Advances in Intelligent Systems and Computing

Volume 978

The series “Advances in Intelligent Systems and Computing” contains publications on theory, applications, and design methods of Intelligent Systems and Intelligent Computing. Virtually all disciplines such as engineering, natural sciences, computer and information science, ICT, economics, business, e-commerce, environment, healthcare, life science are covered. The list of topics spans all the areas of modern intelligent systems and computing such as: computational intelligence, soft computing including neural networks, fuzzy systems, evolutionary computing and the fusion of these paradigms, social intelligence, ambient intelligence, computational neuroscience, artificial life, virtual worlds and society, cognitive science and systems, Perception and Vision, DNA and immune based systems, self-organizing and adaptive systems, e-Learning and teaching, human-centered and human-centric computing, recommender systems, intelligent control, robotics and mechatronics including human-machine teaming, knowledge-based paradigms, learning paradigms, machine ethics, intelligent data analysis, knowledge management, intelligent agents, intelligent decision making and support, intelligent network security, trust management, interactive entertainment, Web intelligence and multimedia.

The publications within “Advances in Intelligent Systems and Computing” are primarily proceedings of important conferences, symposia and congresses. They cover significant recent developments in the field, both of a foundational and applicable character. An important characteristic feature of the series is the short publication time and world-wide distribution. This permits a rapid and broad dissemination of research results.

**** Indexing: The books of this series are submitted to ISI Proceedings, EI-Compendex, DBLP, SCOPUS, Google Scholar and Springerlink ****

More information about this series at http://www.springer.com/series/11156

Rozaida Ghazali · Nazri Mohd Nawi ·
Mustafa Mat Deris · Jemal H. Abawajy
Editors

Recent Advances on Soft Computing and Data Mining

Proceedings of the Fourth International Conference on Soft Computing and Data Mining (SCDM 2020), Melaka, Malaysia, January 22–23, 2020

Springer

Editors
Rozaida Ghazali
Faculty of Computer Science
and Information Technology
Universiti Tun Hussein Onn Malaysia
Batu Pahat, Johor, Malaysia

Nazri Mohd Nawi
Faculty of Computer Science
and Information Technology
Universiti Tun Hussein Onn Malaysia
Batu Pahat, Johor, Malaysia

Mustafa Mat Deris
Faculty of Computer Science
and Information Technology
Universiti Tun Hussein Onn Malaysia
Batu Pahat, Johor, Malaysia

Jemal H. Abawajy
School of Information Technology
Deakin University
Geelong Waurn Ponds Campus, VIC
Australia

ISSN 2194-5357 ISSN 2194-5365 (electronic)
Advances in Intelligent Systems and Computing
ISBN 978-3-030-36055-9 ISBN 978-3-030-36056-6 (eBook)
https://doi.org/10.1007/978-3-030-36056-6

This Springer imprint is published by the registered company Springer Nature Switzerland AG
The registered company address is: Gewerbestrasse 11, 6330 Cham, Switzerland

Conference Organization

Patron

Wahid Razzaly	Vice Chancellor, Universiti Tun Hussein Onn Malaysia

Advisory Committee

Ajith Abraham	Machine Intelligence Research Labs, USA
Junzo Watada	Universiti Teknologi Petronas, Malaysia
Nikola Kasabov	KEDRI, Auckland University of Technology, New Zealand
Rajkumar Buyya	University of Melbourne, Australia
Witold Pedrycz	University of Alberta, Canada

Steering Committee

Jemal H. Abawajy	Deakin University, Australia
Mustafa Mat Deris	Universiti Tun Hussein Onn Malaysia
Nazri Mohd Nawi	Universiti Tun Hussein Onn Malaysia
Rozaida Ghazali	Universiti Tun Hussein Onn Malaysia

Chair

Nazri Mohd Nawi	SMC, Universiti Tun Hussein Onn Malaysia

Organizing Committee

Hairulnizam Mahdin	Universiti Tun Hussein Onn Malaysia
Muhaini Othman	Universiti Tun Hussein Onn Malaysia
Noor Zuraidin Mohd Safar	Universiti Tun Hussein Onn Malaysia
Norhalina Senan	Universiti Tun Hussein Onn Malaysia
Shafry Salim	Universiti Tun Hussein Onn Malaysia

Sofia Najwa Ramli	Universiti Tun Hussein Onn Malaysia
Zubaile Abdullah	Universiti Tun Hussein Onn Malaysia

Program Committee Chair

Nureize Arbaiy	Universiti Tun Hussein Onn Malaysia

Proceeding Chair

Rozaida Ghazali	Universiti Tun Hussein Onn Malaysia

Special Session Chair

Mohd Helmy Abd Wahab	Universiti Tun Hussein Onn Malaysia

Sponsorship

Suhaimi Abd Ishak	Universiti Tun Hussein Onn Malaysia

Media and Promotion Chair

Mohd Norasri Ismail	Universiti Tun Hussein Onn Malaysia

Program Committee

Nureize Arbaiy	Universiti Tun Hussein Onn Malaysia
Chuah Chai Wen	Universiti Tun Hussein Onn Malaysia
Mohd. Najib Mohd. Salleh	Universiti Tun Hussein Onn Malaysia
Mohd. Farhan Md. Fudzee	Universiti Tun Hussein Onn Malaysia
Mohd Helmy Abd Wahab	Universiti Tun Hussein Onn Malaysia
Ruhaidah Samsudin	Universiti Teknologi Malaysia
Aida Mustapha	Universiti Tun Hussein Onn Malaysia
José M. Merigó	University of Chile
Norfaradilla Wahid	Universiti Tun Hussien Onn Malaysia
Alwis Nazir	State Islamic University of Sultan Syarif Kasim Riau
Elena Benderskaya	St. Petersburg State Polytechnical University
Gai-Ge Wang	School of Computer Science and Technology, Jiangsu Normal University
Saima Anwar Lashari	Saudi Electronic University, Saudi Arabia
Riswan Efendi	State Islamic University of Sultan Syarif Kasim, Riau
Mohd Amin Mohd Yunus	Universiti Tun Hussein Onn Malaysia

Mohd Zainuri Saringat	Universiti Tun Hussein Onn Malaysia
Mustafa Mat Deris	Universiti Tun Hussein Onn Malaysia
Nazri Mohd Nawi	Universiti Tun Hussein Onn Malaysia
Hairulnizam Mahdin	Universiti Tun Hussein Onn Malaysia
Rozaida Ghazali	Universiti Tun Hussein Onn Malaysia
Isredza Rahmi Ab Hamid	Universiti Tun Hussein Onn Malaysia
Nurul Hidayah Ab Rahman	Universiti Tun Hussein Onn Malaysia
Azizul Azhar Ramli	Universiti Tun Hussein Onn Malaysia
Mohamad Aizi Salamat	Universiti Tun Hussein Onn Malaysia
Rosziati Ibrahim	Universiti Tun Hussein Onn Malaysia
Noraini Ibrahim	Universiti Tun Hussein Onn Malaysia
Muhaini Othman	Universiti Tun Hussein Onn Malaysia
Noor Azah Samsudin	Universiti Tun Hussein Onn Malaysia
Yana Mazwin Mohmad Hassim	Universiti Tun Hussein Onn Malaysia
Noorhaniza Wahid	Universiti Tun Hussein Onn Malaysia
Norhalina Senan	Universiti Tun Hussein Onn Malaysia
Noryusliza Abdullah	Universiti Tun Hussein Onn Malaysia
Sapi'ee Jamel	Universiti Tun Hussein Onn Malaysia
Nurul Azma Abdullah	Universiti Tun Hussein Onn Malaysia
Kamaruddin Malik Mohamad	Universiti Tun Hussein Onn Malaysia
Pei-Chun Lin	Feng Chia University
Hazlina Hamdan	Universiti Putra Malaysia
Zhiang Wu	Jiangsu Provincial Key Laboratory of E-Business, Nanjing University of Finance and Economics, Nanjing, P.R. China
Hamidah Ibrahim	Universiti Putra Malaysia
Dayang N. A. Jawawi	Universiti Teknologi Malaysia
Maslina Zolkepli	Universiti Putra Malaysia
Haza Nuzly Abdull Hamed	Universiti Teknologi Malaysia
El-Sayed M. El-Alfy	King Fahd University of Petroleum and Minerals
Katsuhiro Honda	Osaka Prefecture University
Jose Santos Reyes	University of A Coruña
Siti Yuhaniz	Universiti Teknologi Malaysia
Mohd Saberi Mohamad	Universiti Teknologi Malaysia
Zaidah Ibrahim	Universiti Teknologi MARA
Paulus Insap Santosa	Gadjah Mada University
Elpida Tzafestas	University of Athens
Nadjet Kamel	University Ferhat Abbes Setif1
Rahmat Hidayat	Politeknik Negeri Padang Indonesia
Fatima Zahra Fagroud	Hassan II University, Casablanca, Morocco
Sanaa El Filali	Hassan II University, Casablanca, Morocco
Ahmed A. Elngar	Beni-Suef University, Egypt
Rajdeep Chowdhury	West Bengal, India
El Habib Benlahmar	University of Hassan II, Morocco

Oumaima Hourrane	University of Hassan II, Morocco
Adnan Abid	University of Management and Technology, Pakistan
Saoud Sahar	National Business School, Ibn Zohr University, Morocco
Fairouz Zendaoui	Institut National de la Poste et des TIC, Algérie
Radzi Ambar	Universiti Tun Hussein Onn Malaysia
Hanaa Hachimi	Ibn Tofail University, Morocco
Vitalii Nitsenko	Odessa I.I.Mechnikov National University, Ukraine
Shamsollah Ghanbari	Islamic Azad University, Ashtian Branch, Iran

Special Session Committee

Computational Intelligence and Its Applied Applications

Ahmed A. Elngar	Beni Suef University, Egypt
Rajdeep Chowdhury	Autonomous University, West Bengal, India

Applied Artificial Intelligence

El Habib Benlahmar	University of Hassan II, Morocco
Oumaima Hourrane	University of Hassan II, Morocco

Soft Computing and Data Mining in Economics and Engineering

Vitalii Nitsenko	Odessa I.I.Mechnikov National University, Ukraine

Emerging Trends in Information Technology and Data Science

Dr. Muhammad Faheem Mushtaq	Khwaja Fareed University of Engineering and IT, Pakistan
Dr. Saleem Ullah	Khwaja Fareed University of Engineering and IT, Pakistan
Engr. Dr. Attaullah Buriro	Khwaja Fareed University of Engineering and IT, Pakistan

Conference Logo

SCDM-2020

Conference Banner

Preface

Data science is one of the fastest growing research and application fields. With rapid advancements in data collection and storage technology, access to a vast amount of data is increasing. Organizations are starting to recruit more and more data scientists to help them leverage data analytics. However, extracting useful information has proven extremely challenging. This is due to the fact that the existence of complex systems often remained intractable to conventional mathematical and analytical methods. For this, data mining, which supports a wide range of business intelligence applications, has opened up exciting opportunities for exploring and analyzing the various types of data. With the deployment of data and soft computing techniques to scour extensive database, various novel and useful patterns—which, otherwise, remain unknown—can be found. As a result, new techniques, technologies, and theories are continually being developed to run advanced analysis. Soft computing which refers to a consortium of computational techniques in computer science can deal with imprecision, uncertainty, partial truth, and approximation to achieve tractability, robustness, and low solution cost. Soft computing tools, individually or in an integrated manner, are turning out to be strong candidates for performing tasks in the area of data mining, decision support systems, supply chain management, medicine, business, financial systems, automotive systems and manufacturing, image processing and data compression, etc. It provides the challenge of transforming data into innovative solutions perceived as a new value by customers.

After the success of our three previous SCDM conferences in 2014 until 2018, we hope to continue this journey of achievements with a fourth international conference. This year, the SCDM 2020 was held in Melaka, Malaysia on January 22–23, 2020. We received 75 paper submissions from 23 countries around the world. The conference also approved four special sessions that are Computational Intelligence and its Applied Applications, Applied Artificial Intelligence, Soft Computing and Data Mining in Economics and Engineering, and Emerging Trends in Information Technology and Data Science. Each paper in regular submission and special sessions were screened by the proceeding's chair and carefully peer-reviewed by two experts from the Program Committee. Finally, only 44 papers with the highest quality and merit were accepted for oral presentation and publication in this volume proceeding, giving an acceptance rate of 59%.

The papers in this proceeding are grouped into five tracks:

- General Track,
- Computational Intelligence and its Applied Applications,
- Applied Artificial Intelligence,
- Soft Computing and Data Mining in Economics and Engineering, and
- Emerging Trends in Information Technology and Data Science.

On behalf of SCDM 2020, we would like to express our highest gratitude to the conference organizer, Faculty of Computer Science & Information Technology, UTHM, and also to Soft Computing & Data Mining research group, Steering Committee, Conference Chair, Program Committee Chair, Organizing Chairs, Special Session Chair, and all Program and Reviewer Committee members for their valuable efforts in the review process that helped us to guarantee the highest quality of the selected papers for the conference.

We would also like to express our thanks to the keynote speakers, Prof. Dr. Rajkumar Buyya from University of Melbourne, Australia, Prof. Dr. Jemal H. Abawajy from Deakin University, Australia, and Prof. Dr. Mustafa Mat Deris from Universiti Tun Hussein Onn Malaysia. Our special thanks are also due to Mr. Suresh Rettagunta and Dr. Thomas Ditzinger for publishing the proceeding in Advances in Intelligent Systems and Computing, Springer. We wish to thank the members of the Organizing and Student Committees for their very substantial work, especially those who played essential roles.

Lastly, we cordially thank all the authors for their valuable contributions and other participants of this conference. The conference would not have been possible without them.

Batu Pahat, Malaysia — Rozaida Ghazali
Batu Pahat, Malaysia — Nazri Mohd Nawi
Batu Pahat, Malaysia — Mustafa Mat Deris
Victoria, Australia — Jemal H. Abawajy

Contents

General Track

An Enhanced Model for Digital Reference Services

Asim Shahzad(✉), Nazri Mohd Nawi(✉), Hairulnizam Mahdin, Sundas Naqeeb Khan, and Norhamreeza Abdul Hamid

Soft Computing and Data Mining Center (SMC), Faculty of Computer Science and Information Technology, Universiti Tun Hussein Onn Malaysia, Parit Raja, Malaysia
asim.sahz@gmail.com, nazri@uthm.edu.my

Abstract. Digital Reference Service (DRS) play a vital role in the Digital Library (DL) research. DRS is a very valuable service provided by DL. Unfortunately, the reference service movement towards digital environment begins late, and this shift was not model based. So, a journey towards a digital environment without following a proper model raises some issues. A few researchers presented a general process model (GPM) in the late 1990s, but this process model could not overcome the problems of DRS. This paper proposes an enhanced model for DRS that use the storage and re-use mechanism with other vital components like DRS search engine and ready reference for solving the issues in DRS. Initially, storage and re-use mechanism are designed and finally, DRS search engine is designed to search appropriate answers in the knowledge base. We improved the GPM by incorporating the new components. The simulation results clearly states that the proposed model increased the service efficiency by reducing the response time from days to seconds for repeated questions and decreased the workload of librarians.

Keywords: DRS model · Virtual reference services · Digital reference services · Model for digital reference services

1 Introduction

In the past few years, one of the critical development in reference services is the emergence of digital reference service (DRS) [1]. Just like traditional libraries moved towards the digital libraries [2], the service of traditional libraries i.e. reference desk service also shifted towards the DRS [3] and unfortunately, this movement towards DRS was not model based. Academic libraries launched the earliest DRS [4], and they implemented this service without following any standard model. Based on the working mechanism of existing DRS, a general process model (GPM) was presented by Lankes in 1998 [5, 6] but there were limitations in this GPM. A conference was arranged in 2002, to bring practitioners and scholars together to share the existing findings and propose potential research question in digital reference [7]. This conference contributed an agenda highlighting a series of investigation areas and the perspectives from which these research areas can be examined [7]. Researchers worked in several different fields

R. Ghazali et al. (Eds.): SCDM 2020, AISC 978, pp. 3–14, 2020.
https://doi.org/10.1007/978-3-030-36056-6_1

and extensively examined some research areas, for instance, efficiency and effectiveness, digital reference information systems, and questions and answers [8]. Researchers also investigated the issue of integrating digital libraries (DLs) with DRS and a good amount of research work has been done in this field [9].

Furthermore, multiple techniques have been investigated for interaction between librarian and patron, for instance, video conferencing, text-based chat, voice chat, and still, this is a hot research topic [10, 11]. Some people worked on librarian-teacher collaboration [12, 13]. Researchers also compared, analyzed and evaluated several existing DRS [14], they explored the framework to analyze and evaluate the DRS and analyzed the librarian's work nature [15, 16] and how a user communicates with DRS [17–21]. The fundamental purpose of digital reference service is to provide timely help to a user [22], to keep an appropriate and relevant collection of resources for reference, to provide the assistance to a user in finding the genuine source of information [23], to assist in searching resources online, to provides the live links to trustworthy webpages and keep the questions and answers in database for use in future. Some researchers [24, 25] provided a comprehensive survey and evaluated different reference services. In DRS the knowledge base, has high importance because, question and answers can be stored in the knowledge base for re-use in future, and information seekers can get the help and benefits from these resources. But the storage mechanism of question and answers for re-use in future is not addressed yet. A great amount of research is needed to be done in this area. In this article, we proposed a model-based approach and introduced the concept of storage and re-use of questions and answers in DRS.

2 Problems in Existing General Process Model (GPM)

An investigational study on digital reference service (DRS) was conducted by Lankes [6], he developed a general digital reference model described in Fig. 1. This model provides a way of understanding the DRS as information systems either as a separate self-contained service or as a part of digital library (DL). There are five fundamental steps in the GPM [7]:

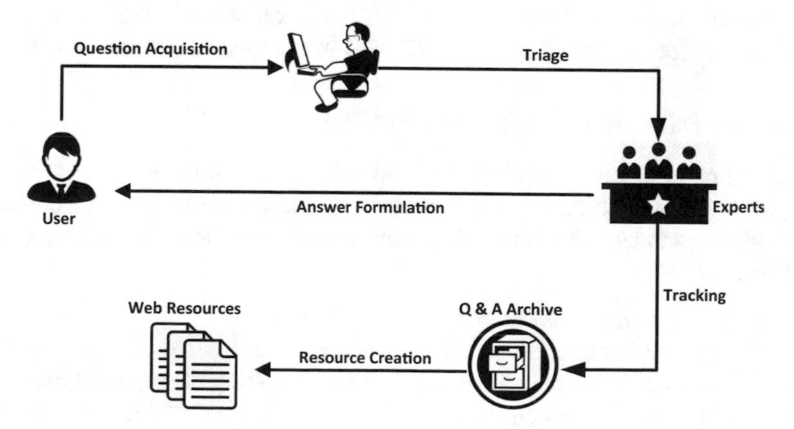

Fig. 1. General process model for digital reference services

(1) The initial step is used to receive questions from web forms, email, live chat.
(2) The second step is responsible for handling and routing the user's question to DL and subject specialist in the service. To filter the repeated questions Triage is applied.
(3) To create the answer for the user's questions, Answer formulation is used.
(4) To identify the latest research areas and gaps in it, tracking is used.
(5) The final step is resource creation which is responsible for expanding or building the collection of data to satisfy the information needs of users. This simple model is followed by every asynchronous DRS i.e. web and email-based DRS [6, 7, 26].

We identified the following shortcomings of the general process model:

(a) There should a storage mechanism of questions.
(b) The model should be able to re-use the already stored questions.
(c) The ready reference service is missing in the GPM. It is an important service for providing the factual answers by reference librarian in a few minutes [27, 28].

3 Proposed DRS Model

The resources and services of digital reference service (DRS) must satisfy the requirements of users [29]. This paper proposes an enhancement to the digital reference service (DRS) model by investigating several academic and public digital libraries and their reference services from around the world. After examining the various digital reference services, we identified that not even a single digital reference service (DRS) is flawless where each service has problems. One of the most critical and common problem in all digital reference services is that there is no procedure defined or discussed for storing and re-using the content. In addition, a critical service of the traditional reference desk known as ready reference is also missing in digital reference service. In which ready reference helps the users to get the answers for factual and short questions in a short period [28]. In order to enhance the general process model (GPM) for obtaining the excellent results, we incorporated some new components, i.e., Digital Reference Service Search Engine (DRSSE), knowledge base, ready reference service, and frequently asked questions (FAQs) in the general process model.

3.1 Incorporating New Components in GPM

Based on some limitations identified by Lankes [7] as explained above in Sect. 2, this paper proposes the enhancement of DRS model by incorporating new vital components in this existing model. The detail discussion about incorporated components is presented below.

1. *Incorporating storage and re-use technique in DRS*
   The storage mechanism is not present neither discussed by the general process model. It can minimize the efforts required for answering the repeated question and will also improve the response time of the system and will reduce the workload of the reference librarian.

2. *Incorporating ready reference service in DRS*
   Ready reference service performs a vital role in the traditional library [30]. The structure of a reference desk in the traditional library is beneficial for people who are asking the library for help, but unfortunately, this service is mainly missing or less famous in DRS [31]. A new enhanced DRS model is required to overcome this issue, which helps and provides the reference desk structure in DRS.
3. *Incorporating search engine in DRS*
   After adding the ready reference service and storage component in general process Model (GPM), we realized the importance of the searching mechanism in digital reference service (DRS) for searching the solutions for repeated queries. So, we incorporated the DRSSE in DRS. The complete working mechanism of digital reference service search engine (DRSSE) can be seen in Fig. 2. The core responsibility of the DRSSE is to receive the user's question and before forwarding the query to reference librarian search for the answer in the knowledge base and other resources. For repeated queries, digital reference service search engine will make it possible to reply to the user's question quickly. By incorporating digital reference service search engine, it improves the efficiency of the service and reduced the librarian's workload. As search engines increase the availability of resources [32] by providing easy techniques to find, identify, and search the required content [23].

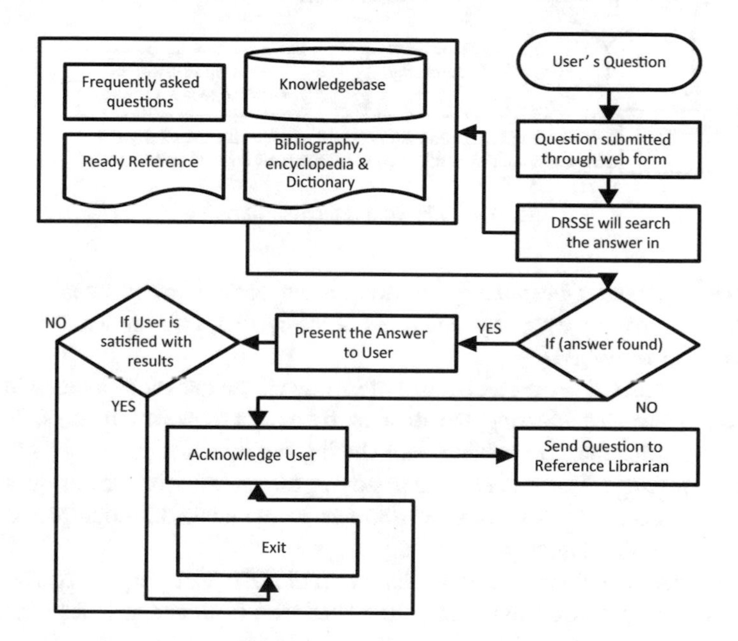

Fig. 2. Digital reference services search engine

3.2 Components of DRS Model

1. *User Interface:* UI is the mean by which a user interacts with DRS. UI is an integral part of DRS because all the components of DRS are linked to UI.

2. *Reference Librarian:* A person employed in a DRS section, and accountable for presenting useful information in answer to a user's question [33]. The tasks of the librarian are depicted in Fig. 3.

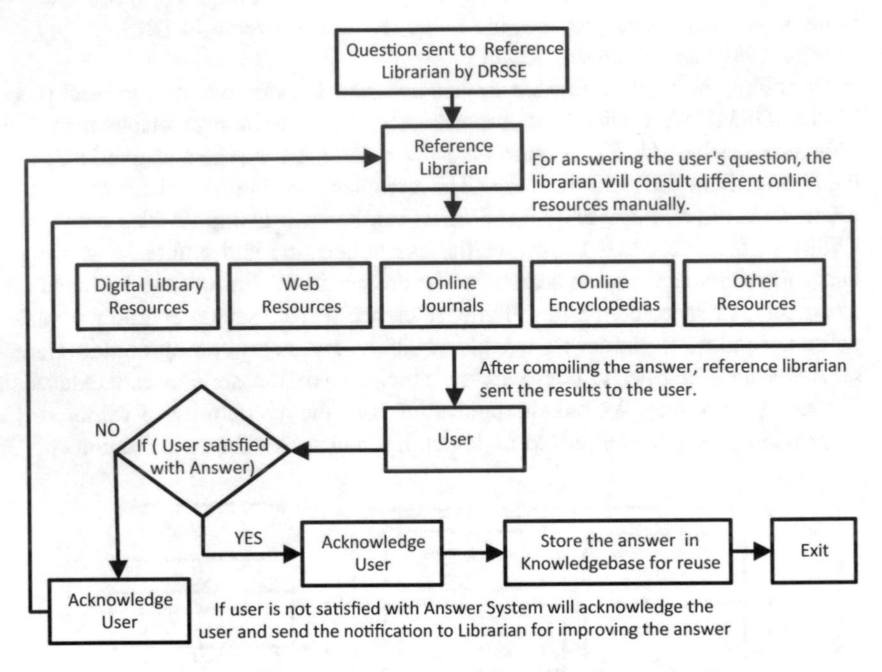

Fig. 3. Tasks of a reference librarian

3. *Frequently Asked Questions:* The FAQ is an archive of answers for a list of questions frequently asked by new users of the system, also known as Question and Answers (Q & A) [34].
4. *Knowledge base:* It is the mechanism of storing all the questions along with answers in a repository for re-using the data in future for repeated queries. It supports retrieving, collecting, organizing, and sharing the information.
5. *Ready Reference:* The repository for storing the answers of factual questions is a ready reference. It consists of three different components, bibliographic dictionary, encyclopedia, and language dictionary.
6. *Digital Reference Services Search Engine (DRSSE):* The purpose of DRSSE is to get the user's question submitted on the DRS web from and search for the solution in stored contents in the knowledge base, ready reference, and FAQs.

3.3 Overall Structure and Interaction of the Components in the DRS Model

In the digital reference service (DRS) model for interaction with a librarian, the users are using DRS user's interface. The interface consists of four different sections which are; knowledge base, ask a reference librarian service, ready reference service and frequently asked question's section. User can search for the answer manually by browsing the content in different sections. For contacting the reference librarian user can use ask a reference librarian service in DRS. After the user submits the question digital reference service search engine (DRSSE) will automatically search the user's question in the knowledge base, ready reference, and FAQs for a solution. If digital reference service search engine found an appropriate solution for the question it will present the results to the user with the option if the user is not satisfied with the results, the DRSSE will forward the question to the reference librarian and will acknowledge the user with a message that the user's question is submitted, and the reference librarian will get back soon with an answer. In case if DRSSE could not find an appropriate answer in the archives, it will forward the question directly to the reference librarian and will acknowledge the user with the message above. After reference librarian received the question from DRSSE, the librarian will consult different web resources (web sites, research papers, web articles, journals, etc.) for compiling the answer and will send back the answer to the user. The system will automatically store the user's question along with the answer compiled by the reference librarian in the knowledge base for re-use in future for repeated questions. The interaction among all components and the overall structure of the DRS Model can see seen in Fig. 4.

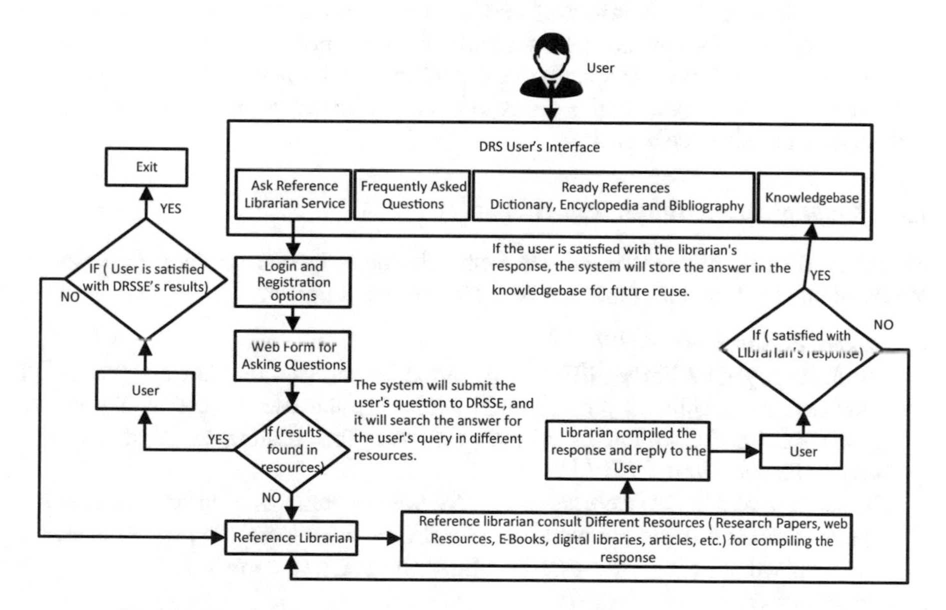

Fig. 4. Components interaction and overall structure of the DRS model

3.4 DRS Model Evaluation

The performances of the proposed digital reference service (DRS) model are compared with several other public and academic reference services easily accessible on the internet. For evaluation, we have selected the reference services of those twenty-seven digital libraries which we explored and analyzed during our survey. Out of these twenty-seven digital libraries (DLs), eleven provides the services to their students and staff only, so we eliminated these eleven services from the comparison. Two more DLs are eliminated from the comparison because these two DLs do not offer any digital reference service. So for the comparison purpose, the remaining of fourteen DL reference services was selected for comparison. A huge number of reference services are easily available on the internet [35], so for the users, it is very difficult to decide which reference service to select for searching a question [20]. We have selected these DRS for comparison because we already gathered enough information about these services during our survey, for instance, who are the users of these services and what kind of questions they can answer. How do our DRS as compared to other services in terms of user's friendliness, librarian's workload, time is taken to compile response, the precision of reply and performance of the service? Keeping these questions in our mind we conducted the experiment to find the answers for the questions above. We designed ten questions for this experiment, the first five were factually based questions and can easily be replied using ready reference service, while the next five questions were a bit technical and based on the academic subject of DRS. The answer for the next five questions can vary from four lines answers to more complex replies. During our experiment, we asked these questions form all DRS twice because the difference between our DRS and other reference services is the knowledge base. Our service can automatically reply for repeated questions which were asked by users in the past, while other reference services do not have this capability. So, by asking the same questions twice from each reference service will show the performance difference between our DRS and other reference services.

3.5 Findings of the Initial Experiment

We received some different responses from selected reference services, the response details of our DRS and all other services are presented below.

1. *Internet Public Library (IPL)*
   The internet public library (IPL) [36] started its services in March 1995. In IPL volunteers from different countries are answering the user's queries. To ask the questions from IPL we have used their web form for question submission.
2. *Toronto Public Library (TPL)*
   The services of the TPL were started in 1999. Some students of library sciences and experts in several fields are working with IPL as volunteers. To ask the questions from TPL we submitted the questions through their web form.
3. *Digital Reference Service (DRS)*
   To evaluate our model, we developed the prototype of the DRS model. The implementation of DRS prototype was necessary for comparison and analysis of the DRS model with other services. We submitted the same ten questions to DRS also.
4. *Other DRS*
   The library reference services (LRS) mentioned below; University of South Florida LRS, North Carolina State University LRS, British Columbia Public LRS,

Ryerson University LRS, and Cornell University LRS did not reply to the factual and research-based questions. Instead of responding to these questions, they answered to inform us that they cannot compile the answers for these complex research questions. The other remaining seven reference services never responded.

5. *Experimental Results*

   Results received from all the services can be seen in Table 1.

Table 1. Responses received from all digital reference services for the initial experiment.

	Response rate (%)	Average time (Hours)	Question's answered (%)	Correct answered (%)	Answer stated (%)	Source provided (%)
Internet public library	70%	41.5 h	50%	50%	20%	50%
Toronto public library	80%	9 h	70%	60%	40%	50%
Digital reference service	100%	4 h	100%	100%	100%	0%

Based on the responses, we evaluated the services on the following factors. The response rate of service, the average time took to compile the answers, out of ten how many queries were answered, how many replies were enough or correct, for how many questions the answers were provided and the percentage of solutions with a link to the source. The average response time of service in hours can be seen in Fig. 5. Figure 6 shows the comparison of reference services based on the factors discussed above. As compared to other services, our service replied the user's questions with real answers instead of just providing them the links to the information available on different websites. The response time of all the digital reference services for the initial experiment is in hours. As we introduced the reuse and searching mechanism in our model so our model can perform better for the repeated questions.

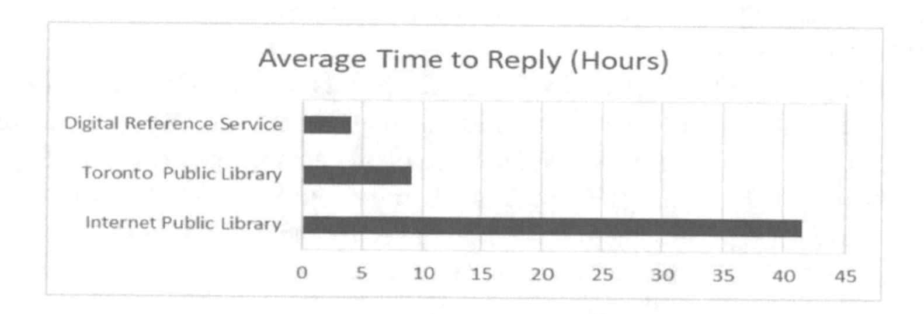

Fig. 5. Average response time of a service in hours

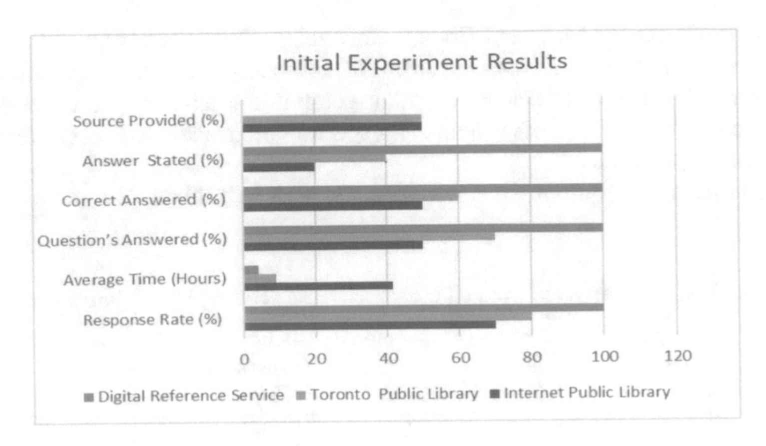

Fig. 6. Comparison of services based on various factors

3.6 Findings of Second Experiment

After six weeks we repeated the same experiment again. The services which responded during the first experiment, we submitted the same ten queries again to those services. The second experiment proved that by incorporating new components we have reduced the workload for reference librarian and enhanced the efficiency of service by providing the solutions for repeated questions within few seconds.

1. *Internet Public Library (IPL)*
   We submitted the same ten questions to IPL again. As compared to the first experiment the response rate and response time of the second experiment are different. The response rate of IPL during the second experiment is a hundred percent.
2. *Toronto Public Library (TPL)*
   Similarly, we submitted the same ten questions again to TPL, the number of answers provided, and response time was different as compared to the first experiment. This time most of the answers were correct and entirely different from their previous answers.
3. *Digital Reference Service (DRS)*
   Knowledge base for storing content and searching mechanism in DRS greatly affected the response time. As all these questions were previously submitted to DRS and it automatically stored all these questions along with answers in the knowledge base, so DRS replied all the questions within a few seconds
4. *Experimental Results*
   After completing the second experiment, the results of each service can be seen in Table 2. Results clearly show that for repeated questions, the response time of DRS is much better than other services. DRS search the answer for the repeated question from knowledgebase in few seconds while other services took hours to recompile the answers for repeated queries. Storage and reuse mechanism in DRS reduced the workload of the reference librarian and enhanced the efficiency of the system. The average response time of service is shown in Fig. 7, while Fig. 8 shows the comparison of reference services based on the various factors.

Table 2. Responses received from all digital reference services for repeated questions.

	Response rate (%)	Average time (minutes)	Question's answered (%)	Correct answered (%)	Answer stated (%)	Source provided (%)
Internet public library	100%	450 min	70%	70%	50%	70%
Toronto public library	100%	90 min	50%	50%	50%	50%
Digital reference service	100%	0.1 min	100%	100%	100%	0%

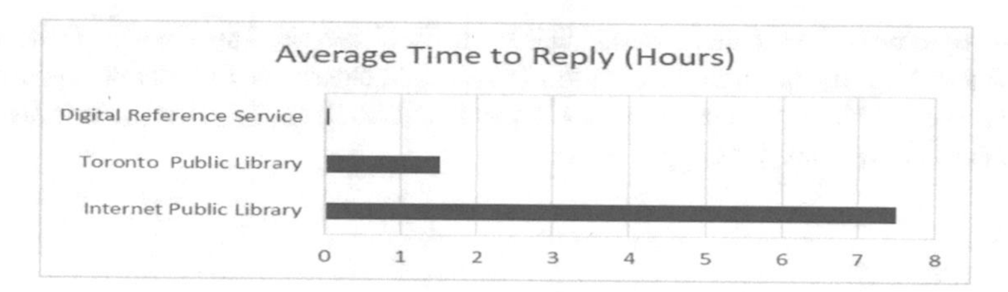

Fig. 7. Average response time of a service in hours

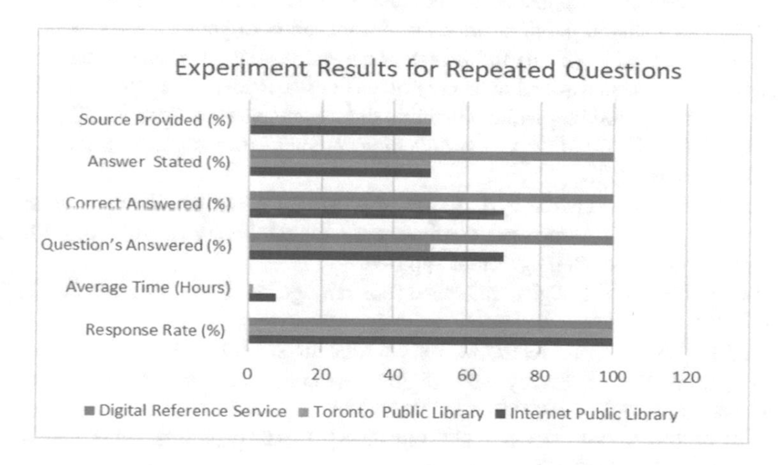

Fig. 8. Comparison of services based on various factors after second time

4 Conclusions

Digital reference service is an essential service from digital libraries. Unfortunately, the service was ignored, and digital reference services movement towards the digital environment was late. This movement was not model based, and after this movement, a general process model was derived from existing services. The general process model was slow, and it also increased the workload of the reference librarian. We incorporated some new vital components in general process model, which enhanced the performance of the new model. We introduced the storage and re-use mechanism along with a digital reference search engine in DRS model; this increased the overall efficiency of the system, reduced the answering time from days to seconds for repeated queries which directly decreased the reference librarian's workload. Our experiments show that the knowledge base is a vital part of DRS, which helps in storing the questions and answers, and all these contents can be reused in the future for repeated queries.

Acknowledgments. The authors would like to thank Universiti Tun Hussein Onn Malaysia (UTHM) and Ministry of Higher Education (MOHE) Malaysia for financially supporting this Research under IGSP grants note U420 and under Trans-displinary Research Grant Scheme (TRGS) vote no. T003.

References

1. Ramos MS, Abrigo CM (2012) Reference 2.0 in action: an evaluation of the digital reference services in selected Philippine academic libraries. Libr Hi Tech News 29(1):8–20
2. Noiret S (2018) Digital public history. In: A companion to public history, p 111
3. Malik A, Mahmood K (2013) Infrastructure needed for digital reference service (DRS) in university libraries: an exploratory survey in the Punjab, Pakistan. Libr Rev 62(6/7):420–428
4. Wasik JM (1999) Building and maintaining digital reference services. ERIC Digest
5. Lankes RD (1998) Building and maintaining internet information services: K-12 digital reference services. ERIC
6. Lankes RD (2004) The roles of digital reference in a digital library environment. In: Proceedings of the international conference of digital library-advance the efficiency of knowledge utilization, Beijing, China, pp 6–8
7. Lankes RD (2004) The digital reference research agenda. J Am Soc Inf Sci Technol 55 (4):301–311
8. Lankes RD (2004) Digital reference. Encycl Libr Inf Sci 1(1):1–3
9. Liu J (2008) Digital library and digital reference service: integration and mutual complementarity. Policy Futur Educ 6(1):59–76
10. Booth C (2008) Video reference and the library kiosk: experimentation and evaluation. J Access Serv 5(1–2):47–53
11. Bakeri Abu Bakar A (2012) Myths and realities of digital reference services: perspectives of libraries from developing countries. Libr Manag 33(3):136–141
12. Huang L, Zhou F (2010) Study on teacher-librarian collaborative distance education model. In: 2010 International conference on intelligent computing and cognitive informatics, pp 138–140
13. Peters T (2018) Online students' use of virtual reference services. J Electron Resour. Librariansh 30(1):1–8

14. Smith LC, Wong MA (2016) Reference and information services: an introduction. ABC-CLIO
15. Arndt TS (2010) Reference service without the desk. Ref Serv Rev 38(1):71–80
16. Collard S, Whatley K (2011) Virtual reference/query log pairs: a window onto user need. Ref Serv Rev 39(1):151–166
17. White MD (2001) Digital reference services: framework for analysis and evaluation. Libr Inf Sci Res 23(3):211–231
18. Dee C, Allen M (2006) A survey of the usability of digital reference services on academic health science library web sites. J Acad Librariansh 32(1):69–78
19. Desai CM (2003) Instant messaging reference: how does it compare? Electron Libr 21 (1):21–30
20. Cloughley K (2004) Digital reference services: how do the library-based services compare with the expert services? Libr Rev 53(1):17–23
21. McCrea R (2004) Evaluation of two library-based and one expert reference service on the Web. Libr Rev 53(1):11–16
22. Kani-Zabihi E, Ghinea G, Chen SY (2006) Digital libraries: what do users want? Online Inf Rev 30(4):395–412
23. Vakkari P (2011) Comparing Google to a digital reference service for answering factual and topical requests by keyword and question queries. Online Inf Rev 35(6):928–941
24. David AT (2001) What is the best model of reference service. Libr Trends 50(2):P183–197
25. Bertot JC, Real B, Jaeger PT (2016) Public libraries building digital inclusive communities: data and findings from the 2013 digital inclusion survey. Libr Q 86(3):270–289
26. Pomerantz J (2004) Question taxonomies for digital reference. SIGIR Forum 38(1):79
27. Jacsó P (2006) Open access ready reference suites. Online Inf Rev 30(6):737–743
28. Cassell KA, Hiremath U (2019) Reference and information services: an introduction. American Library Association
29. Stemper JA, Butler JT (2001) Developing a model to provide digital reference services. Ref Serv Rev 29(3):172–189
30. Weiss A (2016) Examining massive digital libraries (MDLs) and their impact on reference services. Ref Libr 57(4):286–306
31. Buckland MK (2007) The digital difference in reference collections. J Libr Adm 46(2):87–100
32. Layer RM, Pedersen BS, DiSera T, Marth GT, Gertz J, Quinlan AR (2018) GIGGLE: a search engine for large-scale integrated genome analysis. Nat Methods 15(2):123
33. Loesch MF (2017) Librarian as professor: a dynamic new role model. Educ Libr 33(1):31–37
34. Sindhgatta R, Marvaniya S, Dhamecha TI, Sengupta B (2017) Inferring frequently asked questions from student question answering forums. In: EDM
35. Agosti M, Ferro N, Silvello G (2018) Digital libraries: from digital resources to challenges in scientific data sharing and re-use. In: A comprehensive guide through the Italian database research over the last 25 years, pp 27–41. Springer, Cham
36. Internet public library. http://www.ipl.org/. Accessed 02 Jan 2015

Fuzzy Random Based Mean Variance Model for Agricultural Production Planning

Mohammad Haris Haikal Othman[1], Nureize Arbaiy[1(✉)], Muhammad Shukri Che Lah[1], and Pei-Chun Lin[2]

[1] Faculty of Computer Science and Information Technology, Universiti Tun Hussein Onn Malaysia, Batu Pahat, Malaysia
nureize@uthm.edu.my

[2] Department of Information Engineering and Computer Science, Feng Chia University, No. 100, Wenhwa Rd., Taichung, Taiwan

Abstract. Observation and measurement data are the basis of an analysis which usually contains uncertainties. The uncertainties in data need to be properly described as they may increase error in the prediction model. The collected data which contains uncertainty should be adequately treated before analysis. In the portfolio selection problem, uncertainty involves are characterized as fuzzy and random. Hence fuzzy random variables are accounted as input values in the portfolio selection analysis. It is important to preprocess the data sufficiently due to the uncertainties issue. However, only a few studies discuss the systematic procedure for data processing whereby the uncertainties exist. Hence, this study introduces a structure for fuzzy random data processing which deals with fuzziness and randomness in data for building a portfolio selection model. The fuzzy number is utilized to treat the fuzziness and the probability distribution used to treat randomness. The proposed model is applied for agricultural planning. Five types of industrial plants are assessed using the proposed method. The result of this study demonstrates that the proposed method of fuzzy random based data Pre-processing can treat the uncertainties. The systematic procedure of fuzzy random data Pre-processing in this study is important to enable data uncertainties treatment and to reduce error in the early stage of problem model building.

Keywords: Fuzzy random variable · Fuzzy random data · Data Pre-processing · Mean-Variance

1 Introduction

The uncertainties that exist in the real world have a significant effect on forecasting and decision-making. Uncertainty creates risk and may complicate the decision maker [1] because it aggregates decision parameters in real-world problems. For example, production planning is characterized by several critical parameters such as cost, demand, price, and capacity which involved uncertain values. So, it is important to focus on the way uncertainty affects the planning process. In the agriculture industry, there are many kinds of risk and uncertainty that could affect agriculture data such as endogenous, weather variability are the most uncontrollable in the agriculture industry [2].

R. Ghazali et al. (Eds.): SCDM 2020, AISC 978, pp. 15–24, 2020.
https://doi.org/10.1007/978-3-030-36056-6_2

Uncertainty like nature can make the plant a total failure with hail and heavy rain causes low production.

Portfolio selection or portfolio optimization is established to be an important technique among practitioners and researchers in finance. Researchers are currently focusing on the former issue, comparing several risk cases when using portfolio selection [3]. Classical portfolio selection model is developed based on the mean-variance approach [4] and probability theory to handle uncertain situations [5]. Markowitz's portfolio selection approach allows investors to build portfolios that provide the best risk investors. In order to obtain the fuzzy data from the agriculture data, the confidence interval method is applied to transform the single data to interval data where it forms a minimum and maximum data. Based on [6], the confidence interval distance can identify the minimum and maximum interest benefits at the boundary of the confidence area based on the probability for the overall parameter vector.

To analyze uncertain phenomena in the real world, the concept of random variables is widely used in model buildings. The probability theory was previously used to handle random events, and the fuzzy theory provides solutions to address fuzzy data. These two theories play a major role in treating uncertainty in the real world. In addition, existing models treat random events separately from fuzzy data, even in the real world, random events occur simultaneously with fuzzy data. This random situation is represented by a fuzzy random variable. However, both theories are used separately based on one uncertainty. While previous studies have used fuzzy data to address uncertainty, few provide systematic procedures that transform actual data to fuzzy data; and random.

Therefore, this study proposes a procedure based on the fuzzy random variable to address the fuzzy random uncertainty inherent in the data for building a portfolio selection model. The fuzzy random data pre-processing handles a procedure involves treatment to fuzzy data and a random value. The improvement in the data pre-processing proposed in this study to handle the fuzzy random uncertainty is expected to benefit in improving accuracy in the prediction model; i.e. to reduce errors in the initial stages of modeling. Motivated by the uncertainty phenomenon [7], it is appropriate that the data pre-processing should employ a fuzzy random variable concept to overcome the simultaneous uncertainties problem in agricultural production data.

The rest of the paper is organized in the following. Section 2 describes prior knowledge of the related studies. Section 3 describes the main method and Sect. 4 illustrates numerical example by using agriculture data. Section 5 gives concluding remarks.

2 Literature Review

This section provides prior knowledge for portfolio selection by mean-variance model and fuzzy numbers.

2.1 Mean-Variance Model

Portfolio selection involves the process to find an optimal portfolio. Markowitz [4] proposed a method to determine optimum portfolio that provides the highest return and the lowest risk; to which such a situation is considered as portfolio selection problem. The portfolio selection is widely employed in the real-world problem and has been extended to include the solution to deal with uncertainties [8, 9].

The portfolio selection model [10] is defined as follows:

$$\min\left\{\sum_{i=1}^{n}\sum_{j=1}^{n} x_i\sigma_{ij}x_j\right\}$$

$$s.t. \sum_{i=1}^{n} E(r_i)x_i = \rho \tag{1}$$

$$\sum_{i=1}^{n} x_i = 1$$

$$x_i \geq 0, i = 1, \ldots, n$$

where r_i is the random variable for the return, $E(r_i)$ is the expected value for r_i, σ_{ii} is the variance of r_i, σ_{ij} is the covariance of r_i and r_j, x_i is the proportion of capital to be invested in security i and ρ is the parameter of expected return.

2.2 Fuzzy Set

The fuzzy set were initiated and defined by Zadeh [11] as follows:

Definition 1: Let U denote the universal set of the discourse. Then a fuzzy set A on U, is defined in terms of the membership function m_A that assigns to each element of U a real value from the interval [0, 1]. A fuzzy set A in U can be written as a set of ordered pairs in the form $A = \{(x, m_A(x)) : x \in U\}$, where $m_A : U \to [0, 1]$. The value m_A (x), known as the membership degree (or grade) of x in A, expresses the degree to which x verifies the characteristic property of A.

Fuzzy number (FN) is a special form of fuzzy set a of real number set at R. FNs plays an important role in fuzzy math, analogous to the role played by ordinary numbers in classical mathematics. The fuzzy number of x can be written as follows:

$$m_A(X) = \sum_{i=1}^{n} m_i(X)I_{Aj}(X) \tag{2}$$

where $I_{A_i}(X) = 1$ when $x \in A_i$ and $I_{A_i}(X) = 0$ if $x \notin A_i$.

Definition 2: A fuzzy set A on U with membership function $y = m(x)$ is said to be normal, if there exists x in U, such that $m(x) = 1$.

2.3 Fuzzy Random Variables

Fuzzy sets [11] has lead to the data hybridization where the fuzzy data is combined with the randomness. Kwakernaak [12] proposed the notation of fuzzy random variables. While Zadeh [13] introduces a fuzzy random variable with the new definition of exception to generalize the integral of a set-value function.

Theorem 1 Central Limit Theorem

Let $X_1, \ldots, X_n$ be the random variables with mean m and variance σ^2. Let

$$Z_n = \frac{n^{\frac{1}{2}}(\bar{X}_n - m)}{\sigma} = \frac{W_n - nm}{n^{\frac{1}{2}}\sigma}$$

Then, Z_n converges in distribution to Z as $n \to \infty$. We denote $Zn \to Z \sim N(0,1)\, as\, n \to \infty$,

where Z distribute according to a standard normal distribution function *N(0,1)*.

3 Fuzzy Random Portfolio Selection Model for Agricultural Production Planning

The proposed formation for building a fuzzy random portfolio selection model for agricultural production planning is presented in this section. The model contained two important phases. The first phase gives the data pre-processing procedure to deal with the uncertainties by means of fuzzy random data. The second phase exhibits the portfolio selection model using mean-variance model (1) with pre-processed fuzzy random data. The structure is explained as follows.

3.1 Fuzzy Random Data Pre-processing

The fuzzy random data pre-processing is explained.

The yield of production data is obtained from agriculture production and is selected as the predictor. The yield potential production is calculated as Eq. (3).

$$yield\, potential = \frac{Production\,(mt)}{Harvested\; area\,(ha)} \tag{3}$$

The 1% of measurement error [14–16] method is used to fuzzify the data. The fuzzy data contain the maximum production, and the minimum production based on the percentage error calculation. Let us assume that minimum potential is A_i and the maximum potential is B_i. Equation (4) shows the formulation to generate the single value data into the fuzzy data.

$$A_i = yield\,potential - (yield\,potential * 1\%)$$
$$B_i = yield\,potential + (yield\,potential * 1\%) \quad (4)$$

The fuzzified data is presented in an interval form of $F_i = [A_i, B_i]$ where F_i is the Fuzzy data in an interval form. The central point and the width of the data are then calculated. Let central point and width for fuzzified data be a $X_i = [c_i, w_i]$. X_i is a fuzzy parameter, c is the center of the fuzzy number, and w is the width of the fuzzy number. The formula to calculate the central point and width are given as follows;

$$c_i = \frac{A_i + B_i}{2} \forall i = 1, 2, \ldots n$$
$$w_i = \frac{B_i - A_i}{2} \forall i = 1, 2, \ldots n \quad (5)$$

The randomness in data is dealt with probability distribution method in this study. The probability distribution function used are Normal, Weibull, Gamma, and Logistic distribution. Evaluation is based on p-value obtained from four distributions. Distribution with a higher p-value is selected for the next process. The fuzzy random data preprocessing resulted in an interval form $Y_i = [C_i, W_i]$ where C_i is central point and W_i is the width where Y_i is a parameter of fuzzy interval number from the probability distribution result.

3.2 Fuzzy Random Based Portfolio Selection Model for Agricultural Production Planning

Let us assume $Y_i = (C_i, W_i)$ is the set of an interval fuzzy number where $\forall i = 1, 2, \ldots, n$. Expected value and variance of fuzzy interval data are determined as follows:

$$Expected\ value: \quad E(A_i) = (E(c_i), E(w_i)) \quad (6)$$

$$Variance: \quad \sigma^2(A_i) = \left(\sigma^2(c_i), \sigma^2(w_i)\right) \quad (7)$$

where E is the expected value of interval fuzzy number and σ^2 is the variance. Portfolio formulation is given as follows:

$$\left.\begin{array}{l}\max \sum_{i=1}^{n} E(c_i)x_i \\ \min \sum_{i=1}^{n} E(w_i)x_i \\ s.t. \sum_{i=1}^{n} \sigma(c_i)x_i \\ \quad \sum_{i=1}^{n} \sigma(w_i)x_i \\ \quad \sum_{i=1}^{n} x_i \leq 1 \\ \quad \sum_{i=1}^{n} \sigma(w_i)x_i \\ x_i \geq 0 \, \forall i = 1, 2, \ldots, n\end{array}\right\} \tag{8}$$

$max \sum_{i=1}^{n} E(c_i)x_i$ *and min* $\sum_{i=1}^{n} E(w_i)x_i$ in (8) uses expected value in center and width.

Based on 3.1, 3.2 and 3.3, the step in this proposed method can be simplified as follows:

Step 1 Collect agriculture data and change the data into fuzzy data in form of interval F_i parameter.
Step 2 Identify the central point and width of the interval fuzzy data X_i.
Step 3 Identify the underlying distribution of the center and width by using the probability density function.
Step 4 Calculate the expected value and variance using the data from Y_i parameter.
Step 5 Apply expected value and variance into the portfolio selection model in Eqs. (6) and (7).
Step 6 Test the portfoio selection model in Eq. (8) with different k value until the optimal solution is obtained. The optimal solution is indicated with only one portfolio and $\sum_{i=1}^{n} x_i = 1$.

Providing this procedure, it is expected to provide an efficient approach for decision makers to develop the selection model and simultaneously resolve the uncertainty in the data.

4 Numerical Experiment

Five types of agricultural data (Spinach, Mustard Leaf, Cucumber, Chili, and Brinjal) based on Johor farm are used in this experiment. This data is obtained from https://pertanian.johor.gov.my/en/node/528. The data period is 3 years; starting from year 2015–2017. The objective is to identify a portfolio which is the type of vegetable plants that maximizes the return on the investment.

Fuzzy random data pre-processing starts here. The raw data obtained in Table 1 is then fuzzified by using 1% percentage measurement error based on Eq. (4). This resulted in fuzzy interval data in a form of $[a, b]$ where a is minimum data and b is the maximum data. Table 2 shows the fuzzified data.

Table 1. Single form vegetables data A

Vegetable/Year	2015	2016	2017
Spinach	95.21	104.10	98.53
Mustard	164.86	149.92	140.60
Cucumber	245.46	223.67	191.53
Chili	144.75	129.59	191.53
Brinjal	207.53	183.16	184.13

Table 2. Fuzzy interval data $[c_i, w_i]$

Vegetable/Year	2015	2016	2017
Spinach	[95.21, 0.95]	[104.10, 1.04]	[98.53, 0.99]
Mustard	[164.86, 1.65]	[149.92, 1.50]	[140.60, 1.41]
Cucumber	[245.46, 2.45]	[223.67, 2.24]	[191.53, 1.92]
Chili	[144.75, 1.45]	[129.59, 1.30]	[191.53, 1.92]
Brinjal	[207.53, 2.08]	[183.16, 1.83]	[184.13, 1.84]

After collecting the data, the center point and width of the fuzzy data is identified as in Eq. (5). Table 2 shows the center point and width of the fuzzy data. The probability distribution function is then performed to treat the randomness. Each of the vegetables data will provide 4 type of data which each of them will generate Normal, Log, Gamma and Weibull distribution. Each of the distribution will provide the p-value. In this paper, the biggest p-value indicate as the best result to treat the randomness. Table 3 show five types of vegetables are considered and represents in the form of $x_n = (x_1, \ldots, x_5)$ respectively. Table 3 shows the probability distribution function that has been selected based on the highest p-value. Note that the *LOG* denotes logistic distribution, W as the Weibull Distribution and Î³ as gamma distribution. The moment estimator is utilized to approximate the expected value and variance each of the vegetable. Table 4 presents the result. The expected value and variance for each of the vegetable is computed. The data has now completed the pre-processing phase. This pre-processed data is then presented to the portfolio selection model to identify the best portfolio.

The portfolio selection model in Eq. (8) is used to represent the in the Model (9).

$$
\begin{aligned}
max \quad & 99.2806x_1 + 151.7940_{x2} + 218.7535_{x_3} + 155.2864x_4 + 189.9310_{x5} \\
min \quad & 0.9928x_1 + 1.5179x_2 + 2.1875x_3 + 1.5529x_4 + 1.8993x_5 \\
s.t. \quad & \sqrt{20.2x_1} + \sqrt{149.7578x_2} + \sqrt{1006.5329x_3} + \sqrt{1042.4143x_4} + \sqrt{153.9907x_5} = k \\
& \sqrt{0.002} + \sqrt{0.0150x_2} + \sqrt{0.1007x_3} + \sqrt{0.1042x_4} + \sqrt{0.0154x_5} \leq k \\
& x_1 + x_2 + x_3 + x_4 + x_5 \leq 1 \\
& x_i \geq 0 \forall i = 1, 2, \ldots, n
\end{aligned}
\tag{9}
$$

Table 3. Interval number of the probability distribution function

	Center, C	Width, W
Spinach	Î³ (487.952, 0.203)	Î³ (487.952, 0.002)
Mustard	Î³ (153.868, 0.982)	Î³ (155.892, 0.005)
Cucumber	w (230.858, 8.197)	w (1.165, 8.235)
Chili	Î³ (23.134, 6.679)	Î³ (23.432, 0.033)
Brinjal	log (188.989, 6.808)	log (0.955, 0.034)

Table 4. Expected value and variance

		Center, C		Width, W	
		Expected value	Variance	Expected value	Variance
Spinach	x_1	99.2806	20.2000	0.9928	0.0020
Mustard	x_2	151.7940	149.7578	1.5179	0.0150
Cucumber	x_3	218.7535	1006.5329	2.1875	0.1007
Chili	x_4	155.2864	1042.4143	1.5529	0.1042
Brinjal	x_5	189.9310	153.9907	1.8993	0.0154

Equation (9) is then solved by using a linear programming method. We assume that k is the risk level and $x_n = (1,0,0,0,0)$' or $x_n = (0,1,0,0,0)$' or $x_n = (0,0,1,0,0)$' or $x_n = (0,0,0,1,0)$' or $x_n = (0,0,0,0,1)$' is the optimal solution. The calculation to complete the model is stopped when the optimum solution is obtained.

5 Result and Discussion

Table 5 shows the optimal solution results. From the results table, the risk of $k = 4.5$ indicates the optimal solution where the expected return is (99.23, 0.99) for a 1% error. The optimum result based on the five-vegetable data is spinach.

Risk level k is important to support management decision making. The k value is used to indicate the risk level to select certain portfolio. Based on the results in Table 5, the optimal value x^* for risk level of k is 4.5. This indicates that the spinach plant has the potential to produce maximum yield compared to other plants.

The goal of achieving optimal results is to select plants that can provide the maximum potential for maximum profit. In general, $k = 3$, 4, 4.4 produces positive expected return. However, $k = 4.5$ can produce maximum and best returns for the

Table 5. The result with optimal solution

Risk, k	3.0	4.0	4.4	4.5
x_n^*	[0.667, 0, 0, 0, 0]	[0.89, 0, 0, 0, 0]	[0.979, 0, 0, 0, 0]	$[1, 0, 0, 0, 0]^*$
Expected value	(66.27, 0.66)	(88.36, 0.88)	(97.19, 0.97)	(98.30, 0.98)

vegetable yield potential. Therefore, this model can help management focus on the plants with the optimum expected return to produce the best return.

6 Conclusions

In this paper, a procedure to build a portfolio selection model is presented with fuzzy random data pre-processing which dealt with fuzziness and randomness in data. This paper also introduces the fuzzification of the real data technique where the measurement error of 1% is used to fuzzified the data. In order to apply the fuzzy data into the mean-variance model, the fuzzy center c_i and width w_i is identified. The proposed fuzzy random data pre-processing procedure uses the fuzzified data with probability distribution to deal with randomness. The application of the proposed method is applied to production of vegetable data from agriculture sector. The experimental result reveals that the potential vegetables with optimal production yield can be obtained by using this method and used for planning and decision making.

Acknowledgments. The author would like to extend its appreciation to the Ministry of Higher Education (MOHE) and Universiti Tun Hussein Onn Malaysia (UTHM). This research is supported by Fundamental Research Grant Scheme (FRGS) Vote No FRGS/1/2019/ICT02/UTHM/02/7 and Geran Penyelidikan Pascasiswazah (GPPS) grant (Vote H332 & Vote U975). The author thanks the anonymous reviewers for the feedback.

References

1. Espinoza WHI Fernandez, A, Torres D (2017) The semantic web as a platform against risk and uncertainty in agriculture. In: Working conference on virtual enterprises, pp 753–760. Springer, Cham
2. Paulson ND (2007) Three essays on risk and uncertainty in agriculture
3. Righi MB, Borenstein D (2018) A simulation comparison of risk measures for portfolio optimization. Finan Res Lett 24:105–112
4. Markowitz H (1952) Portfolio selection. J Finan 7(1):77–91
5. Shapiro A, Dentcheva D, Ruszczyński, A (2009) Lectures on stochastic programming: modeling and theory. Society for industrial and applied mathematics
6. Uusipaikka E (2008) Confidence intervals in generalized regression models, 1st edn. Chapman and Hall/CRC
7. Chuliá, H, Guillé, M, Uribe JM (2017) Measuring uncertainty in the stock market. Int Rev Econ Finan 48:18–33
8. Sefair JA, MÃ©ndez CY, Babat O, Medaglia AL, Zuluaga LF (2017) Linear solution schemes for mean-semi variance project portfolio selection problems: an application in the oil and gas industry. Omega 68:39–48
9. Lin PC, Watada J, Wu B (2013) Risk assessment of a portfolio selection model based on a fuzzy statistical test. IEICE Trans Inf Syst 96(3):579–588
10. Aouni B, Doumpos M, Pérez-Gladish B, Steuer RE (2018) On the increasing importance of multiple criteria decision aid methods for portfolio selection. J Oper Res Soc 69(10):1525–1542
11. Zadeh LA (1965) Fuzzy sets. Inf Control 8(3):338–353

12. Kwakernaak H (1978) Fuzzy random variables-1. Definition and theorems. Inf Sci 29:1–29. https://doi.org/10.1016/0020-0255(78)90019-1
13. Puri ML, Ralescu DA, Zadeh L (1993) Fuzzy random variables. In: Readings in fuzzy sets for intelligent systems, pp 265–271. Morgan Kaufmann
14. Lah MSC, Arbaiy N, Efendi R (2019) Stock market forecasting model based on AR (1) with adjusted triangular fuzzy number using standard deviation approach for ASEAN countries. In: Intelligent and interactive computing. Springer, Singapore, pp 103–114
15. Rahman HM, Arbaiy N, Wen CC, Efendi R (2019) Autoregressive modeling with error percentage spread based triangular fuzzy number. (2):36–40. https://doi.org/10.35940/ijrte.B1007.0782S219
16. Rahman HM, Arbaiy N, Efendi R, Wen CC, Info A (2019) Forecasting ASEAN countries exchange rates using auto regression model based on triangular fuzzy number. J Electric Eng Comput Sci 14(3):1525–1532. https://doi.org/10.11591/ijeecs.v14.i3

Residual Neural Network Vs Local Binary Convolutional Neural Networks for Bilingual Handwritten Digit Recognition

Ebrahim Al-wajih[1,2](✉), Rozaida Ghazali[1], and Yana Mazwin Mohmad Hassim[1]

[1] Faculty of Computer Science and Information Technology, Universiti Tun Hussein Onn Malaysia, Batu Pahat, 86400 Parit Raja, Johor, Malaysia
ebrahim.q.alwajih@gmail.com, {rozaida,yana}@uthm.edu.my

[2] Society Development & Continuing Education Center, Hodeidah University, 3114 Alduraihimi, Hodeidah, Yemen

Abstract. Most of the public and government documents in Arabic states are typed or written in bilingual forms; Arabic and Latin languages; such as railway reservation slips, bank withdrawal slips, etc. Using a bilingual system is better than using two systems for every language that is not a practical solution. In this paper, a bilingual digit recognition system is developed using Residual Neural Network (ResNet) and Local Binary Convolutional Neural Networks (LBCNN). The proposed systems are evaluated using a bilingual dataset generated from AHDBase and MNIST datasets. The recognition rate of ResNet or LBCNN is 99.38%. In addition, the proposed systems are applied to MNIST and AHDBase datasets separately. The obtained accuracies for MNIST are 99.27% and 99.51% and for AHDBase are 99.29% and 99.38%, respectively. The resulting performance of ResNet and LBCNN are the highest when they are compared against several state-of-the-art techniques.

Keywords: Bilingual digit recognition · Deep learning · Residual Neural Network · Local binary convolutional neural network

1 Introduction

Arabic and English handwriting recognition is a challenging problem because the writing style differs from writer to others as well as the variation of style at different instances of the same writer. Many researches were proposed to address either Arabic [1–4] or Latin [5–9] character/digit recognition problems, while none of researches focused on bilingual Arabic-Latin character/digit systems except those were developed for discriminating between the languages of documents/scripts [10, 11]. A bilingual Arabic-Latin digit recognition is one of the pattern recognition problems that is not concerned by researchers in deep, although it is a significant issue especially in the middle east. Most of the public and government documents existed in Arabic states are typed or written in bilingual forms (i.e., mixed of Arabic and English) such as application forms, railway reservation slips and cheques that need applications which support bilingual involving handwriting recognition systems.

R. Ghazali et al. (Eds.): SCDM 2020, AISC 978, pp. 25–34, 2020.
https://doi.org/10.1007/978-3-030-36056-6_3

Using a bilingual system is better than using two systems for every language that is not a practical solution. This approach reduces the search space in the database.

Nowadays, instead of using the common machine learning classifiers for addressing machine learning based problems such as gender recognition [12, 13], spam detection [14], etc., deep learning techniques are proving their efficiency for image-based pattern recognition such as face recognition [15], Latin handwritten recognition [7], object classification [8], etc. The common technique of deep learning is called Convolution Neural Network or CNN proposed by LeCun [16] in 1989. Many studies have examined the performance, in terms of accuracy, of CNN to Latin digit recognition problem [7], but few of them investigated Arabic digits [17]. Nevertheless, CNN suffers from a large time complexity due to the need of many hidden layers that motivates researchers to suggest many versions or modifications of CNN. In this paper, a fusion of local binary pattern (LBP) and CNN; called LBCNN, which is a CNN variation, has been applied to a bilingual dataset generated from MNIST [18] and AHDBase [4] datasets. The main goal of this paper is to propose a bilingual Arabic-Latin digit system as well as to enhance the performance of LBCNN and ResNet, in term of accuracy, by adjusting the number of convolutional layers.

The rest of the paper is organized as follows. In Sect. 2, some related works are presented and in Sect. 3, the proposed approaches are described. After that, the experimental setup and design are illustrated in Sect. 4. The results are discussed and compared against the state-of-the-art systems in Sects. 5 and 6, respectively. Finally, the conclusion of this study and future work are presented in Sect. 7.

2 Related Works

Many algorithms and approaches have been proposed in handwritten digit recognition either in Arabic [1–4] or Latin [5–9]. In addition, few studies have been done in offline bilingual handwriting recognition, but none of them considered Arabic-Latin digit recognition. On the other hand, many studies classified a script or document to the language written.

For bilingual digit recognition, Dhandra et al. [19] proposed a handwritten/printed recognition system for Kannada and English Digit in 2010. Local features were extracted by dividing images into 64 zones and then the pixel density of each zone was calculated to generate 64 features. After a year, the same features were applied on digits and characters to build a bilingual system for Kannada and English languages [20] and Kannada and Telugu digits [21]. Chaudhari and Gulati [22] calculated the pixel density to classify mixed English and Gujarati Digits. Lehal and Bhatt [23] identified the language of a script by proposing a bilingual handwritten digit recognition system for Devnagri (Hindi) and English language.

For classifying a bilingual script, Elgammal and Ismail [24] proposed a script identification for Arabic and English languages. Runlength Histogram and horizontal projection profiles were used as feature extraction techniques. Meanwhile, Guo et al. [25] proposed a model for printed recognition system for Chinese and English languages. The model was developed using micro structure feature (MSF), peripheral feature (PF) and peripheral + stroke density feature(P + SDF). Jawahar et al. [26] suggested a bilingual printed recognition system for Hindi and Telugu languages. Whole image pixels were used as features and then reduced by PCA to be 150 and 50 features for Telugu and Hindi, respectively. In addition, Kanoun et al. [10] classified printed and handwritten Arabic and Latin documents using a set of geometrical and morphological features extracted from whole scripts. Haboubi et al. [11] proposed an approach to discriminate between Arabic and Latin documents. The words were extracted from documents and then the structural features were extracted from the words. Moreover, Wang et al. [27] produced a model for printed characters of Chinese and English languages. Center-equidistance property-based region partition was used as a feature and segment-based/probability-based classifier was applied to determine the class of the language. In addition, Hangarge and Dhandra [28] produced a bilingual and trilingual handwritten recognition systems for identifying English, Devnagari and Urdu scripts.

Furthermore, some systems were developed for discriminating between the languages of documents/scripts. Pal and Chaudhuri [29] classified a textline of printed documents into five languages: English, Chinese, Arabic, Devanagari and Bangla. Shape-based features, statistical features, and water-reservoir-based features were extracted from 25000 textlines and classified by separation approaches. In 2011, Dhandra et al. [30] proposed a script independent automatic numeral recognition system using digit of Kannada, Telugu and Devanagari languages. Further, Singhal et al. [31] suggested a multilingual system to classify handwritten scripts into four languages; Roman (English), Devanagari (Hindi), Bangla, and Telugu. Gabor filters were used to generate the 40-feature vector from whole script. Meanwhile, Rajput and Anita proposed a system to classify a script into one of six categories [32]. Each category represents a tri-script that contains English script, Hindi script, and a local language script.

3 Proposed Method

In this study, a bilingual digit recognition models are presented by applying deep learning approaches developed using Residual Neural Network (ResNet) [33] and Local Binary Convolutional Neural Network (LBCNN) [8]. These techniques have been applied to a bilingual digit datasets created from Arabic and Latin digit dataset.

3.1 Residual Neural Network (ResNet)

Convolutional Neural Network (CNN) is one of the most used deep learning techniques. CNN uses mainly three architectural ideas, *local receptive fields, weight-sharing* and *pooling layers* [34]. CNN is able to recognize visual pattern directly from image pixels with minimal preprocessing. ResNet is a variation of CNN that enhances the performance of CNN and reduces the time consuming for training deeper networks [33]. The architecture of ResNet contains of traditional convolutional feed-forward networks that connect the output of the ℓth layer as input to the next layer $(\ell+1)^{th}$. There is a transition layer $x_\ell = H_\ell(x_{\ell-1})$ that adds a skip-connection layer that bypasses the non-linear transformations with an identity function: $x_\ell = H_\ell(x_{\ell-1}) + x_{\ell-1}$. The gradient in ResNet flows directly through the identity function from later layers to the earlier layers makes it the benefit of ResNet.

3.2 Local Binary Convolution Neural Networks (LBCNN)

LBCNN [8] is a variation of CNN, which uses LBP technique [35]; as a Local Binary Convolution (LBC). Using CNNs with LBC layers is called Local Binary Convolutional Neural Networks. The LBC layer has several parts including a set of fixed scattered binary convolutional filters, a non-linear activation function and a set of learnable linear weights. The filters are defined in which these filters are not updated in the training phase and the linear weights merge the responses of the activated filter to approximate the corresponding responses of the activated filter of a standard convolutional layer. The significant difference between the LBC and CNN is that LBC has less number of learnable parameters than CNN [8]. The theoretical analysis of this modification presents that the LBC layer is a good approximation for the non-linear activations of standard convolutional layers.

In LBCNN, the encoding using a 3×3 window can be applied using convolution filters in which eight filters of 2-spare difference are used as filters for the first layer in CNN producing 8-bit maps (filtered image). The original LBP is formulated by computing the weighted sum of all the bit maps using a pre-defined weight vector $V = [2^7, 2^6, 2^5, 2^4, 2^3, 2^2, 2^1, 2^0]$. Based on this, Eq. (1) can be reformulated that forms the basis of the LBC layer, as,

$$y = \sum_{i=1}^{8} \sigma(b_i.X).V_i \quad (1)$$

where $X \in \mathbb{R}^d$ is vectorized version of the original image, b_i is i^{th} sparse convolutional filter, σ is the non-linear trinary operator, the Heaviside step function in this case, and $y \in \mathbb{R}^d$ is the resulting LBP image. In this equation, by changing the linear weight V, the *base encoding* and the *ordering* of the encoding can be generalized, and by appropriately changing the non-zero (+1 and −1) support in the convolutional filters' *threshold selection* can be changed. Figure 1 shows the reformulation of LBP encoding using convolution filters and the difference between the architectures of CNN and LBCNN was described in [8].

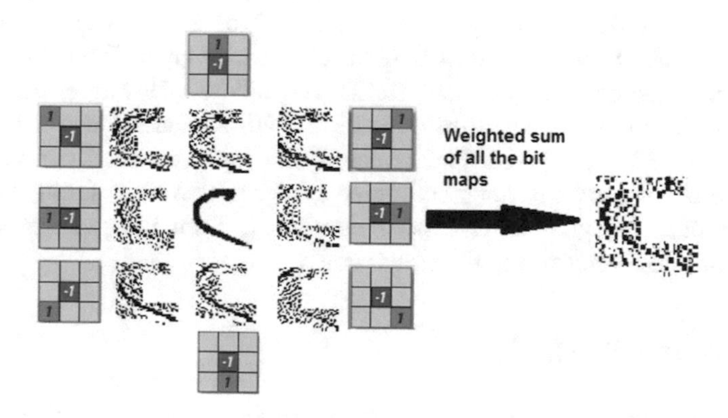

Fig. 1. Reformulation of LBP encoding using convolution filters

4 Experimental Setup and Design

A Ubuntu-based machine (3.6 GHz, i7 CPU, 12 GB of RAM, and NVIDIA GeForce 940 M of 2 GB) was used to implement all experiments. The Digit datasets used are MNIST and AHDBase. Each dataset is composed of 70,000 images; 60,000 for training and 10,000 for testing. The number of classes are 10 for each. However, when they were combined to create the bilingual dataset, the Digit classes that have the same shape were fused as one class. Then, distinguishing between the intra-classes variation is a postprocessing problem that is not in this paper scope. The combining step results the bilingual dataset of 120,000 and 20,000 images for training and testing sets, respectively, and 16 classes. Table 1. displays shapes of Arabic and Latin digits with samples of each class. The rows that show the bilingual datasets illustrates the combined classes which are {0, 5}, {1, 1}, {7, 6}, and {9, 9}. Lua and Torch have been utilized as a platform for evaluating the proposed systems. The performance of each classifier has been evaluated in terms of accuracy.

Table 1. Shapes of Latin and Arabic digits with samples of MNIST and AHDBase datasets.

Latin	Class	0	1	2	3	4	5	6	7	8	9						
	MNIST Sample																
Arabic	Class	0	1	2	3	4	5	6	7	8	9						
	AHDBase Sample																
Bilingual	Class	0	1	2	3	4	5	6	7	8	9	10	11	12	13	14	15
	Samples	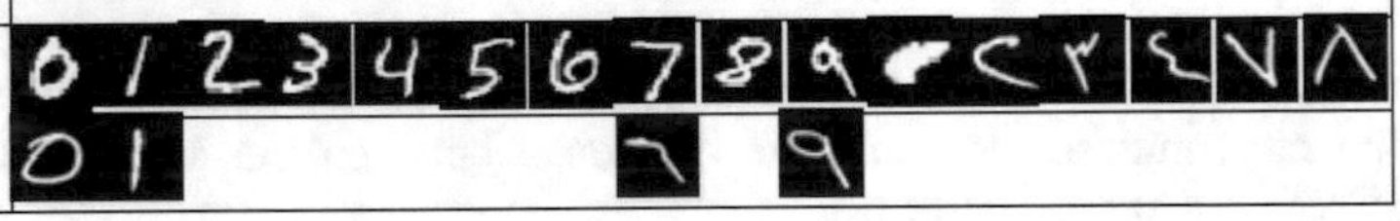															

Moreover, the architectures of ResNet and LBCNN have been configured same as the architecture of [33] with 3 × 3 filter size, 512 of anchor weights with 0.9 of sparsity, 16 of output channels, and 128 of fully connected layer. The learning rate, and the non-linear function are 1e-3, and ReLU, respectively. However, the number of convolutional layers is examined using 5, 10, 15, 20, 25 and 30 layers. Images have been resized to 32 × 32 pixels that represent the input vector. To investigate the performance of the proposed approach, three systems for Arabic, Latin and bilingual handwritten digit recognition have been developed. Then the performances of all systems have been compared against each other.

5 Results and Discussion

The performance of both ResNet and LBCNN techniques have been discussed in term of accuracy which is the main factor for digit recognition systems. The top-1 accuracy on 10-class of MNIST and AHDBase datasets and on 16-class of the bilingual dataset is reported in Table 2. Table 2 shows the accuracies of the techniques using several numbers of convolutional layers or depth. In addition, the number of epochs required to train systems have been included to show the accelerator of building systems as another factor.

Using ResNet approach, the highest acquired accuracies are 99.27%, 99.29%, and 99.38% for MNIST, AHDBase, and the bilingual datasets, respectively. Moreover, it can be noted that the accuracy has been enhanced when both Arabic and Latin datasets have been combined because of the enlarged dataset size. On the other hand, the highest accuracies obtained by LBCNN technique are 99.51%, 99.38%, and 99.38% using MNIST, AHDBase, and bilingual datasets, respectively. Generally, the performance of LBCNN using both Arabic and Latin datasets is better than ResNet and is equal to ResNet when bilingual dataset is used. Further, the top-5 accuracy is presented in the Fig. 2 that shows that most of the accuracies are close to 100%.

Table 2. Accuracies of ResNet and LBCNN techniques.

Tech	Datasets	Depth	5	10	15	20	25	30
ResNet	MNIST	Acc	96.02	98.86	98.48	**99.27**	98.39	98.75
		#epoch	*4*	*12*	*8*	***80***	*11*	*5*
	AHDBase	Acc	99.03	98.86	99.20	99.24	99.05	**99.29**
		#epochs	*11*	*8*	*27*	*24*	*8*	***28***
	BILINGUAL	Acc	99.24	98.97	**99.38**	99.27	98.31	98.41
		#epochs	*80*	*15*	***80***	*80*	*4*	*3*
LBCNN	MNIST	Acc	99.40	99.45	99.34	98.92	98.92	**99.51**
		#epochs	*80*	*80*	*80*	*6*	*9*	***80***
	AHDBase	Acc	**99.38**	99.27	98.91	98.64	99.15	98.96
		#epochs	***80***	*80*	*4*	*4*	*9*	*6*
	BILINGUAL	Acc	**99.38**	99.35	99.33	99.00	98.49	98.45
		#epochs	***80***	*80*	*80*	*17*	*5*	*3*

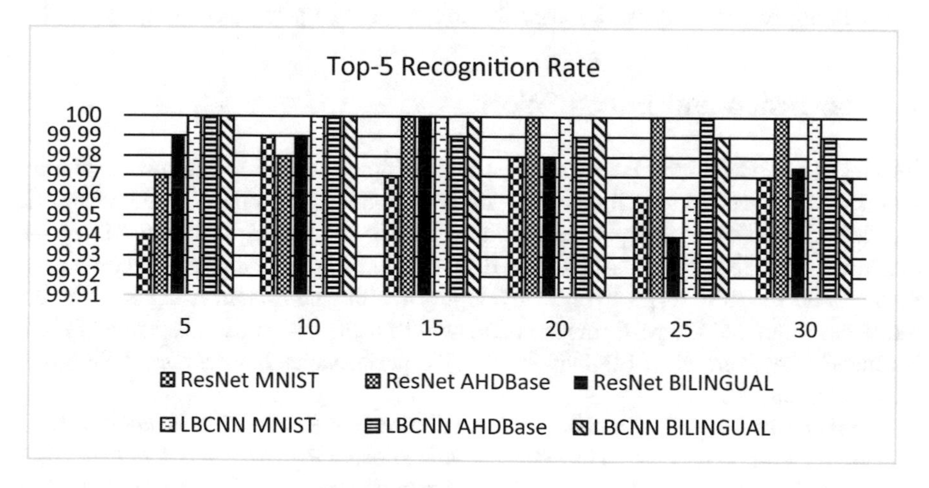

Fig. 2. The top-5 accuracies

Furthermore, the number of convolutional layers has been evaluated, using 5, 10, 15, 20, 25 and 30 layers, to enhance the performance of the systems. For Latin digit dataset, it needs 20 and 30 layers for ResNet and LBCNN techniques, respectively. Moreover, 30 and 5 layers have been needed to get the highest accuracies for Arabic digit dataset to develop ResNet and LBCNN based systems. However, when both datasets were combined, the number of convolution layers has reduced to 15 and 5 for ResNet and LBCNN techniques, respectively.

On the other hand, the number of epochs utilized can be affected by the technique applied, its architecture and dataset used. For example, when ResNet is applied to the bilingual dataset, the number of needed epochs differs based of the number of convolution layers. Also, when LBCNN is trained using 5 convolution layers, the performance varies depend on the datasets. For building Arabic digit classification systems, the highest obtained accuracies require 28 and 80 epochs for ResNet and LBCNN, respectively

6 Comparison Against Previous Studies

The performance of this study has been compared against state-of-the-art techniques. First, the proposed techniques give significant recognition rates for AHDBase dataset, as shown in Table 3. It is clear in Table 3, that ResNet and LBCNN produce the highest accuracies compared others. This implies that ResNet and LBCNN can select the significant features better than using the classic ones.

Although a comparison using MNIST was done in [8], we have discussed another comparison in this section because the number of convolutional layers or depth in our study is less than that of used in [8]. Table 4 shows that the accuracies obtained in the proposed system is the highest against the other systems except the accuracy in [8] which is equal to ours. However, the number of depth in LBCNN in the proposed systems is less than those used in [8] by 100 layers. Unfortunately, to the best of author knowledge, there is no study which addressed the bilingual (Arabic-Latin) digit recognition problem that can be used for comparing our performance against with.

7 Conclusion and Future Work

ResNet and LBCNN have been proposed to develop a system for recognizing bilingual Arabic-Latin handwritten digits. The highest obtained accuracy is 99.38% using ResNet or LBCNN technique. However, when the suggested techniques have been evaluated on MNIST and AHDBase datasets, the highest acquired accuracies are 99.51% and 99.38%, respectively. Developing a bilingual system using ResNet technique has improved the performance compared to using two separate systems for each language. However, when LBCNN is used, the performance has decreased slightly for only Latin dataset.

Furthermore, selecting number of convolution layers or depth is not significant because the highest accuracies have been obtained using ResNet when depth are 20, 30, and 15 for MNIST, AHDBase and bilingual datasets, respectively; and using LBCNN when depth are 30, 5, and 5 for MNIST, AHDBase and the bilingual datasets, respectively. Moreover, the performance of the proposed techniques has been compared against some state-of-the-art systems that shows the significance of the proposed systems.

However, the limitation of this study is hardware resources, i.e. the size of the GBU of 2 GB. Because of this, some restriction has been enforced in all experiments such as the models have been trained with a small size of depth and a mini-batch size of 10; and trained for 80 epochs.

Table 3. A comparison using AHDBase dataset

References	Feature extraction technique	RR %
[1]	2-D DCT	85.26
[2]	DAISY descriptor	98.75
[3]	DWT + DCT	98.32
[4]	Hetero features	99.15
Ours	ResNet	***99.29***
	LBCNN	***99.38***

Table 4. A comparison using MNIST dataset.

References	Feature extraction technique	RR %
[5]	R-STDP	97.2
[6]	CNN with feature map reduction	99.19
[7]	CNN	99.21
[8]	LBCNN-150	99.51
[9]	Histogram of oriented gradients	99.29
Ours	ResNet	*99.27*
	LBCNN-30	***99.51***

Acknowledgment. The authors would like to thank Universiti Tun Hussein Onn Malaysia (UTHM) and Ministry of Education Malaysia for financially supporting this research under the Fundamental Research Grant Scheme (FRGS), Vote No. 1641.

References

1. AlKhateeb JH, Alseid M (2014) DBN-based learning for Arabic handwritten digit recognition using DCT features. In: 2014 6th international conference on computer science and information technology (CSIT). IEEE, pp 222–226
2. Chatterjee A, Malakar S, Sarkar R, Nasipuri M (2018) Handwritten digit recognition using DAISY descriptor: a study. In: 2018 Fifth international conference on emerging applications of information technology (EAIT). IEEE, pp 1–4
3. Lawgali A (2015) Handwritten Digit Recognition based on DWT and DCT. Int J Database Theory Appl 8:215–222
4. El-Sherif EA, Abdelazeem S (2007) A two-stage system for Arabic handwritten digit recognition tested on a new large database. In: Artificial intelligence and pattern recognition, pp 237–242
5. Mozafari M, Ganjtabesh M, Nowzari-Dalini A, Thorpe SJ, Masquelier T (2018) Bio-inspired digit recognition using spike-timing-dependent plasticity (stdp) and reward-modulated stdp in deep convolutional networks. ArXiv Preprint. https://arxiv.org/abs/180400227

6. Chakraborty S, Paul S, Sarkar R, Nasipuri M (2019) Feature map reduction in CNN for handwritten digit recognition. In: Recent developments in machine learning and data analytics. Springer, Singapore, pp 143–148
7. Siddique F, Sakib S, Siddique MAB (2019) Handwritten digit recognition using convolutional neural network in Python with tensorflow and observe the variation of accuracies for various hidden layers
8. Juefei-Xu F, Boddeti VN, Savvides M (2016) Local binary convolutional neural networks. ArXiv Preprint. https://arxiv.org/abs/160806049
9. Das A, Kundu T, Saravanan C (2018) Dimensionality reduction for handwritten digit recognition
10. Kanoun S, Ennaji A, LeCourtier Y, Alimi AM (2002) Script and nature differentiation for Arabic and Latin text images. In: 2002 Proceedings of eighth international workshop on frontiers in handwriting recognition. IEEE, pp 309–313
11. Haboubi S, Maddouri SS, Amiri H (2011) Discrimination between Arabic and Latin from bilingual documents. In: 2011 International conference on communications, computing and control applications (CCCA). IEEE, pp 1–6
12. Al-wajih E, Ahmed M Extended statistical features based on gabor filters for face-based gender classification: classifiers performance comparison. 10
13. Al-wajih E, Ghouti L (2019) Gender recognition using four statistical feature techniques: a comparative study of performance. Evol Intell. https://doi.org/10.1007/s12065-019-00264-z
14. Waheeb W, Ghazali R, Deris MM (2015) Content-based sms spam filtering based on the scaled conjugate gradient backpropagation algorithm. In: 2015 12th International conference on fuzzy systems and knowledge discovery (FSKD). IEEE, pp 675–680
15. Lawrence S, Giles CL, Tsoi AC, Back AD (1997) Face recognition: a convolutional neural-network approach. IEEE Trans Neural Netw 8:98–113
16. LeCun Y (1989) Generalization and network design strategies. Connect Perspect 143–155
17. El-Sawy A, Hazem EB, Loey M (2016) CNN for handwritten arabic digits recognition based on LeNet-5. In: International conference on advanced intelligent systems and informatics. Springer, Cham, pp 566–575
18. LeCun Y (1998) The MNIST database of handwritten digits. Httpyann Lecun Comexdbmnist
19. Dhandra BV, Hangarge M, Mukarambi G (2010) Spatial features for handwritten Kannada and English character recognition. IJCA Spec Issue RTIPPR 3:146–151
20. Dhandra BV, Mukarambi G, Hangarge M (2011) A recognition system for handwritten Kannada and English characters. Int J Comput Vis Robot 2:290–301
21. Dhandra BV, Mukarambi G, Hangarge M (2011) A script independent approach for handwritten bilingual Kannada and Telugu digits recognition. Int J Mach Intell 3
22. Chaudhari SA, Gulati RM (2013) An OCR for separation and identification of mixed English—Gujarati digits using kNN classifier. In: 2013 International conference on intelligent systems and signal processing (ISSP). IEEE, pp 190–193
23. Lehal G, Bhatt N (2000) A recognition system for Devnagri and English handwritten numerals. In: International Conference on Multimodal Interfaces—ICMI 2000, pp 442–449
24. Elgammal AM, Ismail MA (2001) Techniques for language identification for hybrid Arabic-English document images. In: 2001 Proceedings of sixth international conference on document analysis and recognition. IEEE, pp 1100–1104
25. Guo H, Ding X, Zhang Z, Guo F, Wu Y (1995) Realization of a high-performance bilingual Chinese-English OCR system. In: 1995 Proceedings of the third international conference on document analysis and recognition. IEEE, pp 978–981

26. Jawahar CV, Kumar MP, Kiran SR (2003) A bilingual OCR for Hindi-Telugu documents and its applications. In: 2003 Proceedings of seventh international conference on document analysis and recognition. IEEE, pp 408–412
27. Wang K, Jin J, Wang Q (2009) High performance Chinese/English mixed OCR with character level language identification. In: 2009 10th International conference on document analysis and recognition, ICDAR'09. IEEE, pp 406–410
28. Hangarge M, Dhandra BV (2010) Offline handwritten script identification in document images. Int J Comput Appl 4:6–10
29. Pal U, Chaudhuri BB (2001) Automatic identification of English, Chinese, Arabic, Devnagari and Bangla script line. In: 2001 Proceedings of sixth international conference on document analysis and recognition. IEEE, pp 790–794
30. Dhandra BV, Benne RG, Hangarge M (2011) Kannada, Telugu and Devanagari handwritten numeral recognition with probabilistic neural network: a script independent approach. Int J Comput Appl 26:11–16
31. Singhal V, Navin N, Ghosh D (2003) Script-based classification of hand-written text documents in a multilingual environment. In: 2003 Proceedings of 13th international workshop on research issues in data engineering: multi-lingual information management, RIDE-MLIM 2003. IEEE, pp 47–54
32. Rajput GG, Anita HB (2010) Handwritten script recognition using DCT and wavelet features at block level. IJCA Spec Issue RTIPPR 3:158–163
33. He K, Zhang X, Ren S, Sun J (2016) Identity mappings in deep residual networks. In: European conference on computer vision. Springer, Cham, pp 630–645
34. LeCun Y, Bottou L, Bengio Y, Haffner P (1998) Gradient-based learning applied to document recognition. Proc IEEE 86:2278–2324
35. Ojala T, Pietikainen M, Harwood D (1994) Performance evaluation of texture measures with classification based on Kullback discrimination of distributions. In: 1994 Proceedings of the 12th IAPR international conference on pattern recognition, conference a: computer vision & image processing, vol 1. IEEE, pp 582–585

Incorporating the Markov Chain Model in WBSN for Improving Patients' Remote Monitoring Systems

Rabei Raad Ali[1], Salama A. Mostafa[1(✉)], Hairulnizam Mahdin[1], Aida Mustapha[1], and Saraswathy Shamini Gunasekaran[2]

[1] Faculty of Computer Science and Information Technology, Universiti Tun Hussein Onn Malaysia, Batu Pahat, Johor, Malaysia
rabei.aljawary@gmail.com, {salama, hairuln, aidam}@uthm.edu.my

[2] College of Computer Science and Information Technology, Universiti Tenaga Nasional, Kajang, Selangor 43000, Malaysia
sshamini@uniten.edu.my

Abstract. Wireless body sensor network (WBSN) allows remote monitoring for different types of applications in security, healthcare and medical domains. Medical applications involve monitoring a large number of patients in real-time environments. The WBSNs in such environments have to be efficient and reliable in terms of data transfer rate, accuracy, latency, and power consumption. This work focuses on studying the slotted access protocol variables in the Contention Access Period (CAP) with the acknowledged uplink traffic (nodes-to-coordinator) under the WBSN channel. This paper proposes a Markov Chain model in WBSN (MC-WBSN) for improving the efficiency and reliability of patients' remote monitoring systems. The application of the model includes propagating human arm sensory data and analyzing the latency, power consumption, throughput, and higher path loss channel of the WBSN. The results show that the hidden nodes have a great impact on WBSNs performance and throughput. This issue is highly associated with the capacity of the transmitted power.

Keywords: Wireless body sensor networks · Human arm simulation · Media access control · Markov chain

1 Introduction

Wireless Body Sensor Networks (WBSN) in recent years gained great attention due to the innovative wearable, wireless and implantable biosensors [1, 19]. The applications of the WBSN are extended to cover medical, healthcare, in-vivo monitoring, fitness, sport and security sectors. The architecture of the WBSN contains in general wireless Personal Device (PD), wireless nodes, wireless sensors, wireless centre unit and actuators [2]. There are several types of PD such as smartphones. The WBSN has three distinct evaluation dimensions, which are connectivity, sensing modality, size and cost. However, power consumption is still one of the main WBSN issues [3]. The WBSNs

R. Ghazali et al. (Eds.): SCDM 2020, AISC 978, pp. 35–46, 2020.
https://doi.org/10.1007/978-3-030-36056-6_4

usually work on batteries and battery quality and usage life affect the performance of the network [4].

Technically, in the star network topology of the WBSN applications data is collected from monitoring the human body and send to a coordinator through a series of nodes. Most of the communications of the WBSN are uplink (nodes-to-coordinator) using IEEE 802.15.4 Media Access Control (MAC) [5, 6]. IEEE 802.15.4 Media Access Control (MAC) has received much attention to evaluating its performance including the parameters of latency, power consumption, throughput, and higher path loss channel.

There are different researcher contributions to WBSN improvement. Javaid et al. [5] propose an innovative routing protocol for Intra-WBSN that is named as ATTEMPT. This routing scheme uses multi-hop communication approach between the wireless nodes. The ATTEMPT also includes a Body Node Coordinator (BNC) to improve the energy efficiency of the network. Ndih et al. [6] propose a new kind of Media Access Control (MAC) sub-layer schemes known as Slotted Carrier Sense Multiple Access with Collision Avoidance (CSMA/CA). Al Rasyid et al. [7] evaluate the slotted and unslotted CSMA/CA protocol in medical systems. They analyze the impact of the CSMA/CA protocol parameters such as Beacon Order, Backoff Exponent, Maximum Backoff Exponent and Superframe Order. Barakah et al. [8] make an inclusive study of WBSN research. Subsequently, they propose a Virtual Doctor Server (VDS) in the WBSN architecture to assist users.

The main objective of this work is to evaluate the IEEE 802.15.4 standard of the WBSN applications. For this purpose, the Markov Chain is modelled in the WBSN (MC-WBSN) to incorporate beacon-enabled mode of Contention Access Period (CAP) in access protocol of the CSMA/CA [11]. It focusses on studying the slotted access protocol variables in the CAP with the acknowledged uplink traffic (nodes-to-coordinator) under the WBSN channel. It adopts and re-extends the Markov chain based analytical model of [9] and follows the methodology of [11].

2 Literature Review

Now a day, embedded electronics are pervasive or widespread in medical devices, cars and Personal Digital Assistant (PDA) such as mobile phones alarms and clocks [12]. Ambient intelligence technology can be embedded and invisible in our natural surroundings and provides different types of services [1, 13]. Wireless Sensor Networks (WSN) play a key factor in building such technology. In recent years, WSNs are widely adopted in the development of smart sensors in the Internet of Things (IoT) systems [12]. The computing and processing resources of these sensors are limited and can be adopted in different types of appliances. Figure 1 shows a simple sensor node architecture of an ambient intelligence that consists of a processor, sensor, wireless communication modules, Media Access Control (MAC) and power supply. The processor module is responsible for controlling and processes its own data in the storage and, facilitates the operation of sensor nodes to be received by other nodes. The sensor node has a power supply module that usually includes a micro-battery to provide the required energy. The wireless communication module works on sending/receiving data along with their control messages.

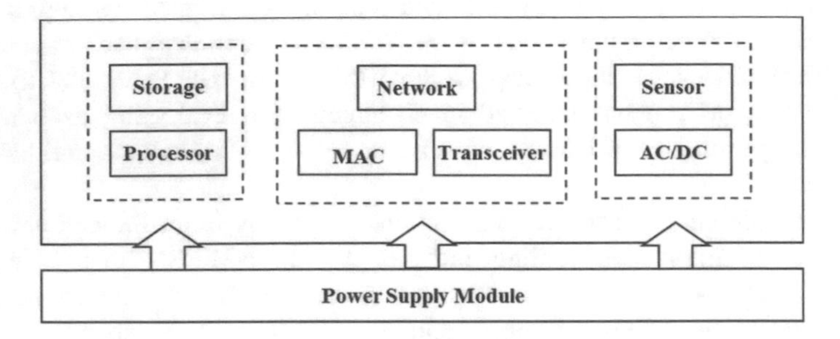

Fig. 1. A simple sensor node architecture [3]

Many areas have been using the WSN technology, such as healthcare monitoring, process management, geographic information system, pollution monitoring, landslide detection and forest fire detection [4]. Medical devices that contain body sensors benefit from the WSN technology which emerges the development of the WBSN.

The energy-efficient sensing devices and energy-efficient Medium Access Control (MAC) protocols are critical in the WBSN [12]. The IEEE 802.15.6 standard mainly stipulates the basic structure of the WBSN physical layer and the MAC protocol [13]. It represents the global standard of low power consumption and low data rate for WBSNs. It is implemented in a wide range of application areas such as industrial process surveillance, surroundings monitoring and personal and health monitoring. The IEEE 802.15.4 defines the MAC of a sensor network in order to connect devices through the Physical layer (PHY) over the wireless channel.

The PHY protocol data unit consists of the length of the packet, a synchronization header and a PHY header that indicate the payload [14]. It is made through the MAC protocol that communicates channels when a device is accessed. The major functions of the MAC include the management of channel access through CSMA-CA, generation of the beacon, synchronization, support for optional device security, disassociation and facilitation of Personal Area Network (PAN), acknowledgement messages, validation messages and maintains of Guaranteed Time Slot (GTS) communication [15].

Various applications and network topologies entail the need for maintaining a flexible MAC frame structure and a simple protocol. The 2.4 GHz PHY supports 16 channels with the range of 2.4–2.4835 GHz and sufficient channel spacing (5 MHz). It is essential in fulfilling the requirements of the reception filter and easing the process of transmission.

According to the IEEE 802.15.4 standard, data transmission occurs in three ways, which are the coordinator of a device transmission, a device to coordinator transmission and transmission between two devices of any kind. Nevertheless, the WSBN which is considered in this study basically involves transmissions between the device and coordinator. More so, the uplink data traffic of the WBSN requires sensing devices in a star topology scheme [16].

3 The MC-WBSN Model

The WBSN application of this work adopts the one-hop star topology. The WBSN is assumed to operate by a common coordinator that has M sensing devices. The nodes are within the required range and do not interrupt ongoing transmissions (no hidden terminal problem). In order to assure consistent data transfer, ACKs are applied in WBSN. The average duration referring to MAC latency (L) is defined as the duration of linking two transmissions less the time of IDLE state, provided in the formula (1):

$$L = \frac{NM\left(1 - p_{idle}^{n}\right)}{S} \tag{1}$$

RTS has been determined based on the average energy utilization of the sensing nodes within transmitting and receiving stages of three radios. The first radio has been set as Idle with ready to receive a command, where the second radio has been set into Transmit mode and the third radio in receive mode. In this sense, the average spending power is set as follows:

$$\pi(idle) + \sum_{i=1}^{5} \pi(bo_i) + \sum_{i=1}^{5}\sum_{j=1}^{2} \pi\left(cs_{ij}\right) + N\pi(tx) + T_{ACK}^{n}\pi\,(ack) \tag{2}$$

The productivity, expressed through channel S, is set as a sequence of time spent in transmission as presented in the following formula:

$$S = \frac{N\pi_S^C}{\pi_{II}^c + T_{B,I}^c\pi_{BI}^c + N\pi_S^c + N\pi_f^c} = \frac{N\beta}{T + T_{B,I}^c(1-\propto) + N(\beta+\delta)} \tag{3}$$

Here, the assumption is that neither an inactivity period or Contention Free Period (CFP) is possessed by the super-frame. Nonetheless, the values of Active Superframe Duration (SD), macBeaconOrder (BCO), macSuperframeOrder (SFO) and Beacon Interval (BI) are used in determining the delay period during the inactive time. The following formulas represent the relation:

$$BI = \left(aBaseSuperframeDuration \times 2^{BCO}\right), 0 \le BO \le 14 \tag{4}$$

and

$$SD = \left(aBaseSuperframeDuration \times 2^{SFO}\right), 0 \le SO \le 14 \tag{5}$$

The time interval between two constant beacon frames is defined by the BI, while the active part within the BI is defined by SD. The length of the base superframe duration is equal to a Base Super Drame Duration equals to 960 symbols or 15.36 ms.

The WBSN applications are interesting particularly when network lifetime and energy consumption are the main concerns. This work assumes that the entire super-frame duration is active in which SFO equals to BCO. Firstly, in the acknowledged

CAP analysis, it is assumed that no packets are lost and that the only possible cause of failure in the transmission is collisions. Each time data has been transmitted successfully, MAC level ACKs are and will reach the destination node. In an event that collision occurs, the packet data will be resent. When a packet with an ACK is received in the scheme, the sensing node is being notified by the coordinator. In case, an ACK is not received by the node, it is assumed that collisions occurred, rather than resending.

The probability p of a node getting a packet to transmit at the subsequent slot is p = λ = N. More so, the implication of this assumption of unsaturated traffic entails neglecting the buffering at the nodes [11]. When nodes are trying to initiate a transmission or transmitted, new packets are not accepted for transmission (p = 0). The individual node behaviour is modelled by means of a Markov chain as shown in Fig. 2.

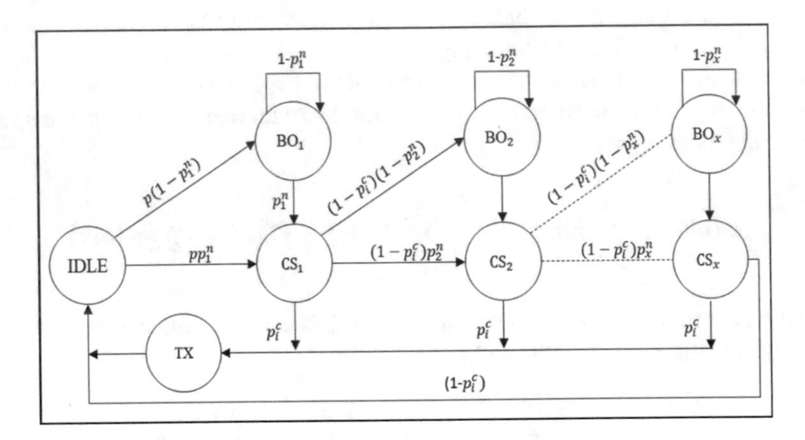

Fig. 2. The MC-WBSN model

The number of back-off slots the node spends in BO1 by X1, an equivalent geometrical distribution is defined in (6).

$$P[X_1 = K] = (1 - P_1^n)^k P_1^n, k = 0, 1, \ldots\infty \tag{6}$$

where $P_1^n = 1/4 : 5$ so that the same meaning will be possessed by the geometric distribution and the uniform distribution is specified by E[X1] = 3:5. Markov equation scheme is solved to obtain the chain for steady-state occupancy as shown in Table 1.

The probability that transmission is initiated by any node since the channel has not been active for the consecutive back-off-slots, is represented by $p_{t=ii}^n$ and computed as follows:

$$p_i^c = \frac{p_{ii}^c}{p_{i/i}^c} \tag{7}$$

Table 1. The Markov equations scheme

No.	Parameter	Equation
1	$\pi(\text{idle})$	$(1-\text{p})\pi(\text{idle})+\pi(\text{ack})+(1-\text{p}_\text{i}^\text{c})\pi(\text{cs}_{51})+\left(1-\text{p}_{\text{i/i}}^\text{c}\right)\pi(\text{cs}_{52})$
2	$\pi(\text{bo}_1)$	$(1-\text{p}_1^\text{n})[\text{p}\pi(\text{idle})+\pi(\text{bo}_1)]$
3	$\pi(\text{cs}_{11})$	$\text{p}_1^\text{n}[\text{p}\pi(\text{idle})=\pi((\text{bo}_1)]$
4	$\pi(\text{cs}_{12})$	$\text{p}_\text{i}^\text{c}\pi(\text{cs}_{11})$
5	$\pi(\text{bo}_2)$	$(1-\text{p}_2^\text{n})[(1-\text{p}_\text{i}^\text{c})\pi(\text{cs}_{11})+\left(1-\text{p}_{\text{i/i}}^\text{c}\right)\pi(\text{cs}_{12})+\pi(\text{bo}_2)]$
6	$\pi(\text{cs}_{21})$	$\text{p}_2^\text{n}[(1-\text{p}_\text{i}^\text{c})\pi(\text{cs}_{11})+\left(1-\text{p}_{\text{i/i}}^\text{c}\right)\pi(\text{cs}_{12})+\pi(\text{bo}_2)]$
7	$\pi(\text{cs}_{22})$	$\text{p}_\text{i}^\text{c}\pi(\text{cs}_{21})$
8	$\pi(\text{bo}_\text{x})$	$(1-\text{p}_\text{x}^\text{n})[(1-\text{p}_\text{i}^\text{c})\pi(\text{cs}_{21})+\left(1-\text{p}_{\text{i/i}}^\text{c}\right)\pi(\text{cs}_{22})+\pi(\text{bo}_\text{x})]$
9	$\pi(\text{cs}_{31})$	$\text{p}_\text{x}^\text{n}[(1-\text{p}_\text{i}^\text{c})\pi(\text{cs}_{21})+\left(1-\text{p}_{\text{i/i}}^\text{c}\right)\pi(\text{cs}_{22})+\pi(\text{bo}_\text{x})]$
10	$\pi(\text{cs}_{31})$	$\text{p}_\text{i}^\text{c}\pi(\text{cs}_{\text{x}1})$

On the other hand, β can be computed as $= \text{Mp}_{\text{t=ii}}^\text{n}(1-\text{p}_{\text{t=ii}}^\text{n})^{\text{M}-1}$. As observed in Fig. 3, the probability of the channel remains idle for two consecutive back-off slots $\text{p}_\text{ii}^\text{c}$ can be determined by solving the Markov chain for the channel.

The probability of the channel is in an idle state consecutively for two back-off slots,$\text{p}_\text{ii}^\text{c}$ can be calculated using (7) and (8) as follows:

$$p_{ii}^{c}\frac{1}{1+T_{B,I}^{C}(1-\propto)+n(\beta+\delta)} \tag{8}$$

These nonlinear equations can be solved by following numerical approximation methods. The Markov chain model is used for the determination of the channel state as presented in Fig. 3. In the beginning, the channel state is IDLE. If any node begins transmission, the Idle changes into Success noting successful transmission.

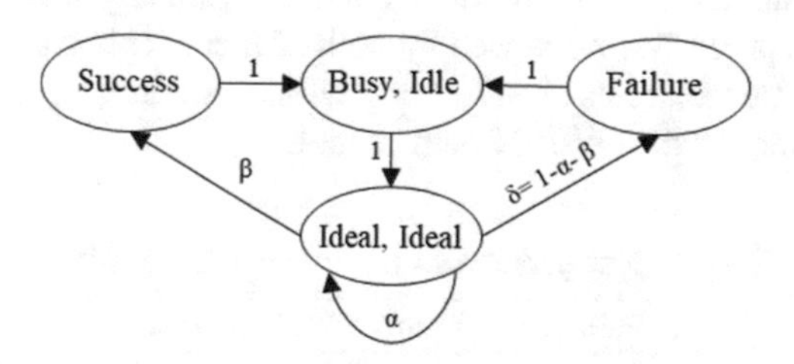

Fig. 3. The Markov chain model determination of the channel state

The channel back to a state "Busy; Idle" out of an intermediate state of "Busy; Idle" at the end of the transmission, if the situation Success or Failure. if the state "busy; Idle" where the node will wait for the ACK coordinator, which reaches when the previous state is a Success. If it is a Failure, and packet data does not arrive at the coordinator, this implies that the corresponding remote sensing node is kept waiting for ACK MacAckWaitDuration. The p_{ack}^{c} calculation requires the ACK of a Success in order to be executed. The complete formulation of the MC-WBSN is presented in [11].

4 Simulation and Results

In the computer networks field, simulation is the most common evaluation methodology. It is frequently used for the development of communication architectures and new network protocols. Simulation tools allow designing a model of computer network via classifying nodes, routing behaviour, communication channels and some other related activities assets of simulation parameters. Different topologies can be implemented in order to study routing behaviour [13].

The simulation model of this work is built by the MATLAB 2010a platform. The radio signals of the WBSN has encountered a higher path loss associated with the empty space as humans' body have a higher packet loss rate [11]. The path loss increases with antenna separation. It is almost ($n \sim 3.35$) in the arm model. Table 2 shows the parameters and the corresponding values of the path loss for the simulated arm model.

Table 2. The path loss parameters of the arm model

Parameter	Arm
d_0[cm]	10
P_0.m[dB]	32–2
N	3.335
a[Db]	4.1

The path-loss in decibels at a distance d, $P_{0,dB}$ represents the path loss at a reference distance d_0 and N is the path loss exponent. The local diffracting components and reflected components are measured and modelled separately using different methods due to the different properties possessed by both of them. This way, extant propagation measurements can be used close to the body within any indoor environment. Table 3 shows the parameters of this WBSN path model.

Table 3. The path loss parameters for different antenna-body separations

Parameter	0 mm	5 mm	10 mm
n	5.8	5.9	6.0
d_0[cm]	0.1		0.1
P_0.m[dB]	56.1	48.4	45.8

The following formula is used for calculating path loss computation:

$$P_{dB} = P_{o,dB} + 10_n log(\frac{d}{d_o}) \quad (9)$$

where n denotes the exponent of path loss, $P_{o,dB}$ is the path loss at a reference distance d_o and P_{dB} represents the path loss in decibels at a distance d.

The certain parameter value is defined in Table 4. The power states and transitions to the radio CC2420 power consumption are illustrated in [11]. It is associated with the transition from state to state S1–S2 by the means of three parameters which are: time of transition, T, current equivalent in the target, and I (S2) and supply current electricity (VDD) = 1.8 V.

Table 4. The path loss parameters for different antenna-body separations

aMinBE	4	aMaxBE	5
macMaxCSMABackoffs	5	CW	2
BCO	5	SFO	5
Number of sensing nodes, M	12	nbeacon	2backoff slots
Ldata	10 backoff slots	N = Ldata	

The simulation results are discussed in terms of latency, power consumption, throughput, and higher path loss channel of the WBSN. The latency is equivalent to the average time used for two broadcasts that are successful minus the average time used during the state of idleness. This is maintained below 25 back off slots that begin with a value of 0:02, and afterwards starts increasing rapidly as a result of congestion of channel. Thus, within this time frame, a packet is available for transmission till the transmission is successfully completed as illustrated in Fig. 4.

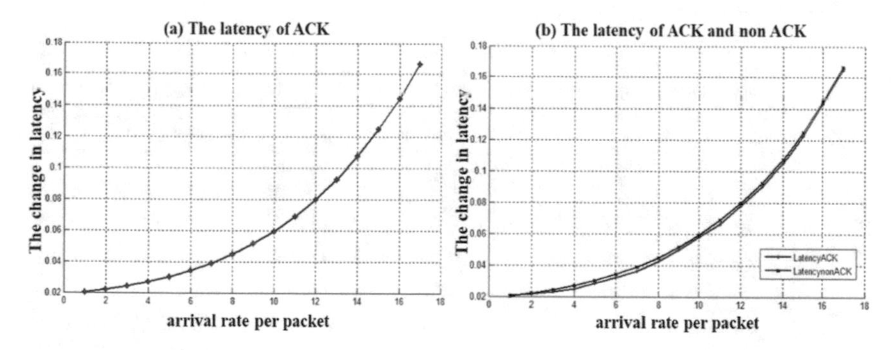

Fig. 4. The latency of packet transition ACK and non ACK over time

Below is a description of the time frame within which a packet becomes ready for transmission and the successful completion of the transmission. When the ACK is included in the model it leads to an obvious increase in latency. However, as an increase occurs in the rate of packet arrival, this increase continues to progress until it gets a maximum increase of 17.4% with regards to latency. The difference in the energy consumption of the ACK and non ACK models is relatively small. The packet arrival rate for the channel maximum load reaches a supreme variance of 5%. Figure 5 describes the energy consumption performance of both the ACK and non ACK models.

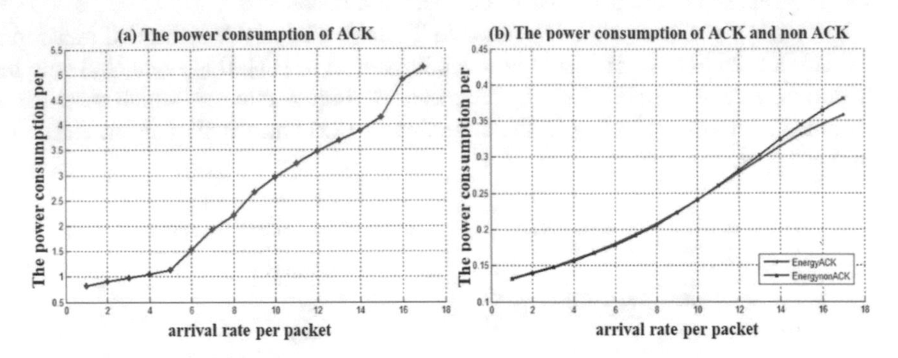

Fig. 5. The energy consumption performance of ACK and non ACK

In the sensing nodes, the power consumption's behaviour of the MAC shows that the ACK rises the power consumption due to waiting and receiving the ACK. Nevertheless, it is not linear due to the fact that the channel throughput makes adjustments to it. The maximum difference that is achieved is about 0:04, thereby leading to the derivation of a difference of 17.6% as shown in Fig. 5(b). Moreover, there is around 0.05 tradeoff between energy consumption and throughput. Figure 6(a) shows the throughput of ACK. The overall throughput performance tends to decline due to consequent collisions and channel congestion. This means, to reduce collision, the traffic load need to be less than 0.1.

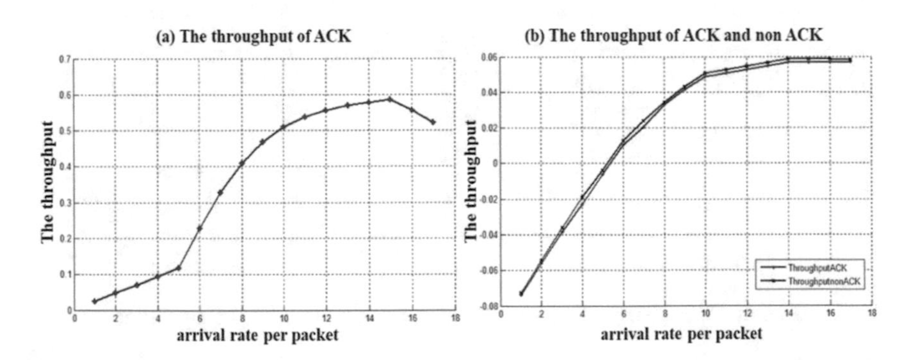

Fig. 6. The overall throughput performance of ACK and non ACK

On the other hand, Fig. 6(b) shows the throughput of non ACK and ACK models. When collisions occur, the highest throughput that can be achieved is often less than the highest bit rate as demonstrated by the results of these models. It is observed that the packet arrival rate of non ACK traffic and ACK traffic is equal to 0.02. Yet, in relation to higher rates of data, a decline occurs for some extent in the throughput of the ACK model. Comparably, the non ACK case produces nearly linearly results, getting a supreme variance of 13:4%. Because of the decrease in coherence, channel congestion occurs thereby leading to an increase in network load.

It is easier to compute the higher path loss if the topology remains as it was previously. Thus, the diffraction around the body of the human leads to the monitoring of higher path loss. In the actual sense, in the higher path loss channel, there is a short distance between the coordinator distance and the sensing nodes in which the distance is often with a circle of 0.3 m. For instance, in this channel, it is possible to have a situation in which the sensing nodes are placed behind the arm of a person with the coordinator in the front arm. Accordingly, there is a real need to decrease the traffic load in order to mitigate power consumption and increase the throughput of the network. Figure 7 shows the throughput of the higher path loss channel.

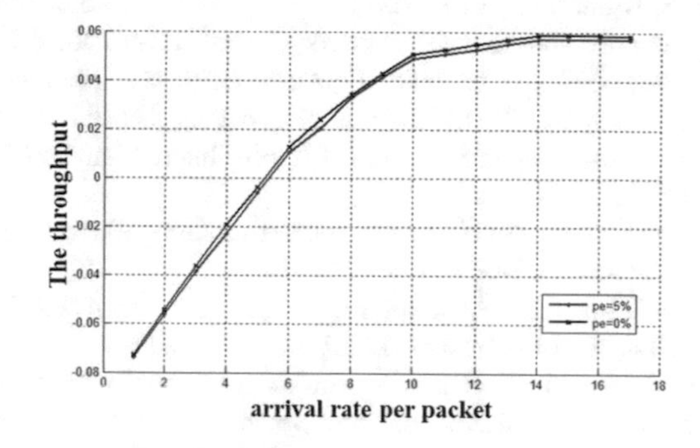

Fig. 7. The throughput of the higher path loss channel

The signification of this study is to assess the performance of the CAP mechanisms of IEEE 802.15.4. It is useful for examining the contention parameters to traffic prioritization and observing service differentiation. In addition, the proposed analytical model serves as a guide to appropriately configuring these parameters and mechanisms for each priority class in the future. The tested parameters of latency, energy consumption and throughput indicate the ability of the protocol and the overall MC-WBSN and possible points of improvements.

5 Conclusion

The transfer packet from the coordinator and the node of the WBSN is needed to ensure reliable data in the applications. There is a need to investigate the performance of the Contention Access Period (CAP) protocol of the IEEE 802.15.4 standard in the WBSN. The simulation study of this paper introduces a WBSN topology based on the Markov chain, MC-WBSN. The evaluation of the MC-WBSN consider the parameters of latency, power consumption, throughput, and higher path loss channel. The evaluation results show that the hidden nodes have a great impact on WBSNs performance and throughput. Furthermore, there is a real need to reduce the traffic load to mitigate the power consumption problem and increase the throughput of the network. The future research considers studying different types of topologies such as the multi-hop topology to ensure comprehensive results. Additionally, this work uses a sensing node that has an ideal disposition, using data of real sensing nodes can ensure more significant results.

Acknowledgements. This work is sponsored by the Malaysia Ministry of Education (MoE) under FRGS grant scheme vote number 1611. It is also supported by University Tenaga Nasional (UNITEN) under the UNIIG Grant Scheme No. J510050772.

References

1. Rashid T, Kumar S, Kumar, A (2017) Effect of Body Node Coordinator (BNC) positions on the performance of intra-body sensor network (Intra-WBSN). In: 2017 4th International conference on power, control & embedded systems (ICPCES), pp 1–6
2. Lall S, Alfa AS, Maharaj BT (2016) The role of queueing theory in the design and analysis of wireless sensor networks: an insight. In: IEEE 14th International conference on industrial informatics (INDIN), pp 1191–1194
3. AboZahhad M, Farrag M, Ali A (2015) A comparative study of energy consumption sources for wireless sensor networks. IJGDC J 8(3):65–76
4. Tiwari P, Saxena VP, Mishra RG, Bhavsar D (2015) A survey of localization methods and techniques in wireless sensor networks. IJTIR 14:2321–1814
5. Javaid N, Abbas Z, Fareed M, Khan Z, Alrajeh NM (2013) M-attempt: a new energy-efficient routing protocol for wireless body area sensor networks. In: International conference on ambient systems, networks and technologies, pp 224–231
6. Ndih EN, Khaled N, De Micheli G (2009) An analytical model for the contention access period of the slotted IEEE 802.15. 4 with service differentiation. In: 2009 IEEE International conference on communications, ICC'09, pp 1–6
7. Al Rasyid MUH, Nadhori I, Sudarsono A, Luberski R (2014) Analysis of slotted and unslotted CSMA/CA wireless sensor network for E-healthcare system. In: 2014 International conference on computer, control, informatics, and Its Application (IC3INA), pp 21–23
8. Barakah DM, Uddin MA (2012) A survey of challenges and applications of Wireless Body Area Network (WBAN) and role of a virtual doctor server in existing architecture. In: International conference on intelligent systems modelling and simulation. IEEE
9. Ramachandran I, Das AK, Roy S (2007) Analysis of the contention access period of IEEE 802.15. 4 MAC. ACM Trans Sensor Netw (TOSN) 3(1):4
10. Shaban H, Abou El-Nasr M (2013) Amplify-and-forward cooperative diversity for green UWB-based WBSNs. Sci World J
11. de Franceschi MA (2009) Performance analysis of the contention access period in the slotted IEEE 802.15. 4 for wireless body sensors networks, Doctoral dissertation, Master thesis, Charles III University of Madrid, Spain
12. Taneja K, Taneja H, Kumar R (2018) Multi-channel medium access control protocols: review and comparison. J JIOS 39(1):239–247
13. Wang J, Xie Y, Yi Q (2015) An all dynamic MAC protocol for WBSN

14. Mohanty S (2010) Energy efficient routing algorithms for wireless sensor networks and performance evaluation of quality of service for IEEE 802.15.4 networks (thesis)
15. Li X, Bleakley CJ, Bober W (2012) Enhanced beacon-enabled mode for improved IEEE 802.15. 4 low data rate performance. Wirel Netw 18(1):59–74
16. Yang X, Wang L, Zhang Z (2018) Wireless body area networks MAC protocol for energy efficiency and extending lifetime. IEEE Sens Lett 2(1):1–4
17. Kanagachidambaresan GR, Chitra A (2015) Fail safe fault tolerant mechanism for wireless body sensor network (WBSN). Wirel Pers Commun 80(1):247–260
18. Mamaghanian H, Khaled N, Atienza D, Vandergheynst P (2011) Compressed sensing for real-time energy-efficient ECG compression on wireless body sensor nodes. IEEE Trans Biomed Eng 58(9):2456–2466
19. Ali RR, Mostafa SA, Mahdin H, Mustapha A, Gunasekaran SS (2019) Incorporating the Markov Chain model in WBSN for improving patients' remote monitoring systems. AUS (Valdivia) n. 26-1

Designing Deep Neural Network with Chicken Swarm Optimization for Violence Video Classification Using VSD2014 Dataset

Ashikin Ali[1(✉)], Norhalina Senan[1], and Iwan Tri Riyadi Yanto[2]

[1] Faculty of Computer Science and Information Technology, Universiti Tun Hussein Onn Malaysia, 86400 Batu Pahat, Johor, Malaysia
gi150038@siswa.uthm.edu.my, halina@uthm.edu.my

[2] Faculty of Mathematics and Natural Sciences, Ahmad Dahlan University, Yogyakarta 55164, Indonesia
yanto.itr@is.uad.ac.id

Abstract. In this paper, a violence video classification model based on deep neural network (DNN) with chicken swarm optimization (CSO) is proposed. Violence is a self-sufficient attribute, the contents that one would not be favorable to see in movies or web videos. This is a challenging problem due to strong content variations among the positive instances of violence. Currently, deep neural network has shown its efficiency in various field that relevant to its implementation and attracts plenty of researchers in awe. However, conservative deep neural network has limitations and a tendency to easily fall into local minima. Regardless of the conventional methods applied to overcome this issue, but these techniques seem insufficiently accurate and does not adopt well to certain webs or user needs. Therefore, the purpose of this study is to assess the classification performances on violence video using Deep Neural Network with Chicken Swarm Optimization (DNNCSO). Hence, in this paper different architectures of hidden layers in DNN have been implemented using the try-error method and the importance of the parameters to examine the effect of the number of hidden layers to the classification performance. The algorithm is evaluated based on error convergence and accuracy. The results have proved the effectiveness of the proposed method up to 77–79% as compared to the conventional DNN which holds 63%. Based on the promising outcome of proposed method, in future the study intended to work on improvised bio-inspired algorithm possibly with different domains and relevant features.

Keywords: Deep neural network · Chicken swarm optimization · Violence classification

1 Introduction

The violence characterization is subjective, and this creates difficulty in defining violent content [1, 2]. Violence possibly will occur any situation or action that may cause harm mentally or physically to users. Violence scenes in video documents regard the content that includes such actions, it also evolves the characteristic of audio signals as well.

R. Ghazali et al. (Eds.): SCDM 2020, AISC 978, pp. 47–56, 2020.
https://doi.org/10.1007/978-3-030-36056-6_5

Provided many web-filtering systems are commercially available and possibilities for users to download from the internet is high. However, these techniques seem insufficiently accurate for classification tasks to meet the user needs. Presently, deep neural network is widely used and popular in practice [3–6]. In particular, the focus is on the architecture of deep neural network as shown in Fig. 1, which is constructed with at least of minimum 3 hidden layers [7].

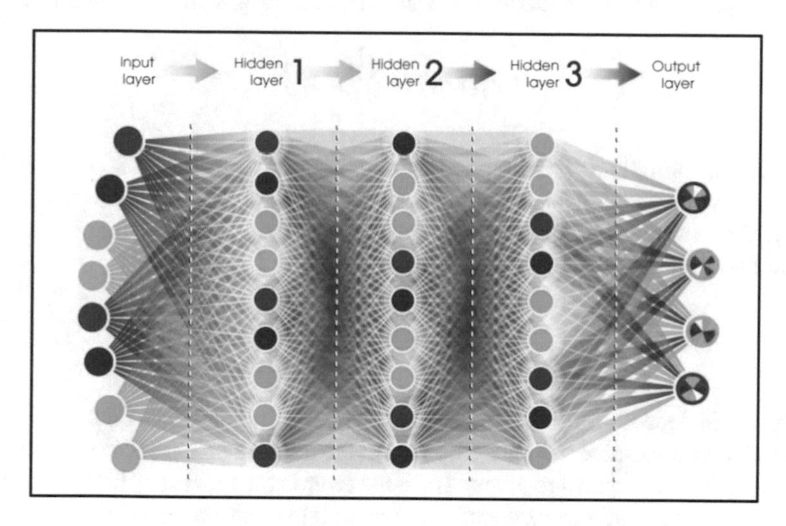

Fig. 1. The Architecture of Deep Neural Network [19].

However, the ability in practical problems causes performance disability to DNN in term of getting optimum outcome to the network. This is basically due to the parameters that been used during the learning phase of DNN such as learning rate, momentum, also not allow computing the desired set of the hidden layers for the network [8].

However, in the arena of evolutionary computing [9–11] there are algorithms that aids to obtain optimum outcome for a specific problem. Thus, some algorithm does not incline with deep neural network. Therefore, Chicken Swarm Optimization (CSO) is one of the new bio-inspired based algorithms [12], it mimics the hierarchical order and behaviors of chicken swarm. The chicken swarm segregated to several category which consists of one rooster, many hens and chicks.

Thus, in this paper, the audio-visual features have been implemented and all the attributes of audio-visual features have been used to fed into the DNNCSO classifier. To examine the accuracy performance, the try-error method was implemented. The experiments were conducted using the generalized set of VSD2014 dataset of 86 YouTube videos consist of violence and non-violence video, segregated uniformly.

The rest of the paper is organized as following. Section 2 explains the basic theory deep neural networks, chicken swarm optimization. Section 3 visualizes the implementation of the proposed method. Section 4 describes the experimental results. Section 5 concludes the whole paper also discuss the future work.

2 Deep Neural Network

The backpropagation algorithm [13] is used in layered feed forward DNN. The backpropagation algorithm uses supervised learning, where the algorithm is provided with the inputs and outputs which the network should compute and then the error is calculated [14]. The training begins with random weights, and the goal is to adjust them so that the error will be minimal. The weighted sum of a neuron is written as:

$$A_j(x, w) = \sum_{i-0}^{n} X_i W_{ji} \tag{1}$$

where the sum of input X_i is multiplied by their respective weights, W_{ji}. The activation depends only on the inputs and the weights. If the output function would be the identity, then the neuron would be called linear. The most used output function is sigmoid function [15]:

$$O_j(x, w) = \frac{1}{1 + e^{-A}(x, w)}. \tag{2}$$

The output depends only in the activation, which in turn depends on the values of the inputs and their respective weights. The goal of the training process is to obtain a desired output when certain inputs are given. Since the error is the difference between the actual and desired output, the error depends on the weights and preferred to be adjusted to minimize the error. The error function for the output of each neuron can be defined as:

$$E_j(x, w, d) = \left(O_j(x, w) - d_j\right)^2. \tag{3}$$

The output will be positive, and the desired target will be greater if the difference is big and lesser if the difference is small. The error of the network will simply be the sum of the errors of all the neurons in the output layer:

$$E(x, w, d) = \sum_j \left(O_j(x, w) - d_j\right)^2. \tag{4}$$

where O_j is the target output and d_j is the target output of the experiment, after finding this, the weight adjusted using the method of gradient descent, the equation is as following:

$$\Delta w_{ji} = -\eta \frac{\partial E}{\partial w_{ji}} \tag{5}$$

This equation can be inferred in the following way: the adjustment of each weight (Δw_{ji}) will be the negative of a constant eta (η), where η is the learning rate. Multiplied by the dependence of the previous weight on the error of the network, which is derivative of E in respect to w_{ji}.

The size of the adjustment will depend on η, and on the contribution of the weight to the error of the function. This is, if the weight contributes a lot to the error, the adjustment will be greater than if it contributes in a smaller amount. Equation (5) is used until appropriate weights with minimal error founded.

Henceforth, derivative of E in respect to w_{ji} discovered. This is the goal of the backpropagation algorithm, since the backwards need to be achieved. First, calculate the error depends on the output, which is the derivate of E in respect to O_j from Eq. (3).

$$\frac{\partial E}{\partial O_j} = 2(O_j - d_j) \tag{6}$$

The reliance of the output on the activation depends on the weights from Eqs. (1) and (2). Can be seen that from Eqs. (6) and (7):

$$\frac{\partial O_j}{\partial w_{ji}} = \frac{\partial O_j}{\partial A_j}\frac{\partial A_j}{\partial w_{ji}} = O_j(1 - O_j)x_i \tag{7}$$

$$\frac{\partial E}{\partial w_{ji}} = \frac{\partial E}{\partial O_j}\frac{\partial O_j}{\partial w_{ji}} = 2(O_j - d_j)O_j(1 - O_j)x_i \tag{8}$$

The adjustment to each weight will begin from Eqs. (5) and (8):

$$\Delta w_{ji} = -2\eta(O_j - d_j)O_j(1 - O_j)x_i \tag{9}$$

Equation (9), for training the network with one more layer, some considerations are needed particularly on training time which can be affected by the architecture of the network.

2.1 Chicken Swarm Optimization

Chicken Swarm Optimization (CSO) algorithm is developed by [12]. The chicken's various movements can be conducive for the algorithm to strike a good balance between the randomness and determination for finding the optima outcome. Recently, CSO has aroused great concern. For example, analysis of the convergence of CSO [16, 17] and the algorithm is demonstrated to meet two convergence criteria, which ensures the global convergence. The chicken swarm optimization mimics the hierarchal order in the chicken swarm and the behaviors of the chicken swarm, which stems from the observation of the birds foraging behavior. The chicken swarm can be divided into several groups, each of which consists of one rooster, many hens, and several chicks. When in foraging, the roosters can always find food differently [18]. The hens always follow the roosters to look for food, and the chicks followed the mother in search of food. The different individuals within the chicken swarm follow different laws of motions.

There exist competitions between the different individuals within the chicken swarm under the hierarchal order. First define the following variables before describing location update formula of the individuals within the chicken swarm: RN, HN, CN and MN are the number of roosters, hens, chicks and mother hens, respectively; N is the number of the whole chicken swarm, D is the dimension of the search space; $xi,\ j(t)\ (i \in [1, \cdots, N],\ j \in [1, \cdots, D])$ is the position of each individuality time t. In the chicken swarm, the best RN chickens would be assumed to be the roosters, while the worst CN ones would be regarded as the chicks.

The rest of the chicken swarm is viewed as the hens. Roosters with best fitting values have priority for food access than the ones with worse fitting values. In the chicken swarm, the best RN chickens would be assumed to be the roosters, while the worst CN ones would be regarded as the chicks. The rest of the chicken swarm is viewed as the hens. Roosters with best fitting values have priority for food access than the ones with worse fitting values. The position update equations of the roosters can be formulated as below:

$$\Delta x_{i,j}^{t+1} = x_{i,j}^{t} * \left(1 + Randn\left(0, \sigma^2\right)\right) \tag{10}$$

$$\sigma^2 = \begin{cases} 1, \ if\, f_i \leq f_k, \\ \exp\frac{(f_k - f_i)}{|f_i| + \varepsilon} \end{cases}, otherwise, k \in [1, N], k \neq i. \tag{11}$$

Where $Randn(0, \sigma^2)$ is a Gaussian distribution with mean 0 and standard deviation σ^2. ε, which is used to avoid zero-division-error, is the smallest constant in the computer. k, a rooster's index, is randomly selected from the rooster group, f is the fitness value of the corresponding x. As for the hens, they can follow their group-mate roosters to search for food. Moreover, they would also randomly steal the good food found by other chickens, though they would be repressed by the other chickens. The more dominant hens would have advantage in competing for food than the more submissive ones. This phenomenon is formulated as below:

$$x_{i,j}^{t+1} = x_{i,j}^{t} + S1 * Rand * \left(x_{i,j}^{t+1} - x_{i,j}^{t}\right) + S2 * Rand * \left(x_{i,j}^{t+1} - x_{i,j}^{t}\right) \tag{12}$$

$$S1 = \frac{\exp(f_i - f_{r1})}{(abs(f_i))} + \varepsilon \tag{13}$$

$$\Delta S1 = \exp((f_{r2} - f_i)) \tag{14}$$

The bigger the difference of the two chicken's fitness values, the smaller S2 and the bigger the gap between the two chickens' positions is. Thus, the hens would not easily steal the food found by other chickens. The reason that the formula form of S1 differs from that of S2 is that there exist competitions in a group. For simplicity, the fitness values of the chickens relative to the fitness value of the rooster are simulated as the competitions between chickens in a group. Hence the more dominant hens would be more likely than the more submissive ones to eat the food. The chicks move around their mother to forage for food, it is formulated as following:

$$x_{i,j}^{t+1} = x_{i,j}^{t} + FL * \left(x_{m,j}^{t} - x_{i,j}^{t}\right) \tag{15}$$

Where $x_{i,j}^{t+1}$ stands for the position of the ith chick's mother ($m \in [1, N]$). $FL(FL \in (0, 2))$ is a parameter, which means that the chick would follow its mother to forage food. The individual difference is considered, the FL of each chick would randomly choose between 0 and 2. Further is the explanation of the implementation of proposed method.

3 Implementation of Proposed DNNCSO

The algorithm is clear indicator of the implementation process of proposed method DNNCSO, Fig. 2 shows scheme of DNN + CSO. For a winning process, objective of the algorithm is to improvise the weights to minimize MSE and increase the accuracy. It will adjust its weight more than a neuron that is closer to the global set. An efficient trial-and-error procedure is applied to tune the DNNCSO parameters.

Algorithm of Deep Neural Network with Chicken Swarm Optimization (DNNCSO)

```
1: Initialization Phase
2: repeat
3: Employed and rank the chicken fitness value using Equation
   (10) and (11)
4: Divide group and determine the relationship between chicks
   and then using Equation (12), (13) and (14)
5: Update the chicken solution until chicken swarm finds the
   best solution using Equation (15)
6: Memorize the best solution achieved so far
7: Until (Cycle = Maximum Cycle Number), Save best solution
8:  Assign all network inputs and output to DNN backpropagation
9: Initialize all weights from step 7
10:  repeat
11:  Present the pattern to the network
Propagated the input forward through the network
    12:     for each layer in the network
    13:      for every node in the layer
    14:      Calculate the weight sum of the inputs to the node
    15:      Add the threshold to the sum
    16:       Calculate the activation for the node
    17:    end
    18:   end
   Propagate the errors backward through the network
    19:     for every node in the output layer
    20:    calculate the error signal
    21:   end
    22: for all hidden layers
    23:     for every node in the layer
    24:       Calculate the node's signal error
    25:       Update each node's weight in the network
    26:     end
    27:    end
    Calculate Global Error
    28:     Calculate the Error Function
29:    end
30: While ((maximum number of iterations < than
    specified))
```

Fig. 2. Scheme of the DNNCSO Algorithm.

From the algorithm, from step 3–5 evolves the CSO algorithm fulfilling the Eqs. 2.10–2.15 to improve the weights in DNN. To be fair, the parameters were set accordingly for this proposed method. Population size is set to 100, max. iteration = 1000, dimension of 20. Chicken, hen, rooster respectively is set to 0.2, 0.6 and 0.1 with different number of hidden layers. That will be discussed further in Sect. 4.

3.1 Experimental Design

The experimental design in this paper consists of feature extraction process and the classification of deep neural network. The program is developed in MATLAB R2019a using MATLAB programming language. The simulations have been conducted in GPU type device MacPro 6, macOS High Sierra operating system, the processor of this is 8 Core Intel Xeon E5 with the speed of 3 GHz and 32 GB RAM.

3.2 Dataset - VSD2014

Benchmark VSD2014 dataset obtained from Technicolor Group has been used as a domain in this study. 86 videos have been used with 434 attributes and 68, 830 instances. This data consists of two classes, namely, violence represented as 1 and non-violence represented as 0. Basically, total of 29 audio features and 405 visual features. For the audio features, consist of 8 features. Each feature is provided by per-video-frame-basis. The 8 features are, accordingly, amplitude envelope (AE), root-mean-square energy (RMS), zero-crossing rate (ZCR), band energy ratio (BER), spectral centroid (SC), frequency bandwidth (BW), spectral flux (SF) and mel-frequency cepstral coefficients (MFCC). For each window, 22 MFCC is computed while all other features are computed in 1-dimensional, by total of 29 features. As for the visual features, it includes 4 features, namely color naming histogram (CNH) with 99-dimensional, local binary patterns (LBP) with 144-dimensional, both color moments (CM) and histogram of oriented gradients (HOG) with 81-dimensional, by total of 405 features.

4 Results and Discussion

In the experiment, Violence Scene Dataset (VSD2014) lone was set up with different parameters for experiment purpose. The results of the simulation indicate that it meets the adequate desired outcome with the aid of try-error-methods. Thus, to quantify the performance of the classifier, using 5-fold cross validation the learning rate is set to 0.01 momentum of 0.7, inertia weight of 0.0002, with 80:20 segregation of data. Acc. represents classification accuracy and Inc. represents the incorrectly classified percentage. Below is the performance summary of the experiments:

Table 1. Summary of performance with 3 hidden layers.

Approach	MSE	Acc. (%)	Inc. (%)
DNNCSO	1.16	42	58
DNN	1.00	50	50

Table 2. Summary of performance with 4 hidden layers.

Approach	MSE	Acc. (%)	Inc. (%)
DNNCSO	0.41	79	21
DNN	0.86	57	43

Table 3. Summary of performance with 5 hidden layers.

Approach	MSE	Acc. (%)	Inc. (%)
DNNCSO	0.45	77	23
DNN	0.82	58	42

Table 4. Summary of performance with 6 hidden layers.

Approach	MSE	Acc. (%)	Inc. (%)
DNNCSO	0.80	60	40
DNN	0.97	51	49

Table 5. Summary of performance with 7 hidden layers.

Approach	MSE	Acc. (%)	Inc. (%)
DNNCSO	0.70	65	32
DNN	0.87	57	43

As a discussion, it is significant that arrangement of a proper hidden layers helps in getting a desired accuracy. As in Table 2, with 4 hidden layers DNNCSO have reached 79% overpowering DNN by 22% and in Table 3, with 4 hidden layers DNNCSO have reached 77% increased by 19% from DNN. The remaining set of experiment,, using 5-fold cross validation the learning rate is set to 0.1 momentum of 0.9, inertia weight of 0.0002, with 80:20 data segregation. Below is the summary of the performances:

Table 6. Summary of performance with 3 hidden layers.

Approach	MSE	Acc. (%)	Inc. (%)
DNNCSO	0.70	65	35
DNN	1.12	44	56

Table 7. Summary of performance with 4 hidden layers.

Approach	MSE	Acc. (%)	Inc. (%)
DNNCSO	0.52	73	27
DNN	0.73	63	37

Table 8. Summary of performance with 5 hidden layers.

Approach	MSE	Acc. (%)	Inc. (%)
DNNCSO	0.49	75	25
DNN	0.82	59	41

Table 9. Summary of performance with 6 hidden layers.

Approach	MSE	Acc. (%)	Inc. (%)
DNNCSO	0.59	71	29
DNN	0.92	54	46

Table 10. Summary of performance with 7 hidden layers.

Approach	MSE	Acc. (%)	Inc. (%)
DNNCSO	0.85	56	44
DNN	0.89	55	45

In Table 7, DNNCSO have reached 73% from 63% of DNN. Meanwhile, in Table 8 DNNCSO scored 75% increased by 16% from DNN. At this point with this focused domain dataset, the algorithm managed to improvise in a better way as compared to the conventional DNN. This shows that the trial-error method of the parameters and layers is the greatest involvement. It can be also concluded that too less or too much of hidden layers can cause the chaos in the training process due to the connecting weights. By optimizing it using the bio-inspired algorithm it has aided to reach a desired output.

5 Conclusion and Future Works

As a conclusion, classification is an intervention to initialize the process of labelling in selected domains. The results presented in this paper may facilitate improvements in this case, it is proven that modified algorithm performs better when compared with algorithm that been implemented individually. The proposed work is considered as an improvement on identifying the significance of using CSO over DNN.

In addition, based on the result, this study obtainable to new view on automatic violence classification. The algorithm enables the learning of information that has been feed to it. Thus, in this paper experimentally proven that the proposed classification algorithm harvests better accuracy compared with the conventional deep neural network. However, further work of the study prefers to focus on most prominent features of the current domain and will consider working with variants of bio-inspired algorithm and domain.

Acknowledgments. This research funded by Ministry of Higher Education (MOHE) under the Fundamental Research Grant Scheme (FRGS) – Vot. No. 1608. Besides, the research is also backed by Universiti Tun Hussein Onn Malaysia.

References

1. Martinez VR, Somandepalli K, Singla K, Ramakrishna A, Uhls YT, Narayanan S (2019) Violence rating prediction from movie scripts
2. Ali A, Senan N (2018) Violence video classification performance using deep neural networks. In: International conference on soft computing and data mining. Springer, Cham, pp 225–233
3. LeCun Y, Bengio Y, Hinton G (2015) Deep learning. Nature 521(7553):436
4. Goodfellow I, Bengio Y, Courville A (2016) Deep learning. MIT Press

5. Zhang C, Bengio S, Hardt M, Recht B, Vinyals O (2016) Understanding deep learning requires rethinking generalization. *arXiv preprint* arXiv:1611.03530
6. Simonyan K, Zisserman A (2014) Very deep convolutional networks for large-scale image recognition. *arXiv preprint* arXiv:1409.1556
7. Skymind (2019) A beginner's guide to neural networks and deep learning. https://skymind.ai/wiki/neural-network
8. Garro BA, Sossa H, Vázquez RA (2011) Evolving neural networks: a comparison between differential evolution and particle swarm optimization. In: International conference in swarm intelligence. Springer, Berlin, pp 447–454
9. Arshad H, Khattak H A, Ameer Z, Abbas A, Khan SU (2019) Estimation of fog utility pricing: a bio-inspired optimisation techniques' perspective. Int J Parallel Emerg Distrib Syst 1–14
10. Chen S, Peng GH, He XS, Yang XS (2018) Global convergence analysis of the bat algorithm using a markovian framework and dynamical system theory. Expert Syst Appl 114:173–182
11. Wang GG (2018) Moth search algorithm: a bio-inspired meta heuristic algorithm for global optimization problems. Memet Comput 1–14
12. Meng X, Liu Y, Gao X, Zhang H (2014) A new bio-inspired algorithm: chicken swarm optimization. In: International conference in swarm intelligence. Springer, Cham, pp 86–94
13. Benevenuto F, Rodrigues T, Almeida V, Almeida J, Gonçalves M (2009) Detecting spammers and content promoters in online video social networks. In: Proceedings of the 32nd international ACM SIGIR conference on research and development in information retrieval. ACM, pp 620–627
14. Eyben F, Weninger F, Lehment N, Schuller B, Rigoll G (2013) Affective video retrieval: Violence detection in Hollywood movies by large-scale segmental feature extraction. PLoS ONE 8(12):e78506
15. Karlik B, Olgac AV (2011) Performance analysis of various activation functions in generalized MLP architectures of neural networks. Int J Artif Intell Expert Syst 1(4):111–122
16. Wang B, Li W, Chen X, Chen H (2019) Improved chicken swarm algorithms based on chaos theory and its application in wind power interval prediction. Math Probl Eng
17. Zarlis M, Yanto ITR, Hartama D (2016) A framework of training ANFIS using chicken swarm optimization for solving classification problems. In: 2016 International conference on informatics and computing (ICIC), pp 437–441. IEEE
18. Wu D, Xu S, Kong F (2016) Convergence analysis and improvement of the chicken swarm optimization algorithm. IEEE Access 4:9400–9412
19. Techsparks (2019) Machine Learning. https://www.techsparks.co.in

Header Based Email Spam Detection Framework Using Support Vector Machine (SVM) Technique

Siti Aqilah Khamis[3], Cik Feresa Mohd Foozy[1,3(✉)], Mohd Firdaus Ab Aziz[1,3], and Nordiana Rahim[2,3]

[1] Applied Computing Technology (ACT), Universiti Tun Hussein Onn Malaysia, Batu Pahat, Johor, Malaysia
feresa@uthm.edu.my

[2] Information Security Interest Group (ISIG), Universiti Tun Hussein Onn Malaysia, Batu Pahat, Johor, Malaysia

[3] Faculty of Computer Science and Information Technology, Universiti Tun Hussein Onn Malaysia, Batu Pahat, Johor, Malaysia

Abstract. Email spam is continuously a major problem, especially on the Internet. Spam consists of malicious malwares which attack user's machine to steal information, destroy the user's machine and trick the user into buying their products. Although spam detection or email spam filtering was developed, there is still a rising number of emails that contain spam. The study of this research is to identify the potentially useful email header features for email spam detection by analyzing two (2) email datasets, the Anomaly Detection Challenges and Cyber Security Data Mining from website. By analyzing the datasets, the main objective of this research is to extract the suitable features of the email header and examine the features to classify the features using Support Vector Machine (SVM) using RapidMiner Studio and Weka 3.9.2. There are five (5) phases in the methodology which are Data Collection, data Pre-Processing, Features Selection, Classification and Detection. Classification of the email header using Support Vector Machine (SVM) for CSDM2010 is higher than the Anomaly Detection Challenges datasets at 88.80% and 87.20% respectively. Thus, SVM proves as a good classifier which produced above 80% accuracy rate for both datasets.

Keywords: Detection · Email spam · Machine learning

1 Introduction

Spam is defined as unsolicited bulk email [1]. There are two ways to recognize spam emails either from its content or the sender's behavior. Email spam sent to the user has a variety of purposes, first is phishing email. Phishing email is used by a company to steal the customer's personal information for a specific purpose. Some steal information like passwords, bank account numbers, credit card numbers and others. Secondly, some spam emails contain commercial purposes where some companies take this opportunity to promote the sales of products at cheap and affordable prices. Lastly,

R. Ghazali et al. (Eds.): SCDM 2020, AISC 978, pp. 57–65, 2020.
https://doi.org/10.1007/978-3-030-36056-6_6

spam email that contains malicious code is the most dangerous. This is because the malicious code may be harmful and be likely to damage the user's machine. The attackers also will always try to create other ways to send spam in emails to achieve their purpose [2].

There are four (4) methods to stop and detect spam emails from entering a user's inbox such as IP Address Blacklisting, Statistics, Heuristics and Phrase Matching. Phrase matching or content based is used to detect header email spam. Phrase matching examines incoming email, looking for phrases that have been used in the mail's header that has already been identified as spam [3] Keywords of spam emails will be analyzed based from its email headers only. There are two (2) datasets that are used in this research such as the Anomaly Detection Challenges [4] and Cyber Security Data Mining [5]. This research is used to detect email spam by using SVM technique based on email header features.

2 Literature Review

This part covers the theory from this research. It explains the terminology needed for understanding the email spam detection framework using email header. It will also give an overview of email's header structure that can be used for feature selection, classification and detection. Other than that, review on the current machine learning techniques will be analyzed.

2.1 Email's Header Structure

Every email consists of two parts such as a header and a body. The content for both header and the body are created by the sender. This is usually done within a Mail User Agent (MUA) [18]. Header usually describes which email address the email has been received from and which Internet Protocol (IP) address the email is coming from [16, 17].

2.2 Spam Emails

Spam is defined as unsolicited bulk email [1]. There are two ways to recognize spam emails either from its content or the sender's attitude. Email spam sent to the user has a variety of purposes, first is phishing email. Phishing email is used by a company to steal the customer's personal information for a specific purpose.

2.3 Comparison of Existing Research Studies

Many techniques have been developed to solve email spamming or detecting email spam. However, these problems still exists even though many techniques have been used because this problem cannot be solved and those techniques were developed to prevent or control the spam emails from entering user's inbox [14, 15]. In email spamming, there are three ways or bases to detect email spamming such as signature based, image based and text based. Table 1 shows the existing study for email spam detection framework by using the email header. Therefore, in this research it is more focused on header base of emails.

Table 1. Existing email detection framework.

No	Author	Technique	Framework
1.	[6]	Naïve Bayes, Random Forest and Support Vector Machine	Data collection, pre-processing, feature selections and classification
2.	[7]	Support Vector Machine	Emails Collection, Training set, pre-processing, feature selections and classification
3.	[8]	Naïve Bayes, Logistic Regression and Support Vector Machine	Pre-processing, email representation using VSM, feature selections and classification
4.	[9]	Support Vector Machine	Emails collection in image input based, converting to text, feature selections and classification

Table 1 above shows the existing studies for email spam detection using different frameworks. Thus, based on the existing framework, the proposed framework was assembled from the first authors which is Al-Jarrah et al. [6]. The framework was chosen because it is more suitable for this research paper in using email header base. Furthermore, based on the table also, SVM technique will be used in this research because it shows good accuracy which is more than 80% for every existing studies. So, SVM was chosen and would be analyzed in the experiments.

2.4 Email Spam Data Collection

Every researcher used different datasets for its experiments. The datasets from other researchers may contain text and also image. Therefore, each dataset consists of ham and spam datasets that will be used in the experiments.

2.5 Email Spam Features Selection

There are many selection type of features that can be extracted in header based email spam detection. Based on Walaa and Sherine [8], Kumar and Biswas [9], the authors use bag of words where the email of datasets were created in Vector Space Model (VSM) after Pre-Processing process. There are formulas to use this method that was proposed by [8, 9].

Hence, Al-Jarrah et al. [6] and Miao et al. [7] the authors use binary numbers when extracting the features. The authors analyzed the email's header structure in the outlook files. Keywords was an important feature in spam detection [10]. Subsequently, authors [6] used keywords such as “dear”, “hey” and “hi” as spam words. For authors [7], X-Mailer is one of the most important structure that is needed to determine spam and ham emails. Therefore, the authors would analyze all the emails whether the path exists. To categorize spam and ham, it would be determined as 1 to represent spam and 0 to represent ham. Hence, Table 2 below shows the feature selection that every author [6–9] chose for their experiments.

Table 2. Features selection for existing framework.

No	Author	Features selection				
		Bags of words	Keyword	From & to field	Subject field	X-mailer
1.	[6]		✓	✓	✓	✓
2.	[7]			✓	✓	✓
3.	[8]	✓				
4.	[9]	✓				

Referring to Table 2 above, it shows that both authors [6, 7] used almost similar features selection but author [7] did not use the keywords as feature selection. For [8, 9], both authors used bag of words that was created in VSM by using their owns formulas.

2.6 Email Spam Classification

Classification is an important phase in detection research. In this phase, classifier is being chosen to run features selected by researcher to identify the result. In the research conducted by Al-Jarrah et al. [6], Miao et al. [7], Walaa and Sherine [8] and Kumar and Biswas [9], classification techniques was used. For authors [6, 8] they used three (3) classifiers. [6] used Naïve Bayes (NB), Random Forest (RF) and Support Vector Machine (SVM) whereas [8] used NB, Logistics Regression (LR) and SVM. For authors [7, 9] they used only one classifier which was SVM. Therefore, Table 3 below shows the classifiers each author used for their framework detection.

Table 3. Classifiers for existing framework.

No	Author	Classifiers machine learning			
		SVM	NB	RF	LR
1.	[6]	✓	✓	✓	
2.	[7]	✓			
3.	[8]	✓	✓		✓
4.	[9]	✓			

From Table 3, it shows the researchers [6–9] used SVM for their classification phase to test whether it was a good classifier. Based on the performance results of their experiments, SVM shows good Accuracy which is more than 80% Accuracy. Hence,

SVM will be used in this research because existing studies showed that it is a good classifiers. SVM also proved that it is suitable for many features selection from email header structure either from text based or image based.

3 Methodology

Hence, the proposed framework for this research paper is generated from the frameworks above. There are five (5) phases in this framework that consist of Data Collection, Data Pre-processing, Features Selection, Classification and Detection.

3.1 Data Collection and Pre-processing

For the dataset, 500 samples from the Anomaly Detection Challenges [4] and 500 samples from CSDM 2010 [5] had been downloaded and used for experiments. In the pre-processing step, features were first extracted from the email subject. The subject field contains the data that need to be pre-processed. Furthermore, there are three (3) steps in this phase which are tokenization, stops words and stemming process. These process are to remove noise, redundant and also common English words that is used that would affect the detection phase [19].

3.2 Feature Selection

Feature Selection is a process on how to determine a feature to be a subset that can be used in the experiments. Therefore, the feature selection is the most important step which affects the detection phase [12]. There are four (4) type of features that will be extracted in this phase. All features were extracted from the email's header structure which comes from "Subject", "From" and "To" fields. Hence, the features also were extracted from the properties of the emails by analyzing its X-Mailer and the return path of the emails.

3.3 Classification

Three (3) classifiers would be tested in this experiment and the performance results of Naïve Bayes (NB), K-Nearest Neighbour (K-NN) and Random Forest (RF) would be compared to SVM performance results.

Support Vector Machine (SVM)
There is no need to include page numbers. If your paper title is too long to serve as a running head, it will be shortened. Your suggestion as to how to shorten it would be most welcome.

Support Vector Machine (SVM) is a classifier that uses hyper plane in distinguishing or separating between a set of objects with a different class. SVM algorithm is an optimal hyper plane with the maximal margin to separate two classes, which require solving the following optimization problem [13].

Naïve Bayes (NB)
Bayesian classifier works on independent events or used on the probability of an event that might occur in the future [21]. This can be detected from the previous occurrance

of the same event. NB is suitable for predicting email spam but not in ham because probably the incoming email or data collection of most of the samples is spam. NB would also limit the total of words if it achieves a higher number than it could be in NB and then it will mark the email to either category.

K-Nearest Neighbor (K-NN)
The k-nearest neighbor (K-NN) classifier is considered an example-based classifier, that means that the training documents are used for comparison rather than as an explicit category representation, such as the category profiles used by other classifiers.

Random Forest (RF)
Random Forest (RF) classifiers is an ensemble from the Decision Tree (DT) classifiers. This is because RF is used as a bootstrap sample. Bootstrap sample was drawn from the labelled data.

3.4 Detection

In this phase, the features that have been extracted would be tested with machine learning classifiers to produce performance results. The performance results were based by Accuracy. Accuracy is a value that measures the level of email spam detection by using machine learning techniques.

4 Result and Discussion

This chapter discusses the experiments that were conducted on 500 samples from the Anomaly Detection Challenges [4] and 500 samples from CSDM 2010 [5] that was run in a tool called RapidMiner Studio. The technique that was used in this experiment was Support Vector Machine (SVM) that would determine whether the data or samples run in the tool were either spam or ham.

4.1 Features Selection

In this phase, the features selection that have been chosen in section three Research Methodology would be tested in several experiments by using two (2) tools which are RapidMiner Studio and Weka 3.9.2. So, there were three (3) experiments results using the different features. Table 4 shows the results of the three (3) experiments for Anomaly Detection Challenges [4] and Table 5 shows the results of three (3) features selection experiments for Cyber Security Data Mining [5].

Table 4. Comparison of 3 features for anomaly detection challenges [4].

Tools features	RapidMiner studio	Weka 3.9.2
	Support vector machine (Accuracy)	Support vector machine (Accuracy)
6	81.20%	81.10%
8	84.50%	83.50%
10	87.20%	87.00%

Table 5. Comparison of 3 features for cyber security data mining [5].

Tools features	RapidMiner studio	Weka 3.9.2
	Support vector machine (Accuracy)	Support vector machine (Accuracy)
6	80.50%	81.20%
8	85.07%	85.50%
10	88.80%	88.80%

Based on Tables 4 and 5 above, every size of features selection gives different Accuracy outcomes. Ten (10) features were used in the classification and detection phases.

4.2 Classification and Detection

The chosen classifier is Support Vector Machine (SVM), SVM was chosen based on the existing framework done by researchers in literature reviews. To validate SVM as a better classifier, both datasets would be tested using three (3) other classifiers. The chosen classifiers were Naïve Bayes (NB), K-Nearest Neighbour (K-NN) and Random Forest (RF). Hence, the Accuracy for every classifiers would be compared. Table 6 shows the results of three (3) classifier experiments for Anomaly Detection Challenges [4] and the results of three (3) classifier experiments for Cyber Security Data Mining [5].

Table 6. Comparison of 3 classifiers for dataset [4, 5].

Classifiers	Tools			
	RapidMiner studio (Accuracy)		Weka 3.9.2 (Accuracy)	
	[4]	[5]	[4]	[5]
SVM	87.20%	88.80%	87.00%	88.80%
Naïve Bayes (NB)	78.60%	76.75%	83.80%	80.40%
Random Forest (RF)	82.60%	82.96%	83.50%	84.60%
K-NN	81.40%	80.56%	84.50%	82.70%

Based on Table 6 above, every classifier gives different Accuracy outcomes. The highest Accuracy showed in Table 6 is SVM with 87.20% and 88.80% using Rapid-Miner. Meanwhile by using Weka the SVM result 87.00% and 88.80%. It shows that the performance results has a slight difference with the second dataset and that SVM gives the highest accuracy among other techniques.

5 Conclusion

There were several experiments that had been done to validate the email spam detection framework using header emails. From the result, it is shown that the Support Vector Machine (SVM) is one of the technique that can produce the best performance results with two datasets. Based on the performance results shown (refer to Table 6), it shows that SVM also has good performance in both data mining tool when using three (3) other machine learning techniques.

In addition, the extracting features also help in this research where there are ten (10) list of features that were used. By using these features, it shows that such features are suitable with this framework as the techniques used shows good performance results. The features chosen were also tested in two (2) tools to validate the results.

The features selection and classification experiments also prove that the SVM classifier is suitable for both datasets and list of features that have been extracted to detect spam, as also suitable for email spam detection framework where the performance results showed over than 80% Accuracy.

Acknowledgments. This research is supported by Universiti Tun Hussein Onn Malaysia under TIER 1 Vot H237 and Ministry of Education Malaysia under the Fundamental Research Grant Scheme (FRGS) Vot K075.

References

1. Koprinska I, Poon J, Clark J, Chan J (2017) Learning to classify e-mail. Inf Sci 177 (10):2167–2187
2. Mbah KF, Lashkari AH, Ghorbani AA (2017) A phishing email detection approach using machine learning techniques, Doctoral dissertation, University of New Brunswick
3. Idris I, Selamat A (2014) Improved email spam detection model with negative selection algorithm and particle swarm optimization. Appl Soft Comput 22:11–27
4. Kaggle Inc (2017) https://www.kaggle.com/c/adcg-ss14-challenge-02-spam-mails-detection. Accessed 14 Sept 2017
5. DMC (2010) http://csmining.org/index.php/spam-email-datasets-.html. Accessed 1 Oct 2017
6. Al-Jarrah O, Khater I, Al-Duwairi B (2013) Identifying potentially useful email header features for email spam filtering. In: The sixth international conference on digital society (ICDS)
7. Ye M, Tao T, Mai FJ, Cheng XH (2008) A spam discrimination based on mail header feature and SVM. In: 2014 4th International conference on wireless communications, networking and mobile computing, WiCOM'08. IEEE, pp 1–4
8. Gad W, Rady S (2015) Email filtering based on supervised learning and mutual information feature selection. In: 2015 Tenth international conference on computer engineering & systems (ICCES). IEEE, pp 147–152
9. Kumar P, Biswas M (2017) SVM with Gaussian kernel-based image spam detection on textual features. In: 2017 3rd International conference on computational intelligence & communication technology (CICT). IEEE, pp 1–6

10. Ozarkar P, Patwardhan M (2013) Efficient spam classification by appropriate feature selection. Int J Comput Eng Technol (IJCET) 4(3). ISSN 0976–6375
11. Shoaib M, Farooq M (2015) USpam—a user centric ontology driven spam detection system. In: The 48th Hawaii international conference on system sciences, pp 3661–3669

A Mechanism to Support Agile Frameworks Enhancing Reliability Assessment for SCS Development: A Case Study of Medical Surgery Departments

Abdulaziz Ahmed Thawaba[1](✉), Azizul Azhar Ramli[1], Mohd. Farhan Md. Fudzee[1], and Junzo Wadata[2]

[1] Faculty of Computer Science and Information Technology, Universiti Tun Hussein Onn Malaysia, Parit Raja, Batu Pahat 86400, Johor Darul Takzim, Malaysia
hi170059@siswa.uthm.edu.my, {azizulr, farhan}@uthm.edu.my

[2] Computer and Information Sciences Department Center for Research in Data Science (CERDAS), Universiti Teknologi PETRONAS, Seri Iskandar 32610, Perak Darul Ridzuan, Malaysia
junzo.watada@utp.edu.my

Abstract. Safety-critical systems (SCSs) are distributed in many areas such as flight, railway, medical, nuclear and defense, and failure of these systems can harm human life or damage the environment. There are several ways to reduce or possibly eliminate failure, according to researchers improving the development processes of SCSs reduces failure by 40%. Therefore, developers have invented many development methods and measurement techniques to reduce failures during development processes. Agile Software Development (ASD) has entered the field of safety-critical system development (SCSD) in the past few years. Agile has produced reliable and high-quality products by expanding its principles and adapting new measurement techniques. However, there is still a gap in improving measurement techniques and stopping failures. This paper proposed a new measurement mechanism called Package Metrics for Improving Software Development (PM-ISD). The proposed measurement enhances reliability assessment and decision-making during agile development processes. PM-ISD assists the Agile management team throughout the Medical Surgery Department (MSD) project phases to track the completion of tasks.

Keywords: Safety-critical systems (SCSs) · Safety critical-system development (SCSD) · Agile software development (ASD) · Scaled agile framework (SAFe) · Area: system development · Performance

1 Introduction

Safety-Critical Systems are distributed in many different fields in our daily lives such as medical, military, nuclear, defense. SCSs are susceptible to failure, and failures in these systems can lead to significant threats to humans or the environment [1]. Developers

R. Ghazali et al. (Eds.): SCDM 2020, AISC 978, pp. 66–76, 2020.
https://doi.org/10.1007/978-3-030-36056-6_7

faced significant challenges in building reliable SCSs due to the complexity and distribution of the system [2]. Therefore, developers are working hard to introduce appropriate measurement techniques to help reduce defects during the SCSD life cycle. ASD changed the way most systems were developed. Agile now comes with a new framework that can control the complexity and requirements of SCSs while continuing to apply agile principles. Scaled Agile Framework (SAFe) is a freely exposed knowledge base for proven and integrated models of enterprise-wide Lean-Agile development [3]. The SAFe framework will adapt to SCSD with strong management control because it divides the project into three levels, portfolio, program, and team [4]. There are many metrics used to measure the development processes. According to Silva, in order to avoid failures, the metric must measure every small chunk in the project of the SCS [5]. ASD follow some predefined metrics and a range of rigorous practices such as; burn-down-charts, test-pass rates, and sustainable pace [6]. However, SCSD still needs more evaluation methods to obtain product safety and reliability. This paper focuses on how to support the management team in enhancing the evaluation of SCS reliability through the agile development framework. The remaining sections of this paper are organized as follows; Sect. 2 presents the PM-ISD mechanism. Section 3 discusses the implementation of PM-ISD throughout the agile development of the MSD project. Section 4 is the conclusion.

2 Package Metrics for Improving Software Development (PM-ISD)

There are a large number of measurement techniques, such as Key performance indicators (KPIs) and the metrics of SAFe [7]. Most measurement techniques focus on specific parts of the development life cycle, such as KPI metrics, but some focus on more areas such as SAFe metrics [8]. The project management team decompose the project into several tasks. Each task has a set of sub-tasks that must be followed up to ensure they are implemented based on the project plan and standards. The task or sub-task in a project needs a time plan, cost, manpower, requirements, functions to be implemented, and a standard to meet. SCSD requires a metric that is capable of measuring most aspects of development processes and integrate standards into the measurement. PM-ISD uses task indicators to trace implementation, resources and predict the task completion. PM-ISD provides accurate figures that support management decisions to improve productivity through early measurements during SCSD phases.

2.1 PM-ISD Components

PM-ISD components are the project plan, project testing table, and five measurements. The five measurements are Achievement metrics, Standards achievement metrics, Resources metrics, Prediction metrics for task completion, and Prediction metrics for the achievement of the final standards. This paper is not enough to cover the discussion of the five measurements, so we only discuss achievement metrics and resource metrics.

2.2 Project Plan (PP)

The SCSD project plan contains indicators identified by the management team, such as cost, time, functions, manpower, and standards. PM-ISD Metric divided the manpower into four categories to facilitate management and distribution. Manpower categories are team A has five or less, team B has ten or less, team C has less than 20, and team D has more than 20. The purpose of the dividing mechanism is to be suitable for large projects, which have many tasks and each task has different manpower distribution.

2.3 Testing Table (TT)

In the testing table, there are several tests per sub-task or main task. The test team should enter the data and results required for the task, for example; task ID, test data, cost expenditure, functions done, and standards achieved.

2.4 Achievement Metrics (M1)

Achievement metrics use standards and indicators to ensure that the sprint scope is completed according to plan. M1 is more efficient than Agile Sprint Burndown because it measures the completion of the sprint scope by using measurement results of subtask achievement. M1 has two metrics, Achievement metric for all tasks and Achievement metric for main tasks using subtasks achievement.

2.5 Resources Metrics (M2)

M2 assists the management team in verify task resources when tests conducted and identify the available or required resources during the project lifecycle. The resources indicators of M2 are time, cost, and manpower. M2 consists of three sub-matrices; Recourse metric, Available resources metric, and Required resources metric.

3 Discussions of Medical Surgery Department (MSD) Project

The selected case study is the MSD project. This project has implemented according to the standards of the Yemeni Ministry of Health. The project designed by the Medical Projects Section of the Ministry of Health and supervised by the Medical Standards and Quality Control Department. MSD was implemented by the Ministry of Public Works. MSD project has two main components; construction works and medical equipment. Construction works have five main tasks and thirty subtasks. There are 24 subtasks for the supply and installation of medical devices and instruments.

3.1 Failures of Medical Project

The medical project management team must be careful during the construction stages in dealing with elements that may lead to defects in buildings and facilities. Deficiencies in the performance of medical devices pose an unacceptable risk of injury to the patient. According to Haryati's studies on four hospitals, most defects are identified in architectural works, followed by electrical works, mechanical works, civil works, and biomedicine [9].

3.2 Applying Agile Development Framework

The choice of development methods depends on the developer's experience, the culture of the company and the way the final product is delivered. Developers often prefer the Agile framework these days instead of the classical framework due to the speed and incremental method in system delivery and low cost.

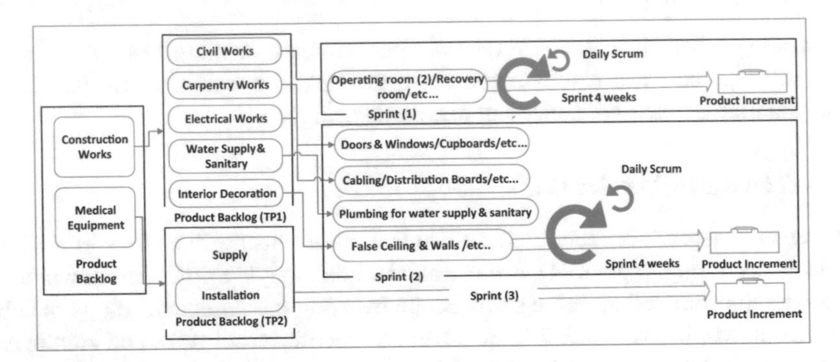

Fig. 1. MSD project (Using SAFe)

As shown in Fig. 1, the main product backlog contains civil works and medical equipment. Product backlog (TP1) contains five tasks and two Sprints. Product backlog (TP2) contains two tasks and one Sprint. PM-ISD starts measuring from Sprint activities, and these activities are subtasks of the main task in Product Backlog. As shown in Fig. 1, activities in Sprint (1) are a subtask of the main task (Civil Works) in TP1. PM-ISD uses Sprint's daily or weekly subtask test results to measure the completion of the main task.

3.3 Measurements Using PM-ISD

Project Plan (PP)

The PP of MSD project contains five indicators; cost, time, functions, manpower, and standards. PM-ISD added two other indicators as shown in Eqs. (1) and (2).

$$\text{Cost for Each Function} = \frac{\textit{Task Cost}}{\text{Task Functions}} \tag{1}$$

Equation (1) calculates the cost required to achieve each function by divide the cost of the task by the number of the task's functions.

$$\text{Time for Each Function} = \frac{\text{Task Duration}}{\text{Task Functions}} \tag{2}$$

Equation (2) calculates the time required to achieve each function by divide the cost of the task by the number of the task's functions.

Project Plan for Construction of Surgery Department

Task ID	Task Name	Start Date	Cost	Duration (Day)	Man-power	Standard (Points)	Standard (%)	Functions (Count)	Cost required for each function	Time required for each function (Day)	Time required for each standard points (in Day)	Time required for each standard % (in Day)
TP0	Construction of Surgery Department	2/1/2014	$577,771	270	D	264	100	220	2626.23	1.23	1.02	2.7
TP1-CW	Construction Works (Part One)	2/1/2014	$457,662	210	D	191	100	159	2878.38	1.32	1.10	2.1

Fig. 2. Project plan for construction of MSD (48 tests and sub-tests)

Figure 2 shows the project plan data according to the eight indicators and adds the task ID, task name, and start date for each task or sub-task. The PP contains the data for 47 tasks and sub-tasks, started with TP0 task and ended with TP2-ME9 sub-tasks.

Testing Table (TT)

The data presented in the testing table was entered by the testing team during project implementation.

Testing Table (118 tests)

ID	Task ID	Task Name	Test Date	Time Spent (Day)	Spend Cost	Functions Done	Standards Achieved
1	TP1-CW1-CI-01	Location adjustment	2/6/2014	5	$5,295.00	1	95
2	TP1-CW1-CI-02	Drilling of additional cement bases	2/13/2014	5	$4,475.00	1	95
3	TP1-CW1-CI-03	Soil reclamation works	2/20/2014	5	$4,000.00	2	95

Fig. 3. Testing table for construction of MSD Project (118 tests)

Data were taken in Fig. 3 from the implementing company at Al Hazm Hospital (Al Jouf)/Yemen. The data presented in Fig. 3 was entered by the testing team during project implementation. The TT contains 118 sub-tasks and main tasks tests.

Achievement Metric (M1)

I. **Achievement Metric for All Tasks (M1-1)**

M1.1displays the testing details of tasks or sub-tasks such as; time spent, cost expenditure, functions already performed, standards achievement and the remaining of time, cost, function, and standards. M1.1 has eight equations as following:

$$Remaining\ Time = \frac{Task\ Duration}{Testing\ Date - Starting\ Date} \quad (3)$$

Equation (3) calculates the remaining time for task achievement by dividing the task duration by the testing date subtracted from the task starting date.

$$\mathit{Time\ Spent} = \frac{\mathit{Testing\ Date} - \mathit{Starting\ Date}}{\mathit{Task\ Duration}} * 100 \tag{4}$$

Equation (4) calculates the time spent during task implementation, by subtracted task starting date from task testing date and dividing by the task duration and multiplied by 100.

$$\text{Remaining Cost} = \mathit{Task\ Cost} - \mathit{Spending\ Cost} \tag{5}$$

Equation (5) calculates the remaining cost for task achievement by task cost minus cost spending.

$$\text{Spending Cost } (\%) = \frac{\text{Spending Cost}}{\text{Task Cost}} * 100 \tag{6}$$

Equation (6) calculates the spending cost percentage during task implementation by the spending cost divided by task cost and multiplied by 100.

$$\text{Remaining Functions } = \mathit{Task}\,\text{Functions } - \text{Functions Done} \tag{7}$$

Equation (7) calculates the remaining functions for the task by the task's functions minus the functions already done.

$$\text{Functions Done } (\%) = \frac{\text{Functions Done}}{\mathit{Task}\text{Functions}} * 100 \tag{8}$$

Equation (8) calculates the functions done percentage during task implementation by the functions already done divided by the task's functions and multiplied by 100.

$$\text{Remaining Standards } = \text{Task Standards} - \text{Standards Achieved} \tag{9}$$

Equation (9) calculates the remaining standards for the task by the task's standards minus the standards already achieved.

$$\text{Standards Achieved } (\%) = \frac{\text{Standards Achieved}}{\text{Task Standards}} * 100 \tag{10}$$

Equation (10) calculates the standards achieved percentage during task implementation by the standards already achieved divided by the task's standards and multiplied by 100.

M1-1: Achievement Metrics for All Tasks (118 Tests)

ID	Task ID	Task Name	Test Date	Remaining Time (Day)	Time Spent (%)	Remaining Cost	Spending Cost (%)	Functions Remaining	Functions Done (%)	Standards (%) Remaining	Achieved Standards (%)
1	TP1-CW1-CI-01	Location adjustment	2/6/2014	2	71.4%	261.00	95.3%	0	100.0%	5	95
2	TP1-CW1-CI-02	Drilling of additional cement bases	2/13/2014	2	71.4%	14.00	99.7%	0	100.0%	5	95

Fig. 4. Achievement metric for all tasks of construction of MSD project

Figure 4 shows 118 tests conducted during the project life cycle. Test No. 1 for sub-task (TP1-CW1-CI-01) shows that the time spent is 71%, the costs spent 95% and the completed functions of the task are 100% and the achievement of the current working standards 95%.

II. **Achievement Metric for Main Tasks Using Subtasks Achievement (M1-2)**
M1.2 uses the achievement of sub-tasks to identify the achievement of its main tasks. M1-2 contains two procedures, (1) using the results of completing the subtask, and (2) selecting the latest test data for each subtask. M1.2 has five equations as follows:

$$\text{Number of } subtask = subtask1 + subtask\ldots\ldots + subtaskN \tag{11}$$

Equation (11) calculates the number of sub-tasks by sum all sub-tasks for the main task.

$$Time\ \text{Spent}\ (\%) = \frac{Sum(Time\ Spent\ of\ Subtasks)}{Number\ of\ Subtasks} \tag{12}$$

Equation (12) calculates the percentage of the time spent during the main task implementation by sum the percentage of the time spent of its sub-tasks and divided by the number of its sub-tasks.

$$Spend\ Cost(\%) = \frac{Sum(Spend\ Cost\ of\ Subtasks)}{Number\ of\ Subtasks} \tag{13}$$

Equation (13) calculates the percentage of the cost spending during the main task implementation by sum the percentage of the spending cost of its sub-tasks and divided by the number of its sub-tasks.

$$\text{Functions } Done(\%) = \frac{Sum(Functions\ Done\ of\ Subtasks)}{Number\ of\ Subtasks} \tag{14}$$

Equation (14) calculates the percentage of the functions already done during the main task implementation, by sum the functions already done of its sub-tasks and divided by the number of its sub-tasks.

$$\text{Standards Achieved } (\%) = \frac{Sum(\text{Standards Achieved } of\ Subtasks)}{NumberofSubtasks} \tag{15}$$

Equation (15) calculates the percentage of the standards achieved during the main task implementation by sum the percentage of the standards achieved of its sub-tasks and divided by the number of its sub-tasks.

M1-2: Achievement Metrics for Main Tasks

Main Task: Civil Works (TP1-CW1-CI-0)

Sub-Task		Main Task	
Test Number	1	Number of last tests of sub-tasks	6
Task ID	TP1-CW1-CI-01	Current Date	10/7/2019
Task Name	Location adjustment	Time Spent of the main task(%)	84.05%
Last Test Date	2/6/2014		
Time Spent (%)	71.43%	Spending Cost (%)	99.10%
Spending Cost (%)	95.30%	Functions Done (%)	100.00%
Functions Done (%)	100.00%	Achieved Standards (%)	95.00%
Achieved Standards (%)	95		

Fig. 5. Achievement metric for TP1-CW1-CI-0 Task using its subtasks achievement

Figure 5 shows on the left side the completion of six sub-tasks for the main task (TP1-CW1-CI-0). On the right side shows the main task achievement. The main task achieved by the sum of the sub-task results of each indicator and then divided by the number of sub-tasks.

Resources Metrics (M2)

I. **Resources Metric (M2-1)**

The Resource Metric (M2.1) displays the current state of all resources and assist the management team to review the resource status after each task. M2.1 has four equations as following:

$$\text{Available Time } = Task\ Duration - (Testing\ Date - Starting\ Data) \tag{16}$$

Equation (16) calculates the available time of sub-tasks or main tasks by identifying the difference between testing date and start date, and then minus the result from the duration of sub-tasks or tasks.

$$\text{Available Cost} = Task\ Cost - Spending\ Cost \tag{17}$$

Equation (17) calculates the available cost of sub-tasks or main tasks by minus the spend cost from the cost of sub-tasks or tasks.

M2-1: Resources Metrics

ID	Task ID	Task Name	Test Date	Available Time(day)	Available Cost	Manpower	Remaining Function
1	TP1-CW1-CI-01	Location adjustment	2/6/2014	2	261	A	0
2	TP1-CW1-CI-02	Drilling of additional cement bases	2/13/2014	2	14	A	0

Fig. 6. Resources metric for all tasks of construction of MSD project

Figure 6 shows the available resources and the remaining functions. According to Fig. 6, the resources available for sub-task (TP1-CW1-CI-01) after the test on June sound are two days and 261,00 dollars.

II. **Available Resources Metric (M2-2)**
The available resources metric identifies only the available resources during the implementation of the sub-tasks or main tasks. There are two procedures in M2-2, first using data from M2.1 and second checking whether the available time is greater than zero and the remaining functions equal to zero and printing available resources data.

M2-2: Available Resources Metrics

ID	Task ID	Test Date	Task Name	Available Time(day)	Available Cost	Manpower	Remaining Functions
108	TP1-CW	8/28/2014	Construction Works (Part One)	2	$0.00	D	0
107	TP1-CW5-ID-0	8/21/2014	Interior Decoration	3	$0.00	D	0

Fig. 7. Available resources metrics during project lifecycle

M2-2 identifies the available resources by using data of 118 tests performed for the MSD Project and only 29 tests showed that there are some resources available. According to Fig. 7, the resources available for the completed task (TP1-CW5-ID-0) after test No. 107 on August is D team for three days.

III. **Required Resources Metric (M2-3)**
M2-3 Metric identify only the resources required during the implementation of the sub-tasks or main tasks. There are two procedures, first using data from M2.1 and second checking whether the available time is lesser than zero and the remaining functions equal or greater than one and printing required resources data.

M2-3: Required Resources Metrics

ID	Task ID	Test Date	Task Name	Available Time(day)	Available Cost	Manpower	Remaining Function	Time Required (day)	Cost Required
76	TP1-CW2-EW-012	7/17/2014	Other Appliances (Fans / Stockist)	-3	$10.00	A	2	2	$259.50
77	TP1-CW2-EW-013	7/17/2014	Lighting	-3	$0.00	A	1	2	$293.33

Fig. 8. Required resources metric during project lifecycle

M2.3 specifies the required resources using data from 118 tests carried out for the Construction of MSD Project and only 4 tests showed that there are some resources needed for four sub-tasks. Figure 7 shows the time and cost required to complete the four sub-tasks. According to Fig. 8, the resources required to complete the sub-task (TP1-CW2-EW-012) after test number 76 on July 17 are team A for three days and $10.00.

3.4 Discussions Results

After applying PM-ISD to the MSD project, we found that the PM-ISD mechanism takes into account the failure of MSD and thus measures the standards in each test. PM-ISD provides project management with accurate figures for each task or subtask throughout project phases, as shown in Figs. 4 and 5. PM-ISD tracks resources during the project phases, as shown in Fig. 6, and determines the status of these resources, as shown in Figs. 7 and 8. PM-ISD helps management track implementation and mitigate errors by improving decision-making mechanisms.

4 Conclusion

The development of SCS becomes a major challenge for developers because of the consequences of any failure. Software Engineering has invented many measurement techniques to assist developers specifically for SCSD. Agile has expanded its principles in recent years to accommodate the complexities of SCSs. This paper proposed a measurement mechanism called PM-ISD, which is a management measure that contains a number of sub-metrics. PM-ISD helps to track progress and reduce SCS failure by measuring all tasks in the early stages of the project. M1-1 and M1-2 Metrics assisted the management team of the MSD project to track the achievement of all subtasks and use this achievement to determine the achievement of main tasks. M2.1, M2.2, and M2.3 provided figures on the availability of resources during the project lifecycle, and management can use these figures to reallocate resources. For future research, PM-ISD is still required to add some algorithms such as Random Forest to check for errors during development processes.

Acknowledgments. This work is sponsored by Universiti Tun Hussein Onn Malaysia under TIER1 FASA 1/2007, UTHM Research Grant (VOT U896) and Gates IT Sdn. Bhd.

References

1. Martins LEG, Gorschek T (2017) Requirements engineering for safety-critical systems: overview and challenges. IEEE Softw 34(4):49–57
2. Kramer S, Raab P, Mottok J, Racek S (2014) Comparison of enhanced markov models and discrete event simulation: for evaluation of probabilistic faults in safety-critical real-time task sets. In: 2014 17th Euromicro conference on digital system design, Verona, Italy, pp 591–598
3. Leffingwell D (2016) SAFe 4.0 reference guide: scaled agile framework for lean software and systems engineering, 1st edn. Addison-Wesley Professional
4. Stettina CJ, Hörz J (2015) Agile portfolio management: an empirical perspective on the practice in use. Int J Project Manag 33(1):140–152
5. Silva N, Vieira M (2014) Towards making safety-critical systems safer: learning from mistakes. In: 2014 IEEE international symposium on software reliability engineering workshops, Naples, Italy, pp 162–167
6. Padmini KVJ, Dilum Bandara HMN, Perera I (2015) Use of software metrics in agile software development process. In: 2015 Moratuwa engineering research conference (MERCon), Moratuwa, Sri Lanka, pp 312–317

7. Kylili A, Fokaides PA, Lopez Jimenez PA (2016) Key performance indicators (KPIs) approach in buildings renovation for the sustainability of the built environment: a review. Renew Sustain Energy Rev 56:906–915
8. Leffingwell D (2011) Agile software requirements: lean requirements practices for teams, programs, and the enterprise
9. Isa HM, Hassan FP, Mat MC, Isnin ZB, Sapeciay Z (2011) Learning from defects in design and build hospital projects in Malaysia

Link Bandwidth Recommendation for Indonesian E-Health Grid

Sri Chusri Haryanti[1(✉)], Ummi Azizah Rachmawati[1], Elan Suherlan[1], Nabylla Azahra[1], Sarah Syahwenni Utari[1], and Heru Suhartanto[2]

[1] Faculty of Information Technology, Universitas YARSI, Jakarta, Indonesia
sri.chusri@yarsi.ac.id
[2] Faculty of Computer Science, Universitas Indonesia, Jakarta, Indonesia

Abstract. Today, almost all the agencies and institutions, including government agencies and hospitals, utilize information technology to improve efficiency and effectiveness. The application of information technology in the health sector is required to enhance the performance of health services to the community. In this research, we investigate the e-Health Grid based on the provincial hospital in Indonesia using a Mininet emulator. Actual distances between hospitals have been applied. The packet rate is determined based on the assumption that it is proportional to the population served by the hospital. The e-Health Grid consolidates 34 hospitals, four controllers, and four switches. The result of the simulation yields a recommendation of link bandwidth that provides minimum round trip time from each node in the grid.

Keywords: E-Health grid · Indonesia · Link bandwidth

1 Introduction

Eysenbagh in [1] defined e-Health as "an emerging field in the intersection of medical informatics, public health, and business, referring to health services and information delivered or enhanced through the Internet and related technologies. In a broader sense, the term characterizes not only a technical development, but also a state-of-mind, a way of thinking, an attitude, and a commitment for networked, global thinking, to improve healthcare locally, regionally, and worldwide by using information and communication technology".

E-Health also describes as an application Information and Communication Technology (ICT) technology in healthcare. Increasing the efficiency and quality of healthcare are the objectives. E-Health brings new interactions between patients and healthcare workers. Nevertheless, it focuses on the patient's empowerment. Information and communication exchanges take place in a standardized model in e-Health [1]. Services in e-Health include electronic health records, consumer health informatics, telemedicine, health knowledge management, healthcare teams, healthcare information systems, m-Health, and medical research [2].

The development of e-Health commits the hospitals, health centers, universities, drug manufacturers, the pharmaceutical industry, Government, and also community. The environment needs resource sharing between those components. Grid computing

R. Ghazali et al. (Eds.): SCDM 2020, AISC 978, pp. 77–87, 2020.
https://doi.org/10.1007/978-3-030-36056-6_8

systems are adequate to resolve the problem of resource sharing on information, data structures, databases, processor, and storage resources on separate locations dynamically employing open-standard protocols [3]. It is mandatory to use an open-source to meet the requirement of the e-Health. The grid computing system has designed a set of middleware to promote e-Health applications. Grid computing for e-Health cut down the cost and fully employ the existing IT resources.

According to the Law of Health in Indonesia (UU No. 36/2009), the government is responsible for the health sector. Attaining equitable and affordable health care services demand proper planning, actuating, controlling, supervising, and monitoring [4]. Indonesia is the vastest archipelagic country in the world, consists of 17,508 islands with a population of more than 237 million [5]. Sumatra, the most significant island, covers 25.2% of the entire Indonesian region and has 21.3% of the population. Papua that covers 21.8% of the territory has 1.5% of the population, and Java that covers only 6.8% of the land has 57.5% of the population [6]. Consequently, information and communication are critical aspects of improving the resilience of the country, including in the health area.

The integration of data and information is critical to health organizations in Indonesia. According to the reports, Indonesia is a country for several epidemic diseases such as tuberculosis, malaria, dengue fever, and others, but the data still could not be obtained quickly and accurately. Collaboration is required to find new methods and techniques for data interaction to improve healthcare services. The interoperability system is necessary for national e-Health development in Indonesia. It could attain integrated, secure, efficient data, and information. Interoperability also upholds a system to share and integrate information and processes using a set of standard work. Grid computing technology is one solution to the interoperability problem. The grid computing system integrates services across distributed and heterogeneous virtual organizations on different resources and relationships [7].

In the previous work, we have simulated e-Government Grid in Indonesia based on provinces using GridSim toolkit [8]. We carry out this study to support the Government for giving excellent services to the community and also to assist the researchers, practitioners, Universities, pharmacists, and government agencies.

As a part of the study in adopting Grid technology on e-Health services in Indonesia, we have suggested an e-Health Grid topology model based on the provinces in Indonesia and simulated the topology using NS-3. In this study, we use the topology as a reference in building the e-Health Grid of the country.

2 E-Health Grid

2.1 E-Health

E-Health is expected to increase the efficiency and quality of healthcare. E-Health is evidence-based and focused on consumers' or patients' empowerment. It brings a new relationship between patients and healthcare workers using online sources. E-Health supports communication and information exchange in a standardized manner [7].

2.2 Grid Computing

According to Foster et al. [9], a grid computing system, is a collection of hardware and software infrastructure that provides reliable, consistent, pervasive, and inexpensive access to high-end computational capabilities. The grid computing or the Grid integrates and coordinates resources and users and built-in multifunctional protocols, interfaces, and open standards. Grids enable sharing resources to achieve the quality of services associated with response time, throughput, availability, and security. Various types of resource allocations can be used to meet user demand for complex computation.

Berman et al. [10] stated that a grid computing infrastructure and data management would yield electronic foundation aggregation for the global community regarding business, government, research, science, and entertainment. The grid connects and integrates resources such as networks, communication, computing, and information systems to create a virtual platform that supports computing and data management.

Buyya et al. defined Grids as infrastructures that owned and regulated by several organizations [11]. Those organizations work in an integrated and collaborative nature.

The grid is an established concept in the IT community that facilitates the collaboration of distributed resources with secure and consistent access irrespective of users' physical location. It is verified to share, select, and aggregate distributed heterogeneous resources, for instance, storage, data, instruments, and software systems, as a single and consolidated resource [12].

2.3 E-Health Grid

Brenton et al. define the health grid as the application of grid technologies to the health area [13]. E-Health Grid integrates distributed medical datasets. It aligns a large number of distributed data, a large-scale statistics capacity, and vast epidemiology [14].

E-Health changes the decision-making process and health outcomes. E-Health also needs a reliable online system for education and consultations. A reliable connected system is vital to improve productivity, give more access to medical services, and provide a high quality of diagnostic for patient safety. Technology should be able to break in some existing barriers to access e-Health in the world [15].

3 E-Health in Indonesia

In resolving the UN Millennium Development Goals (MDGs), the Indonesian government announced a Healthy Indonesia 2010 goal. The health-related targets are achievements in the development of national health. Indonesia has been prosperous in collecting resources for developing public health, including the Global Fund to overcome AIDS, Tuberculosis, and Malaria (GFATM). The health sector must strengthen the capability to develop and utilize resources mobilization [16].

Health services are decentralized in unintegrated health information systems that make the share-out of reporting responsibilities is not clear. Indonesia does not have comprehensive data on the health condition of the entire nation that could be used to monitor health programs in the whole country [16]. Indonesia needs to build an interoperable system that supports handling health issues in the country.

The primary purpose of this works is modeling an Indonesia E-Health Grid and obtaining link bandwidth recommendation using Mininet. In 2015, the Ministry of Health of the Republic of Indonesia listed the number of hospitals in Indonesia as 2,228, consists of 1,718 General Hospital and 510 Specialist Hospital [17]. Each province has a central hospital in the capital of the province. Table 1 shows the list of provincial hospitals in Indonesia.

Table 1. Provincial Hospital in Indonesia

No	Province	Name of Hospital
1	Aceh	RSUD Idi Rayeuk
2	Sumatra Utara	RSUD Pandan
3	Sumatra Barat	RSUD Dr. Achmad Darwis
4	Riau	RSUD Provinsi Kepulauan Riau Tanjungpinang
5	Kepulauan Riau	RS Otorita Batam
6	Jambi	RS Siloam
7	Sumatra Selatan	RSU Lahat Palembang
8	Bengkulu	RS Bhayangkara Jitra
9	Lampung	RS Bhayangkara
10	Belitung	RSUD Kabupaten Belitung
11	Jakarta	RS Dr. Cipto Mangunkusumo
12	Jawa Barat	RSUD Kab. Sumbawa Barat
13	Banten	RSUD Banten
14	Jawa Tengah	RSU Ambarawa
15	Daerah Istimewa Yogyakarta	RS Bethesda Yogyakarta
16	Jawa Timur	RSUD Kab Seram Bagian Timur
17	Bali	RS Trijata Polda Bali
18	Nusa Tenggara Barat	RSUD Kab. Sumbawa Barat
19	Nusa Tenggara Timur	RSUD Kota Kupang
20	Kalimantan Barat	RSUD Landak
21	Kalimantan Tengah	RSUD Lamandau
22	Kalimatan Selatan	RSU Datu Sanggul Rantau
23	Kalimantan Timur	RSUD Am Parikesit Tenggarong
24	Kalimantan Utara	RSUD Tarakan
25	Sulawesi Utara	RSU Datoe Binangkang
26	Sulawesi Barat	RSU Majene
27	Sulawesi Tengah	RSU Anutapura Palu
28	Sulawesi Tenggara	RSUD H.M Djafar Harun
29	Sulawesi Selatan	RSU Labuang Baji
30	Gorontalo	RSU Prof Dr H Aloei Saboe
31	Maluku	RSUD Dr. M. Haulussy
32	Maluku Utara	RSU Ternate
33	Papua Barat	RSUD Kaimana
34	Papua	RSUD Tiom

We considered the government of Indonesia, consisting of 34 provinces. These 34 provinces are by reason of the agreement between the Indonesian Parliament and the Government to include North Kalimantan as a new province on October 22, 2012 [8]. The network topology form on the provinces in Indonesia shown in Fig. 1. Indonesian e-Health Grid Network.

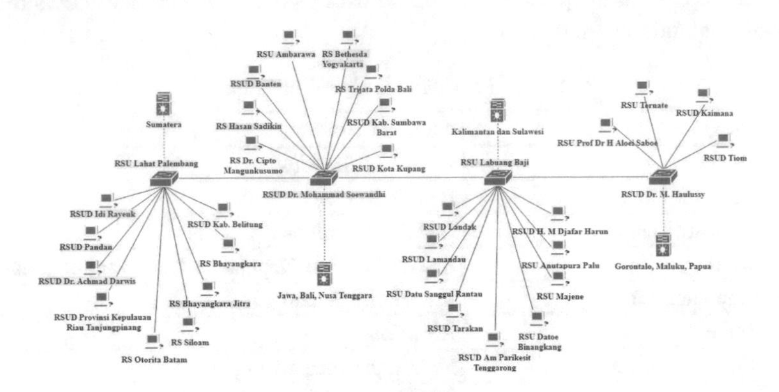

Fig. 1. Indonesian e-Health grid network

We divided the provinces into four clusters. Cluster I consists of hospitals located in provinces at Sumatera Island at the provinces as follows: Nanggroe Aceh Darussalam (NAD), North Sumatera, West Sumatera, Riau, Jambi, South Sumatera, Bengkulu, Lampung, Bangka Belitung, and Riau archipelago. Cluster II covers hospitals located in the provinces of Java, Bali, and Nusa Tenggara that are Special Capital Region Jakarta, West Java, Banten, Central Java, Special Region Yogyakarta, East Java, Bali, West Nusa Tenggara, and East Nusa Tenggara. Cluster III contains hospitals located in the province of Kalimantan and Sulawesi Islands that are West Kalimantan, Central Kalimantan, South Kalimantan, East Kalimantan, North Kalimantan, North Sulawesi, Central Sulawesi, South Sulawesi, Southeast Sulawesi, and West Sulawesi. Cluster IV consists of hospitals placed in the provinces of Gorontalo, Maluku archipelago, and Papua Island that are, Maluku, North Maluku, Papua, and West Papua. Figure 1 displays the network topology based on the province in Indonesia

4 Simulation

Examining a Grid in a real system is very expensive, limited, and time-consuming. In addition to the difficulties of carrying out the heterogeneous and dynamic system, we have to consider different administrative arrangements at each resource. In this research, we evaluate and analyze the performance of topology proposed in the simulation. We use Mininet to build and simulate the topology. We performed scenarios to see the round-trip time (*rtt*) by way of a package rate set to adjust the number of persons in each region in each Province.

We set the packet rate in the simulation, 5 Kbps for the population of 0–10000, 10 Kbps for the populations of 10000–20000, 15 Kbps for the populations of 20000–30000, 20 Kbps for the populations >30000. The link rate is set at 50 Mbps and 100 Mbps. The result of this simulation is a graphic of the *rtt* link bandwidth. Figure 2 shows the comparison of *rtt* between nodes in Sumatra province for a bandwidth of 50 Mbps and 100 Mbps. Tables 2, 3, 4, and 5 shows the packet rate of data from each province, based on the province's population. Population by provinces referred to the Central Bureau of Statistics [18].

Table 2. Population and Packet Rate (Sumatera)

Province	Population in 2015 ($\times$1000)	Packet rate (Kbps)
Aceh	5002	5
Sumatera Utara	13937	10
Sumatera Barat	5196	5
Riau	6344	5
Jambi	3402	5
Sumatera Selatan	8052	5
Bengkulu	1875	5
Kepulauan Riau	1973	5
Kepulauan Bangka Belitung	1373	5

Table 3. Population and packet rate (Java, Bali, and Nusa Tenggara)

Province	Population in 2015 ($\times$1000)	Packet Rate (Kbps)
DKI Jakarta	10178	10
Jawa Barat	46709	20
Jawa Tengah	33774	20
DI Yogyakarta	3679	5
Jawa Timur	38847	20
Bali	4153	10
Nusa Tenggara Barat	4835	5
Nusa Tenggara Timur	5120	5

Table 4. Population and packet rate (Kalimantan and Sulawesi)

Province	Population in 2015 ($\times$1000)	Packet rate (Kbps)
Kalimantan Barat	4789	5
Kalimantan Tengah	2495	5
Kalimantan Selatan	3990	5
Kalimantan Timur	4068	5
Sulawesi Utara	2412	5
Sulawesi Tengah	2877	5
Sulawesi Selatan	8520	5
Sulawesi Tenggara	2499	5
Sulawesi Barat	1282	5

Table 5. Population and Packet Rate (Gorontalo, Maluku, and Papua)

Province	Population in 2015 ($\times$1000)	Packet rate (Kbps)
Gorontalo	1133	5
Maluku	1686	5
Maluku Utara	1162	5
Papua Barat	871	5
Papua	3149	5

The results of the simulations are depicted in Figs. 2, 3, 4 and 5. We run the simulation and compare the result for bandwidth 50 Mbps and 100 Mbps. Figure 2 shows the comparison graph of *rtt* between nodes in Sumatra provinces. Figure 3 shows the comparison graph of *rtt* between nodes in Java, Bali, and Nusa Tenggara provinces. Figure 4 shows the comparison graph of *rtt* between nodes in Kalimantan and Sulawesi provinces. Figure 5 shows the comparison graph of *rtt* between nodes in Gorontalo, Maluku, and Papua provinces.

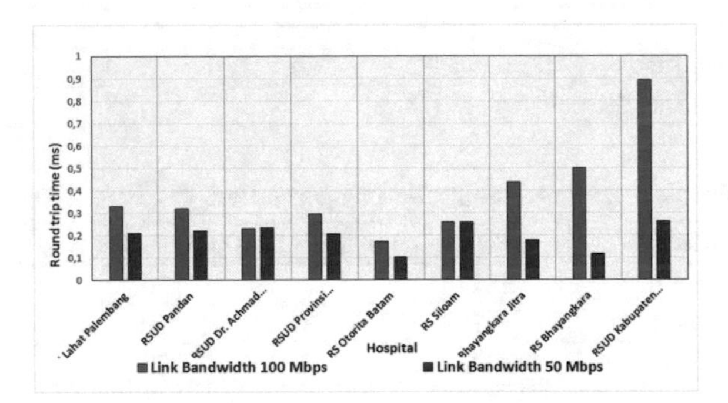

Fig. 2. The comparison of *rtt* between nodes in Sumatera to RSU Lahat Palembang

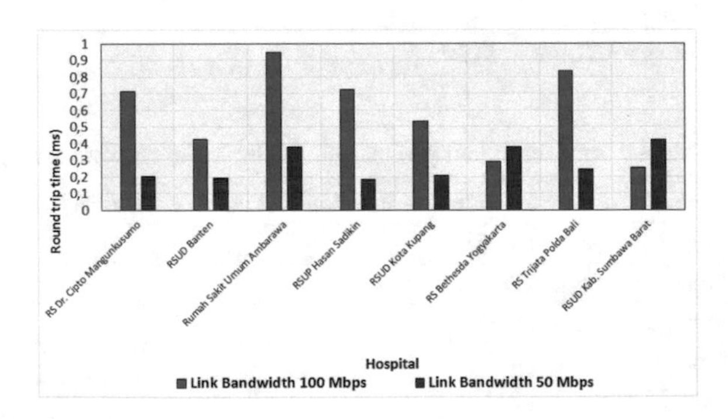

Fig. 3. The comparison of *rtt* between nodes in Jawa, Bali, & Nusa Tenggara to RSUD Dr. M. Soewandie

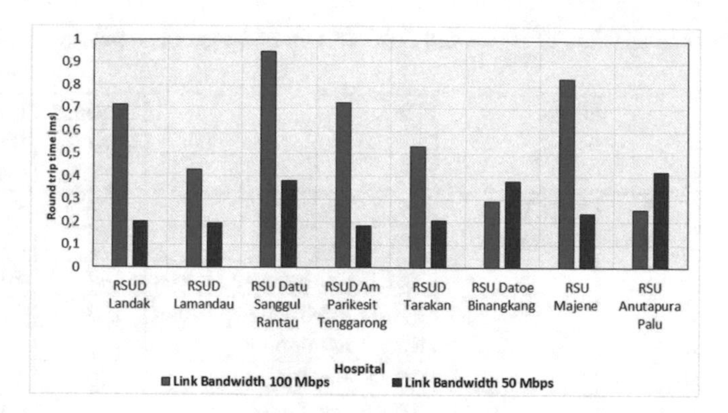

Fig. 4. The comparison of *rtt* between nodes in Kalimantan and Sulawesi to RSU Labuang Baji

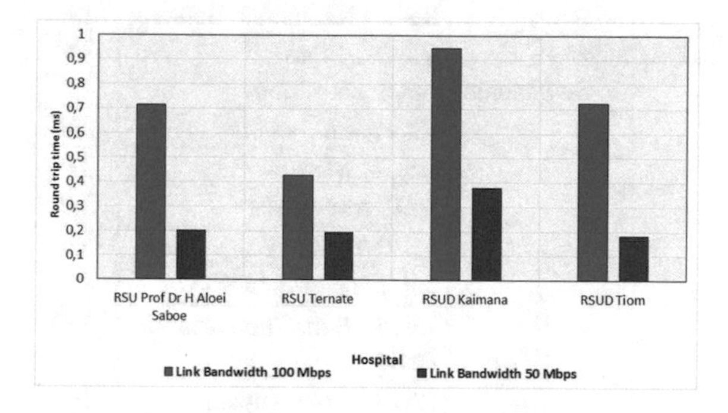

Fig. 5. The comparison of *rtt* between nodes in Gorontalo, Maluku, and Papua to RSUD Dr. M. Haulussy

Of all the measurements performed, the link bandwidth between nodes that generates the minimum *rtt* is summarized in Table 6. These results provide bandwidth recommendations for each link that can provide a smaller round trip time for the e-Health Grid Indonesia network.

We use the frequent pattern mining [19] to recommend the link bandwidth with support 20% and confidence 80%. The result shows that if the distance is more than 600 km, the best link bandwidth is 50 Mbps.

Table 6. Recommendation of bandwidth link between nodes on the topology of e-Health Grid in Indonesia

	From	To	Distance (km)	Link Bandwith (MBps)
Region 1 Sumatera	RSU Lahat Palembang	RSUD Idi Rayeuk	1632	50
		RSUD Pandan	1112	50
		RSUD Dr. Achmad Darwis	649	50 or 100
		RSUD Provinsi Kepulauan Riau Tanjungpinang	742	50
		RS Otorita Batam	756	50
		RS Siloam Jambi	305	50 or 100
		RS Bhayangkara Jitra	295	50
		RS Bhayangkara	349	50
		RSUD Kabupaten Belitung	1123	50
Region 2 Jawa, Bali & Nusa Tenggara	RSUD Dr. M Suwandhie	RS Dr. Cipto Mangunkusumo	835	50
		RS Hasan Sadikin	726	50
		RSUD Banten	974	50
		RSU Ambarawa	380	50
		RS Bethesda Yogyakarta	336	100
		RS Trijata Polda Bali	422	50
		RSUD Kab. Sumbawa Barat	162	100
		RSUD Kota Kupang	141	50
Region 3 Kalimanatan & Sulawesi	RSU Labuan Baji	RSUD Landak	2478	50
		RSUD Lamandau	1400	50
		RSU Datu Sanggul Rantau	1017	50
		RSUD Am Parikesit Tenggarong	1415	50
		RSUD Tarakan	1612	50
		RS Datoe Binangkang	236	100
		RSU Majene	358	50
		RSU Anutapura Palu	213	100
		RSUD H. M Djafar Harun	134	50
Region 4 Gorontalo, Maluku & Papua	RSUD Haulussy	RSU Prof Dr. H Aloei Saboe	1335	50
		RSU Ternate	620	50
		RSUD Kaimana	892	50
		RSUD Tiom	2090	50

5 Conclusion and Future Works

This research produced an e-Health Grid network design for a hospital in Indonesia by province. Total connected hospitals 34 with four controllers and four switches by region-based division. Measurements of e-Health Grid network performance in Indonesia based on province provides the recommended bandwidth of each link that can provide a shorter round-trip time.

Research on e-Health Grid in this research can be continued by testing alternative topology forms and other scenarios. The performance measurement of the e-Health Grid based on the province network in Indonesia on Mininet emulator recommends the bandwidth of each link, which can give smaller *rtt* (round trip time). The use of Mininet emulators facilitates research on e-Health Grid to be continued implementation to the testbed.

Acknowledgements. This work under the support of the Indonesian Ministry of Research and Technology and Higher Education, Directorate General of Higher Education Excellent Research Grants Number 105/K3/KM/2017.

References

1. Eysenbach G (2001) J Med Internet Res 3(2):e20
2. Blum HL (1981) Planning for health development and application of social change theory. Human Sciences Press
3. Rachmawati UA, Suhartanto H, Sensuse DI (2012) Function group based of Indonesian e-Government grid services topology. Int J Comput Sci 9(2), 2
4. Health Law of Indonesia Number 36/2009
5. Statistics Indonesia (2012) 2010 Population Census. Accessed 20 Sept 2012
6. Rachmawati UA, Suhartanto H (2014) Analysis of Indonesia e-Government grid services simulation based on population. Int J Soft Eng Appl 8(11):89–100
7. Silva V (2005) Grid computing for developers. Charles River Media, Hingham, MA
8. Rachmawati UA et al (2016) E-health grid network topology based on province in Indonesia. Int J Bio-Sci Bio-Technol 8(2):307–316
9. Foster I, Kesselman C, Tuecke S (2001) The anatomy of the grid: enabling scalable virtual departments. Int Supercomput Appl 15(3)
10. Berman F, Fox G, Hey AJG (2003) Grid computing: making the global infrastructure a reality. Wiley, New York, NY, USA
11. Buyya R, Murshed M (2002) Gridsim: a toolkit for the modeling and simulation of distributed management and scheduling for grid computing. J Concurr Comput: Pract Exp 14:13–15
12. Kldiashvili E (2005) Grid-New perspective of e-Health applications for georgia (2005). http://georgia.telepathology.org
13. Breton V, Dean K, Solomonides T (2005) The healthgrid white paper. In: Healthgrid 2005. Ios Press, pp 249–318
14. De Vlieger P et al (2009) Grid-enabled sentinel network for cancer surveillance. Stud Health Technol Inf 147:94–289

15. Kldiashvili E (2010) The application of virtual organization technology for E-Health. In: Grid technologies for E-Health: applications for telemedicine services and delivery. IGI Global, USA, ch 1, pp 1–17
16. World Health Organization (WHO) Country Report (2015). http://www.who.int/countries/idn/en/
17. Department of Health, Republic of Indonesia (2015). http://sirs.buk.depkes.go.id/rsonline/report/report_by_catrs_2013.php)
18. Central Bureau of Statistics, Population projection by Province. https://www.bps.go.id/statictable/2014/02/18/1274/proyeksi-penduduk-menurut-provinsi-2010—2035.html
19. Han J, Kamber M (2006) Data mining concepts and techniques, 2nd edn. Morgan Kauffman Publishers

Investigating the Optimal Parameterization of Deep Neural Network and Synthetic Data Workflow for Imbalance Liver Disorder Dataset Classification

Nova Eka Diana[1](✉), Andi Batari Ahmad[1], and Zwasta Pribadi Mahardika[2]

[1] Informatics Department, Faculty of Information Technology, Universitas YARSI, Jakarta 10510, Indonesia
nova.diana@yarsi.ac.id

[2] Faculty of Medicine, Universitas YARSI, Jakarta 10510, Indonesia

Abstract. DNN (Deep neural network) has emerged as one of the standard methods to create a classification model. The most common issue affecting DNN performance is the class-imbalanced distribution dataset. This research designed two workflows for generating synthetic dataset using SMOTE algorithm, SDS-1, and SDS-2 dataset. We further investigated the optimal DNN parameters that generate the best optimum performance over those datasets. We used Indian Liver Patient Dataset (ILPD) from the oldest source, UCI Machine Learning Repository, with a total of 583 records, consist of 416 positives and 167 negatives data. We measured the DNN performance using sensitivity and F-score metric following the nature of the medical domain that mainly focused on identifying a particular disease. The experiment results revealed that DNN model with the learning rate of 1E-1, TanH activation function, Xavier weighting, the epoch of 40, and the hidden layers of 10, delivered the best sensitivity and F-score value, 98.40% and 99.18%, respectively. The results suggested that the workflow for generating the class-balanced dataset will leverage the DNN performance.

Keywords: Deep neural network · Imbalanced dataset · SMOTE · Sensitivity · F-score

1 Introduction

Liver disorder remains a high-burden disease worldwide with nearly 2 million deaths each year, due to complexities of cirrhosis, viral hepatitis, and hepatocellular carcinoma (HCC) [1]. WHO monitoring report for the SDGs (Sustainable Development Goals) in 2018 affirmed that many countries were encountering a "double burden" of malnutrition, in which the under-nutrition infirmity coexisted with overweight and obesity. Obesity in childhood could lead to future risk diseases, such as liver disease, respiratory problems, diabetes, and low self-confidence emotional problems. The reports insinuated that most of the liver disease burden in the world due to the severe progression of

R. Ghazali et al. (Eds.): SCDM 2020, AISC 978, pp. 88–97, 2020.
https://doi.org/10.1007/978-3-030-36056-6_9

hepatitis B (HBV) or hepatitis C (HCV) virus inside the body [2]. Earlier diagnosis of liver disorder could prevent these viruses from escalating into severe liver disease.

Machine learning had attracted countless researchers to design a classifier model that automatically diagnosed a liver disorder, either working on clinical or image dataset. Some algorithms were Quantum-based Binary NN [3], Multilayer Perceptron NN [4], Random Forest Model [5], and Deep Learning [6, 7]. DNN is one of deep learning algorithm that consistently adopted to produce the optimal classifier model, either in the medical or non-medical domain. Lee et al. [8] developed a DNN model to estimate LDL Cholesterol risk for CVD (cardiovascular disease) from two detached datasets of the Korean population. Their research suggested that DNN could accurately estimate LDL-C risk compared to the Novel and Friedewald method. Kannadasan et al. [9] proposed a stacked autoencoders-based DNN to classify Prima Indians dataset and achieved high accuracy. Singaravel et al. [10] also selected deep-learning NN to predict the energy used to explore space in building design. Their research indicated that DNN could advance the performance over other simple ANN models. Many studies had also extended the DNN model to get a higher classification rate, such as Aung et al. [11] research. They implemented DCNN (deep convolutional NN) for classifying multi-label label cover indices of satellite images. Their proposed prototype was tantamount with multi-class SVM for distinguishing multi-label land cover. Their system was valuable to measure the land cover tampering detection on the satellite images. Besides, other studies had also applied Deep Feedforward Neural Network [12] and Fully Conditional Deep Neural Network (FCDNN) [13] to process estimation and segmentation [13] accurately.

Machine learning researchers have continuously perfected the classifier model to attain the most robust performance. Nonetheless, they still hold the standard issue for the imbalanced dataset that poorly represented classes' distribution. Many methods had been introduced to handle the class-imbalanced dataset, such as RUS (Random Under Sampling), Random Over Sampling (ROS), and SMOTE (Synthetic Minority Oversampling Technique). Among these methods, numerous data scientists were experimenting SMOTE to create the balanced-synthetic dataset by oversampling the minority class instead of just creating copies of the data. Zhang et al. [14] had successfully applied SMOTE to answer the dimension disaster on their dataset and successfully raised the prediction capability of their proposed model on identifying ion channels-targeted conotoxins. Numerous studies also had recommended the variant of SMOTE technique to produce the class-balanced and noise-free dataset better [15–18].

Most researches were concentrating on enhancing the distribution of the dataset; they did not reveal the leverage of the distinctive scheme on generating synthetic data toward the classifier performance. This research aimed to examine two SMOTE-based data synthetics workflows toward DNN classifier performance. We assessed the DNN performance based on the variation of learning-rate, epoch, and the number of hidden layer parameters.

2 Materials and Methods

2.1 Liver Dataset Preprocessing

We used ILPD dataset from UCI Machine Learning Database with a whole of 583 data (416 positive and 167 negative data, respectively). Table 1 compiled the variables of ILPD dataset in which each row illustrated the attribute name, data type, and the interval of dataset distribution (e.g., Total Bilirubin Numeric [0.4, 7.5] had the lowest and the highest value of 0.4 and 75, respectively, for TB attribute). Figure 1 depicted the two-dimensional ranking of ILPD dataset features using Pearson-rank correlation to show how well the attributes correlated with the diagnose result. The darker red color meant that those attributes were highly correlated to each other and toward the diagnose result, e.g., DB and TB variables.

Table 1. ILPD data format; *1: positive, 2: negative

Variable	Information	Data type
Age	Age	Numeric [4, 90]
Gender	Gender	String [Female, Male]
TB	Total Bilirubin	Numeric [0.4, 75]
DB	Direct Bilirubin	Numeric [0.1, 19.7]
ALP	Alkaline Phospatase	Numeric [63, 2110]
SGPT	Alamine Aminotransferasae	Numeric [10, 2000]
SGOT	Aspartate Aminotransferase	Numeric [10, 4929]
TP	Total Proteins	Numeric [2.7, 9.6]
ALB	Albumin	Numeric [0.9, 5.5]
A/G	Albumin/Globulin Ratio	Numeric [0.3, 2.8]
CLASS	Data label	Numeric [1, 2]*

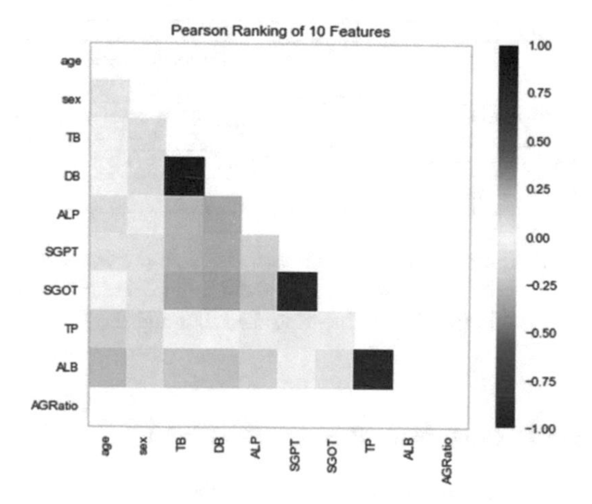

Fig. 1. Pearson-ranking visualization of ILPD dataset.

As aforementioned, we had an imbalance dataset that was a prevalent difficulty met by the data scientists. Goodfellow et al. [19] inferred that dataset synthesis could produce artificially contrived data for verifying mathematical models and enhanced machine learning models. In this research, we employed a popular method, SMOTE, to resolve the imbalance ILPD dataset. SMOTE applied a K-Nearest Neighbor algorithm to produce synthetic data. Figure 2 displayed the implementation of KNN in SMOTE algorithm to create new instances. We generated two synthetic datasets and then analyzed their performance to the classification model. Figure 3 represented the workflow to produce the dataset employed for this research.

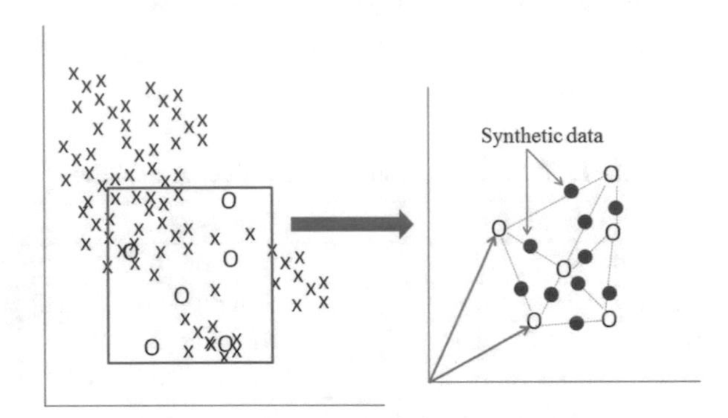

Fig. 2. SMOTE applies KNN to select k nearest neighbor, computes the distance between vectors and multiply the difference by a random number (0, 1) to create a synthetic data

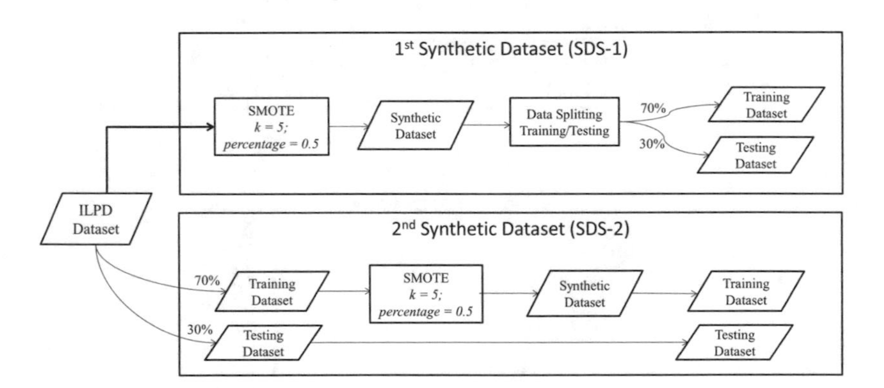

Fig. 3. Approach to create synthetic datasets with $k = 5$ and $percentage = 0.5$. 1st workflow applied SMOTE to all ILPD dataset and then split them into 0.7 training and 0.3 testing data. 2nd workflow split the dataset into training and testing data first and then performed SMOTE toward the training data only

2.2 Configuration of Deep Neural Network (DNN)

Figure 4 demonstrates the structure of a fully-connected DNN adopted in this research. DNN is a stacked of neural networks that consist of several layers, wherein every layer is composed of several nodes. A node is like a neuron in the brain that arouses whenever encounters with stimuli. A node assigns a set of weights to the inputs, sum all these input-weighted products, and then passes it to the activation function to decide whether the signal should proceed further within the network. Our DNN model employed Xavier algorithm to give the initial weights of all layers. Xavier will randomize the initial weights using Eq. (1), and keep the signal of each activation function to fall into the sound range through many layers. We applied TanH activation function to determine when a neuron should be stimulated. TanH or Hyperbolic Tangent is a non-linear function that transforms inputs into high dimensional data, which is essential for classification purpose. This function will transform the real number input into an output value that is close to zero and in the range of [−1, 1] using Eq. (2). Hence, the function will evenly distribute all weights values between positive and negative number through the layers.

$$w^l \sim N(0, \sqrt{\frac{2}{n^l + n^{l-1}}}). \tag{1}$$

$$\tanh(x) = 2\delta(2x) - 1. \tag{2}$$

In this research, we examined the DNN model based on sensitivity and F-score metrics, as shown in Eqs. (3) and (4), respectively. In the medical field, it was more important to evaluate the classifier model on how it correctly identified the actual data in the training set than focused on the accuracy metric. Sensitivity or recall metric measures the exact positive rate of the model to correctly identify the actual positive

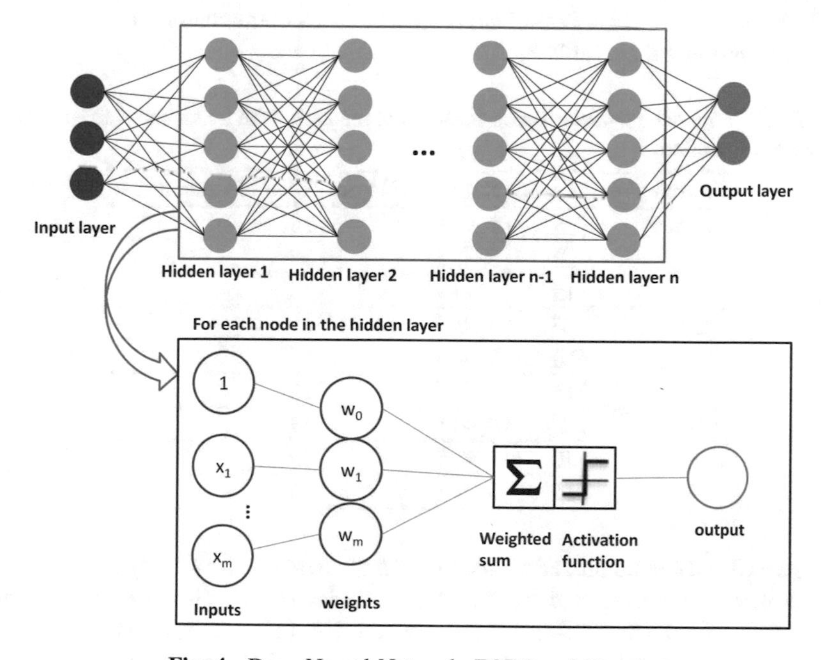

Fig. 4. Deep Neural Network (DNN) architecture

data. Whereas, F-score calculates the harmonic average of recall and precision to measure the achievement of a classifier model for the positive class in the dataset.

$$sensitivy = recall = \frac{true\,positive}{true\,positive + false\,negative}. \quad (3)$$

$$F - score = 2 * \frac{precision * recall}{precision + recall}. \quad (4)$$

3 Results and Discussion

As aforementioned, we analyzed the success of DNN model based on the sensitivity and F-score metrics. We evaluated the model according to three parameters: learning-rate, epoch, and the quantity of DNN hidden layers. The learning rate is critical to determine how effective and converge the learning process. The selection of epoch value is also essential to avoid the under-fitting or over-fitting regime in identifying the correct data. Finally, we examined the number of hidden layers that yielded the most favorable outcome.

In this research, we divided the experiment into two scenario stages. The First scenario aimed to determine the best performed-dataset toward the same DNN models. After that, we determined the best DNN model configuration based on sensitivity and F-score metrics. Table 2 summarized the DNN configurations applied to SDS-1 and SDS-2 dataset, with the learning-rate variation of 1E-1, 1E-2, and 1E-3. We executed each learning-rate for a total of 45 experiments with a combination of epoch and layer number, as presented in Table 2.

Table 2. DNN model parameterization for the first scenario, with a number of five experiments for each configuration.

Learning rate	Epoch	Number of layers	Total of experiments
1E-1	5	3, 4, 5	15
	10	3, 4, 5	15
	15	3, 4, 5	15
1E-2	5	3, 4, 5	15
	10	3, 4, 5	15
	15	3, 4, 5	15
1E-3	5	3, 4, 5	15
	10	3, 4, 5	15
	15	3, 4, 5	15

Figures 5 and 6 summarized the recall and F-score metric for all experiments stated in Table 2, in which the first and the second row on each figure displayed the experiment results upon SDS-1 and SDS-2 dataset, respectively. In most experiments, the second dataset, SDS-2 always had a higher performance than SDS-1 dataset. Furthermore, the

DNN model with a learning-rate of 1E-1, a layer number of 5, and an epoch number of 15 gave the best recall value of 82.00% and the F-score of 78.35%, respectively. Thorough analysis exhibited that the higher learning-rate and epoch number would leverage the DNN model to swiftly and steadily converge to the optimal solution. Whereas, the expanding number of hidden layer did not warrant that the designed model would produce a more robust performance. The figures also showed that the DNN configuration mentioned above could correctly and accurately identify the higher particular liver disease in the SDS-2 than the SDS-1 dataset. The SDS-2 dataset had the difference recall and F-score value of 13.64% and 15.48%, respectively, which were more leading than the DNN model execution toward SDS-1 dataset. We found that by firstly splitting the data and then applying the SMOTE algorithm, then the distribution of the input dataset (training and testing) would be more balanced. We could highly trust the accuracy result because we wholly extracted the testing data from the original dataset.

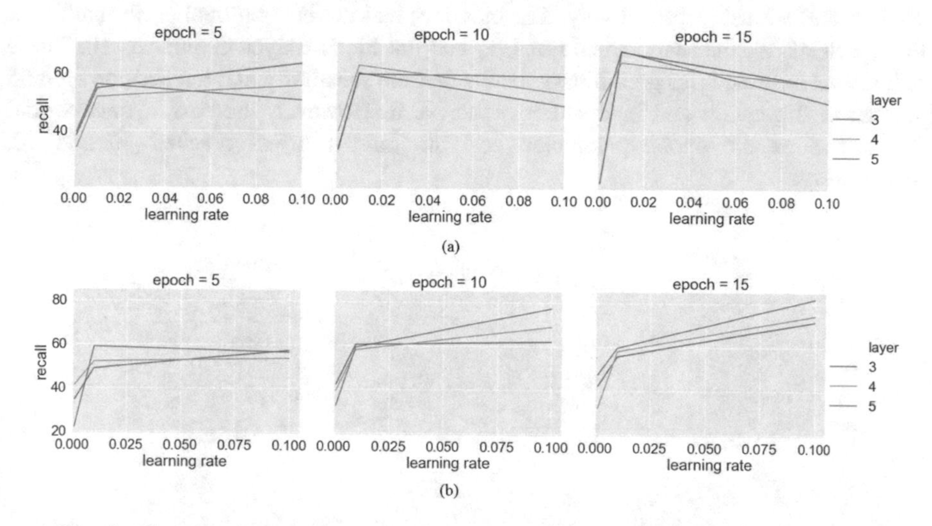

Fig. 5. The experiment values of recall metric on (a) SDS-1 and (b) SDS-2 dataset

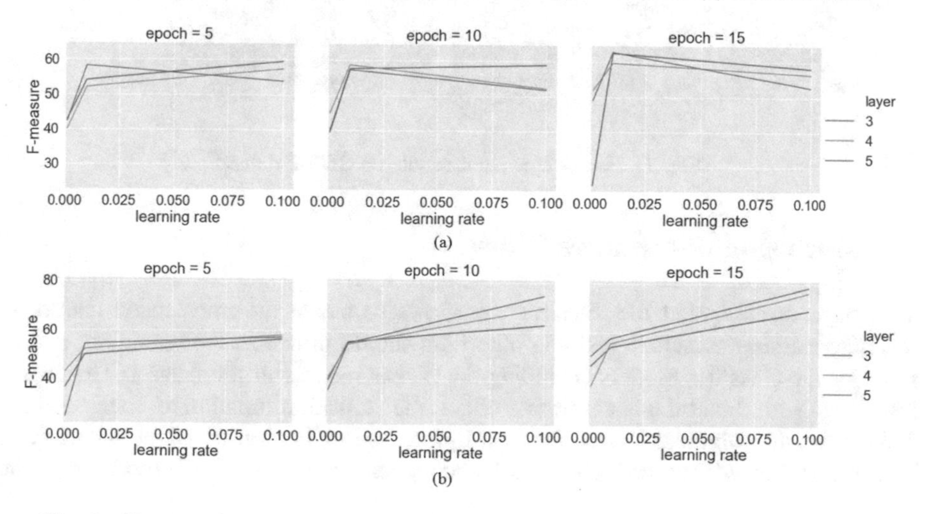

Fig. 6. The experiment values of F-score metric on (a) SDS-1 and (b) SDS-2 dataset

The ILPD dataset classified the liver disease according to enzymatic parameters. We needed to create a classifier model that was correctly and highly recognizing the positive records in the dataset. Based on the results above, we further investigated the optimum properties of DNN architecture only on SDS-2 dataset. Our initial investigations revealed that the current optimal properties of DNN model were five hidden layers and an epoch number of 15. Then, we started new experiments with these parameters as the baselines. We repeated a total of 60 experiments for each learning-rate by updating the epoch parameter to the value of 20, 30, 40, and 50. Moreover, we also began the experiment with five hidden layers, and then repeated the same process with 10 and 15 hidden layers.

Figure 7 illustrated the DNN architecture performance in identifying the real dataset using recall and F-score metric. Further experiments showed that the classifier model had a significant improvement in the recall and F-score metric, which were 98.40% and 99.18%, respectively. The classifier had the best optimal performance for the epoch of 40, the learning-rate of 0.1, and the hidden layer number of 10. These outcomes confirmed our preliminary results that the sensitivity and the accuracy would elevate as the epoch and layer number increased. However, the performance would decrease after the epoch parameter, and the hidden layers reached 40 and 10, respectively.

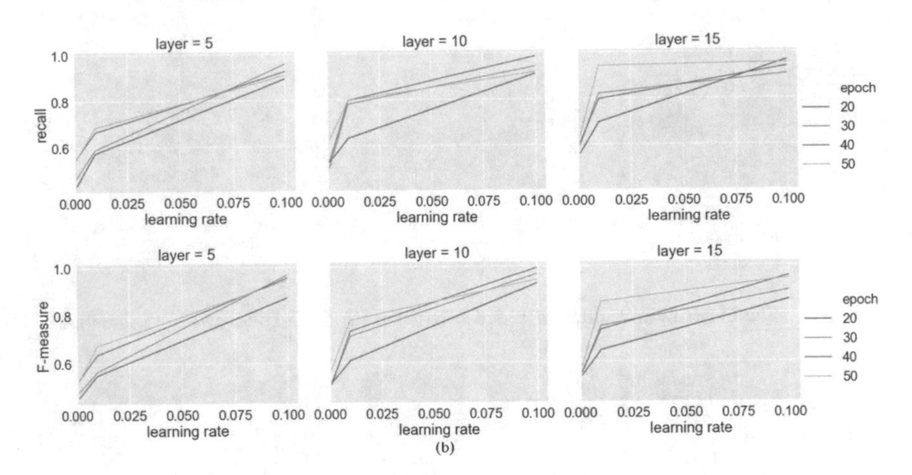

Fig. 7. Final experiment results on SDS-2 dataset

4 Conclusion and Future Work

This paper investigated the different workflows to tackle the imbalanced dataset in machine learning. Machine learning algorithm should prioritize the sensitivity and F-score of the classifier model for finding the higher records in the dataset. The experiments suggested that the learning-rate of 0.1, the activation function of TanH, and the Xavier weight parameters were the best optimal configuration of DNN architecture. The results showed that the higher epoch and hidden layers in the DNN model, then the performance would increase. The research tackled the common imbalance dataset and optimization parameters of DNN model faced by researchers.

Acknowledgments. The authors wish to thank Universitas YARSI for funding this research (No. 183/INT/UM/WRII/UY/VIII/2016).

References

1. Asrani SK, Devarbhavi H, Eaton J, Kamath PS (2019) Burden of liver diseases in the world. J Hepatol 70:151–171
2. World Health Organization (2018) World health statistics 2018: monitoring health for the SDGs. Sustainable Development Goals, Geneva
3. Patel OP, Tiwari A (2015) Liver disease diagnosis using quantum-based binary neural network learning algorithm. In: Proceedings of fourth international conference on soft computing for problem solving, advances in intelligent systems and computing, vol 336. Springer, New Delhi, pp 425—434
4. Abdar M, Yen NY, Hung JCS (2018) Improving the diagnosis of liver disease using multilayer perceptron neural network and boosted decision trees. J Med Biol Eng 38(6):953–965
5. Wu CC et al (2019) Prediction of fatty liver disease using machine learning algorithms. Comput Methods Programs Biomed 170:23–29
6. Hassan TM, Elmogy M, Sallam ES (2017) Diagnosis of focal liver diseases based on deep learning technique for ultrasound images. Arab J Sci Eng 42(8):3127–3140
7. Das A, Rajendra Acharya U, Panda SS, Sabut S (2019) Deep learning based liver cancer detection using watershed transform and Gaussian mixture model techniques. Cogn Syst Res 54:165–175
8. Lee T, Kim J, Uh Y, Lee H (2019) Deep neural network for estimating low density lipoprotein cholesterol. Clin Chim Acta 489:35–40
9. Kannadasan K, Edla DR, Kuppili V (2018) Type 2 diabetes data classification using stacked autoencoders in deep neural networks. Clin Epidemiol Glob Health
10. Singaravel S, Suykens J, Geyer P (2018) Deep-learning neural-network architectures and methods: using component based models in building-design energy prediction. Adv Eng Inform 38:81–90
11. Aung SWY, Khaing SS, Aung ST (2019) Multi-label land cover indices classification of satellite images using deep learning. In: ICBDL 2018: big data analysis and deep learning applications, vol 744. Springer, Singapore, pp 94–103
12. Chemali E, Kollmeyer P, Preindl M, Emadi A (2018) State-of-charge estimation of Li-ion batteries using deep neural networks: a machine learning approach. J Power Sour 400:242–255
13. Bazrafkan S, Thavalengal S, Corcoran P (2018) An end to end deep neural network for iris segmentation in unconstrained scenarios. Neural Netw 106:79–95
14. Zhang L, Zhang C, Gao R, Yang R, Song Q (2016) Using the SMOTE technique and hybrid features to predict the types of ion channel-targeted conotoxins. J Theoret Biol 403:75–84
15. Guo H, Zhou J, Wu C-A (2018) Imbalanced learning based on data-partition and SMOTE. Information 9:238–250
16. Raghuwanshi BS, Shukla S (2019) SMOTE based class-specific extreme learning machine for imbalanced learning. Knowl-Based Syst (2019)

17. Maldonado S, Lopez J, Vairetti C (2019) An alternative SMOTE oversampling strategy for high-dimensional datasets. Appl Soft Comput 76:380–389
18. Douzas G, Bacao F, Last F (2018) Improving imbalanced learning through a heuristic oversampling method based on k-means and SMOTE. Inf Sci 465:1–20
19. Goodfellow I et al (2016) Deep learning (Adaptive Computation and Machine Learning Series). The MIT Press

Genetic Algorithm Based Parallel K-Means Data Clustering Algorithm Using MapReduce Programming Paradigm on Hadoop Environment (GAPKCA)

Sayer Alshammari, Maslina Binti Zolkepli(✉), and Rusli Bin Abdullah

Universiti Putra Malaysia, 43400 UPM Selangor, Malaysia
sayer.r.alsayer@gmail.com, masz@upm.edu.my

Abstract. Data clustering algorithm has been receiving considerable attention in many application areas such as data mining, document retrieval, image processing and pattern classification. A hybrid data clustering algorithm using the combination of genetic algorithm (GA) with a popular variant of K-Means clustering algorithm, parallel k-Means clustering algorithm (PKCA) is proposed in this paper. The objective of the proposed algorithm is to combine the search process of GA to generate new data clusters and apply parallel K-Means to further speed up the quality of the search process during clusters formation. The proposed approach is implemented using the popular MapReduce programming model on Hadoop framework. Experiments were conducted with multiple synthetic datasets to evaluate the performance of the proposed algorithm. Results show that the proposed algorithm was able to speed up document clustering process by 0.54 s on average and outperformed PKCA. Data analysts in marketing and finance, telecommunication and transport companies and researchers in academia can use this algorithm to make sense out of their huge volume of data.

Keywords: K-Means · Parallel K-Means · Clustering algorithm · MapReduce · Hadoop · Genetic algorithm

1 Introduction

The rise of different big data frameworks such as Apache Hadoop and, more recently, Spark, for massive data processing based on the MapReduce paradigm has allowed for the efficient utilization of data mining methods and machine learning algorithms in different domains. Document clustering is regarded as one of the most widely used data mining technique with the help of K-Means algorithm as a result of its simplicity and efficiency. The algorithm groups N objects into K clusters maintaining high intra group similarity and low inter group similarity of the objects. Initially provided with k cluster centers (centroids), the k-means algorithm places data object into a group by finding closest cluster center using a distances measure.

R. Ghazali et al. (Eds.): SCDM 2020, AISC 978, pp. 98–108, 2020.
https://doi.org/10.1007/978-3-030-36056-6_10

This measure calculates similarity distance between each data objects and centroids. Among many available distance measures, Euclidean distance is the widely used measure for distance calculation between objects and their centroids [1]. K-means is an iterative algorithm that converges after a finite number of iterations or if a prior condition for converge is met. Each iteration provides a set of objects to a centroid whose distance is lower. A new centroid is then calculated based on the mean of the data objects of each group. This new centroid is fed to the next iteration and so on. The classical clustering algorithms lack in performance which causes enterprises to lose time and the marketing advantage [2]. This situation requires an efficient computing platform that can exploit this potential to the maximum level, keeping in mind the current challenges associated with it. Obviously, we need to transform classical clustering algorithms for the efficient computing platform designed for large data processing. Enterprises have experienced that existing centralized architectures requires to be replaced with the distributed architecture in order to efficiently handle huge volume of data [3].

Apache Hadoop is arguably the most influential, established and efficient distributed computing framework for large data processing [4, 5]. Hadoop has two key components: Hadoop distributed file system (HDFS) and MapReduce distributed programming paradigm. While distributing the computing process, HDFS takes responsibility to divide dataset and send them to multiple computing nodes and keep track of them whereas MapReduce process algorithmic steps in each computing node. Consequently, this paper proposed an evolutionary parallelized k-means algorithm using MapReduce programming paradigm and run on Hadoop framework.

2 Cluster Analysis

Cluster analysis have been a hot topic of research as a result of its applicability in many disciplines that includes cloud computing, image processing, data mining and information retrieval [6]. The document cluster analysis problem was defined by [7] as a tuple (X, A, R, ϕ) where:

- X is a finite set of n data objects: $X = \{O1, O2, \ldots On\}$
- A is a finite set of attributes or feature: $A = \{Ao1, Ao2, \ldots Ao2\}$. Thus, each data object $Oi \in X$ has a corresponding discrete set of attributes AOi from which it can be identified.
- R is the relationship defining the constrain on the resulting clusters. Or any two clusters Ci, Cj where $Ci \cap Cj = \Phi$, for $i, j = 1, 2, \ldots k$. The relationship is to abide by all clusters formed that no pairs of clusters should have a data object in common.
- Φ is the objective function used for cluster evaluation

A solution to a clustering problem requires the partition of the n data objects into a set of k clusters Ck such that Φ is optimized. The number of possible partitions for the n objects in k clusters is given by the Stirling number of the second kind:

$$S_n^k = \frac{1}{k!}\sum_{i=1}^{k}(-1)^{k-1}\binom{k}{i}i^n \tag{1}$$

Equation (1) leads to a combinatorial explosion as a result of multiple number of solutions that would make it impossible to search all the alternative clusters [8]. The problem more complex when the number of clusters to search is not known from the initial stage. In that instance, the number of different portions is sum of the Stirling number numbers of the second kind given by:

$$\sum_{i=1}^{k_{max}} S_n^i \tag{2}$$

Where k_{max} denotes the maximum number of clusters such that $k_{max} \leq n$.

As a result of the aforementioned reasons, there is increasing need to propose an evolutionary strategy to solve clustering problem using cost function that quantifies the efficacy of a clustering algorithm on the basis of similarity and dissimilarity measures of the data objects. One of the commonly used objective function is the sum of the squared distance of the data objects within their clusters representatives.

Given a partition P and a cluster representative C. The cluster function is usually assessed by Eq. (3).

$$\Phi = \sum_{i=1}^{n} Dist(Oi, Cpi) \tag{3}$$

Where *Dist* is a distance function between the data objects and the criterion in the cluster. In clustering scenario, Euclidean Distance (*ED*) is the most widely used distance function. This function is represented by Eq. (4).

$$ED = Dist(O1, O2) = \sqrt{\sum_{i=1}^{k}\left(O_1^i - O_2^i\right)^2} \tag{4}$$

The interest of researchers and complexity that lies within big data analytics has stimulated the search for efficient data clustering approximation algorithm [3, 8, 13] broadly grouped under three categories: hierarchical, partitional and heuristics search methods [14].

Hierarchical clustering methods [4, 5, 15] cluster data objects by using either divisive or agglomerative style. The agglomerative style starts with each data object on its own cluster and merges the closest pair of clusters at each step with the help of cluster proximity metrics. Different agglomerative data algorithms differ from one another on how clusters of data objects are merged. Partitional data clustering algorithms [2, 16, 17] operates by iterative repositioning of data objects between clusters. Data objects are divided into a set of non-overlapping clusters in such a way that each data objects resides inexactly one cluster. Clustering criteria is use for measuring of a given solution on this clustering method. The value of the criteria is increased by the algorithm after each iteration until a convergence criterion is achieved. This class of algorithms are considered among the fastest algorithms yet, present a result of inferior quality. The third category of these algorithms, are metaheuristic clustering algorithm presented in [18–21], are formed as an alternative to the traditional clustering methods. These techniques are presented with the advantage of being flexible. Moreover, this strategy can be applied for solving clustering problem. These techniques used the principle of encoding possible candidate solution through a so-called objective function. They have the advantage that they could efficiently escaped local optimum convergence. Class of these algorithms start by generating initial population and iteratively replace current solution with a better solution by considering the objective function of each solution in a well-defined strategy. Their performance largely depends on finding the best possible clusters of objects within the solution space. That is, the ability of exploring different region of search space for obtaining high quality solutions within those regions of the solution space. Example of these algorithms include particle swarm optimization (PSO) [11, 12], genetic algorithm (GA) [7, 20], Tabu search [15].

K-Means clustering algorithm is unsupervised data clustering algorithm used for clustering unlabeled data. The goal of the algorithm is to find clusters in data such that the number of clusters is represented by the variable K. the algorithm iteratively works to assign each data point to one of the K clusters according to a criterion provided. Clusters are formed based on feature and similarities. The literature looks at two variant of K-Means clustering algorithms; classical and parallel K-Means clustering algorithms. Classical data clustering algorithm has been used in both academia and in the industries for developing application that deals with unsupervised data clustering in a big data environment.

A hybrid GA-based K-Means clustering algorithm [7] was proposed to improve the quality of clusters formed in order to increase the search process, this algorithm has limitation as the algorithm did incorporate any distributed file system mechanism in the implementation. Tsai et al. [22] proposed a document clustering algorithm by implementing metaheuristic-based data clustering algorithm in a cloud environment to reduce the response time delay on data analytics systems. Moreover, a parallel K-Means data clustering algorithm using MapReduce programming paradigm on Hadoop environment for analyzing remote sensing images which generates direct view of objects was proposed by [23]. The speed of convergence rate of Genetic K-Means data clustering algorithm is enhanced [8] by incorporating the search process of an evolutionary algorithm and clustering process of a classical K-Means algorithms. Sardar and Ansari [25] proposed a document clustering approach that implemented a parallel K-Means clustering algorithm using MapReduce programming paradigm on a Hadoop framework.

3 GAPKCA Algorithm

Parallel K-Means algorithm [17] implemented by GAPKCA is a simple and well-known algorithm used for solving the clustering problem. The goal of the algorithm is to find the best partitioning of n (number of) objects into k (a constant, i.e. number of) clusters, so that the total distance between the cluster's members and its corresponding centroid, representative of the cluster is minimized. The algorithm uses the mapper function to perform the task of assigning each data object to its closes center while the reduce function performs the task of updating newly formed centers. To reduce the cost of communication, a function that deals with the partial combination of intermediate objects with the same key within the same map operation. The main operations of GAPKCA are carried out by three function; the map function, combine function and reducer function. The algorithm uses an iterative refinement strategy in the following steps:

1. This step determines the starting cluster centroids. A very commonly used strategy is to assign random k, different objects as being the centroids.
2. Assign each object to the cluster that has the closest centroid. In order to find the cluster with the most similar centroid, the algorithm must calculate the distance between all the objects and each centroid.
3. Recalculate the values of the centroids. The values of the centroid are updated by taking as the average of the values of the object's attributes that are part of the cluster.
4. Repeat 2 and 3 iteratively until objects can no longer change clusters

Also, the optimization and belong to the group of Evolutionary Algorithms. They simultaneously examine a set of possible solution. Input to GAs is an initial population of solutions called individuals or chromosomes. The most fundamental unit of genetic information is the gene which is one of the constituents of a chromosome. Every gene can assume different values called allele. All genes of an organism form a genome which affects the appearance of an organism called phenotype. Chromosomes represent a point among the candidate solutions' search space and they are encoded using a unique representation. In other to access the quality of individuals, they are assigned scores (i.e., fitness values). A random generation procedure can be used to generate the initial population and the population could also be generated using more advanced techniques to ensure the resulting chromosomes are of very high quality. Chromosomes are selected by the reproduction operator in a random fashion or based on fitness values from the population of potential parents. This population is then made to enter into a mating pool. Note that in order to ensure efficient exchange of information using the cross-over operator probability, parents are care-fully drawn and combined from the mating pool. The resulting population enters into an intermediate population after they are subjected to mutation. It is essential to ensure diversity within the population without disrupting the cross-over results. This is achieved by applying the mutation operator with a low probability. In order to populate a new generation, the initial population is updated using a selection scheme. During the course of the genetic 'operation', individuals from a population pass through an evolution process (from generation to generation) using an evaluation procedure. After a number of generations, the population becomes uniform and an optimal or near-optimal solution is obtained. Next, we elaborate the application of the genetic algorithm for clustering problem. Algorithm 1 shows the various steps used in the proposed genetic algorithm.

Algorithm*: *GAPKCA
input : *Problem P*0
output: *Solution Cfinal*(*P*0)
1 *begin*
2 *Generate initial population;*
3 *Evaluate the fitness of each individual in the population;*
4 *while* (*Not Convergence reached*) *do*
5 *Select individuals according to a scheme to reproduce;*
6 *Breed each selected pairs ofindividuals through crossover;*
7 *Apply parallel K* − *Means if necessary to each offspring*
8 *Evaluate the fitness of the intermediate population;*
9 *Replace the parent population witha new generation* ;
10 *end* while
11 *end*

- Initial population

The initial population consists of individuals generated randomly in which each gene's allele is assigned randomly a label from the set of cluster labels.

- Fitness function

Fitness function is an objective function fundamental to the application of genetic algorithms. It is a numerical value that expresses the performance of an individual so that different individuals can be compared. The fitness function used by our genetic algorithm is simply by Eq. (4).

- Crossover

The task of the cross-over operator is to reach regions of the search space with higher average quality. New solutions are created by combining pairs of individuals in the population and then applying a crossover operator to each chosen pair. The individuals are visited in random order. An unmatched individual is matched randomly with an unmatched individual. Thereafter, the two-point crossover operator is applied using a crossover probability to each matched pair of individuals. The two-point crossover selects two randomly points within a chromosome and then interchanges the two parent chromosomes between these points to generate two new off-spring.

- Mutation

Mutation operator is a genetic operator used to maintain genetic diversity from one generation of a population of genetic algorithm chromosomes to the next. It is analogous to biological mutation. Mutation alters one or more gene values in a chromosome from its initial state. In mutation, the solution may change entirely from the previous solution.

4 ADS Dataset

Five synthetic datasets [26] (ADS1-ADS5) are selected for running the experiment. These datasets were used by researchers for evaluating the efficacy of their proposed clustering algorithms such as [27–30]. Two of these datasets (ADS1 and ADS2) are partially synthesized [31] from a comma separated value (.csv) format of titanic dataset [32] retrieved from https://github.com/datasciencedojo/datasets on 15 July 2019. This dataset is having 893 rows and 12 columns. The remaining three datasets (ADS3-ADS5) are fully synthesized data derived from the US election voters registration available for download at https://catalog.data.gov/dataset?tags=elections. This dataset was retrieved on 15 June 2019. The dataset represents a section of the state of Oregon voters record of over forty thousand organized in six different columns.

5 Experiment

The experiment was conducted on a Dell PC running Windows 10 professional edition with 8 GB ram and 3.24 GHz processor. JDK 11, Hadoop 3.20 and NetBeans IDE version 1.8.2 were used for the experiment. Five artificial datasets of considerable sizes were used for running the experiment. The smallest sized dataset contains 100 randomly selected records, while the largest is having 10000 randomly selected records from the chosen datasets (see Table 1).

Table 1. Experimental data

Dataset	Size	# Clusters	# Trial	Execution time (m/s)	
				GAPKCA	PKCA
ADS1	100	10	10	2.25	2.51
ADS2	500	20	25	2.28	2.58
ADS3	1000	50	50	2.75	3.16
ADS4	5000	100	50	2.81	3.60
ADS5	10000	100	100	3.09	4.07
Average	3320	5	32600	2.64	3.18

Table 1 presents the dataset used for running the experiment (Dataset) and their properties (Size). The table also presents the initial number of clusters formed from each dataset (# Cluster) and the number of iteration (# Trial = Size/# Cluster) performed on each dataset. The last column of the table reports the time taken by the algorithms (GAPKCA and PKCA) to identify relevant clusters within the specified number of trials on each dataset with the help of the objective function defined in Eq. (4).

6 Result and Discussion

Execution time presented in Table 1 was used to compare the proposed clustering algorithm (GAPKCA) and the classical Parallel K-Means Clustering Algorithm (PKCA). Bar chart was used to describe the results for comparing the performance of the two algorithms with respect to their execution time. Figures 1 and 2 present the execution time and the average execution time of the two clustering algorithms.

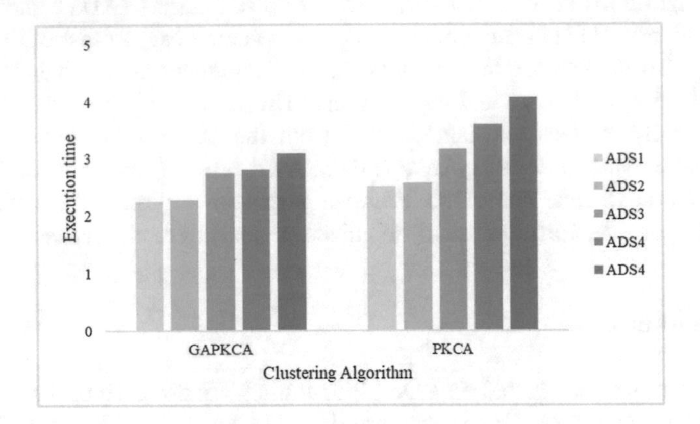

Fig. 1. Execution time

From the chart presented by Fig. 1, it can be seen that, on all the datasets PKCA spend the highest execution time. This indicated that our proposed clustering algorithm outperforms the classical parallel K-Means clustering algorithm on all the dataset. Next is to compare the average execution time of the two clustering algorithms.

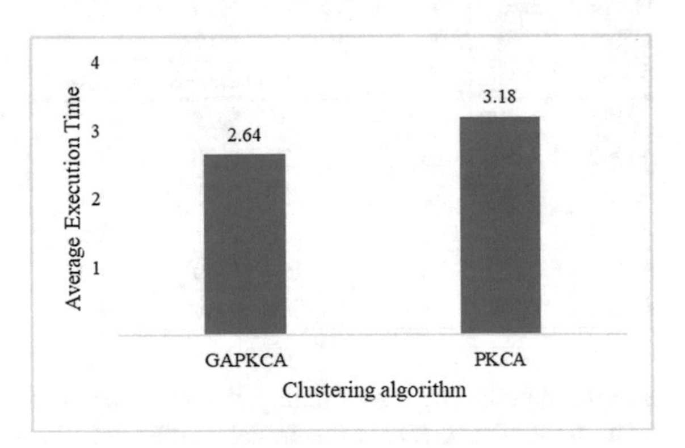

Fig. 2. Average execution time

Figure 2 presents the average execution time of the clustering algorithms. The results indicated that on average, GAPKCA was faster than PKCA by spending 2.64 s on average on the five datasets. While PKCA spent 3.18 s on average on the considered datasets.

7 Conclusion

A GA-based parallel K-Means data clustering algorithm using MapReduce programming model on Hadoop framework was proposed to aid document clustering process. The proposed algorithm is able to increase the efficacy of data clustering process of unsupervised learning by speeding up the process of cluster formation. The algorithm outperformed the classical parallel K-Means clustering algorithm on all the five datasets considered during the experiment and on average by 0.54 s.

The proposed GAPKCA was purposely developed to enhance the processes of identifying meaningful information the resides in big data environment. This algorithm uses GA with parallel K-Means clustering and apply MapReduce model on Hadoop framework to identify set of objects within a cluster. Results of the experiment indicated that our proposed algorithm was able to reduce the execution time during the process of unsupervised learning. In future we intend to extend the capability of the proposed algorithm to handle large volume of data and consider addition metrics such as precision and recall. The proposed algorithm can used in both industries and academia by data analysists and researchers dealing with smaller dimension of continuous numeric data. Such as marketing and customer segmentation, banking and proud detection, public transport analysis and telecommunication companies.

References

1. Garg N, Singla S, Jangra S (2016) Challenges and techniques for testing of big data. Procedia Comput Sci 85:940–948
2. Cuzzocrea A, Darmont J, Mahboubi H (2009) Fragmenting very large XML data warehouses via K-means clustering algorithm. Int J Bus Intell Data Min 4(3/4):301–328. Inderscience, Genèva
3. Zhao W, Ma H, He Q (2009) Parallel K-Means clustering based on MapReduce. In: CloudCom 2009: cloud computing. Springer, Heidelberg, pp 674–679
4. Cohen-Addad V, Kanade V, Mallmann-Trenn F, Mathieu C (2017) Hierarchical clustering: objective functions and algorithms, no. 1
5. Chatziafratis V, Niazadeh R, Charikar M (2018) Hierarchical clustering with structural constraints, pp 1–23
6. Singh K, Malik D, Sharma N (2011) Evolving limitations in K-means algorithm in data mining and their removal. Int J Comput Eng Manag 12:2230–7893
7. Bouhmala N, Viken A, Lønnum JB (2015) Enhanced genetic algorithm with K-means for the clustering problem. Int J Model Optim 5(2):150–154
8. Lu Y, Lu S, Fotouhi F, Deng S, Brown SJ (2004) FGKA: a fast genetic K-means clustering algorithm. In: Proceedings of the 2004 ACM symposium on applied computing, pp 622–623

9. Alswaitti M, Albughdadi M, Isa NAM (2018) Density-based particle swarm optimization algorithm for data clustering. Expert Syst Appl 91:170–186
10. Wu X, Zhu X, Wu G-Q, Ding W (2014) Semana 07-data mining with big data. Knowl Data Eng IEEE Trans 26(1):97–107
11. Goel L, Jain N, Srivastava S (2017) A novel pso based algorithm to find initial seeds for the k-means clustering algorithm. In: Communication and computing systems: proceedings of the international conference on communication and computing systems ICCCS 2016, pp 159–163, November
12. Younus ZS et al (2015) Content-based image retrieval using PSO and k-means clustering algorithm. Arab J Geosci 8(8):6211–6224
13. Bandyopadhyay S, Maulik U (2002) An evolutionary technique based on K-means algorithm for optimal clustering in RN. Inf Sci 146:221–327
14. Oussous A, Benjelloun FZ, Ait Lahcen A, Belfkih S (2018) Big data technologies: a survey. J King Saud Univ Comput Inf Sci 30(4):431–448
15. Shukri S, Faris H, Aljarah I, Mirjalili S, Abraham A (2018) Evolutionary static and dynamic clustering algorithms based on multi-verse optimizer. Eng Appl Artif Intell 72(April):54–66
16. Deepali AP, Varshney S (2016) Analysis of K-means and K-medoids algorithm for big data. Phys Procedia 78:507–512 (2016)
17. Ramírez-Gallego S, Fernández A, García S, Chen M, Herrera F (2018) Big data: tutorial and guidelines on information and process fusion for analytics algorithms with MapReduce. Inf Fusion 42:51–61
18. Hua Y, Jin Y, Hao K (2019) A clustering-based adaptive evolutionary algorithm for multiobjective optimization with irregular pareto fronts. IEEE Trans Cybern 49(7):2758–2770
19. Zhang X, Tian Y, Cheng R, Jin Y (2018) A decision variable clustering-based evolutionary algorithm for large-scale. IEEE Trans Evol Comput 22(1):1–17
20. Sinha A, Jana PK (2018) A hybrid MapReduce-based k-means clustering using genetic algorithm for distributed datasets. J Supercomput 74(4):1562–1579
21. Garza-Fabre M, Handl J, Knowles J (2018) An improved and more scalable evolutionary approach to multiobjective clustering. IEEE Trans Evol Comput 22(4):515–535
22. Tsai CW, Chang WY, Wang YC, Chen H (2019) A high-performance parallel coral reef optimization for data clustering. Soft Comput 2
23. Lv Z, Hu Y, Zhong H, Wu J, Li B, Zhao H (2010) Parallel K-means clustering of remote sensing images based on MapReduce. In: WISM 2010, vol 6318, pp 254–262
24. Krishna K, Murty NM (1999) Genetic K-means algorithm. IEEE Trans Syst Man Cybern B Cybern 29(3):433–439
25. Sardar TH, Ansari Z (2018) Partition based clustering of large datasets using MapReduce framework: an analysis of recent themes and directions. Futur Comput Inf J 3(2):247–261
26. Drechsler J (2011) Synthetic datasets for statistical disclosure control: theory and implementation, vol 201. Springer Science & Business Media
27. Banerjee S, Choudhary A, Pal S (2016) Empirical evaluation of K-means, bisecting K-means, fuzzy C-means and genetic K-means clustering algorithms. In: 2015 IEEE International WIE Conference on Electrical and Computer Engineering WIECON-ECE 2015, pp 168–172
28. Sathiyakumari K, Preamsudha V, Manimekalai G (2011) Unsupervised approach for document clustering using modified fuzzy C mean algorithm. Int J Comput Organ Trends 1 (3):10–14
29. Hotho A, Staab S, Stumme G (2003) Text clustering based on background knowledge. Inst Appl Inf 1–36

30. Nur'Aini K, Najahaty I, Hidayati L, Murfi H, Nurrohmah S (2016) Combination of singular value decomposition and K-means clustering methods for topic detection on Twitter. In: ICACSIS 2015, pp 123–128
31. Surendra H, Mohan H (2017) A review of synthetic data generation methods for privacy preserving data publishing. Int J Sci Technol Res 6(3):95–101
32. Cicoria S, Sherlock J, Muniswamaiah M, Clarke L (2014) Classification of titanic passenger data and chances of surviving the disaster data mining with weka and kaggle competition data. In: Proceedings of the student-faculty research day, CSIS, Pace University, pp 1–6

Android Botnet Detection by Classification Techniques

Athirah Zaffira Binti Majit[1], Palaniappan Shamala[1](✉), Cik Feresa Mohd Foozy[1], Chuah Chai Wen[1], and Muruga Chinniah[2]

[1] Information Security Interest Group (ISIG), Faculty of Computer Science and Information Technology, Universiti Tun Hussein Onn Malaysia (UTHM), Johor, Malaysia
shamala@uthm.edu.my

[2] Faculty of Business Management, Universiti Teknologi MARA (UiTM), Segamat, Johor, Malaysia

Abstract. Currently, android botnet attacks have shifted from computers to smartphones due to its functionality, ease to exploit, and based on financial intention. Mostly, the Android malware attack increased due to its popularity and high usage among end users. Android botnet is defined as a collection of compromised mobile smartphones and controlled by a botmaster through a command and control (C&C) channel to serve a malicious purpose. Current research are still lacking in terms of their low detection rate due to their selected features. This approach is implemented by extracting two different types of features permissions, software features as well as API calls. Thus, this paper proposes an approach that utilizes ensemble learning for Android malware detection. It combines advantages of static analysis with the efficiency and performance of ensemble machine learning to improve Android malware detection accuracy. For dataset was collected from UNB the Canadian Institute for Cybersecurity dataset and benign from google play. Canadian Institute for Cybersecurity is actually a lightweight method for detection of Android botnet that infers detection patterns automatically and enables identifying botnet directly on the smartphone. The machine learning algorithms used are random forest and naive bayes for classification however while random forest show more accuracy compared with another algorithm. The performance of various classifiers is evaluated by identifying the rate of False Positive and True Positive and accuracy. The results showed that Random Forest Algorithm achieved the highest accuracy rate of 97.1%. In future, more significant approach by using different feature selection such as intent, string and system component will be further explored for a better detection and accuracy rate.

Keywords: Malware · Android botnet · Smartphones · Machine learning · Static analysis

R. Ghazali et al. (Eds.): SCDM 2020, AISC 978, pp. 109–120, 2020.
https://doi.org/10.1007/978-3-030-36056-6_11

1 Introduction

Nowadays, smartphones and tablets are extremely popular not only for adults but also for children [1]. According to [2], because of the daily increase in the number of users and facilities, mobile device technologies have grown fast. According to the latest estimates, the number of smartphone users at the beginning of 2016 was 2 billion and is expected to increase to over 2.50 billion in 2018 [3].

Most of smartphone are created along with android application packages. Malware is used to damage Internet connected devices and gather sensitive information from individuals, or it uses spyware for accessing the most private information on the infected device. Installing and updating anti-virus software, updating latest security patches of mobile operating system and avoid downloading and installing mobile application from third-party application market are examples to detect and handle botnet attacks.

However, based on the statement given above, it contains some drawback from that method such as updating the software will consume a lot of memory spaces. This is because lot of smartphones nowadays having limited memory space to updating their software frequently. Another issue is if their mobile company giving a latest security patch. It can be seen at most of the Android-based platform based on the mobile model. Usually only the latest model that will get the privilege from their respectively company. So, we will be focusing on detection application rather than apps installation.

In recent years android botnet detection has been a major research topic but there is no awareness to user to take more responsibility of this issues. This lacking has motivated to do research on this android botnet malware. So that, user can be exposed more with this type of botnet malware.

Therefore, in this paper prominent permissions and Application Programming Interface (API) calls for mobile malware identification using machine learning techniques random forest and Naive Bayes is extracted. Here, for android malware detection, static analysis will apply to approach in this paper with the aim to examine if ensemble features with respect to individual features minimizes misclassification, thereby improving accuracy of rate.

The rest of the paper is organized as follows: Sect. 2 describes the related work on android malware detection techniques and machine learning algorithms. Section 3 presents the proposed framework research methodology used in completing this research. Last Sect. 4 shows the evaluation methodologies and experimental results based on performance metric.

2 Related Works

This section describes the related works about Android malware detection approaches.

2.1 Machine Learning Algorithm

Based on article by [4], machine learning algorithms are used to develop a classifier to detect patterns. The general method of using the data mining technique to detect malware is to start by generating the feature sets. The feature sets include sequence of instructions, API / system call sequence, hexadecimal byte code sequence [4].

Machine learning algorithms are often categorized as being supervised or unsupervised. There are consist many of algorithms can be applied for any data problem such as Linear Regression, Logistic Regression, Decision Tree, K-Means, Random Forest etc. In this research two algorithms are proposed: Random Forest and Naives Bayes.

2.1.1 Random Forest

Random Forest uses Bagging to create a diverse group of classifiers by introducing randomness into the introduction of learning algorithms. According to [5], random forests are comparable to AdaBoost, but more robust to errors and outliers. The generalization error for a forest converges as long as the number of trees in the forest is large. Thus, overfitting is not a problem. The accuracy of a forest at random depends on the strength of the individual classifiers and the measurement of their dependence. The ideal is to maintain the strength of individual classifiers without their correlation being increased [6].

Random Forest has several advantages that can be leveraged for improved machine learning based detection: no special preprocessing of input is required; can deal with large numbers of training instances, missing values, irrelevant features [7]. Based on researcher [8] reported that their achieved an accuracy of 86.41% and 0.92 of Area Under Curve (AUC) using Random Forest classifier.

2.1.2 Naives Bayes

Naives Bayes is the probabilistic classifiers based on the Bayes's theorem decision rule with strong naives assumption between the features. The features of the subject to be tested can be learnt by machine learning using this algorithm because naives bayes classification are highly scalable.

According [9], in this algorithm we assume that all the features are independent of each other. The classification is based on the calculating the maximum probability of the attributes which belong to a particular class.

2.1.3 K- Nearest Neighbor (KNN)

Based on [10], The K-Nearest Neighbor Algorithm is the simplest of all machine learning algorithms. It is based on the principle that the samples that are similar, generally lies in close vicinity. Based on [11] explains k-Nearest Neighbours algorithm is known as lazy-learning algorithm as it takes less time for training. Its computations are eager-based learning algorithms [11].

Based on the Table 1 below show the commonly algorithms machine learning used of a previously researchers using in their thesis to make more efficient with classification algorithms. According to the previous researcher, it can conclude that in this proposed research would like to apply the classification algorithm: Random Forest and Naives Bayes. It is because these two algorithms commonly and preferably used in other researcher in their thesis.

In this study, two types of classification algorithms which are Random Forest (RF) and Naïve Bayes (NB), are adopted. Random forest is a logic-based algorithm that

Table 1. Machine learning algorithms

Authors	Algorithms			
	Random forest	Decision tree	Naives Bayes	K-NN
[9]	√		√	
[10]				√
[7]	√	√	√	

is proven to produce high-accuracy results for malware detection. Naïve Bayes is widely used in text classification. The features of subject to be tested can be learnt by machine learning using this algorithm because Naïve Bayes classifier are highly scalable. These two are chosen for this research because both are the most popular machine learning algorithms and generally, are able to give results with good accuracy [12].

2.2 Study of Android Malware Detection Techniques

There are mainly two approaches to analyze the Android Malware: Static and Dynamic Analysis approach.

2.2.1 Static Analysis

Based on [13], static analysis is analysis of the applications is done, and the features are extracted without executing the application on an emulator or device. In comparison to other analysis techniques for android malware detection, static analysis consumes fewer resources and time as it does not involve execution of the application. This analysis can detect runtime errors, logical inconsistencies, and possible security violations. The most commonly used static features are the Permission and API calls [13].

Static analysis is usually more efficient, since no code execution is required in providing fast detection and classification for known mobile malware. Static analysis of Android malware can rely on Java bytecode extracted by disassembling an application. The manifest file is also a source of information for static analysis.

2.2.2 Dynamic Analysis

Dynamic analysis is the testing and evaluation of a program by executing data in real-time. The objective is to find errors in a program while it is running, rather than by repeatedly examining the code offline. It is a detection technique which aims at evaluating malware by executing the application and the main advantage of this technique is that determines the application behavior during runtime and loads target data. The resource consumption in this analysis technique is more as compared to static analysis. Dynamic behavioral detection method constructs operation environment by using a sandbox, virtual machine, and other forms, and simulates the execution of the application to acquire the application's behavior model [13].

2.2.3 Hybrid Analysis

According to [13], Hybrid Analysis is a combination of static and dynamic analysis. It is a technology or method that can integrate run-time data extracted from dynamic analysis into a static analysis algorithm to detect behavior or malicious functionality in the applications. The hybrid analysis method involves combining static features obtained while analyzing the application and dynamic features and information extracted while the application is executed. Though it could increase the accuracy of the detection rate, it makes the system cumbersome and the analysis process is time consuming [13].

Table 2. Malware analysis techniques

Authors	Techniques		
	Static analysis	Dynamic analysis	Hybrid analysis
[9]	√	√	
[13]	√	√	√
[7]	√		
[4]	√	√	√

As conclusion, based on the Table 2 above show the android malware analysis techniques approach by previously researchers in their thesis. According to the previous researcher, it can conclude that in this proposed research will apply the malware analysis by using static analysis based on behavior based techniques because it can detect known as well as unknown malwares also with the efficiency and performance of ensemble machine learning to improve Android malware detection accuracy.

2.3 Comparison Table

Based on Table 3 below, it can conclude that the proposed research will apply the following features such as permission and API call/System call where most other researchers have used this of features on their thesis to solve the problem occur with android malware.

As conclusion, this research is proposed to conduct on android botnet malware because mostly researches have been conducted on android malware by researchers. The techniques have been proposed for this research is behavior-based detection techniques. The analysis type is static analysis. Machine learning used: Random Forest and Naives Bayes for the classification, to detect whether the malicious is benign or malware. The dataset was collected from UNB the Canadian Institute for Cybersecurity.

Table 3. Comparison by related work

Related Works	Research Area	Analysis	Features	Machine Learning Algorithm	Dataset	Evaluation Features
[9]	Trojan & Spyware	Static & Dynamic analysis	Static: Permission Dynamic: API call	Random forest Naïve Bayes Simple logistic Sequential minimal optimization J48	Drebin: malware dataset: Google play store: Benign	ROC TPR/FPR AUC ROC Accuracy
[8]	Android Malware	Static analysis	API Call Permission Software/ hardware features	Naives Bayes Random Forest J48	Contagion: Malware dataset Google Play Store: Benign	Accuracy TPR/ FPR
[13]	Android Malware	Static, Dynamic & Hybrid analysis	-	-	-	Accuracy
[18]	Android Malware	Dynamic analysis	Opcodes features	Random forest Naïve Bayes J48	Google play Store: Benign Malacia-project: malware dataset	TPR/ FPR TNR /FNR
[19]	Android Ransom ware	-	Lock detected Encryption detected	Random Forest Clustering: K-Means	VirusTotal & Malgnome	Accuracy ROC TPR / FPR
[20]	Android Malware	Static & Dynamic	Network Traffic	Decision Tress	Drebin & ContagionDump Set: Malware Dataset	Accuracy ROC TPR / FPR
Proposed Work	Android Botnet	Static Analysis	API Call Permission	Random Forest Naïve Bayes	Canadian Institute for Cybersecurity: Botnet malware Google Play Store: Benign	TPR / FPR Accuracy

3 Methodology

3.1 Research Framework

This research process usually begins with identifying a problem, collecting data and analyzing data to achieve the objectives of the research. The feature of static analysis that is used in this case study is requested permission call and responsible API call (Fig. 1).

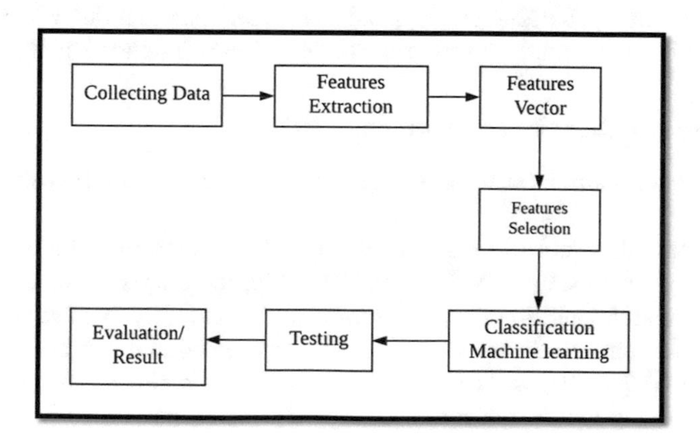

Fig. 1. Research framework

3.2 Dataset

In this phase, this research used two types of dataset which one for malware samples and for benign samples. Malware samples downloaded from UNB Canadian Institute for Cybersecurity and benign samples downloaded from google play. Malware dataset and Benign dataset was used to find the features of android botnet.

Canadian Institute for Cybersecurity dataset contains large collection of Android botnet samples representing 14 botnet families dataset includes 1929 samples dataset that combines some botnet samples from the Android Genome Malware project, VirusTotal and samples provided by well-known anti-malware vendor [21]. Majority of existing android malware samples which appear from 2010 to 2014. In this experiment used for 500 samples for malware dataset. Another 500 samples of Android benign from Google Play store. The whole dataset is available in the android application package (APK) format.

3.3 Features Extraction and Feature Vector

Features extraction is a process of extracting a certain value according to the features lists from processed data. There are many ways to perform this process. Static analysis can be done to determine the specific value of the features. Then, these values will be converted to be used in machine learning. The objective of Feature Extraction is to find that transformation, usually linear, which helps not only to eliminate irrelevant information and redundancy but also provide better separation of classes.

In this phase, result from feature extraction phase is transform into vector in comma separated value (CSV) file for each of 1000 Android applications. The vector is recorded either with value 1 or 0 which is based on the feature that had been extracted from Android permission and ends with the class either botnet or benign.

3.4 Classification Algorithm

In this research, Random Forest and Naive Bayes had been used for the classification algorithms by features selection of permission and API calls is acquired. Both feature selection and classification phases in data analysis module are run using WEKA (Waikato Environment for Knowledge Analysis), a software tool that widely employed for implementing the feature selection methods and the classification algorithms.

3.5 Feature Selection: Information Gain (IG)

Features selection is a process where an appropriate feature is chosen to ensure the best results.

Information gain is aimed to remove redundant or irrelevant features from a given feature vector in previous phase. Thus, in this study, the process of features se-lection will use Information Gain (IG) to get a top-ranked both feature which are top 20 and top 40. The final process in this phase, the features selection using In-formation Gain (top 20 or top 40 features) was tested again and obtained a new accuracy result, which will be explained in the next chapter.

3.6 Performance Matric: Accuracy, True Positive (TP) and False Positive (FP)

Performance Metric is a measurement to evaluate the performance of the model. In this research, the performance metric is used to measure the performance of the detection model using machine learning: Random Forest and Naive Bayes. Will considered on this research two performance metric that are Accuracy and True Positive (TP) and False Positive (FP) rate.

Accuracy of detection model can represent the trueness with systematic error and precision with random errors. Accuracy also can be defined as the combination of trueness and precision. Accuracy equation also shown in equation below can be defined as the probability of correct number prediction divided to the total number of instances.

A True Positive (TP) rate is an outcome that shows a given condition has been satisfied. TP error also can be defined as a single condition is checked and the result is in positive decision and usually designated as true or false. A true positive rate is better when the value is closer to 1.

A False positive rate is an outcome where it shows a given condition has been satisfied but actually has not been satisfied. It checked a single condition and given a negative decision usually designated as true or false. For False positive rate, the best value is when it closer to 0.

4 Experimental Result

4.1 All Experimental Results

In this section, the results obtained from the analysis of Android botnet detection using Random Forest (RF) and Naive Bayes (NB) are presented. This analysis contains 158 features for permission and 132 features for API Calls.

(a) Random Forest (RF)

Table 4 shows the results of Android botnet detection based on static analysis using Random Forest as a Machine Learning classifier.

Table 4. Android botnet detection results using RF

Features	TP rate	FP rate	Accuracy %
Permission	0.947	0.053	94.7%
API calls	0.970	0.030	97.0%
Combine (Permission + API Calls)	0.975	0.025	97.5%

Based on the results in Table 4, True Positive (TP) rate for permission features using Random Forest is 0.947 which means that the system correctly classified 464 botnet as botnet and the False Positive (FP) rate for permission is 0.053, which means that there are 36 benign applications classified as a botnet. The accuracy of botnet detection for permission is 94.7%.

Based on the results in Table 4, True Positive (TP) rate for API features using Random Forest is 0.970 which means that the system correctly classified 479 botnet as botnet, whereas the False Positive (FP) rate for API calls is 0.030 which means that there are 21 benign applications classified as a botnet. The accuracy of botnet detection for API calls is 97.0%.

Next, the results of combined features, where permission features are combined with API calls features. Table 4 shows that the True Positive (TP) rate is 0.975 which means that the system correctly classified 478 botnet as botnet, whereas the False Positive (FP) rate is 0.025 which means that there are 22 benign applications classified as a botnet. The accuracy of botnet detection for combined features is 97.5%.

The results of Android botnet detection using Random Forest indicates that permission gets a low accuracy, which is 94.7% compared to API calls features and combined features which obtained a quite similar accuracy of 97.5%.

(b) Naive Bayes (NB)

Table 5 shows the results of Android botnet detection based on static analysis using Naïve Bayes as a second Machine Learning classifier used.

Table 5. Android botnet detection results using NB

Features	TP rate	FP rate	Accuracy %
Permission	0.909	0.091	90.9%
API calls	0.914	0.086	91.4%
Combine (Permission + API Calls)	0.947	0.053	94.7%

Based on the results in Table 5, the True Positive (TP) rate for permission features is 0.909 which means that the system correctly classified 432 botnet as bot-net and the False Positive (FP) rate for permission is 0.091 which means that there is 68 benign application classified as botnet. The accuracy of botnet detection for permission is 90.9%.

Based on the results in Table 5, the True Positive (TP) rate for API Calls features is 0.914 which means that the system correctly classified 472 botnet as bot-net and the False Positive (FP) rate for API calls is 0.086 which means that there are 28 of benign applications classified as botnet. The accuracy of botnet detection for intent is 91.4%.

Next, the results for combined features, where permission features combined with API calls features in Table 5 show that the True Positive (TP) rate is 0.947 which means that the system correctly classified 468 botnet as botnet and the False Positive (FP) rate is 0.053 which means that there is 32 benign application classified as botnet. The accuracy of botnet detection for combined features is 94.7%.

The results of Android botnet detection using Naïve Bayes show that permission gets a low accuracy, which is 90.9% compared to API calls features and combined features, which get the accuracy of 91.4% and 94.7%.

4.2 Experimental by Ranked Result

Based on NB (40 features), permission gets a lower accuracy which is 90.70% compared to API calls and combination of features that get the accuracy which is 91.02% and 93.30%. Meanwhile, based on RF (40 features), the permission features also get a lower accuracy which is 94.30% compared to API calls and combination of features that get the accuracy which is 96.71% and 97.1%.

Meanwhile, based on NB (20 features), permission gets a lower accuracy which is 88.7% compared to API calls accuracy is 91.22% and the combination of features accuracy is 94.0%. Meanwhile, based on RF (20 features), the permission features also get a lower accuracy which is 92.10% compared to API calls and combination of features that get the accuracy which is 96.71% and 95.80% (Table 6).

Table 6. Results for accuracy of Android ransomware detection using NB & RF based on top 40 and 20 features

Random forest (RF)		
Feature name	Total features use	Accuracy %
Permission	40	94.30
	20	92.1
API Calls	40	96.71
	20	96.71
Combination Features	40	97.1
(Permission + API)	20	95.8
Naïve Bayes (NB)		
Feature Name	Total features use	Accuracy %
Permission	40	90.70
	20	88.7
API Calls	40	91.02
	20	91.22
CombinationFeatures	40	93.3
(Permission + API)	20	94.0

As a conclusion, the technique in this study for static analysis using permission features and API calls features gives the highest accuracy of Android botnet detection by using only top 40 features of combination (permission + api) and applying Random Forest as a machine learning classifier with accuracy 97.1%.

5 Conclusion

Android malware can be detect by various techniques. In this research we used behavior based detection techniques and static analysis to detect the android malware. For android malware classification we used machine learning which include Naïve Bayes and Random Forest algorithm. The process of executing of experiments including data, features extraction, behavior based detection, algorithm classification, model prediction and result. To run the experiment the WEKA tools is using for data classifications.

The study performed in this research was a proof-of-concept. Therefore, several future improvements related to this study are identified:

- Use more features extraction
- Comparative study based on accuracy and time taken for the An-droid Botnet detection
- Comparative study of other techniques for Android Botnet
- Use more Machine Learning classifiers

The other researcher also can use other types of data set to detect the malware.

Acknowledgments. The authors would like to acknowledge the financial support received from the Research Management Centre (RMC) Universiti Tun Hussein Onn Malaysia (UTHM) under the Tier 1 Research Grant Scheme (Code H101).

References

1. Saracino A, Sgandurra D, Dini G, Martinelli F (2018) MADAM: effective and efficient behavior-based android malware detection and prevention 15(1):83–97
2. Simon JP (2016) How Europe missed the mobile wave. Info 18(4):12–32
3. Martín I, Hernández JA, Muñoz A, Guzmán A (2018) Android malware characterization using metadata and machine learning techniques. Secur Commun Netw (2018)
4. Srivastava P, Raj M (2018) Feature Extraction for enhanced malware detection using genetic algorithm 7(4):44–49
5. Belgiu M, Drăguţ L (2016) Random forest in remote sensing: a review of applications and future directions. ISPRS J Photogramm Remote Sens 114:24–31
6. Breiman L (2001) Random forests. Mach Learn 45(1):5–32
7. Yerima SY, Sezer S, Muttik I (2015) High accuracy android malware detection using ensemble learning. IET Inf Secur 9(6):313–320
8. Aswini AM, Vinod P (2015) Towards the detection of android malware using ensemble features. J Inf Assur Secur 10(1)
9. Kapratwar A (2016) Static and dynamic analysis for android malware detection
10. Jadhav SD, Channe HP (2016) Comparative study of K-NN, naive Bayes and decision tree classification techniques. Int J Sci Res (IJSR) 5(1):1842–1845
11. Singh A, Lakshmiganthan R (2017) Impact of different data types on classifier performance of random forest, naive bayes, and k-nearest neighbors algorithms
12. Dietterich TG (2000) Ensemble methods in machine learning. In International workshop on multiple classifier systems. Springer, Berlin, Heidelberg, pp 1–15
13. Rao V, Hande K (2017) A comparative study of static, dynamic and hybrid analysis techniques for android malware detection 5(2):1433–36
14. Sgandurra D, Muñoz-González L, Mohsen R, Lupu EC (2016) Automated dynamic analysis of ransomware: benefits, limitations and use for detection. arXiv preprint arXiv:1609.03020
15. Aswini AM, Vinod P (2014) October. android malware analysis using ensemble features. In: International conference on security, privacy, and applied cryptography engineering. Springer, Cham, pp 303–318

16. Fereidooni H, Conti M, Yao D, Sperduti A (2016) Anastasia: android malware detection using static analysis of applications. In: 2016 8th IFIP international conference on new technologies, mobility and security (NTMS) (pp 1–5). IEEE
17. Yusof M, Saudi MM, Ridzuan FHM (2018) A systematic review analysis for mobile Botnet detection using GPS exploitation. Malays J Sci Health Technol 2(Special)
18. Sahay SK, Sharma A (2016) Grouping the executables to detect malwares with high accuracy. Procedia Comput Sci 78:667–674
19. Khalid NS (2016) Android malware classification using K-means clustering algorithm (Doctoral dissertation, Universiti Tun Hussein Onn Malaysia)
20. Zulkifli A, Hamid IRA, Shah WM, Abdullah Z (2018) Android malware detection based on network traffic using decision tree algorithm. In: International conference on soft computing and data mining (pp 485–494). Springer, Cham
21. Kadir AFA, Stakhanova N, Ghorbani AA (2015) Android botnets: what urls are telling us. In: International conference on network and system security. Springer, Cham, pp 78–91

Android Ransomware Detection Based on Dynamic Obtained Features

Zubaile Abdullah[1(✉)], Farah Waheeda Muhadi[1], Madihah Mohd Saudi[2], Isredza Rahmi A. Hamid[1], and Cik Feresa Mohd Foozy[1]

[1] Information Security Group, Faculty of Computer Science & Information Technology, Universiti Tun Hussein Onn Malaysia (UTHM), Batu Pahat, Johor, Malaysia
zubaile@uthm.edu.my

[2] Information Security Assurance Group, Faculty of Science & Technology, Universiti Sains Islam Malaysia (USIM), Nilai, Negeri Sembilan, Malaysia

Abstract. Along with the rapid development of new science and technology, smartphone functionality has become more attractive. Smartphones not only bring convenience to the public but also the security risks at the same time through the installation of malicious applications. Among these, Android ransomware is gaining momentum and there is a need for effective defense as it is very important to ensure the security of smartphone user. There are various analysis techniques used to detect instances of Android ransomware. In this paper, we proposed the Android ransomware detection using dynamic analysis technique. Two dataset were used which is ransomware and benign dataset. The proposed approach used the system calls as features which obtained from dynamic analysis. The classification algorithms Random Forest, J48, and Naïve Bayes were used to classify the instances based on the proposed features. The experimental results showed that the Random Forest Algorithm achieved the highest detection accuracy of 98.31% with lowest false positive rate of 0.016.

Keywords: Android · Ransomware · Machine learning · Classification · Dynamic analysis

1 Introduction

In the past few years, the popularity of smartphone have risen significantly [1]. Nowadays, smartphone are no longer limited for phone calling or sending messages but also being used for web browsing, social networking, applications downloading and installing and online banking transaction. Smartphone user kept confidential information such as contacts, bank account number, username and password for online banking, credit card number, memorable and private pictures in these devices [2]. Report by McAfee [3], indicated that the hackers are increasingly targeting Android smartphone users through malicious application installation (malware). The number of Android smartphone users affected by malware are increasing during January–March 2016 especially ransomware as illustrated in Fig. 1.

R. Ghazali et al. (Eds.): SCDM 2020, AISC 978, pp. 121–129, 2020.
https://doi.org/10.1007/978-3-030-36056-6_12

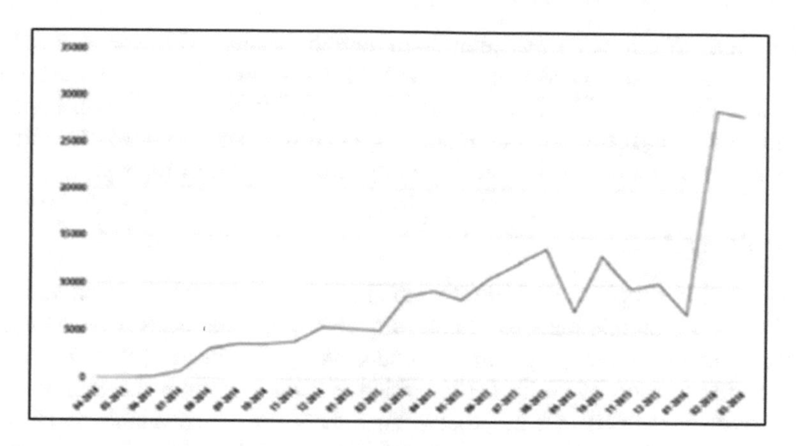

Fig. 1. The number of users encountering mobile ransomware at least once in the period April 2014 to March 2016 [4].

Ransomware is a type of malware that infects the personal files of the user and does not permit to access until ransom is paid. Ransomware operates in two different ways, from simply locking the devices of the infected devices to encrypting all of its files. This is a dangerous because ransomware is designed to extort money, damage personal data, and block infected devices. Once the device is infected, the attacker asks the victim to pay a ransomware order to obtain the key for decrypting the files or restoring the control of the Android smartphone. Recent example of ransomware targeting mobile devices was "Charger", found in January 2017 and bundled with Energy Rescue application. This malicious application was snooped on Google Play Store and targeted Android smartphone before being pulled from the market. "Charger" demanded 0.2 Bitcoins and threatened to sell the victim's personal information on the black market if the ransom is not paid [5]. Kaspersky Lab, on the other hand, had reported that from April 2014 to March 2015, they had protected 35,413 users from mobile ransomware. A year later the number had increased almost 4% to 136, 532 users [4]. Another researcher, Kim et al., also reported that the users attacked with ransomware also increased: from 2.04% in 2014–2015 to 4.63% in 2015–2016 [6]. The growth curve may be less compared with computer ransomware, however it is still a worrying trend.

2 Literature Review

This section contains the related literature, which includes an overview of ransomware and technique used in detection analysis.

2.1 Ransomware

Ransomware intrudes into users' devices after pretending as a normal (benign) application thus smartphone user will not aware of it presence. Ransomware restricts the use of the system in various ways after intruding the system [7]. For example, *DoubleLocker* ransomware encrypt files in Android smartphone thus, user cannot access important file when needed. *Reveton* ransomware in the other hand, will impersonate law enforcement agencies such as FBI and locked user smartphone screen until user contacted a given name and instructed to pay some amount of money.

2.2 Static Analysis

Static analysis is simple and efficient method in providing fast detection and classification for known mobile malware. In static analysis, application is reverse-engineered in order to examine the programming code, strings or method that is suspicious. However, this method has a limitation which it cannot monitor the malicious application behaviour on the run time process since there are no executable process occurs [6]. As a result, there some malicious application is downloaded during run time process. To overcome this limitation, dynamic analysis is used to detect the ransomware.

2.3 Dynamic Analysis

Dynamic analysis on the other hand, examines the application during execution. Some of the code sections that may miss by static analysis can be detected in dynamic analysis. Dynamic analysis relies on runtime observation of memory usage, CPU usage, network behaviour, as well as examines the system calls of application. The main advantage of this technique is it detects dynamic code loading and records the application behaviour during runtime, which is hindered in static analysis.

2.4 Related Work

In this section, some of the techniques used by prior researchers to analyses Android ransomware based on static, dynamic and hybrid analyses are reviewed. Currently, little related work specifically targets ransomware [8]. Hence, related works on ransomware and more general malware-oriented papers are reviewed.

Sgandurra et al. [8] present a machine learning for dynamically analyzing and classifying ransomware called EldeRan. This proposed method monitors a set of actions performed by applications in their first phases of installation checking for characteristic signs of ransomware. In addition, EldeRan is a framework to identify the most significant ransomware dynamic features and uses them to detect ransomware. The machine learning component of EldeRan consists of two phases, which are features selection and classification. EldeRan also has the ability to detect new families of ransomware family that is not included in the features selection and training process of the classifier. The proposed method were able to detect more than 96% of ransomware.

Fereidooni et al. [9] proposed ANASTASIA, a system to detect Android malware using static analysis. This approach analyses large-scale malware datasets which are 18677 malware datasets and 11187 benign datasets. The research extracts many features such as intents, permission, Application Programming Interface (API) calls, system commands and malicious activities. The authors used several machine learning for the classification such as XGboost, Random Forest, Logistic Regression, Support Vector Machine, Naïve Bayes, Deep Learning and Decision Tree Classifier. The authors also designed a tool to extract features from Android application known as uniPDroid. They obtained an accuracy of 97% for the detection of Android malware.

Martinelli et al. [10] proposed a framework called BRIDEMAID (Behavior-based Rapid Identifier Detector and Eliminator of Malware for AndroID) which exploits static and dynamic for accurate detection of Android malware. In their paper, the authors used n-grams matching for static analysis and multi-level monitoring of device, applications and user behaviour for dynamic analysis. The framework is a combination between static and dynamic, known as hybrid technique. BRIDEMAID had been tested against 2794 malicious applications and achieved a detection accuracy of 99.7%.

The summary of the comparison consists of features selection and technique used by prior researchers for Android malware detection are presented in Table 1.

Related works of Android Ransomware Detection.

Table 1. Summary of related work.

Researcher (s)	Type of analysis	Classification algorithm	Features	Accuracy of detection
[8]	Dynamic analysis	Naïve Bayes, Support Vector Machine (SVM), Logistic Regression	Application Programming Interface (API)	96.3%
[9]	Static analysis	XGboost, Random Forest, Logistic Regression, Support Vector Machine, Naïve Bayes, Deep Learning and Decision Tree Classifier	Intent, Permission, API calls, System commands, Malicious activities	97%
[10]	Hybrid analysis	Support Vector machine (SVM)	Permission, API calls, system calls	99.7%

3 Methodology

This section explains on how this research is conducted including methodology used in data collection and analysis, explanation of research tools and environment and overview of proposed research.

3.1 Dataset

There are two datasets used for this research; training dataset and testing dataset. Training dataset is use to build up a detection model, while a testing dataset is to validate the model. 400 training dataset is retrieved from VirusTotal[1] which is among

[1] www.virustotal.com.

the largest malware repository available freely on the Internet. The testing dataset is a benign application downloaded from Google Play Store, an official market for Android. A total of 400 benign apps were downloaded and used for this research.

3.2 Dynamic Analysis

In this process, all the applications from datasets is being installed to obtain features which is system calls. In order to achieve this, an Android emulator which is Genymotion was used. In this process each Android ransomware and benign application are executed in separate emulator and the system calls were recorded as soon as the application is running in the emulator as shown in Fig. 2.

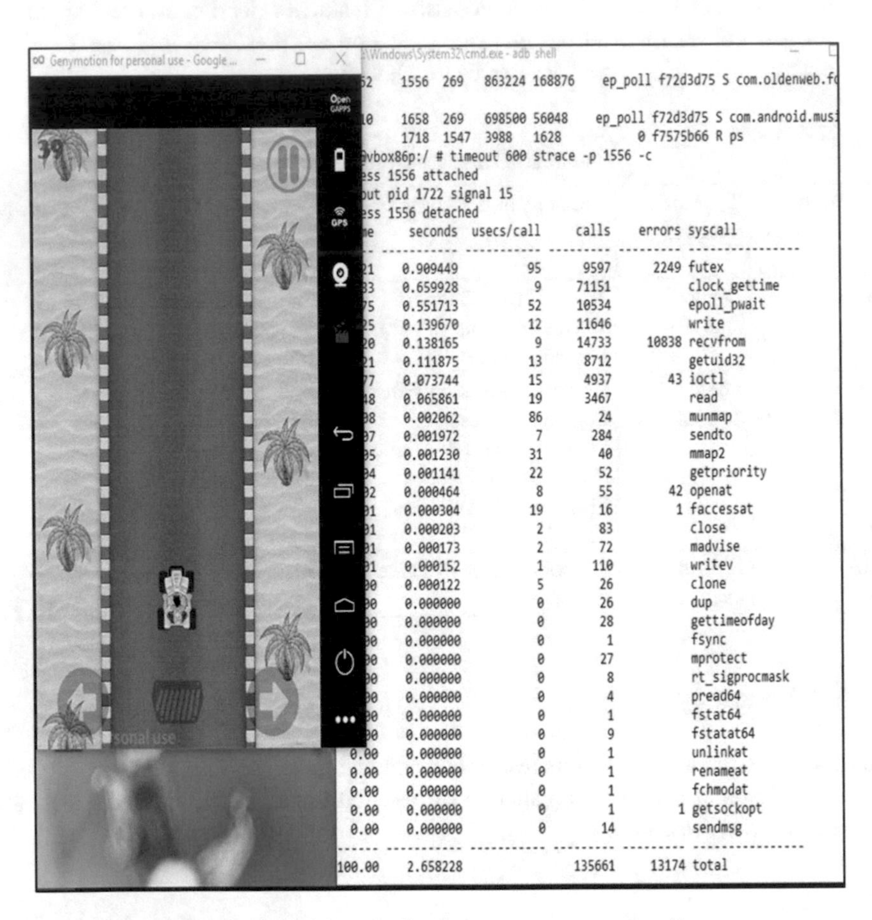

Fig. 2. Executing application to extract system calls.

3.3 Feature Vector

In this phase, extracted and recorded features from previous process are transformed into nominal representation to form features vector of system calls for each Android applications. Feature in application that matched with the feature list from feature selection process, is assigned as 1 indicated it presence in the application while 0 indicated the absence.

Let V be an application vector containing set of features in the application that matched with selected features list. For every $_i$th application in the dataset (ransomware and benign), Vi = {f1, f2, f3,...fn} and f is determined in formula as shown in Eq. 1 below:

$$fn = \begin{cases} 1, & \textit{if feature } n_{th} \textit{ exist and match} \\ 0, & \textit{otherwise} \end{cases} \quad (1)$$

3.4 Feature Selection

There are more than 250 system calls in an Android operating system. Applications use system calls to perform specific tasks such as read, write and open since they cannot directly interact with the Android operating system. Upon issuing a system call in user mode, Android operating system switches to kernel mode to perform the required task. However, in this research, it found that not all of the system calls are requested by the application in dataset, thus irrelevant system calls are not selected. In this features selection phase, 52 systems call are used for the classification as presented in Table 2.

Table 2. List of selected system calls for classification.

Selected system calls		
futex	epoll_ctl	madvise
epoll_wait	sendto	fcntl
getpid	recvfrom	gettid
chmod	read	getsockopt
mprotect	write	bind
ioctl	open	getpriority
munmap	close	rt_sigreturn
clock_gettime	brk	fdatasync
access	prctl	mkdir
dup	rt_sigpromask	fork
umask	epoll_pwait	getsockname
gettimeofday	unlink	stat
clone	lseek	lstat
sigprocmask	rename	fstat
_llseek	fsync	getuid
pread	sched_yield	fchown
writev	nanosleep	pwrite

3.5 Classification

In this research, Random Forest, J48 and Naïve Bayes were used as classification algorithms. These three classifiers were chosen due their usage in others similar researches. Feature selection and classification phases in data analysis are run using WEKA (Waikato Environment for Knowledge Analysis).

4 Result

In order to evaluate the performance of different classification algorithms, the following metrics are used, True Positive Rate (TPR), False Positive Rate (FPR) and Accuracy.

TPR is the number of Android ransomware classified as ransomware. FPR is the number of benign applications classified as ransomware and Accuracy is total percentage of classifier correctly classified the instances. Table 3 shows the comparison between the three classification algorithms.

Table 3. Classification result.

Algorithm	TPR	FPR	Accuracy
Naïve Bayes	0.93	0.19	86.75%
J48	0.95	0.045	95.32%
Random Forest	0.98	0.016	98.31%

From Table 2, it shows that Random Forest algorithm achieved the best performance in term of TPR, FPR and Accuracy as presented in Fig. 3 while Fig. 4 illustrated Random Forest result in WEKA screen shot.

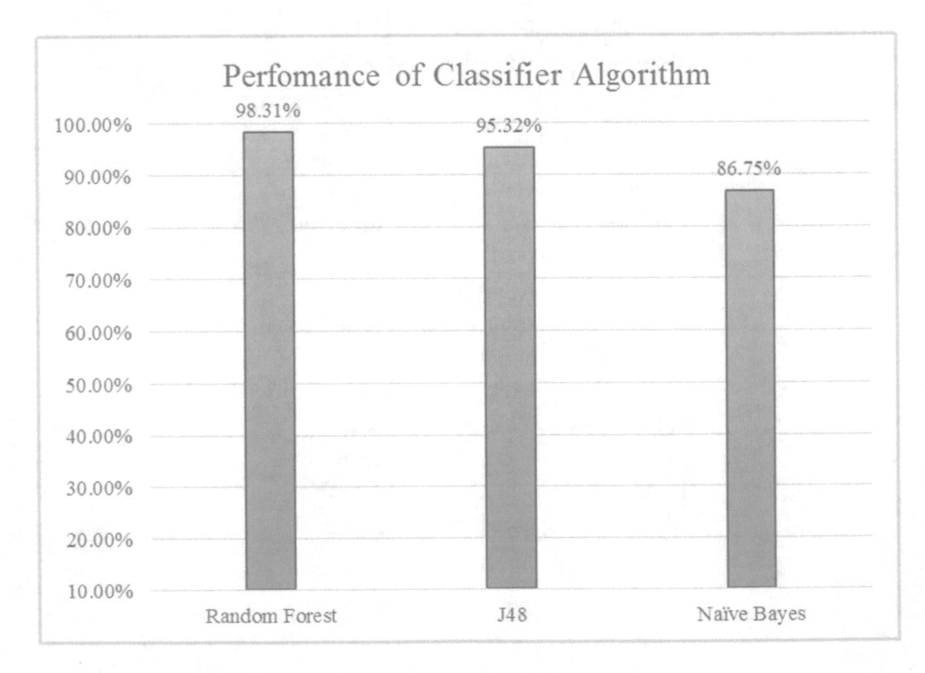

Fig. 3. Performance of classifier algorithm.

```
=== Summary ===

Correctly Classified Instances         757               98.3117 %
Incorrectly Classified Instances        13                1.6883 %
Kappa statistic                          0.9662
Mean absolute error                      0.0602
Root mean squared error                  0.1348
Relative absolute error                 12.0459 %
Root relative squared error             26.959  %
Total Number of Instances              770

=== Detailed Accuracy By Class ===

                 TP Rate  FP Rate  Precision  Recall   F-Measure  MCC      ROC Area  PRC Area  Class
                 0.982    0.016    0.985      0.982    0.983      0.966    0.998     0.999     ransomware
                 0.984    0.018    0.982      0.984    0.983      0.966    0.998     0.998     benign
Weighted Avg.    0.983    0.017    0.983      0.983    0.983      0.966    0.998     0.998

=== Confusion Matrix ===

   a   b   <-- classified as
 383   7 |   a = ransomware
   6 374 |   b = benign
```

Fig. 4. Random forest classifier result.

5 Conclusion and Future Work

Android ransomware are the new threats for smartphones as they possess serious threat. This research uses the system calls in the Android applications as features via dynamic analysis approach, to differentiate ransomware and benign applications. Information Gain is used to select the most significant system calls, then the classification algorithms, Random Forest, J48 and Naïve Bayes are used to classify applications whether the applications is ransomware or benign. The result shows that Random Forest algorithm achieved the highest detection accuracy of 98.31% with the lowest false positive rate of 0.016. For future research, more dataset of Android ransomware and benign applications will be used to achieve better detection accuracy.

Acknowledgments. This research is sponsored by Universiti Tun Hussein Onn Malaysia (UTHM) via UTHM Registrar Office and Tier 1 Research Grant H237. The authors would like to thank Universiti Tun Hussein Onn Malaysia and Ministry of Higher Education Malaysia for the facilities and financially supporting this research.

References

1. Zulkifli A, Hamid IRA, Shah WM, Abdullah Z (2018) Android malware detection based on network traffic using decision tree algorithm. Adv Intell Syst Comput 700:485–494
2. Abdullah Z, Saudi MM, Anuar NB (2017) ABC: android botnet classification using feature selection and classification algorithms. Adv Sci Lett 23(5):4717–4720
3. McAfee, Mobile Threat Report Criminal Quest for Money Could Make 2018 the Year of Mobile Malware

4. Cluley G (2018) The android Ransomware threat has quadrupled in just one year. Tripwire, Inc. Available: https://www.tripwire.com/state-of-security/featured/the-android-ransomware-threat-has-quadrupled-in-just-one-year/. Accessed 10 Aug 2018
5. Yang T, Yang Y, Qian K, Lo DC-T, Qian Y, Tao L (2015) Automated detection and analysis for android ransomware. In: 2015 IEEE 17th International conference on high performance computing 2015 IEEE 7th international symposium on cyberspace safety and security 2015 IEEE 12th international conference on embedded software, no. 1, pp 1338–1343
6. Arshad S, Ali M, Khan A, Ahmed M (2016) Android Malware detection & protection: a survey. Int J Adv Comput. Sci Appl 7(2)
7. Naway A, LI Y (2018) A review on the use of deep learning in android malware detection. Int J Comput Sci Mob Comput 7(12):42–58
8. Sgandurra D, Muñoz-González L, Mohsen R, Lupu EC (2016) Automated dynamic analysis of ransomware: benefits, limitations and use for detection
9. Fereidooni H, Conti M, Yao D, Sperduti A (2016) ANASTASIA: android malware detection using static analysis of applications. In: 2016 8th IFIP international conference on new technologies, mobility and security (NTMS), pp 1–5
10. Ferrante A, Malek M, Martinelli F, Mercaldo F, Milosevic J (2018) Extinguishing ransomware - a hybrid approach to android ransomware detection, 242–258

Training Functional Link Neural Network with Ant Lion Optimizer

Yana Mazwin Mohmad Hassim(✉) and Rozaida Ghazali

Faculty of Computer Science and Information Technology, Universiti Tun Hussein Onn Malaysia (UTHM), 86400 Batu Pahat, Johor, Malaysia
{yana, rozaida}@uthm.edu.my

Abstract. Functional Link Neural Network (FLNN) has becoming as an important tool used in machine learning due to its modest architecture. FLNN requires less tunable weights for training as compared to the standard multilayer feed forward network such as Multilayer Perceptron (MLP). Since FLNN uses Backpropagation algorithm as the standard learning algorithm, the method however prone to get trapped in local minima which affect its performance. This paper proposed the implementation of Ant Lion Algorithm as learning algorithm to train the FLNN for classification tasks. The Ant Lion Optimizer (ALO) is the metaheuristic optimization algorithm that mimics the hunting mechanism of antlions in nature. The result of the classification made by FLNN-ALO is compared with the standard FLNN model to examine whether the ALO learning algorithm is capable of training the FLNN network and improve its performance. From the result achieved, it can be seen that the implementation of the proposed learning algorithm for FLNN performs the classification task quite well and yields better accuracy on the unseen data.

Keywords: Functional link neural network · Learning algorithm · Ant lion optimizer

1 Introduction

In the field of Machine Learning (ML), ANNs is known as a family of statistical learning algorithm inspired by the way of human brain process information [1]. They are data driven self-adaptive method, in which they can change their structure based on external or internal information that flows through the network to model complex relationships between inputs and outputs. The recent vast research activities in neural classification have established that ANNs are a promising tool and have been widely applied to various real-world classification task especially in industry, business and science [2, 3]. One of the best-known types of ANNs is the Multilayer Perceptron (MLP). The application of MLP in classification tasks has shown better performance in comparison to the statistical method due to their nonlinear nature and training capability [4, 5].

Despite the development of various types of ANNs, this research work examines the ability of Higher Order Neural Networks (HONNs) which focusing on Functional

R. Ghazali et al. (Eds.): SCDM 2020, AISC 978, pp. 130–139, 2020.
https://doi.org/10.1007/978-3-030-36056-6_13

Link Neural Network (FLNN) for solving classification problems. FLNN is a class of HONNs that can perform nonlinear mapping by using only single layer of units [6]. HONNs utilize higher order terms to expand their input space into higher dimensional space to achieve nonlinear separability which reduced the complexity of the network. The single layer property of FLNN also makes the learning algorithm used in the network less complicated as compared to other standard feedforward neural networks [7].

In neural classifications, network training is essential in order to build a classification model. The purpose of network training is to find an optimal connection weights values that can minimize the network error and this method can also be viewed as an optimization task [8]. The most widely used error correcting learning method for network training is the Backpropagation (BP) learning algorithm [9–11]. The BP-learning algorithm is a type of gradient descent optimization method. However, network learning algorithm which is based on gradient descent optimization method has several drawbacks that can affect the performance of the neural network model [12–14]. Therefore, this research emphasizes on improving the FLNN network learning algorithm by using metaheuristic optimization algorithm to overcome such drawback.

2 Related Work

This section presents an overview on Functional Link Neural Network, the standard learning algorithm of FLNN and Ant Lion Optimizer Algorithm that will be utilized in this work.

2.1 Functional Link Neural Network

Functional Link Neural Network (FLNN) is a class of Higher Order Neural Networks (HONNs) that utilized higher order of its input introduced by Pao [15, 16]. FLNN is much more modest than MLP as it has a single layer of trainable weights whilst able to handle non-linear separable problems. The flat architecture of FLNN has also make the learning algorithm in the network less complicated [7]. In order to capture non-linear input-output mapping, the input vector of FLNN is extended with a suitable enhanced representation of the input nodes which artificially increase the dimension of input space [15, 16]. Figures 1 and 2 show the network structure of MLP and FLNN both with 2 input nodes. The network structure of FLNN presented in Fig. 2 was enhanced up to 2nd order (the highest order) make it employed only 4 trainable parameters (3 weights + I bias) in its structure. As compared to. Figure 1, the MLP with the same number of input nodes (2 inputs) and with a single hidden layer of 2 nodes (the least numbers of hidden nodes and layers) formed 9 trainable parameters (6 weights +3 biases) in its structure. Comparing of both networks, FLNN need less trainable weights as compared to MLP for training and this make the learning scheme for FLNN less complicated.

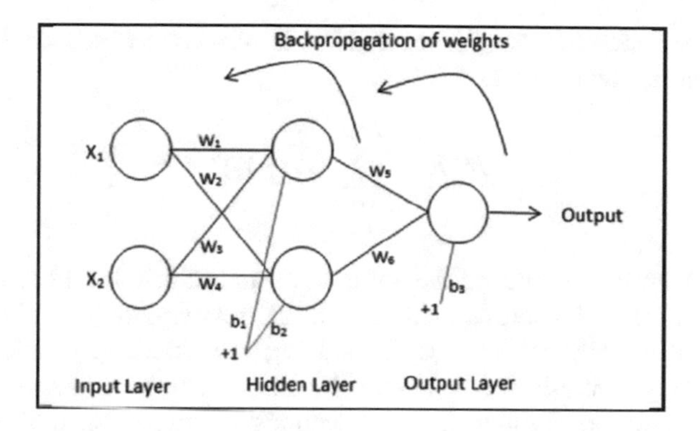

Fig. 1. MLP structure with 2 input nodes

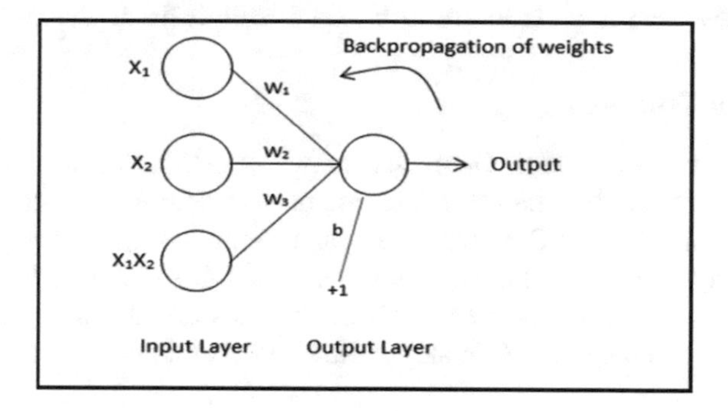

Fig. 2. FLNN structure with 2 input nodes

This work is focused on the FLNN with generic basis architecture which uses a tensor representation. Figure 2 illustrates the FLNN structure of 2 inputs up to second order input enhancement. The first order consists of the original inputs which are x_1 and x_2. To provide the FLNN network with non-linear mapping capability the inputs is enhanced with additional higher order unit by extending them based on the product unit. The product unit of this network is x_1x_2 as depicted is Fig. 2 and is known as the 2nd order input enhancement for the network.

2.2 Learning Algorithm

In the context of FLNN training, the mostly used learning algorithm is the Back-propagation (BP) learning algorithm. Backpropagation algorithm is a supervised learning method used to monitor the neural network learning which was introduced by Rumelhart et al. [17]. The algorithm is widely used in conjunction with an optimization method such as gradient descent [18]. The idea of the BP-learning algorithm is to reduce network error in order to optimize the network performance. In BP-learning algorithm, the FLNN weights are adjusted in accordance with the Delta rule with the purpose to minimize the network error between the desired output and the network output. The network error for FLNN training is measured based on Mean Squared Error, MSE and is calculated as follows;

$$MSE = \frac{1}{n}\sum_{i=1}^{n}(\hat{Y}_i - Y_i)^2 \tag{1}$$

Although BP-learning is one of the most used algorithms for FLNN training, the algorithm, however, has several limitations. The algorithm tends to easily get trapped in local minima especially when used for solving non-linear separable classification problems. Solving a non-linear separable classification is a multimodal problem and may encounter many local minima. As BP-learning is strictly depends on the shape of error surface, this has made the algorithm prone to stuck in some local minima when moving along the error surface during the training phase. Therefore, further investigation to improve learning algorithm in FLNN is still desired.

2.3 Ant Lion Optimizer

Ant Lion Optimizer or Antlion Optimizer known as ALO, is a recent meta-heuristic that mathematically models the interaction of ants and antlions in nature developed by Seyedali Mirjalili [19]. ALO optimization algorithm has been developed to solve optimization problems [20, 21]. The algorithm of ALO simulate interaction between ants in the snare and antlions. In the algorithm, ants are the prey and they moves randomly in the search space, while antlions are hunters that will hunt ants and become fitter using traps [22].

In ALO there are five main steps of hunting prey which are the random walk of ants, building traps, entrapment of ants in traps, catching preys, and re-building traps.

Random Walk of Ants.
ALO model the movement of ants by using the equation of random walk;

$$X(t) = [0, cumsum(2r(t_1) - 1), cumsum(2r(t_2) - 1), \ldots, cumsum(2r(t_n) - 1),] \tag{2}$$

where $r(t) = \begin{cases} 1\ rand > 0.5 \\ 0\ rand \leq 0.5 \end{cases}$

The random walk in Eq. (2) is used in the following equation to update the position of ants;

$$X_i^t = \frac{(X_i^t - a_i) \times (d_i - c_i^t)}{(d_i^t - a_i)} + c_i \tag{3}$$

where a_i is the minimum of random walk of i-th variable, c_i^t is the minimum of i-th variable at t-th iteration, and d_i^t indicates the maximum of i-th variable at t-th iteration.

Building Traps.
ALO employ roulette wheel to represent the building traps. The roulette wheel operator selects antlions based of their fitness during optimization. This mechanism gives high chances to the fitter antlions for catching ants.

Entrapment of Ant in Traps.
The entrapment of ant's movements by antlion are represented as follows;

$$c_i^t = Antlion_j^t + c^t \quad (4)$$

$$d_i^t = Antlion_j^t + d^t \quad (5)$$

Where c_i^t the minimum of all variables for i-th ant, c^t the minimum of all variables at t-th iteration, d_i^t the maximum of all variables for i-th ant and d^t vector including the maximum of all variables at t-th iteration.

Catching preys and Rebuilding Traps.
In ALO, catching prey occurs when ants get fitter than its associated antlion. An antlion is then required to update its position to the latest position of the hunted ant to enhance its chance of catching new prey simulated by following equation;

$$Antlion_j^t = Ant_i^t if, f(Ant_i^t) > f(Antlin_i^t) \quad (6)$$

Where t is the present iteration, $Antlin_i^t$ the position of selected j-th antlion at t-th iteration and Ant_i^t indicate the position of i-th ant at t-th iteration.

3 FLNN Training Using Ant Lion Optimizer

Inspired by the robustness and flexibility offered by the ALO algorithm, we proposed the implementation of the ALO algorithm as the learning algorithm for FLNN training. The proposed pseudo code is presented in Fig. 3. In the initial process, the FLNN architecture (weight and bias) is transformed into objective function along with the training dataset. This objective function will then be fed to the ALO algorithm in order to search for the optimal weight parameters. The weight changes are then tuned by the ALO algorithm based on the error calculation (difference between actual and expected results).

```
Begin
1) Initialize Maximum Cycle Number, MCN
2) Set FLNN weights as objective function: f(X), X = (x1,x2,...xd);
3) Generate an initial population of ants and antlions
4) Evaluate fitness of the antlions
5) Cycle = 1;
   while (Cycle < MCN)
   for i = 1 : n (all n ants)
   select antlion to build trap using roulette wheel
   Update distance of ants and antlions using following equation;
```

$$c^t = \frac{c^t}{I} \quad and \quad d^t = \frac{d^t}{I}$$

```
   Move ants using equation (2) and equation (3)
   end for
   evaluate fitness of all ants
   replace antlions with its corresponding ants using equation (6)
   update the best antlion
   end while
End
```

Fig. 3. ALO algorithm for training FLNN network

4 Experimentation Results

The Simulation experiments were performed on the training of FLNN with standard BP-learning (FLNN-BP), FLNN and with Ant Lion Optimizer (FLNN-ALO). In this work we have considered 6 public benchmark datasets obtained from UCI Machine Learning Repository as presented in Table 1. The result of the FLNN-ALO also was compared with standard model of Multilayer Perceptron (MLP-BP) to assess the classification performance in term of network complexity. Table 2 summarized parameters setting for the experiment.

Table 1. UCI machine learning public benchmark datasets

Datasets	Number of instances	Number of attributes	Number of classes
BUPA Liver Disorders	345	6	2
Breast Cancer Wisconsin	699	9	2
Indian Liver Patient Dataset (ILPD)	583	10	2
Iris	150	4	3
PIMA Indians Diabetes	768	8	2
Thyroid Disease	215	5	3

The reason of conduction a simulation on MLP and FLNN architectures is to provide a comparison on neural complexity between standard ANN and HONN. In this simulation we used a single hidden layer of MLP with the numbers of hidden nodes equal to it input features while for the FLNN architecture we used a second order input enhancement to provide a nonlinear input-output mapping capability. The neural complexity refers to the numbers of trainable parameters (weight and bias) needed to implement good approximation in each neural network architecture. The less numbers

Table 2. The parameters setting for the experiment

Parameter	MLP-BP	FLNN-BP	FLNN-ALO
Learning rate	0.1–0.5	0.1–0.5	–
Momentum	0.1–0.9	0.1–0.9	–
Epoch	1000	1000	1000
Minimum error	0.001	0.001	0.001

Table 3. Result on BUPA liver disorders dataset

BUPA Liver Disorders			
Learning algorithm	Best network structure	Number of tunable weights	Accuracy (%)
MLP-BP	6-6-1	33	64.64
FLNN-BP	21-1	22	60.70
FLNN-FA	**21-1**	**22**	**65.72**

of parameters indicate that the network is less complex as it required a smaller number of weight and bias to be updated at every epoch or cycle. Tables 3, 4, 5, 6, 7 and 8 shows the classification success achieved for each dataset in terms of network structure, and classification accuracy.

Table 4. Result on Breast Cancer Wisconsin dataset

Breast cancer Wisconsin			
Learning algorithm	Best Network structure	Number of tunable weights	Accuracy (%)
MLP-BP	9-9-1	67	95.38
FLNN-BP	**45-1**	**46**	**96.50**
FLNN-ALO	45-1	46	96.45

Table 5. Result on Indian Liver Patient dataset (ILPD)

Indian liver patient dataset			
Learning algorithm	Best network structure	Number of tunable weights	Accuracy (%)
MLP-BP	10-10-1	109	65.75
FLNN-BP	55-1	56	66.98
FLNN-ALO	**55-1**	**56**	**69.58**

Table 6. Result on Iris dataset

Iris dataset			
Learning algorithm	Best network structure	Number of tunable weights	Accuracy (%)
MLP-BP	4-4-3	43	94.60
FLNN-BP	**11-3**	**33**	**92.02**
FLNN-ALO	11-3	33	85.89

Table 7. Result on PIMA Indians diabetes dataset

PIMA Indians diabetes			
Learning algorithm	Best network structure	Number of tunable weights	Accuracy (%)
MLP-BP	8-8-2	91	72.69
FLNN-BP	36-1	37	70.15
FLNN-ALO	**36-1**	**37**	**75.15**

Table 8. Result on thyroid disease dataset

Thyroid disease			
Learning algorithm	Best network structure	Number of tunable weights	Accuracy (%)
MLP-BP	5-5-3	45	94.38
FLNN-BP	15-3	48	91.09
FLNN-ALO	**15-3**	**48**	**91.51**

Based on the results achieved for each dataset presented in Tables 3, 4, 5, 6, 7 and 8, it shows that the proposed FLNN-ALO outperforms the standard FLNN-BP and MLP-BP in almost every dataset. For the case of BUPA, ILPD, PIMA and THYROID, the simulation result shows that the proposed metaheuristic learning algorithm can successfully train the FLNN for solving classification problems with better accuracy percentage on unseen data. This demonstrates that FLNN-ALO network classifies better than MLP-BP network with higher accuracy and smaller number of weight parameters.

5 Conclusion

In this work, we evaluated the FLNN-ALO for classification problems. The experiment has demonstrated that FLNN-ALO performs the classification task quite well. This research work is conducted to discover a better learning algorithm for training the Functional Link Neural Network that can give a promising result. In future work, we will conduct an experiment on high dimensional classification benchmark data in order to explore and to assess the feasibility of the proposed learning algorithm.

Acknowledgments. The authors would like to thank Universiti Tun Hussein Onn Malaysia (UTHM) for supporting this research.

References

1. Michalski RS, Carbonell JG, Mitchell TM (2013) Machine Learning: an Artificial Intelligence Approach. Springer Science & Business Media (2013)
2. Yeatman TJ, Zhou JX, Bloom GC, Eschrich SA (2014) Artificial neural network proteomic tumor classification. Google Patents
3. Al-Jarrah O, Arafat A (2015) Network intrusion detection system using neural network classification of attack behavior. J Adv Inf Technol 6
4. Zabidi A, Lee Yoot K, Mansor W, Yassin IM, Sahak R (2010) Classification of infant cries with asphyxia using multilayer perceptron neural network. In: Computer engineering and applications (ICCEA), 2010 second international conference on, pp 204–208
5. Silva LM, Marques de Sá J, Alexandre LA (2008) Data classification with multilayer perceptrons using a generalized error function. Neural Netw 21:1302–1310
6. Giles CL, Maxwell T (1987) Learning, invariance, and generalization in high-order neural networks. Appl Opt 26:4972–4978
7. Misra BB, Dehuri S (2007) Functional link artificial neural network for classification task in data mining. J Comput Sci 3:948–955
8. Karaboga D, Basturk B (2007) On the performance of artificial bee colony (ABC) algorithm. Elsevier Appl Soft Comput 8:687–697
9. Rojas R (2013) Neural networks: a systematic introduction. Springer Science & Business Media (2013)
10. Eberhart RC (2014) Neural network PC tools: a practical guide. Academic Press
11. Rubio JD, Angelov P, Pacheco J (2011) Uniformly stable backpropagation algorithm to train a feedforward neural network. Neural Netw IEEE Trans 22:356–366
12. Sierra A, Macias JA, Corbacho F (2001) Evolution of functional link networks. Evolut Comput IEEE Trans 5:54–65
13. Abu-Mahfouz I-A (2005) A comparative study of three artificial neural networks for the detection and classification of gear faults. Int J Gen Syst 34:261–277
14. Dehuri S, Mishra BB, Cho S-B (2008) Genetic feature selection for optimal functional link artificial neural network in classification. In: Proceedings of the 9th international conference on intelligent data engineering and automated learning, pp 156–163. Springer, Daejeon, South Korea (2008)
15. Klassen M, Pao YH, Chen V (1988) Characteristics of the functional link net: a higher order delta rule net. In: Neural networks IEEE international conference on, vol 501, pp 507–513

16. Pao YH (1989) Adaptive pattern recognition and neural networks. Addison-Wesley Longman Publishing Co., Inc
17. Rumelhart DE, McClelland JL (eds) (1986) Parallel distributed processing: explorations in the microstructure of cognition, vol 1: foundations. MIT Press
18. Haykin S (2004) Network neural: a comprehensive foundation. Neural Netw 2
19. Mirjalili S (2015) the ant lion optimizer. Adv Eng Softw 83:80–98
20. Dinakara Prasasd Reddy P, Veera Reddy VC, Gowri Manohar T (2018) Ant Lion optimization algorithm for optimal sizing of renewable energy resources for loss reduction in distribution systems. J Electr Syst Inf Technol 5:663–680
21. Tian T, Liu C, Guo Q, Yuan Y, Li W, Yan Q (2018) An improved ant lion optimization algorithm and its application in hydraulic turbine governing system parameter identification. Energies 11:95
22. Abd-AlKareem M, Riyadh T (2018) Exploring ant lion optimization algorithm to enhance the choice of an appropriate software reliability growth model. Int J Comput Appl 182:1–8

Classification of Metamorphic Virus Using N-Grams Signatures

Isredza Rahmi A. Hamid[1(✉)], Nur Sakinah Md. Sani[1], Zubaile Abdullah[1], Cik Feresa Mohd Foozy[1], and Kuryati Kipli[2]

[1] Faculty of Computer Science and Information Technology, Universiti Tun Hussein Onn Malaysia, Parit Raja, Johor, Malaysia
rahmi@uthm.edu.my

[2] Faculty of Engineering, Universiti Malaysia Sarawak, 94300 Kota Samarahan, Malaysia

Abstract. Metamorphic virus has a capability to change, translate, and rewrite its own code once infected the system to bypass detection. The computer system then can be seriously damage by this undetected metamorphic virus. Due to this, it is very vital to design a metamorphic virus classification model that can detect this virus. This paper focused on detection of metamorphic virus using Term Frequency Inverse Document Frequency (TF-IDF) technique. This research was conducted using Second Generation virus dataset. The first step is the classification model to cluster the metamorphic virus using TF-IDF technique. Then, the virus cluster is evaluated using Naïve Bayes algorithm in terms of accuracy using performance metric. The types of virus classes and features are extracted from bi-gram assembly language. The result shows that the proposed model was able to classify metamorphic virus using TF-IDF with optimal number of virus class with average accuracy of 94.2%.

Keywords: Metamorphic virus · Classification · Term frequency inverse document frequency (TF-IDF)

1 Introduction

Currently, security threat has become vicious and countermeasure must be taken seriously. The number of security threat towards the user is increasing each year. The virus inventor becomes more creative in order to penetrate the system. Once the virus was in the system, it will either corrupting the system or remains dormant until it gets to attack the target. Thus, the system must become more alert towards the virus intrusion in order to protect it from the virus attack.

Metamorphic virus has capabilities to change, translate, and rewrite its own code when it infects a system. It is the most viral and if it is not detected earlier the system can be seriously damage. The difference between Polymorphic and Metamorphic virus is that the Polymorphic Virus keeps the original code and only encrypt the code. The Metamorphic Virus is much more complex and requires programming expert to create this virus [1]. This research has three main objectives, which are:

R. Ghazali et al. (Eds.): SCDM 2020, AISC 978, pp. 140–149, 2020.
https://doi.org/10.1007/978-3-030-36056-6_14

- To design a virus detection model on metamorphic virus using static detection.
- To classify metamorphic virus using Term Frequency Inverse Document Frequency.
- To evaluate the proposed model using Naïve Bayes algorithm in terms of accuracy and efficiency.

The rest of the paper is organized as follows: Sect. 2 describes the related work on metamorphic virus detection techniques. Section 3 presents the proposed classification model for metamorphic virus detection based on Term Frequency Inverse Document Frequency (TF-IDF). Section 4 shows experimental setup. Section 5 will discuss about the result from the experiment. Finally, Sect. 6 concludes the work and highlights a direction for future research.

2 Related Work

During the former phases of virus creation, virus programmers tried to infect a large number of victims. Virus was created similar in type of infection, but the malicious actions performed were different. However, the methods employed to infect a host machine and spread to other machines were similar to all virus. Most of the early stage of virus detection was discovered based on its signature files and activities performed by the virus. As virus detection systems managed to detect and stop the infections with increasing strength, virus programmers started to implement new methods to spread the virus infections [2]. The evolution of virus becomes more advanced that it produce virus that used encryption technique to obfuscate their presence. This makes the virus existence unclear to confuse the virus detector.

Metamorphic virus changed its code while propagate. Thus, it can avoid detection by static signature-based virus scanners. This leads to possibility of undetectable breed of malicious programs. Moreover, static analysis metamorphic virus also uses code obfuscation techniques which could beat dynamic analysers, such as emulators. Hence, the metamorphic virus managed to alter its behaviour when discovered executing under a manipulated environment. The metamorphic virus used several metamorphic transformations to differ the visual aspect, such as register usage exchange, code permutation, code expansion, code shrinking, and garbage code insertion [3]. Metamorphic virus also capable to create a new generation that looks different to their parents.

Table 1 shows the comparison between three virus detection approaches. Signature based is the most efficient approach as compared to anomaly based and code emulation in term of detection strength, accuracy and low at cost. However, it is only limited to new malware variant.

Table 1. Comparison of virus detection approaches [4].

Methods / Parameter	Signature based	Anomaly based	Code emulation
Strength	Efficient	New malware	Encrypted virus
Limitation	New malware	Unproven	Complex
Cost	Low	High	High
Accuracy	High if database is updated	Low	High

Qu used behaviour-based features consists of 602 malware from the VCL family to create the signature of the virus family [5]. The algorithm used is backward regression model and regression model. The regression model was used to determine the identification of VCL malware and act as indicator to show the influential of VCL malware. The backward regression model achieved 90.3% accuracy in identifying VCL malware. Kuriakose [6] used feature selection method to detect the presence of metamorphic virus. This research used 3344 malware sample and 1218 benign of 32 win XP. A significant bi-gram of variable lengths is used for constructing learning models using AdaboostM1 (using J48 as base classifier) and Random forest with default settings in WEKA [7]. They managed to achieve 99.8% and 92.6% for benign and virus detection. Shabani [8] used the Bayesian Network features to detect metamorphic virus tested on 600 samples. Bayesian Network learning is known as a NP-hard problem because it's utilizing exploratory research proved that was helpful in many learning approach although it does not guarantee optimistic result. They used Hill climbing algorithm because it is a popular algorithm just because of exchanging between computational demands and the quality of the model. Their approach managed to achieve above 90% accuracy.

Pomorova [9] proposed a metamorphic virus detection method using altered emulators situated on corporate network hosts. The method involves fuzzy logic in order to classify metamorphic viruses. The results of experimental studies demonstrated the detection accuracy of 85%.

Our work differ than other researchers [5, 6, 8, 9] as shown in Table 2 in such a way that we used Second Generation virus dataset. Then, the dataset was classified using Term Frequency Inverse Document Frequency (TF-IDF) algorithm and tested with Naïve Bayes classification algorithm. We used bi-gram features to identify that metamorphic virus classes based on its own features can distinct themselves from each other.

Table 2. Virus classification approach.

Work by	Features used	Sample used	Machine learning algorithm	Results (Acc)
Qu et al. [5]	Behaviour-based feature	602	Backward Logistic Regression Model or Logistic Regression Model	90.3%
Kuriakose et al. [6]	Feature selection	3344 malware and 1218 benign	Adaboost and Random Forest	Benign – 99.8% Virus – 92.6%
Shabani et al. [8]	Bayesian network	600	Hill Climbing	Above 90%
Pomorova et al. [9]	Fuzzy logic	–	Fuzzy Inference System	82%

3 Metamorphic Virus Classification Model

This section explains about the metamorphic virus classification model. The proposed model consists of three important phases including the pre-processing, feature extraction, and model generation with prediction. Figure 1 shows the flow of the proposed virus classification model.

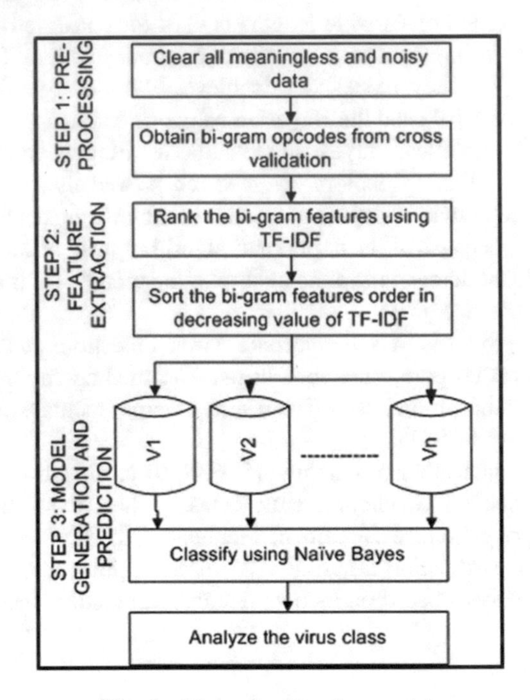

Fig. 1. Virus classification model.

3.1 Pre-processing

The pre-processing phase involves raw data cleansing. The dataset consists of 152 viruses which 96 viruses from Second Generation. Data describing allows the distribution of data values, while data transformation help in performing calculation on existing columns.

The pre-processing involves data mining techniques to train the classification models with set of rule based about the virus function. Then, the model will be trained and used to classify the testing data. Normally, data mining techniques is used for large of datasets for pattern detection [8]. The final process in data preparation is data sampling, which help the creation of training and to validate the datasets.

3.2 Feature Extraction

The feature extraction process is to create set of new features that can be used for classification. First, we obtained the bi-gram feature value. Then, the bi-gram feature was further classified using Term Frequency Inverse Frequency Document (TF-IDF). The TF-IDF algorithm gives weight value for each word in the whole document as shown in Fig. 2. TF-IDF method allows each word to be considered as important and is inversely proportional on how often it occurs in whole document.

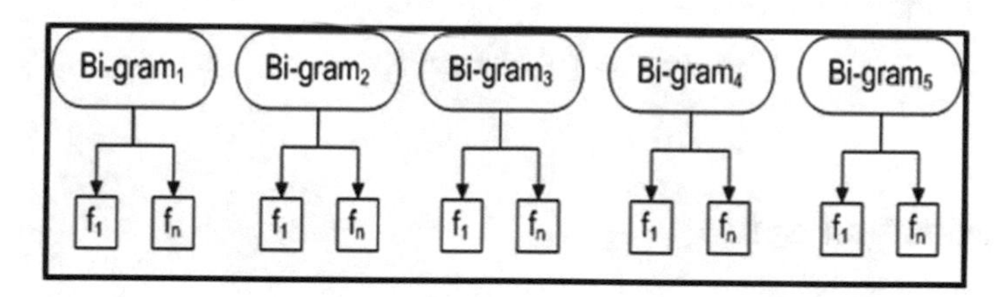

Fig. 2. Classification of dataset.

The classification using Term Frequency Inverse Document Frequency has three main steps:

1. The frequency of bi-gram features was calculated for the whole dataset.
2. The frequency value was normalized to avoid biased result by calculating the bi-gram feature value (Tf) divide by maximum bi-gram value in the dataset as shown in Eq. 1.

$$\mathrm{Tf}/\mathrm{Max}(\mathrm{Tf}\ldots\mathrm{n}) \tag{1}$$

3. Then, calculate the Inverse Document Frequency(IDF) using Eq. 2, where, N is total number of documents in the corpus, {d∈ D: ∈ d} is number of documents where the term appears and (t, d) $\neq 0$ if the term is not in the corpus. This will lead to a division-by-zero.

$$\mathrm{idf}(\mathrm{t},\mathrm{d}) = \log \mathrm{N}/(|\{\mathrm{d} \in \mathrm{D} \sim \mathrm{t} \in \mathrm{d}\}|) \tag{2}$$

3.3 Model Generation and Prediction

Term Frequency is used to categorize the text document. This method does not involve any binary values. Generally, Term Frequency means number of times the word or term, exists in a document, d. To get a better result, the Term Frequency will be divided with maximum number of raw Term Frequency according to the length of the document. This can be simplified by Eq. 3.

$$\mathrm{tf, Normalized} = \mathrm{tf}/(\max/(\mathrm{tf, dl1...tf, dln})) \quad (3)$$

The second step will be the calculation the IDF using Eq. 4.

$$\log(\mathrm{n, d}/\mathrm{countif}(\mathrm{tf, d1...tf, dn})) \quad (4)$$

The last step is to calculate the Term Frequency Inverse Document Frequency by multiplying the values of tf, normalized with idf value. Finally, the TF-IDF value will be obtained through Eq. 5.

$$\mathrm{tf} - \mathrm{idf} = (\mathrm{tf, Normalized} * \mathrm{idf}) \quad (5)$$

4 Experimental Setup

This section discuss the experimental setup for metamorphic virus classification.

4.1 Metamorphic Virus Detection System

In our study, the virus classification was performed using WEKA. The experiment will show the accuracy and effect of classification algorithm used on the dataset. This experiment uses the Second Generation Virus Kits which has 52 features. The dataset contained long string of uni-gram assembly language. However, the uni-gram feature does not have any significant meaning for virus classification. So, the bi-gram feature is selected for this experiment. The bi-gram features are calculated using Term Frequency Inverse Frequency Document (TF-IDF) to classify the virus class. We only used the first five bi-gram features from the whole string to show that the TF-IDF algorithm is the effective method to classify virus.

4.2 Performance Metric

In order to measure the effectiveness of the classification approach, we refer to three possible outcomes as: True Positive (TP) and False Positive (FP) and Receiver Operating Curve (ROC).

True positive is where the number of correctly identified as metamorphic virus.

$$\mathrm{TP} = \mathrm{TP}/(\mathrm{TP} + \mathrm{FN}) \quad (6)$$

While False positive is where the number of wrongly identified as metamorphic virus.

$$\mathrm{FP} = \mathrm{FP}/(\mathrm{TN} + \mathrm{FP}) \quad (7)$$

ROC value is where the classification made and algorithm used can be determine it certainty. The best result of ROC value should be close to one. This shows that the datasets has equal sensitivity and specificity.

5 Result and Discussion

This section discusses number of class and performance for each bi-gram in each datasets.

5.1 TF and FP Rate

The dataset has different types of bi-gram features and total number of classes. We selected the first five bi-gram features from the whole string to show that the TF-IDF algorithm is the effective method to classify virus. Table 3 shows the bi-gram features and number of classes for five bi-gram intended for Second Generation virus kit dataset.

Table 3. TP and FP rate of second generation virus kits.

Bi-gram	Features	Number of class	TP rate	FP rate	Bi-gram	Features	Number of class	TP rate	FP rate
1	Movint	20	0.854	0.01	4	movlea	18	0.854	0.027
	Callpop	13	0.958	0.003		movint	9	0.958	0.052
	Jmpmov	13	0.906	0.05		Intmov	18	0.906	0.015
	Jmpcall	3	0.99	0.24		callpop	9	0.885	0.018
2	Intmov	6	0.969	0.018		intcmp	10	0.927	0.006
	Subpush	6	0.948	0.017		pushpush	17	0.844	0.009
	Movsub	6	0.979	0.016		intpush	11	0.885	0.008
	Movadd	6	0.969	0.007		poppush	10	0.917	0.007
	Movxor	7	0.927	0.067		popsub	6	0.969	0.003
	Addadd	6	0.948	0.012		loopcall	6	0.979	0.057
	Popsub	5	0.948	0.051		intjz	2	0.99	0.99
	Addsub	5	0.958	0.358		pushmov	15	0.844	0.017
	Addinc	4	0.979	0.101	5	leamov	22	0.813	0.024
	Xoradd	4	0.979	0.201		intmov	22	0.833	0.028
	Xorinc	3	0.958	0.958		popmov	16	0.896	0.012
3	Pushpush	10	0.979	0.001		intpush	16	0.865	0.094
	Subpush	7	0.948	0.004		pushpush	18	0.844	0.015
	Submov	7	0.938	0.005		jmov	11	0.917	0.017
	Incloop	4	0.958	0.002		subpush	12	0.927	0.014
	loopcall	5	1	0		movsub	12	0.875	0.01
	Sublea	4	0.854	0.071		cmpjz	12	0.875	0.054
	subloop	4	0.958	0.313		pushlea	10	0.906	0.02
	loopmov	5	0.979	0.146		poppush	15	0.854	0.077
	pushmov	10	0.938	0.004		sublea	4	0.958	0.076
	Incinc	3	0.969	0.001		popsub	7	0.948	0.181
	addloop	3	0.979	0		pushmov	18	0.823	0.017

For the first bi-gram, `jmpcall` has the highest value of True Positive and False Positive rate with 0.99 and 0.24 respectively. However, `callpop` has better result compared to `jmpcall` because it has only slightly lower True Positive rate at 0.958 plus they have lower False Positive rate at 0.003.

As for second bi-gram, `movsub`, `addinnc`, and `xoradd` features have highest True Positive value with 0.979. Moreover, `movadd` has the lowest value for FP rate and `xorinc` has the highest value for False Positive rate with 0.958.

The `loopcall` features in third bi-gram achieved the highest True Positive rate and lowest False Positive rate that are 1 and 0 respectively. Other features also have high value of True Positive which shows that the virus classification is more diverse and correctly classified.

In fourth bi-gram, `intjz` feature has highest True Positive and False Positive value with 0.99 and 0.99 correspondingly. A `posub` feature produce better result as it has high True Positive rate at 0.969 and lower False Positive rate at 0.003.

The `sublea` feature in fifth bi-gram has the highest True Positive value with 0.958. However, `popsub` has better result although it has slightly lower True Positive rate at 0.927 compared to `subpush`, which has much lower False Positive rate at 0.014.

5.2 ROC Value

The ROC value should be close to one to be considered as good. Figures 3, 4, 5, 6 and 7 shows the ROC value for all bi-grams for Second Generation virus kit dataset. All features in fifth bi-grams shows the highest ROC value as compared to other bi-grams. This demonstrates that the dataset has equal sensitivity and specificity when classify using fifth bi-grams.

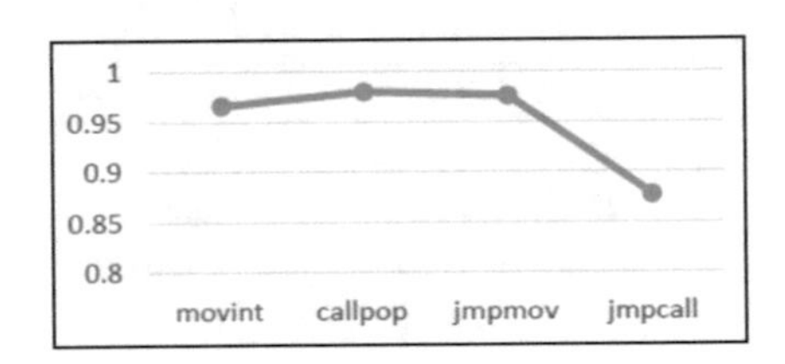

Fig. 3. ROC of first bi-gram features.

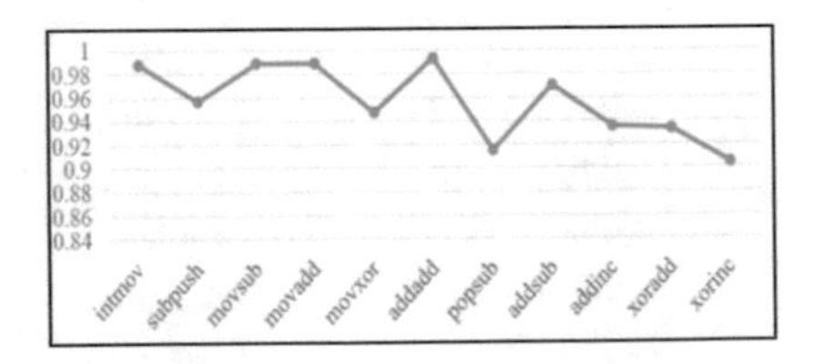

Fig. 4. ROC of second bi-gram features.

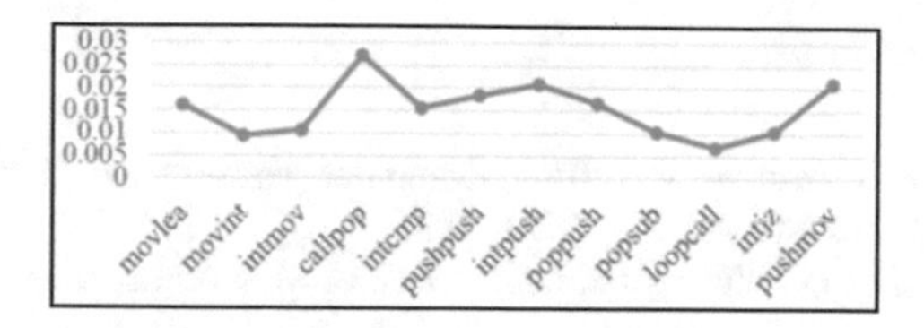

Fig. 5. ROC of third bi-gram features.

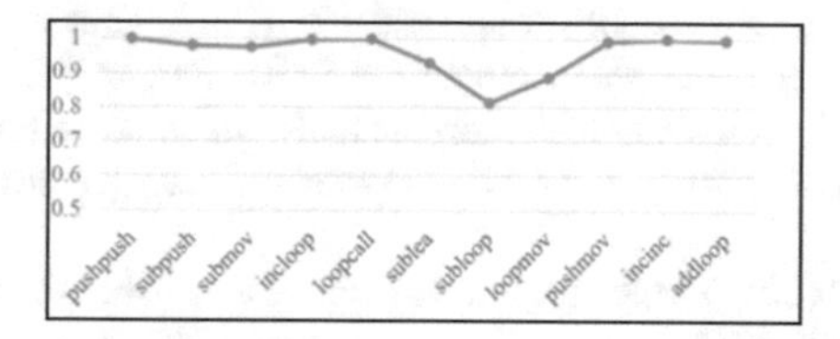

Fig. 6. ROC of fourth bi-gram features.

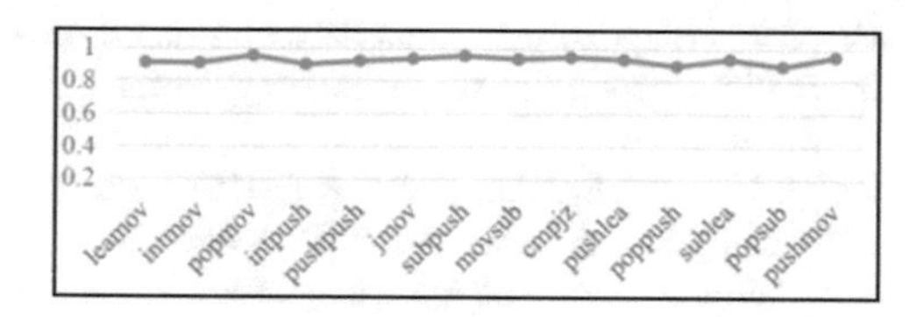

Fig. 7. ROC of fifth bi-gram features.

6 Conclusion

The classification of metamorphic viruses shows that the viruses can be reduced into small group. However, the technique used to cluster the metamorphic viruses is depends on the types of dataset used. The metamorphic virus classification model used Term Frequency Inverse Document Frequency (TF-IDF) to cluster the virus. This technique was widely implement in many research field that had terms or words as their dataset. In addition, this technique gives weight to important terms that need to be highlighted in a document. The proposed model managed to get high True Positive and low False Positive value when classifying the virus based on bi-gram features.

Acknowledgments. This research work is fully funded by Fundamental Research Grant Scheme (FRGS/1/2018/ICT04/UTHM/02/1), Ministry of Education Malaysia (MOE) and supported by the Research Management Center (RMC), Universiti Tun Hussein Onn Malaysia under Vot K047.

References

1. Kumar D, Kumar N, Kumar A (2014) Computer viruses and challenges for anti-virus industry. Int J Eng Comput Sci 3(2):3879–3872
2. Lin D, Stamp M (2011) Hunting for undetectable metamorphic viruses. J Comput Virol 7 (3):201–214
3. Konstantinou E (2018) Metamorphic virus: analysis and detection. Technical Report of University of London. http://www.rhul.ac.uk/mathematics/techreport. Accessed 24 Sep 2018
4. Kakad R, Kamble SG, Bhuvad SS, Malavade VN (2014) Study and comparison of virus detection techniques. Int J Adv Res Comput Sci Softw Eng 4(3):251–253
5. Qu Y, Hughes K (2013) Detecting metamorphic malware by using behavior-based aggregated signature. In: Proceeding of the world congress on internet security (WorldCIS-2013), London, pp 13–18
6. Kuriakose J, Vinod P (2014) Metamorphic virus detection using feature selection techniques. In: Proceeding of 5th IEEE international conference on computer and communication technology, ICCCT 2014, pp 141–146
7. Hall M, Frank E, Holmes G, Pfahringer B, Reutemann P, Witten IH (2009) The WEKA data mining software: an update. SIGKDD Explor 11(1):10–18
8. Shabani N, Jahan MV (2014) Metamorphic virus detection based on Bayesian network. In: Proceeding of 1st international congress on technology, communication and knowledge (ICTCK), pp 1–8
9. Pomorova O, Savenko O, Lysenko S, Nicheporuk A (2016) Metamorphic viruses detection technique based on the modified emulators. In: ICT in education, research and industrial applications, integration, harmonization and knowledge transfer. vol 1614, pp 375–383

Computational Intelligence and its Applied Applications

Experimental Analysis of Tuberculosis Classification Based on Clinical Data Using Machine Learning Techniques

Hery Yugaswara, Muhamad Fathurahman(✉), and Suhaeri

Informatics Department, Faculty of Information Technology, Universitas YARSI, Jakarta 10510, Indonesia
{hery.yugaswara,muhammad.fathurrachman}@yarsi.ac.id

Abstract. The early detection of tuberculosis plays a significant rule to reduce the death rate of tuberculosis. However, the early detection of tuberculosis nowadays has a limitation such as it needs long periods of time to acquire accurate diagnosis because it includes many clinical examinations. To overcome this problem a new diagnosis schema is needed. This study evaluates the common machine learning techniques including Logistic Regression, K-Nearest Neighbour, Naive Bayes, Support Vector Machine, Random Forest, Neural Network and Linear Discriminant Analysis to diagnose tuberculosis using classification methods based on clinical data. The results show that most of machine learning techniques that use in this study have a good performance in classifying tuberculosis based clinical data. Those machine learning techniques have achieved 0.97–0.99 in testing F1-Score.

Keywords: Tuberculosis · Machine learning · Classification · Early detection

1 Introduction

According to a recent result of Basic Health Research by Indonesian Health Ministry in 2018, The prevalency of Tuberculosis (TB) has increased between 2013–2018 [1]. The result shows that 29 of 36 provinces has higher prevalency of TB compared to those in 2013. However, the death rate of TB in Indonesia has decreased since 2000–2017 [2]. The early detection of TB plays an important rule to reduce the death rate of TB. Nevertheless, the early detection of TB has a limitation such as time-consuming to correctly diagnosis of tuberculosis [3] because of it needs for many clinical examinations. Therefore, an accurate and fast early detection of TB is required to help the clinician in selecting an appropriate treatment for the patients.

Recently, machine learning techniques have many common methods that widely use in the identification of diseases. This approach utilizes a set of clinical data to create a model which can be used to automatically detect TB. Several works have been done using a variety of feature in order to detect TB. Olatunji et al. [4] using a genetic neuro-fuzzy inferential model-based neural network for diagnosis of TB. Bobak et al. [5]

R. Ghazali et al. (Eds.): SCDM 2020, AISC 978, pp. 153–160, 2020.
https://doi.org/10.1007/978-3-030-36056-6_15

utilize common machine learning techniques Random Forest (RF), Support Vector Machine (SVM) and Partial Least Square (PLS) for classification of TB based on transcriptional biomarkers. Mithra and Emmanuel [6] proposed Gaussian Fuzzy Neural Network for diagnosis TB based on sputum smear microscopic images. Another approach also introduces by Uçar and Karahoca [3], Adaptive Neuro-Fuzzy Inference System (ANFIS) to detect mycobacterium TB on patients. However, the TB classification using machine learning techniques remains a challenging task because of its accuracy is dependent on what features were used.

This study aims to evaluate the performance of machine learning techniques in particular classification methods to classify TB based on clinical data. The main contribution to this study is to provide the preminary analysis of the use of the implementation of machine learning techniques to the TB classification based on clinical data. This paper is organized as follows. Section 2 presents the datasets and methods. Section 3 provides the results and discussions. Conclusion and future work are given in Sect. 4.

2 Dataset and Methods

2.1 Dataset

The datasets used in this study is obtained from the public health center which is also called (Puskesmas) in Jakarta. The dataset contains 81 samples and have two categories which are positive and negative classes. The positive class (denoted as 1 in the dataset) refers to the patient with confirmed TB whereas the negative class denoted as 0 in the dataset) refers to the patient with confirmed with no TB.

The number of samples that belongs to the positive class are 54 samples and 27 samples are in the negative class. Each sample have 19 clinical features that are composed of numerical and binary features. The numerical features are age, temperature, lower blood pressure, upper blood pressure and pulse per minutes whereas the binary features are composed of sex, cough in two weeks, bleeding cough, sweating at night, fever, feverish, nausea and vomiting, easily tired, epigastric pain, headache, abdominal pain, dyspnea, decreased appetite and weight loss. The distribution of clinical features can be seen in Tables 1 and 2.

Table 1. Mean of numerical values in the dataset

No	Feature name	Mean
1	Age	47.64
2	Temperature	40.69
3	Lower blood pressure	124.32
4	Upper blood pressure	78.89
5	Pulse per minutes	78.02

Table 2. The propotion of binary feature in the dataset

No	Feature name	Values	Proportion of feature values	
			1	0
1	Sex	{Male, Female}	0.59	0.41
2	Cough in two weeks	{Yes, No}	0.63	0.37
3	Bleeding Cough	{Yes, No}	0.37	0.63
4	Sweating at night	{Yes, No}	0.14	0.86
5	Fever	{Yes, No}	0.17	0.83
6	Feverish	{Yes, No}	0.21	0.79
7	Nausea and Vomitus	{Yes, No}	0.23	0.77
8	Easily tired	{Yes, No}	0.25	0.75
9	Epigastric pain	{Yes, No}	0.26	0.74
10	Headache	{Yes, No}	0.33	0.67
11	Abdominal pain	{Yes, No}	0.33	0.67
12	Dyspnea	{Yes, No}	0.56	0.43
13	Decrease appetite	{Yes, No}	0.22	0.78
14	Weight Loss	{Yes, No}	0.19	0.81
15	Class	{TB, Not TB}	0.67	0.33

Table 1 shows the mean of numerical features. Table 2 present the distribution of each binary feature values. In case of a sex feature, the value 1 represents a male whereas 0 represents a female.

2.2 Design Experiment

The design experiment framework to evaluate the performance of the classification of TB have three main processess including pre-processing, training phase, and testing phase, the detail of these processess can be seen in Fig. 1 below

Figure 1 shows the framework of the design experiment that used in this study. In the first step, Preprocessing data is applied to the clinical data including data transformation. Then 10-Fold cross-validation is used to split the set of clinical data into 10 folds which every number of folds have three different sets which are training, validation and testing set. The training and validation sets are used to train and to find the appropriate parameter settings of classification methods in order to achieve high accuracy. To find the appropriate parameter settings for each classification method during the training phase, each number of folds are repeated 10 times with the different number of parameter settings of classification model using a grid search [7]. Every single classification method in the training phase will be evaluated using evaluation methods. The best classification method which produces high accuracy will be selected as a trained model. In the next step, the trained models are used to predict the "unseen"

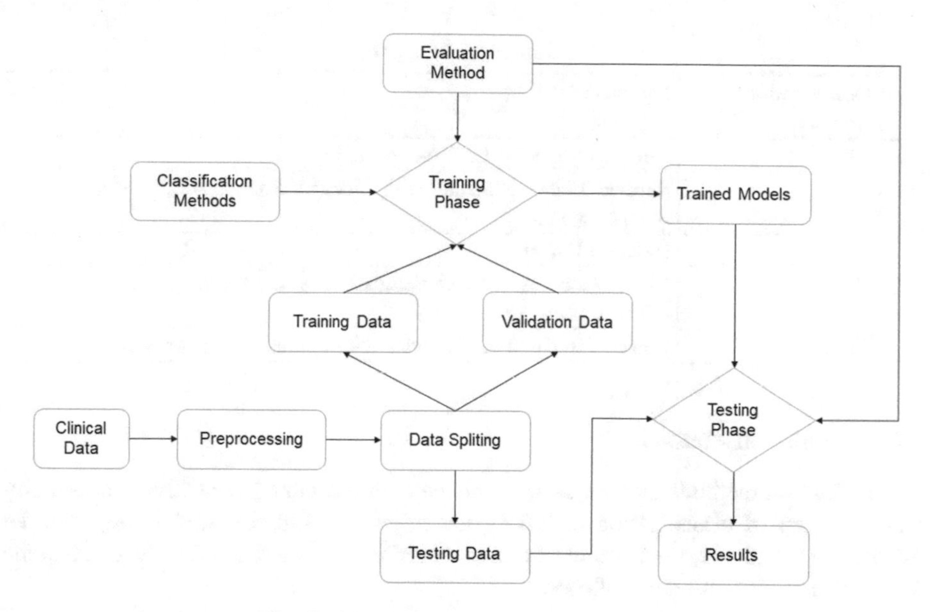

Fig. 1. The design experiment framework

clinical data in the testing set. Then the prediction result of the testing set will be evaluated by the similar evaluation methods as used in the training phase. The details of classification and evaluation methods that use in this paper will be described in the next sub-sections.

2.3 Classification Methods

In this study, several common classification methods are used including Logistic Regression (LR), K-Nearest Neighbour (KNN), Naive Bayes (NB), Support Vector Machine (SVM), Random-Forest (RF), Neural Network (NN) and Linear Discriminant Analysis (LDA). Those methods are available in the caret package that is implemented in the R language. The package provides a training method that supports adjusting parameter settings for each classification methods in order to achieve maximum accuracy [8]. The details of parameter settings for each classification methods can be seen in Table 3 below

Table 3. Parameter settings of classification methods

No	Classification methods	Parameter setting
1	LR	Penalized Logistic Regression with lambda Coeff = 0.01, 0.001, 0.0001
2	KNN	K-Nearest Neighbour with evaluation number of K = 5, 7, 9, 11, 13

(*continued*)

Table 3. (*continued*)

No	Classification methods	Parameter setting
3	NB	Naive Bayes with Laplace correction
4	SVM	Support Vector Machine with Linear Kernel, C = 1, Scaling Sample = True
5	RF	Random Forest
6	NN	Neural Network with hidden unit: 3–9 and weight decay: 0.01, 0.001, 0.0001
7	LDA	Linear Discriminant Analysis with default parameter settings

2.4 Evaluation Methods

The evaluation methods that are used in this paper are accuracy, sensitivity, specificity and F1 score of classification of TB in the training, validation and testing set. To estimate the accuracy, the number of correct classification should be obtained. Thus the accuracy can be written as follows:

$$Accuracy = \frac{TP + TN}{N} \tag{1}$$

The sensitivity and specificity are used to estimate for each class categories. Thus these measures can be defined as follows:

$$Sensitivity = \frac{TP}{TP + FN} \tag{2}$$

$$Specificity = \frac{TN}{FP + TN} \tag{3}$$

To estimate the F1 score, the recall (same as sensitivity) and precision value should be obtained first. In this case, the recall and sensitivity are the same, thus the precision value can be defined as follows

$$Precision = \frac{TP}{TP + FP} \tag{4}$$

Therefore, the F1-Score can be wrap as follows:

$$F1\text{-}Score = 2 \times \frac{sensitivity * precision}{sensitivity + precision} \tag{5}$$

Where TP (True Positive) and TN (True Negative) refer to the number of samples that correctly classify belong to its classes. Whereas N refers to the total number of samples. FP and FN are the number of samples that incorrect classification according to its class.

3 Results and Discussion

The experimental result of TB classification using several classification methods is presented. All methods are applied to the same dataset and evaluated to the same evaluation methods. The evaluation results of the classification method on the training, validation and testing set are provided in Tables 4, 5 and 6. To simplify the comparison of classification model performance, we only consider the F1-Score because of its value also accommodate the values of sensitivity and precision.

Table 4. Evaluation result on the training set

No	Classification method	Accuracy	Sensitivity	Specificity	F1-score
1	LR	1.00	1.00	1.00	**1.00**
2	KNN	0.98	0.95	1.00	0.97
3	NB	0.98	0.99	0.98	0.98
4	SVM	1.00	1.00	1.00	**1.00**
5	RF	1.00	1.00	1.00	**1.00**
6	NN	1.00	1.00	1.00	**1.00**
7	LDA	1.00	1.00	1.00	**1.00**

Table 5. Evaluation result on the validation set

No	Classification method	Accuracy	Sensitivity	Specificity	F1-Score
1	LR	0.99	0.97	1.00	0.98
2	KNN	0.97	0.92	0.99	0.94
3	NB	0.98	0.99	0.98	0.97
4	SVM	1.00	1.00	1.00	**1.00**
5	RF	1.00	1.00	1.00	**1.00**
6	NN	0.99	0.98	1.00	0.99
7	LDA	0.98	0.99	0.98	0.97

Table 6. Evaluation result on the testing set

No	Classification method	Accuracy	Sensitivity	Specificity	F1-Score
1	LR	0.99	0.98	1.00	**0.99**
2	KNN	0.97	0.91	1.00	0.95
3	NB	0.97	0.98	0.96	0.96
4	SVM	0.99	0.98	1.00	**0.99**
5	RF	0.99	0.98	0.99	0.98
6	NN	0.99	0.98	1.00	**0.99**
7	LDA	0.99	0.98	0.99	0.98

According to Table 4, LR, SVM, RF, NN, and LDA produce 1.00 in accuracy sensitivity, specificity, and F1-score which means all data in the training set are correctly classified. Experimental results on the validation set (Table 5) show that SVM and RF achieve the high value in accuracy, sensitivity, specificity, and F1-Score followed by NN, LR, NB, LDA, and KNN. Table 6 shows the classification result on the testing set. Based on the F1-Score in Table 6, SVM, LR, and NN have better performance followed by RF, LDA, NB, and KNN.

Based on the overall experiment, all machine learning techniques perform well in classifying two class categories of TB. One of the causes of getting high accuracy in all experiment is that the sample in the training, validation and testing dataset are quite similar. The limitation in applying machine learning models, in this case, is that these models are lacked of interpretability which means the model cannot explain why a sample is classified to a class. In the other side, the interpretability is an important aspect for the clinician to support their diagnosis.

4 Conclusion and Future Work

This study evaluates the performance of machine learning techniques for classifying TB based on patient clinical data. The classification methods that use in this study are LR, SVM, RF, NN, and LDA. In the training set all classification methods achieve more than $\sim$0.98–1.00 in accuracy, 0.95–1.00 in sensitivity, 0.98–1.00 in specificity and 0.97–1.00 in F1-Score. In the validation set, SVM and RF have slightly better performance to the other methods according to F1-Score. In addition, the evaluation result in the testing set (unseen) shows that SVM, LR, and NN has better performance according to F1-Score compare to those of other methods. In overall performance, the SVM model has consistently achieved the high performance compared to the other methods on training, validation and testing set according to the F1-Score (1.00 in training and validation set and 0.99 in testing set). The high accuracy, sensitivity, specificity, and F1-Score that achieve by several classification methods on the experiment above indicate that this approach is potentially used to support the clinician in TB diagnosis.

In the future work, more clinical data and medical images of TB patients will be used for a multimodal feature in order to classify TB. Another focus to this study is that to improve the interpretability of machine learning model in the TB classification.

Acknowledgments. This research has been fully funded by Internal Grant Research of Universitas YARSI.

References

1. Kementerian Kesehatan RI (2018) Hasil Utama RISKESDAS 2018, Jakarta, Indonesia
2. Frieden TR, Brudney KF, Harries A (2014) Global tuberculosis: perspectives, prospects, and priorities. 312(14)
3. Uçar T, Karahoca A (2011) Procedia computer predicting existence of mycobacterium tuberculosis on patients using data mining approaches. Procedia Comput Sci 3:1404–1411

4. Olatunji M, Williams O, John E (2017) A genetic-neuro-fuzzy inferential model for diagnosis of tuberculosis. Appl Comput Inf 13(1):27–37
5. Bobak CA, Titus AJ, Hill JE (2019) Comparison of common machine learning models for classification of tuberculosis using transcriptional biomarkers from integrated datasets ☆. Appl Soft Comput J 74:264–273
6. Mithra KS, Emmanuel WRS (2018) GFNN : gaussian-fuzzy-neural network for diagnosis of tuberculosis using sputum smear microscopic images. J King Saud Univ—Comput Inf Sci
7. Max A et al (2019) Classification and regression training, p 216
8. Fern M, Cernadas E (2014) Do we need hundreds of classifiers to solve real world classification problems ? 15:3133–3181

Impact of Hyperparameter Tuning on Deep Learning Based Estimation of Disease Severity in Grape Plant

Shradha Verma(✉), Anuradha Chug, and Amit Prakash Singh

University School of Information, Communication and Technology, Guru Gobind Singh Indraprastha University (GGSIP), Delhi, India
verma.shradha@gmail.com, {anuradha,amit}@ipu.ac.in

Abstract. Accurate and quantitative estimation of disease severity in plants is a complex task, even for experienced agronomists and plant pathologists, where incorrect evaluation might lead to the inappropriate use of pesticides. In this paper, the authors have utilized two Convolutional Neural Networks (CNN), namely, AlexNet and ResNet18, for assessing the disease severity in Grape plant. The images for Isariopsis Leaf Spot disease in Grape plant, also known as Leaf Blight, were taken from the PlantVillage dataset, divided into three categories of severity stages (early, middle & end) and used for training the CNN models, via the transfer learning approach. The effects of fine-tuning the hyperparameters such as mini-batch size, epochs and data augmentation were also observed and analysed. For performance comparison, measures such as classification accuracy, mean F1-score, mean precision, mean recall, validation loss, ROC curves and time taken were recorded.

Keywords: Deep learning · Convolutional neural networks (CNN) · Hyperparameter tuning · Plant diseases · Disease severity · Grape plant

1 Introduction

Agricultural productivity is severely hampered by the onset of several plant diseases, leading to huge economic as well as ecological losses. Incorrect or late diagnosis might result in improper use of agrochemicals, thereby negatively affecting the agrobiodiversity and subsequently, the crop yield [1]. Intensification of agricultural activities in recent times, have led to several novel methodologies, resulting in a significant increase in the overall agricultural output. These techniques encompass multi-disciplinary approaches, comprising of advanced computational practices such as artificial intelligence, image processing, genetic algorithms, machine learning etc., and deep learning being the latest and most successfully applied in this arena. Traditional approaches have invariably given an insight into the applications of image processing and feature extraction implementations for plant disease predictions. Pal et al. [2] studied the natural defence mechanisms of a plant and developed a SVM-based tool for identification of resistive (R) genes/proteins in plants. Camargo and Smith [3] showcased a SVM based approach, with the traditional technique of feature extraction and selection,

R. Ghazali et al. (Eds.): SCDM 2020, AISC 978, pp. 161–171, 2020.
https://doi.org/10.1007/978-3-030-36056-6_16

for identification of plant diseases. The authors concluded that textural features are crucial for correct identification of plant diseases. In traditional machine learning approaches, the performance of models is highly dependent on the cumbersome task of hand-engineered features. Author in [4, 5] employ the traditional operations for plant disease identification and demonstrate the labor-intensive processes. In the last few years, there has been a tremendous progress in the applications of deep learning in various domains, especially in agriculture such as plant disease diagnosis, fruit counting, plant species classification, weed identification etc. [6]. Deep learning provides end to end training by combining the feature extraction and classification tasks, but it's performance is proportional to the size and diversity of the dataset.

2 Background

In 2016, Mohanty et al. [7] implemented AlexNet and GoogLeNet on the PlantVillage dataset for correct identification of plant diseases, with the aim of developing a smartphone application to be used by the growers on a massive scale. They achieved a classification accuracy of 99.35%. Similarly, in 2018, Ferentinos [8] utilised a database of 87,848 images and implemented five CNN architectures for identifying the plant-disease pairs, with VGG achieving the highest classification accuracy of 99.53%. Both laboratories as well as real life cultivation images were employed for training. In 2017, Wang et al. [9] investigated the performance of four deep CNNs for assessing the disease severity in apple black rot disease. The best classification accuracy achieved was 90.4% by VGG16.

Liang et al. [10] implemented a new deep learning architecture for evaluating the plant disease severity and classification, as well as plant species identification, with ResNet50 as the base model, achieving the accuracies of 91%, 98% and 99% resp. Sladojevic et al. [11] proposed a plant disease identification model to categorize 13 distinct plant diseases and successfully discern leaves from immediate surroundings. Their model, based on Caffe, achieved a success rate of 96.3%. Verma et al. [12] implemented five CNN models to identify tomato plant diseases, utilizing the images from PlantVillage dataset, with ResNet50 achieving the best classification accuracy of 96.14%, which was used to develop a mobile application for classification of tomato plant diseases successfully. Lu et al. [13] proposed a novel CNN based model for identification of 10 rice plant diseases. The authors utilised 500 natural images of leaves as well as stems, both healthy and diseased, achieving an accuracy of 95.48%. Sun et al. [14] proposed a 26 layer deep CNN for plant identification, utilizing 10,000 natural images of 100 plant species, with a success rate of 91.78%. Similarly, Zhang et al. [15] investigated a three channel CNN architecture, whereby it exploited the RGB characteristics of a colored image to generate distinct feature maps and later fuse them into one for plant disease recognition. Toda and Okura [16] explored several visualization methods such as attention maps, hidden layer output and feature visualization etc. to examine and reveal the internal representations and processes of disease diagnosis by a CNN.

With deep learning implementations achieving promising results, choosing hyperparameters is a crucial task. Potential factors may conflict with each other; our aim is to find a balance/trade-off between them. In this paper, the authors have analyzed the relative impact and importance of three hyperparameters, namely, mini-batch size, epochs and data augmentation, on the overall performance of the two CNN models, AlexNet [17] and ResNet18 [18], for assessing the disease severity in grape leaf blight disease. The remaining paper is organized as: Sect. 2 outlines the materials and methods along with implementation details, Sect. 3 discusses the results and Sect. 4 presents the conclusion of the study.

3 Materials and Methods

3.1 Dataset

PlantVillage dataset [19] is a publicly available collection of 54,306 leaf images of 14 crops and corresponding 26 diseases, annotated and categorized into 38 distinct plant - disease pairs, accessible in three forms: colored, grayscale and segmented. Grape plant is affected by several diseases such as black measles, leaf blight, black rot etc. Disease severity, as mentioned in [20], is distinguished by the plant tissue or leaf area visibly damaged by the disease, in comparison to the total leaf surface. Figure 1 showcases the colored images of grape leaf blight disease in various stages of disease severity. Table 1 gives the overall count of images, selected manually from the PlantVillage dataset, in each category, based on visual inspection of diseased area. In the healthy category, 320 images were selected from the dataset.

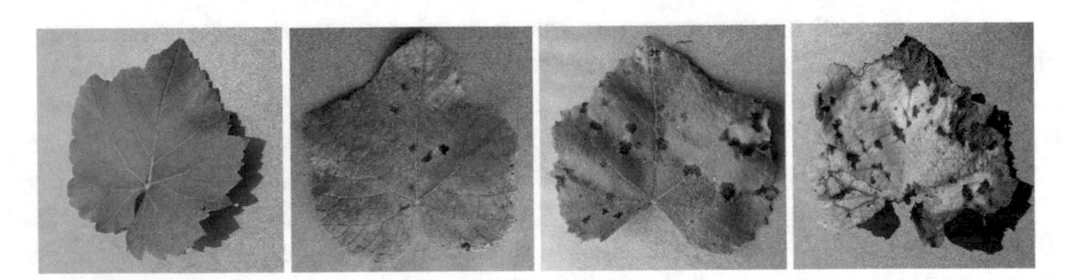

Fig. 1. Sample images of healthy, early, middle and end disease severity stages [19]

Table 1. Dataset

Categories		Total images available	No. of images (Color)
Leaf Blight/Isariopsis Leaf Spot	Early stage	1076	314
	Middle stage		367
	End stage		292

3.2 CNN Architectures

AlexNet, designed by Alex Krizhevsky, consisting of five convolutional, followed by three fully connected layers, was the first large-scale and deep CNN, also the winner of

ILSVRC 2012. It utilized Rectified Linear Unit (ReLU) for the very first time, to introduce non-linearity, instead of tanh or sigmoid, resulting in a significant increase in the training speed. Tanh and sigmoid were affected by the vanishing gradient problem, which was not the case with ReLU. Hence, now ReLU, a non-saturating activation, is a standard for most deep learning architectures. To overcome overfitting and obtain a generalised model, data augmentation (cropping, rotation, flips) was employed and 50% dropout was implemented after every fully connected layer. The network was split in two parts and trained continuously for five to six days on two parallel GTX 580 GPUs for classification of 1.2 million images into 1000 classes, achieving a top-5 error rate of 15.3%. ResNet18 is an 18-layer version of ResNets, or Residual Networks, winner of ILSVRC 2015, employed for several computer vision implementations. Before residual networks were introduced, deeper networks were affected by the vanishing gradient problem. ResNet employs 'identity skip connections', where the original input is added element-wise with the output of the convolution block, followed by ReLU, providing an alternate path for the gradient to pass, while also ensuring that the performance of higher layer is similar to the lower layer. These residual modules are stacked together, followed by the final branch of global average pooling, fully connected and softmax layers. ResNet18 has comparably lower number of parameters than AlexNet. More about the architectures can be found in [17, 18]. Table 2 lists a few details of the implemented CNN models.

Table 2. CNN architectures implemented [21]

	Parameters (Millions)	Size (MB)	Layers	Depth	Image input size
AlexNet	61	227	25	8	227 × 227
ResNet18	11.7	44	72	18	224 × 224

3.3 Implementation

The implementation was performed on a system equipped with GTX1060 6 GB GPU, using MATLAB R2018b. The dataset was loaded as the image datastore and partitioned into the train-validation sets with 80–20 split. Thereafter, a pre-trained CNN network was loaded for training on the dataset. In order to retrain a network on a different dataset to classify new images, the final learnable and classification layers are replaced with new layers, so that the new fully connected layer has the number of outputs same as the number of class labels in the new dataset. As the dataset being utilized is small comparatively, the weights in the initial ten layers for AlexNet and twenty layers for ResNet18, were frozen, hence, during training these weights weren't updated, leading to increased training speed, while also preventing these layers from overfitting to new images. Table 3 lists the fixed as well as variable hyperparameters used for training the CNN architectures. For comparison purposes, performance metrics namely accuracy, mean F1-score, mean precision, mean recall, validation loss, ROC curves and time taken were observed.

Table 3. Hyperparameters used for training

Fixed hyperparameters	
Algorithm for training	SGDM
Momentum	0.9
Initial learning rate	0.001
Learn rate drop factor	0.2
Learn rate drop period	Every 5 epochs
L2Regularization	1.0000e-04
Shuffle	Every Epoch
Freeze weights upto	10 layers for AlexNet 20 layers for ResNet18
Train-validation split	80:20
Tested Hyperparameters	
Epochs	10, 30 & 50
Mini batch size	16, 32, 64 & 128
Augmentation	With augmentation &without augmentation

4 Results and Discussion

The focus of this study was to observe the effects of fine-tuning three hyperparameters, mini-batch size, epochs and data augmentation, for disease severity estimation of grape leaf blight disease. The resultant performance measures, i.e. classification accuracy, mean F1-score, mean precision, mean recall, validation loss, ROC curves and time taken were recorded, as shown in Tables 4 and 5, and demonstrated in Figs. 2, 3, 4, 5 and 6. The classification accuracies observed for AlexNet were between 86.05%–90.31%, and 81.4%–87.6% for ResNet18. The highest accuracies achieved for AlexNet and ResNet18 are highlighted in the tables, vertically indicating the best epoch and horizontally indicating the best mini-batch size. Clearly, with augmentation, AlexNet has performed best with 30 epochs and 32 mini-batch size, while ResNet18 resulted in better performance measures with 10 epochs and 64 mini-batch size. Without the use of augmented dataset for training, it was assumed that classification accuracies would be lower, but that was not observed. AlexNet achieved the highest accuracy of 90.31%, without data augmentation. However, in few instances the models seemed to be overfitting, as training accuracies reached 100% but lower accuracies were recorded on the validation set. In case of AlexNet, without data augmentation, validation loss increased significantly, for all observations. As for ResNet18, the validation loss is decreasing with increasing mini-batch sizes. With increasing epochs and decreasing mini-batch sizes, training time increased notably. Also, time taken for training is slightly higher, in case of augmentation. While training the CNN models, at times it was noted that after 33–35 epochs, the validation accuracy remained flat, concluding that for training pre-trained networks to classify new images, larger epochs aren't necessary. Also, size of the mini-batch is limited by the total available images and GPU memory. It has a notable effect on training speed and time.

Table 4. Performance measures for AlexNet

AlexNet (With augmentation)

Mini-batch size	10 Epochs		30 Epochs		50 Epochs	
	Accuracy% {Mean F-score, mean precision, mean recall, loss}	Time taken	Accuracy% {Mean F-score, mean precision, mean recall, loss}	Time taken	Accuracy% {Mean F-score, mean precision, mean recall, loss}	Time taken
16	87.21%$_{\{0.8755,0.8718,0.8792,0.3606\}}$	249 s	**87.60%**$_{\mathbf{\{0.8804,0.8771,0.8837,0.3829\}}}$	734 s	86.43%$_{\{0.8702,0.8657,0.8747,0.3615\}}$	1195 s
32	**87.98%**$_{\mathbf{\{0.8832,0.8827,0.8838,0.3681\}}}$	150 s	**89.53%**$_{\mathbf{\{0.8982,0.8961,0.9003,0.4112\}}}$	383 s	**88.37%**$_{\mathbf{\{0.8912,0.8890,0.8933,0.3201\}}}$	666 s
64	86.43%$_{\{0.8682,0.8670,0.8694,0.3747\}}$	82 s	**88.76%**$_{\mathbf{\{0.8921,0.8893,0.8950,0.3100\}}}$	245 s	87.60%$_{\{0.8798,0.8764,0.8832,0.3077\}}$	401 s
128	86.82%$_{\{0.8733,0.8698,0.8768,0.3276\}}$	63 s	**89.15%**$_{\mathbf{\{0.8945,0.8911,0.8980,0.3493\}}}$	184 s	87.98%$_{\{0.8846,0.8862,0.8831,0.3753\}}$	307 s

AlexNet (without augmentation)

Mini-Batch Size	**10 Epochs**		**30 Epochs**		**50 Epochs**	
	Accuracy%$_{\{\text{Mean F-score, Mean Precision, Mean Recall, Loss}\}}$	Time taken	Accuracy%$_{\{\text{Mean F-score, Mean Precision, Mean Recall, Loss}\}}$	Time taken	Accuracy%$_{\{\text{Mean F-score, Mean Precision, Mean Recall, Loss}\}}$	Time taken
16	87.21%$_{\{0.8762,0.8763,0.8761,0.5380\}}$	239 s	86.05%$_{\{0.8640,0.8628,0.8652,0.5418\}}$	713 s	84.88%$_{\{0.8551,0.8593,0.8509,0.5877\}}$	1173 s
32	**89.15%**$_{\mathbf{\{0.8955,0.8963,0.8947,0.4537\}}}$	127 s	**87.98%**$_{\mathbf{\{0.8827,0.8816,0.8838,0.5388\}}}$	374 s	**90.31%**$_{\mathbf{\{0.9059,0.9057,0.9061,0.5519\}}}$	623 s
64	86.82%$_{\{0.8733,0.8753,0.8713,0.4757\}}$	77 s	86.82%$_{\{0.8711,0.8681,0.8742,0.5301\}}$	225 s	**89.15%**$_{\mathbf{\{0.8943,0.8946,0.8940,0.4506\}}}$	379 s
128	87.21%$_{\{0.8754,0.8746,0.8762,0.5281\}}$	62 s	86.82%$_{\{0.8715,0.8675,0.8757,0.3715\}}$	179 s	**88.76%**$_{\mathbf{\{0.8906,0.8884,0.8928,0.4195\}}}$	302 s

Table 5. Performance measures for ResNet18

ResNet18 (With augmentation)						
Mini-Batch Size	10 Epochs		30 Epochs		50 Epochs	
	Accuracy% {Mean F-score, Mean Precision, Mean Recall, Loss}	Time taken	Accuracy% {Mean F-score, Mean Precision, Mean Recall, Loss}	Time taken	Accuracy% {Mean F-score, Mean Precision, Mean Recall, Loss}	Time taken
16	**85.66% {0.8606,0.8586,0.8625,0.8046}**	378 s	83.72% {0.8437,0.8453,0.8422,0.7045}	1073 s	83.33% {0.8395,0.8403,0.8387,0.6969}	1771 s
32	**85.66% {0.8632,0.8642,0.8621,0.5377}**	218 s	83.72% {0.8428,0.8435,0.8422,0.6545}	611 s	82.95% {0.8355,0.8346,0.8364,0.5482}	993 s
64	**87.60% {0.8827,0.8818,0.8836,0.4217}**	155 s	**86.82% {0.8737,0.8736,0.8739,0.4226}**	402 s	**85.66% {0.8639,0.8631,0.8647,0.4302}**	659 s
128	**86.43% {0.8696,0.8699,0.8694,0.3706}**	112 s	85.66% {0.8627,0.8617,0.8636,0.3630}	328 s	85.27% {0.8585,0.8591,0.8580,0.3644}	558 s
ResNet18 (Without augmentation)						
Mini-Batch Size	**10 Epochs**		**30 Epochs**		**50 Epochs**	
	Accuracy% {Mean F-score, Mean Precision, Mean Recall, Loss}	Time taken	Accuracy% {Mean F-score, Mean Precision, Mean Recall, Loss}	Time taken	Accuracy% {Mean F-score, Mean Precision, Mean Recall, Loss}	Time taken
16	82.95% {0.8382,0.8441,0.8323,0.7469}	347 s	81.40% {0.8277,0.8426,0.8133,0.8908}	1038 s	**83.72% {0.8454,0.8513,0.8396,0.7698}**	1733 s
32	83.33% {0.8397,0.8414,0.8379,0.5233}	220 s	82.95% {0.8371,0.8416,0.8327,0.5245}	596 s	**86.05% {0.8684,0.8745,0.8623,0.4491}**	991 s
64	84.88% {0.8553,0.8594,0.8513,0.3574}	131 s	**85.66% {0.8640,0.8691,0.8589,0.4333}**	390 s	**86.05% {0.8680,0.8738,0.8623,0.3907}**	653 s
128	**85.66% {0.8624,0.8643,0.8606,0.3642}**	111 s	84.88% {0.8554,0.8588,0.8520,0.3445}	334 s	85.27% {0.8594,0.8635,0.8554,0.3495}	546 s

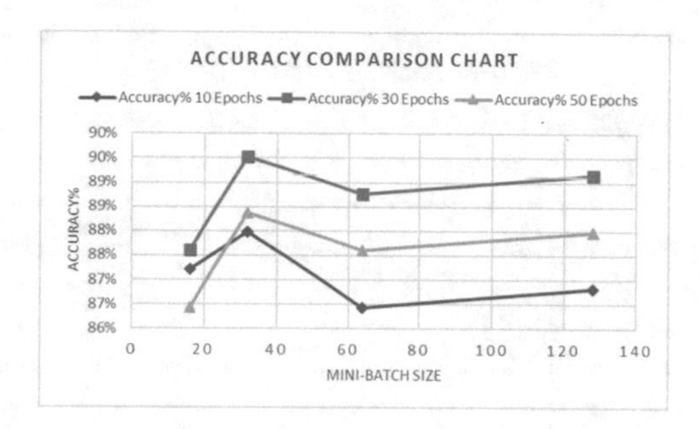

Fig. 2. Accuracy comparison chart for AlexNet (with augmentation)

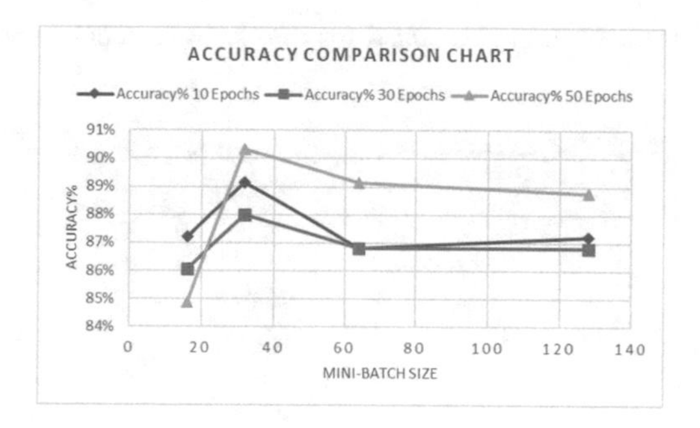

Fig. 3. Accuracy comparison chart for AlexNet (without augmentation)

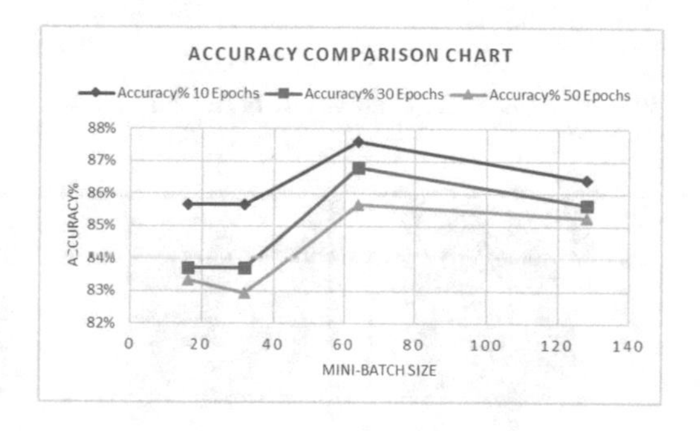

Fig. 4. Accuracy comparison chart for ResNet18 (with augmentation)

Assessment of the disease severity requires fine-grained classification, in order to differentiate between adjoining severity stages. Figure 6 showcases the ROC curves for AlexNet and ResNet18, clearly demonstrating that mostly, the middle stage images are the ones misclassified, while healthy and end stages are correctly classified. Figures 7 and 8 demonstrate a few samples of category-wise classification for both CNN models.

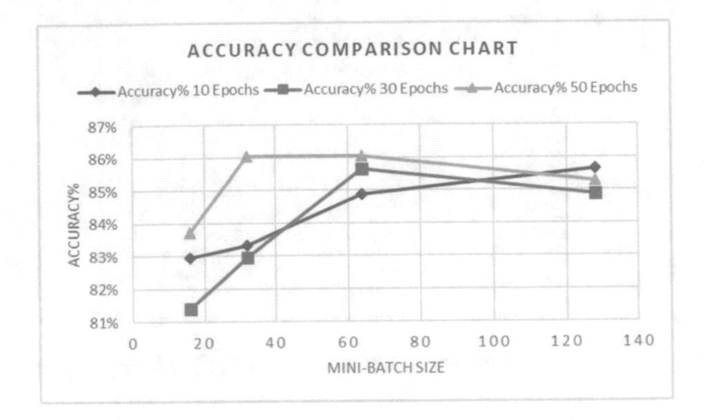

Fig. 5. Accuracy comparison chart for ResNet18 (without augmentation)

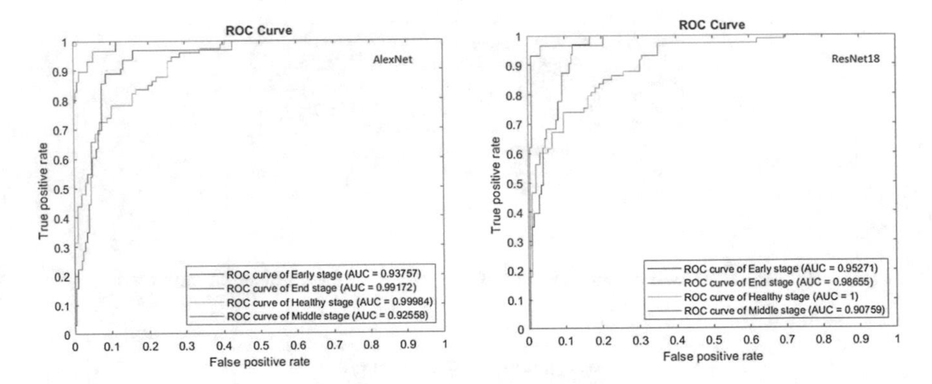

Fig. 6. Sample ROC curves of both CNN models

Fig. 7. Sample images of category-wise classification results for AlexNet

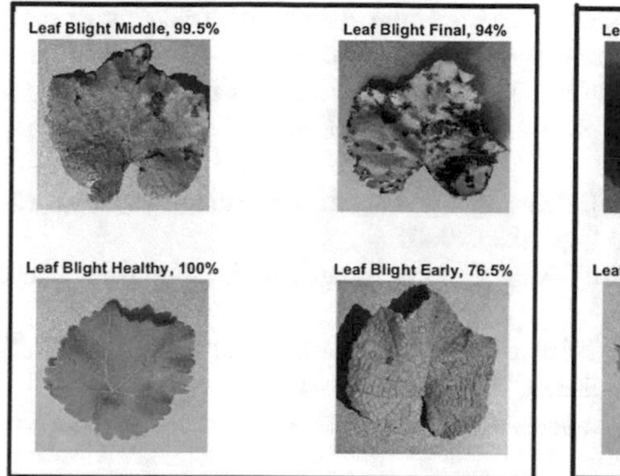

Fig. 8. Sample images of category-wise classification results for ResNet18

5 Conclusion

In this study, two pre-trained CNN architectures, AlexNet and ResNet18, were implemented via transfer learning, to evaluate the disease severity of grape leaf blight disease and effects of tuning three hyperparameters were observed. The diseased as well as healthy leaf images (a total of 1293 images) were utilized from the PlantVillage dataset. The diseased images were categorized into three severity stages (early, middle and end), and impact of varying the mini-batch size, epochs and data augmentation (with/without) were analyzed, for both CNN models. The highest classification accuracy of 90.31% was achieved by AlexNet, with 32 mini-batch size, 50 epochs and without augmentation. The highest classification accuracy given by ResNet18 was 87.6%, with 64 mini-batch size and 10 epochs, with augmented dataset. The effect of varying the mini-batch size and epochs on the performance measures and time taken could be clearly noted. But, there was no clear correlation observed with the use augmented dataset. However, without augmentation, in few instances, the models seemed to be overfitting. With augmented dataset, AlexNet and ResNet18 performed best for 30 and 10 epochs respectively, on all mini-batch sizes. Similarly, AlexNet and ResNet18 resulted in best classification accuracies with 32 and 64 mini-batch sizes respectively, trained for all epochs.

Acknowledgments. This study was supported by Department of Science and Technology (DST), Government of India, New Delhi, under Interdisciplinary Cyber Physical Systems (ICPS) Programme (Reference No. T-319).

References

1. Giomi T, Runhaar P, Runhaar H (2018) Reducing agrochemical use for nature conservation by Italian olive farmers: An evaluation of public and private governance strategies. International journal of agricultural sustainability. 16(1):94–105
2. Pal T, Jaiswal V, Chauhan RS (2016) DRPPP: A machine learning based tool for prediction of disease resistance proteins in plants. Comput Biol Med 78:42–48

3. Camargo A, Smith JS (2009) Image pattern classification for the identification of disease causing agents in plants. Comput Electron Agric 66(2):121–125
4. Al Bashish D, Braik M, Bani-Ahmad S (2011) Detection and classification of leaf diseases using K-means-based segmentation and Neural Networks-based Classification. Inf Technol J 10(2):267–275
5. Camargo A, Smith JS (2009) An image-processing based algorithm to automatically identify plant disease visual symptoms. Biosys Eng 102(1):9–21
6. Kamilaris A, Prenafeta-Boldú FX (2018) Deep learning in agriculture: A survey. Comput Electron Agric 147(February):70–90
7. Mohanty SP, Hughes DP, Salathé M (2016) Using Deep Learning for Image-Based Plant Disease Detection. Frontiers in Plant Science. 7(September):1–10
8. Ferentinos KP (2018) Deep learning models for plant disease detection and diagnosis. Comput Electron Agric 145:311–318
9. Wang, G., Sun, Y., & Wang, J.: Automatic Image-Based Plant Disease Severity Estimation Using Deep Learning. Computational Intelligence and Neuroscience. 1–8, (2017)
10. Liang Q, Xiang S, Hu Y, Coppola G, Zhang D, Sun W (2019) PD2SE-Net: Computer-assisted plant disease diagnosis and severity estimation network. Comput Electron Agric 157:518–529
11. Sladojevic, S., Arsenovic, M., Stefanovic, D., Anderla, A., Culibrk, D.: Deep Neural Networks Based Recognition of Plant Diseases by Leaf Image Classification. Computational Intelligence and Neuroscience. 1–11, (2016)
12. Verma, S., Chug, A., Singh, A. P., Sharma, S., Rajvanshi, P.: Deep Learning-Based Mobile Application for Plant Disease Diagnosis: A Proof of Concept With a Case Study on Tomato Plant. In Applications of Image Processing and Soft Computing Systems in Agriculture. 242–271, IGI Global, (2019)
13. Lu Y, Yi S, Zeng N, Liu Y, Zhang Y (2017) Identification of rice diseases using deep convolutional neural networks. Neurocomputing. 267:378–384
14. Sun Y, Liu Y, Wang G, Zhang H (2017) Deep learning for plant identification in natural environment. Comput Intell Neurosci 2017:1–6
15. Zhang S, Huang W, Zhang C (2019) Three-channel convolutional neural networks for vegetable leaf disease recognition. Cogn Syst Res 53:31–41
16. Toda Y, Okura F (2019) How Convolutional Neural Networks Diagnose Plant Disease. Plant Phenomics. 2019:9237136
17. Krizhevsky, A., Sutskever, I., Hinton, G. E.: Imagenet classification with deep convolutional neural networks. In Advances in neural information processing systems. 1097–1105, (2012)
18. He, K., Zhang, X., Ren, S., Sun, J.: Deep residual learning for image recognition. In Proceedings of the IEEE conference on computer vision and pattern recognition. 770–778, (2016)
19. Hughes, D. P., Salathe, M.: An open access repository of images on plant health to enable the development of mobile disease diagnostics. ArXiv Preprint. https://doi.org/10.1111/1755-0998.12237, (2015)
20. Campbell, C. L., Neher, D. A.: Estimating disease severity and incidence. In Epidemiology and management of root diseases, pp 117–147. Springer, Berlin, Heidelberg, (1994)
21. https://www.mathworks.com/help/deeplearning/ug/pretrained-convolutional-neural-networks.html. Last accessed: May 2019

Multi-modal Feature Based for Phonocardiogram Signal Classification Using Autoencoder

Muhamad Fathurahman(✉), Ummi Azizah Rachmawati, and Sri Chusri Haryanti

Fakultas Teknologi Informasi, Teknik Informatika, Universitas YARSI, Jakarta, Indonesia
muhammad.fathurrachman@yarsi.ac.id

Abstract. Phonocardiogram classification plays an important rule in the diagnosis of heart disease. It can be used in selecting a proper treatment to the patients. However, automated PCG classification has many issues. One of the important issues is the feature extraction process. It is difficult to extract relevant features from PCG signals due to some noises that corrupt the original signal. The noises are included murmur, intestine and breathing sounds. To overcome this problem, several works have been proposed such as performing segmentation on PCG signals before the feature extraction process, using many types of signal features including wavelet, mfcc, spectral, time-frequency and statistical features. These types of features experimentally affect the classification accuracy of PCG signals. This study proposes a feature fusion based using an autoencoder model in order to obtained new repfresentation features. The result shows that this approach provides a competitive result of PCG classification compare to those of the baseline methods.

Keywords: PCG classification · Autoencoder · Feature fussion · Heart sounds

1 Introduction

Phonocardiogram (PCG) Signal classification plays an important rule in the diagnosis of heart disease. It can be used to help the clinician in order to determine the appropriate treatment for the patients. Nevertheless, the automated PCG classification is difficult to apply because of the complexity of PCG signals such as the PCG signal contains other sounds including murmur, intestine and breathing sounds. Theses noises make the identification of Fundamental Heart Sound (FHS) which are S1 and S2 are difficult to find. To overcome these problems, Many segmentation methods are proposed such as Shannon Energy et al. [1] Short Time Fourier Transformation and Wavelet [2], and recently, Logistic-Regression Hidden Semi-Markov Model is introduced by [3]. These segmentation methods are commonly used as a preprocessing method before the feature extraction phase. Another issue related to the PCG Classification is the feature extractions. The Feature extraction in the context of PCG classification is a process to find good signal representation that can be fed to a

R. Ghazali et al. (Eds.): SCDM 2020, AISC 978, pp. 172–180, 2020.
https://doi.org/10.1007/978-3-030-36056-6_17

classification algorithm. Recently many feature extraction methods are used to extract features from PCG Signal including wavelet features [4], deep structured features of CNN [5], spectral features [6], Mel Frequency Cepstral Coefficient (MFCC [7], time-frequency and statistical features [8]. Most of the methods above are difficult to extract features from signals that have varying duration, therefore signal preprocessing is needed. Zhang et al. proposed scaled-spectrogram [9] to handle the problem in extracting features from varying duration of PCG signals. Zhang's approach [9] improves the performance of the PCG classifications. However, the classification of the PCG signal remains a challenge because of its laborious task to perform feature extraction in the complex PCG signal.

Recently, a new architecture of Neural Network called Autoencoder is used to learn the representation of raw data. This approach has widely used in speech recognition including dysarthric speech recognition [10], Speech-Emotion recognition [11], Speaker Recognition [12]. In this paper, we proposed a new schema for PCG signal feature extractions. The neural networks based Autoencoder is used to learn from MFCC and wavelet features of the PCG signals in order to produce the new representation features which improve the classification accuracy. This research aims to evaluate the performance of PCG classification based on the feature obtained from Autoencoder.

The rest of the paper is arranged as follows, Sects. 2 present the framework, methods and materials in order to perform PCG classification. Section 3 provides the results and discussions. The Conclusion and Future work are provided in Sect. 4.

2 Methods and Materials

2.1 Dataset

This study used two benchmark PCG datasets including PASCAL-A and PASCAL-B which are available online on-site [13]. The PASCAL-A contains four major categories which are normal, murmur, extra sound and artifact whereas the PASCAL-B dataset only has three major categories which are normal, murmur and extrasystole. The duration of each signal is varied which is approximately 1–30 s [14]. Both datasets are divided into training and testing data (Table 1).

Table 1. PCG datasets [9, 14].

No	Dataset	Class category	Training	Testing
1	PASCAL A	Normal	31	14
		Murmur	34	14
		Extrasound	19	8
		Artifact	40	16
2	PASCAL B	Normal	200	136
		Murmur	66	39
		Extrasystole	46	20

2.2 Classification Framework

The proposed framework of the PCG classification is comprised of several steps. The detailed framework can be seen in Fig. 1. below.

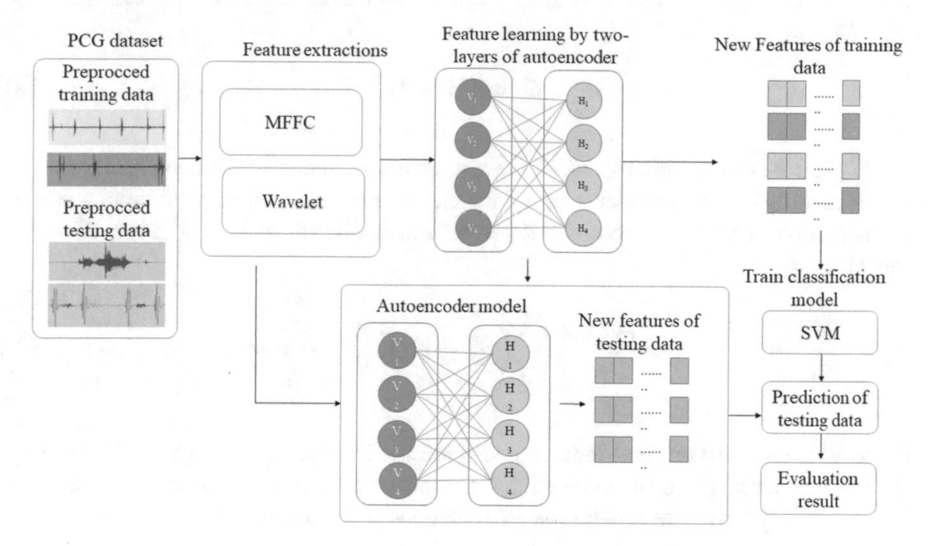

Fig. 1. The proposed framework of PCG classification.

All the training and testing signals are preprocessed that follow the step from [9] including normalization [9] and filtered by the 6-order-Butterworth filter to reduce noise [9]. Then, the preprocessed training and testing signals are extracted separately using both Mel Frequency Cepstral Coefficient (MFCC) and Wavelet to obtain features. In the next step, the MFCC and Wavelet features from the training signal are fed into 2-layers of the autoencoder architecture. Then the autoencoder architecture is trained iteratively for learning new representation features from both MFCC and Wavelet features. After the training of autoencoder architecture, then this autoencoder model is used to obtained new representation features both training and testing data. The new representation features of training data are used to train a support vector machine classifier (SVM), whereas the new representation features of testing data are used to evaluate the classification result of the SVM classifier.

2.3 Autoencoder

Autoencoder is one of the unique types of neural network architecture that uses to learn an input representation [15]. This architecture has two main phases which are encoder $\mathbf{V}$ and decoder $\mathbf{O}$ [16, 15]. In the encoder phase, the input $\mathbf{x}$ is learned through the $\mathbf{V}$ layer according to the following formula [15]

$$V = sigm(W_1 x + b_1) \tag{1}$$

$\mathbf{W_1}$ and $\mathbf{b_1}$ represent the matrix of weights and bias vector at the encoder layer. In the decoder phase, the reconstruction of an original input **x** is estimated using the following term [15, 16]

$$O = W_2 V + b_2 \tag{2}$$

The $\mathbf{W_2}$ and $\mathbf{b_2}$ represent a matrix of weights and bias vector at the decoder layer. The goal of training an autoencoder is to reduce the error between an original input **x** and its reconstruction of **x'**. Then the optimal parameters can be set according to the following term [15, 16].

$$\theta = \frac{argmin}{\theta} \sum_{I=1}^{M} \left(x_2^{(i)} - x^{(i)} \right)^T \left(x_2^{(i)} - x^{(i)} \right) \tag{3}$$

Where V^i is equal to original input **x** and $\boldsymbol{\theta}$ are network parameters $\{\mathbf{W_1}, \mathbf{W_2}, \mathbf{b_1}, \mathbf{b_2}\}$. The optimal parameters of autoencoder can be estimated using Stochastic Gradient Descent (SGD) [15]. The illustration of autoencoder architecture can be seen in Fig. 2 below.

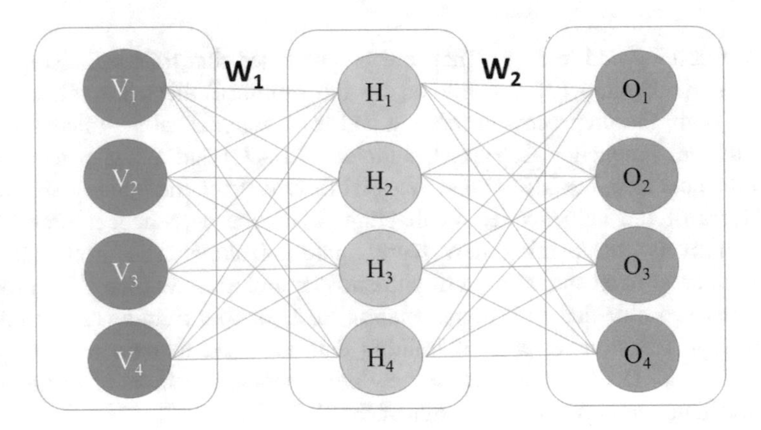

Fig. 2. Autoencoder architecture.

In this study, we evaluate the two-layer of hidden layers with each layer have the same number of hidden units. The total layer of its architecture is four including the input (first encode phase) and output (last decode phase) layer. the detail architectures are provided in Table 2 below.

Table 2. Autoencoder architecture uses in this study.

No	Name of autoencoder archictectures	Hidden layer	Hidden unit
1	AE1	2	250
2	AE2	2	300

2.4 Feature Extraction

2.4.1 MFCC Features

Mel Frequency Cepstral Coefficient (MFCC) is one of the common features that is used in audio analysis. MFCC features can capture the valuable information in an audio or a signal. The MFCC feature can be obtained through the following steps [17].

1. The signal is divided into small-piece of windowing (Hamming Window)
2. Fast Fourier Transform is applied for each window
3. Mapping into a mel-scale form using triangular overlapping windows.
4. Calculate the Log Score for each mel-frequency
5. Then, the Discrete Cosine Transform is applied to extract mel-cepstral coefficients.

The further details of the equations that use in these steps can be seen in [17]. In this study, we use Matlab built-in function [18] to automatically extract the MFCC feature in PCG signals.

2.4.2 Wavelet Features

In this paper, the maximal overlap discrete wavelet transform is used to decompose the PCG signal. On an account of the 6th-order Daubechies wavelet has an identical pattern to the PCG signal, it was used in this experiment [9, 19]. The features from wavelet are the unbiased estimates of the wavelet variance by scale [22]. These features are obtained by using the Matlab built-in function [20, 21].

2.5 Classification Method

In the classification phase, Multi-class Support Vector Machine (SVM) is used to classify the PCG signals based on the new feature representation that is obtained by Autoencoder Model. The parameters settings of the SVM model are configured as follows. Kernel Method = 'Polynomial' with the degree of the polynomial is set to 2. This SVM model is provided by the Matlab [22] built-in function.

2.6 Evaluation Methods

To evaluate the classification results, we evaluate the PCG signal category in both dataseta using the precision, sensitivity, specificity, and Youden Index values. In the case of the PASCAL-A dataset, the F-Score is used to measure the overall performance [23]. Besides, The Discriminant Power is also used for measuring the heart problem categories [9, 23]. All the measurement above can be defined as follows

$$Precision = \frac{TP}{TP + FP} \quad (4)$$

$$Sensitivity = \frac{TP}{TP + FN} \quad (5)$$

$$Specificity = \frac{TN}{FP_{TN} + TN} \quad (6)$$

$$F - Score = \frac{(\beta^2 + 1) + P_r S_e}{\beta^2 + P_r S_e} \quad (7)$$

$$Discriminant\ Power = \frac{\sqrt{3}}{\pi}\left(log\frac{S_e}{1 - S_e} + log\frac{S_e}{1 - S_e}\right) \quad (8)$$

$$Youden\ Index = S_e - (1 - S_p) \quad (9)$$

3 Results and Discussions

3.1 Result in Dataset PASCAL-A

This dataset contains four-different class categories including normal, murmur, extra sound and artifact. to evaluate the classification result, we evaluate the precision for each class categories, artifact sensitivity, specificity and Youden Index of the artifact. then F-Score and total precision are used to evaluate the overall performance. The result can be seen in Table 3. In overall performance, AE2 has a better precision result of normal (0.62), extra sound (0.58) and artifact (0.92) compare to those of AE1. Also, AE2 produces the same result with AE1 in the precision of murmur (0.60), artifact sensitivity (0.94) and F-Score (0.29). Beside that, AE2 has a higher score of total precision compare to that of AE1. In comparison to the previous research [9], both AE1 and AE2 have lower performance in almost all evaluation criterions compare to those of Scaled Spectrogram-Partial Least Square Regression (SS-PLSR) that is proposed by [9]. Our approach only outperforms this method in the precision of extra sound score which achieves 0.57 (AE1) and 0.58 (AE2). Equally important, the AE2 produces a high score of the precision of extra sound, precision of artifact and F-Score which has a little difference with baseline methods (SS-Method and SS-PLSR).

Table 3. Comparison with previous research [9] on Dataset PASCAL A

Evaluation criteria	SS-METHOD	SS-PLSR	AE1	AE2
Precision of normal	**0.67**	0.60	0.60	0.62
Precision of murmur	**0.91**	**0.91**	0.60	0.60
Precision of extra sound	0.37	0.44	0.57	**0.58**
Precision of artifacts	0.76	**0.94**	0.91	0.92
Artifact sensitivity	**1.00**	**1.00**	0.94	0.94
Artifact specificity	0.58	**0.64**	0.58	0.60
Youden index of artifact	0.58	**0.64**	0.52	0.54
F-score heart problem	0.28	**0.30**	0.29	0.29
Total precision	2.71	**2.89**	2.67	2.73

3.2 Result in Dataset PASCAL-B

Differ with the previous dataset, this dataset only contains three major categories which are Normal, Murmur and Extrasystole. To evaluate the performance in this dataset, the precision score for all categories is evaluated. Also, as referred to [9, 14] the discriminant power is also utilized to evaluate the heart problem categories. The experiment results in this dataset can be seen in Table 4 below.

Table 4. Comparison with previous research [9] on dataset PASCAL B

Evaluation criteria	SS-METHOD	SS-PLSR	AE1	AE2
Precision of normal	0.74	0.76	**0.81**	**0.81**
Precision of murmur	**0.66**	0.65	0.59	0.60
Precision of extrasystole	0.24	0.33	0.12	0.12
Heart problem sensitivity	0.24	0.34	**0.55**	**0.55**
Heart problem specificity	0.89	**0.90**	0.81	0.82
Youden index of heart problem	0.13	0.24	**0.36**	**0.36**
Discriminant power	0.24	0.36	0.39	**0.40**
Total precision	1.57	**1.75**	1.53	1.53

Based on Table 3. Both AE1 and AE2 produce the same score of precision of normal (0.81), heart problem sensitivity (0.55), precision of extrasystole (0.12), Youden Index of heart problem (0.36) and total precision (1.53). The AE2 achieves a better score of precision of murmur (0.60), heart problem specificity (0.82) and discriminant power (0.40) compare to that of AE1. In comparison with the previous result [9]. AE2 outperforms the performance of method SS-PLSR in evaluation criteria of precision of normal (0.81 vs 0.76), heart problem sensitivity (0.55 vs 0.34), Youden Index of heart problem (0.36 vs 0.24), discriminant power (0.36 vs 0.40). However, the SS-Method and SS-PLSR have superior performance in precision of murmur, heart problem specificity, and total precision.

4 Conclusion and Future Work

This study proposed the PCG classification based on learned features of multi-modal features through an autoencoder architecture. This approach has successfully classified the PCG signal with a competitive result in both datasets compare to the baseline methods [9]. In the PASCAL-A dataset, our approach has lower performances in almost all evaluation criteria except in the score of precision of murmur. Another fact is that our approach achieves a high score of evaluation of artifact category which has a little different score with baseline methods. Similarly, In the PASCAL-B dataset, the performance of our approach has a better evaluation score in the precision of normal, heart problem sensitivity, Youden Index of heart problem and discriminant power. This study shows that the use of learned features of multi-modal features through autoencoder architecture can improve the several scores of the evaluation criteria of the PCG classification. In future work, some feature extraction methods will be used to be learned through autoencoder architecture. Besides, the deep architecture of an autoencoder and pre-trained deep of neural network model will be evaluated for PCG classification.

Acknowledgments. This research is supported by Internal Research Grant of Universitas YARSI Number: 0007.2/FTI/ST-PN.00/I/2019.

References

1. Saini M (2016) Proposed algorithm for implementation of Shannon energy envelope for heart sound analysis. 7109(3):15–19
2. Deng Y, Bentley P, A robust heart sound segmentation and classification algorithm using wavelet decomposition and spectrogram. Peterjbentley.Com
3. Springer DB, Tarassenko L, Clifford GD (2016) Logistic regression-HSMM-based heart sound segmentation. IEEE Trans Biomed Eng 63(4):822–832
4. Imani M, Ghassemian H (2016) Curve fitting, filter bank and wavelet feature fusion for classification of PCG signals. In: 2016 24th Iran. Conf. Electr. Eng. ICEE 2016, pp 203–208
5. Tschannen M, Kramer T, Marti G, Heinzmann M, Wiatowski T (2016) Heart sound classification using deep structured features. Comput Cardiol Conf (CinC), 2016
6. Leal A et al (2018) Noise detection in phonocardiograms by exploring similarities in spectral features. Biomed Signal Process Control 44:154–167
7. Cheng X, Sun K, Zhang X, She C (2016) Feature extraction and recognition methods based on phonocardiogram, pp 87–92
8. Randhawa SK, Singh M (2015) Classification of heart sound signals using multi-modal features. Procedia Comput Sci 58:165–171
9. Zhang W, Han J, Deng S (2017) Heart sound classification based on scaled spectrogram and partial least squares regression. Biomed Signal Process Control 32:20–28
10. Vachhani B, Bhat C, Das B, Kopparapu SK (2017) Deep autoencoder based speech features for improved dysarthric speech recognition. In: Proceedings annual conference international speech communication association INTERSPEECH, pp 1854–1858
11. Deng J, Zhang Z, Eyben F, Schuller B (2014) Autoencoder-based unsupervised domain adaptation for speech emotion recognition. IEEE Signal Process Lett 21(9):1068–1072

12. Oldˇrich Plchot PM, Burget L, Aronowitz H (2016) Audio enhancing with DNN autoencoder for speaker recognition, pp 5090–5094
13. Peter Bentley RG, Nordehn G, Coimbra M, Mannor S (2012) Classifying heart sounds challenge. [Online]. Available: http://www.peterjbentley.com/heartchallenge/. Accessed 09 Apr 2019
14. Bentley P, Nordehn G, Coimbra M, Mannor S (2011) The PASCAL classifying heart sound challenge 2011 results. [Online]. Available: http://www.peterjbentley.com/heartchallenge/#aboutdata. Accessed 16 Jun 2019
15. Guorong W, Dinggang S, Mert SR (2016) Machine learning and medical imaging. Elsevier
16. Brosch T, Yoo Y, Tang LYW, Tam R (2016) Deep learning of brain images and its application to multiple sclerosis, no. 1. Elsevier Inc
17. Atibi M, Atouf I, Boussaa M, Bennis A (2016) Comparison between the MFCC and DWT applied to the roadway classification. In: Proceedings—CSIT 2016 2016 7th international conference on computer science information technology, pp 0–4
18. Extract mfcc, log energy, delta, and delta-delta of audio signal—MATLAB mfcc. [Online]. Available: https://www.mathworks.com/help/audio/ref/mfcc.html. Accessed 16 Jun 2019
19. Kumar D, Carvalho P, Antunes M, Paiva RP, Henriques J (2011) An adaptive approach to abnormal heart sound segmentation, pp 661–664
20. Maximal overlap discrete wavelet transform—MATLAB modwt. [Online]. Available: https://www.mathworks.com/help/wavelet/ref/modwt.html. Accessed 19 Jun 2019
21. Multiscale variance of maximal overlap discrete wavelet transform—MATLAB modwtvar. [Online]. Available: https://www.mathworks.com/help/wavelet/ref/modwtvar.html#buytgpj-1. Accessed 19 Jun 2019
22. Matlab Documentation, Support Vector Machine Classification—MATLAB & Simulink (2019). [Online]. Available: https://www.mathworks.com/help/stats/support-vector-machine-classification.html. Accessed 09 Apr 2019
23. Marques N, Almeida R, Rocha AP, Coimbra M (2013) Exploring the stationary wavelet transform detail coefficients for detection and identification of the S1 and S2 heart sounds. Comput Cardiol Conf (CinC), 891–894

Implementation of 1D-Convolution Neural Network for Pneumonia Classification Based Chest X-Ray Image

Muhamad Fathurahman(✉), Sri Cahya Fauzi, Sri Chusri Haryanti, Ummi Azizah Rahmawati, and Elan Suherlan

Teknik Informatika, Fakultas Teknologi Informasi, Universitas YARSI, Jakarta, Indonesia
muhammad.fathurrachman@yarsi.ac.id

Abstract. Pneumonia is an infectious disease that attacks the lungs, causing the air sacs in the lungs to become inflamed and swollen. Pneumonia is caused by fungi, bacteria, and viruses. Pneumonia can affect anyone, including children. The most successful type of method for analyzing images to date is the Convolutional Neural Network (CNN). The Convolutional Neural Network classification is implemented based on multiple extraction features. The purpose of this study is to evaluate the performance of ID-Convolutional to classify pneumonia with various CNN architectures to produce the best performance in accuracy and compare it with the other baseline methods that have been made. Experiments are carried out based on the number of hidden layers and configuration parameters. The parameters used are the epoch, kernel, strides, and pool size. The result shows that the proposed method achieves 94% inaccuracy and 0.93 in AUC. Besides, CNN has a competitive result compare to the other baseline methods.

Keywords: Convolutional neural network · Chest X-Ray · Pneumonia

1 Introduction

Pneumonia leading infection cause death among children under five. Based on the UNICEF, pneumonia killing approximately 2,400 children a day and pneumonia accounted for approximately 16 percent or around 880.000 children in 2016 [1]. In other words, every hour there are 2 to 3 children who die from pneumonia in Indonesia. According to the World Health Organization (WHO), pneumonia is the biggest cause of death in children worldwide, there were 9808.894 children under the age of 5 in 2017 [2]. One of the causes of high pneumonia cases is a lack of public understanding of this disease for early detection. Common symptoms caused by pneumonia are common so people ignore this [3]. Handling pneumonia is still difficult because many

R. Ghazali et al. (Eds.): SCDM 2020, AISC 978, pp. 181–191, 2020.
https://doi.org/10.1007/978-3-030-36056-6_18

risk factors have not been resolved. One way to prevent pneumonia is by giving a pneumonia vaccine, but in Indonesia, the vaccine has not yet entered immunization which is included in the government program [3, 4]. Then to detect pneumonia, namely with a chest X-ray, However, getting a diagnosis requires a long time because it has to adjust the practice schedule from the experts. With the existence of an image-based pneumonia classification system that can assist medical staff in diagnosing pneumonia.

The challenge of automatic image classification is difficult because of the noise in the image, to be able to reduce noise in an image can use one technique, namely Filter Adaptive Noise Removal Technique that is by updating each pixel and changing the pixel value with the average value of neighboring pixels that affect pixel value processed. The size of the large image dimensions also affects the time needed to process the image and the value of accuracy in the image. One of the approaches is commonly used for object recognition is using convolutional neural networks [5].

Convolutional Neural Networks (CNN) have been applied as an effective structure of models for understanding image content, giving state-of-the-art results on image detection, segmentation and recognition [6, 7]. The key to enabling the several factors behind those results was techniques for scaling up the networks to tens of millions of parameters and massive labeled datasets that able to support the learning process. Under these conditions, CNN has been shown to learn interpretable and powerful image features [8].

Rahib H Abiyev classified Chest X-ray images using the Convolutional Neural Network (CNN) method resulting in an accuracy value of 92.4%, with the Backpropagation Neural Network (BPNN) method producing an accuracy value of 80.04%, the method of Competitive Neural Network (CpNN) produces an accuracy value of 89.57%. However, research conducted by Abiyev by classifying 12 disease predictions in the chest classification rates for Pneumonia is still small, with a probability value of approximately 0.1 [9].

Bar Y et al. Classifying the CNN with GIST method resulted in an accuracy value of 92% [6, 10]. Islam M.T classifies Pneumonia with VGG16 resulting in an accuracy value of 86% and with VGG19 producing an accuracy value of 92% [6, 11]. Pattar classified the Lung cancer and Pneumonia using the Feed Forward method resulting in an accuracy value of 94.8% and 94.82% using the Basis Function Network [12]. Max Noris classifying Pneumonia with the VGG16 method produces an accuracy of 86% [13]. Saraiva conducted a study using k-fold cross validation to share data and the CNN method for image classification resulted in an accuracy value of 95.30% [14]. In this research, it was implemented the Convolutional Neural Network for the classification of X-ray pneumonia images based on multiple extraction features.

2 Methods and Evaluation

In this section, methods 1-D Convolutional Neural Network, Histogram of Oriented Gradient, Gray Level Co-occurrence, and Evaluation methods for both proposed method and reference methods were introduced.

2.1 1D Convolutional Neural Network

Convolutional Neural Network (CNN) is a development model of Multilayer Perceptron (MLP) designed for 2-dimensional data. CNN is included in the type of Deep Neural Network because of the high network depth and much applied to image data. Because of the nature of the convolution process, CNN can be used on data that has a 2-dimensional structure, such as image and audio [15]. There are 3 main types of layers used on CNN architectures, namely convolutional layers, pooling layers, and fully connected layers [16] (Fig. 1).

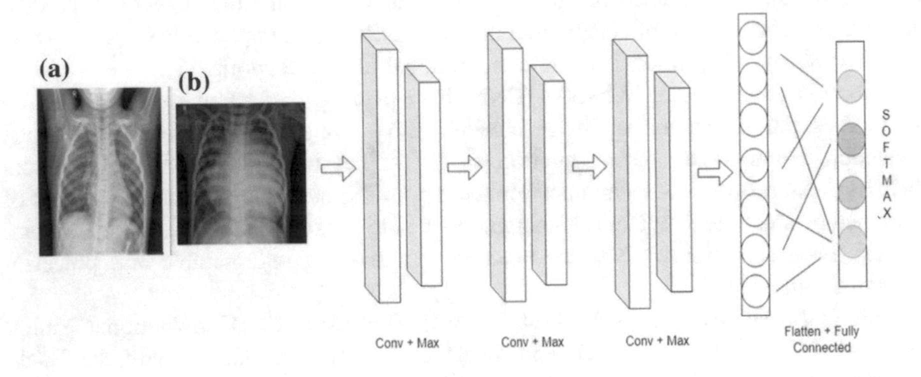

Fig. 1. CNN architecture

On the CNN architecture, there are several layers and parameters. Layer of Input: this layer will save the raw pixel value or change it from the image. Convolutional Layers: this layer combines the input image with a series of fixed-size filters or kernels used to convolve the image data, and each produces a feature map in the output image. Pooling layers: this layer is responsible for down-sampling the spatial dimensions of the input (width and height).

The use of the pooling layer on CNN only focuses on the most important pattern and aims to reduce the size of the image so that it can be easily replaced with a convolution layer with the stride which is the same as the pooling layer in question. Rectified Linear Unit (ReLU) Layers; this layer is responsible for applying the non-linear function to the output x of the previous layer. The ReLU layer applies elementwise activation functions, such as max (0, x).

Fully Connected Layer; this layer is used to understand the pattern generated by the previous layer. The neuron in this layer has a full connection to all activations in the previous layer. After being trained, the transfer learning approach can extract features in this layer to train other classifiers. The second baseline descriptor included in this research is representing feature extraction using Histogram of Oriented Gradient (HOG) and Gray Level Co-occurrence Matrix (GLCM).

2.2 Histogram of Oriented Gradient (HOG)

Histogram of Oriented Gradient (HOG) is a method used to detect objects, this method calculates the gradient value in a particular area of an image. Each image has the characteristics shown by the gradient distribution obtained by dividing the image into small areas called cells. then there is a block feature for calculating normalization in each cell [17] (Fig. 2).

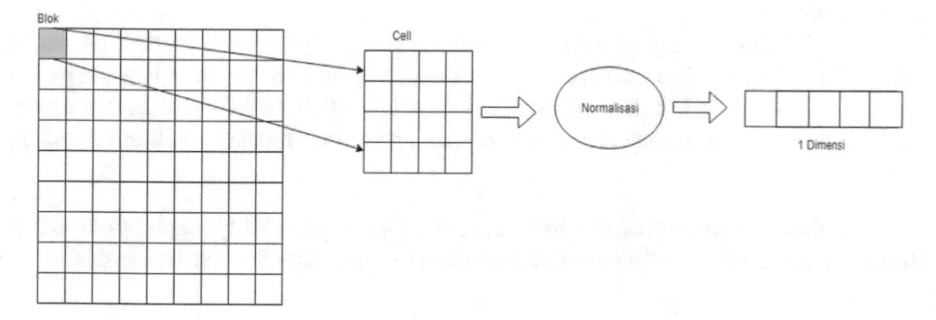

Fig. 2. Histogram of oriented gradient

2.3 Gray Level Co-occurance Matrix (GLCM)

Gray Level Co-occurrence Matrix (GLCM) method is a way of extracting second-order statistical texture features [18]. In this experiment, the features used were contrast, homogeneity, energy, correlation, and dissimilarity.

$$Contrast = \sum_{i,j=0}^{N-1} P_{i,j}(i-j)^2 \tag{1}$$

$$Homogenity = \sum_{i,j=0}^{N-1} \frac{P_{i,j}}{1+(i-j)^2} \tag{2}$$

$$Energy = \sum_{i,j=0}^{N-1} P_{i,j}^2 \tag{3}$$

$$Correlation = \sum_{i,j=0}^{N-1} P_{i,j}\left[\frac{(i-\mu_i)(j-\mu_j)}{\sqrt{(\sigma_i^2)(\sigma_j^2)}}\right] \tag{4}$$

Then combine the two extraction features into one. The use of this feature extraction for the introduction of object categories by converting 2-dimensional image data into 1-dimensional feature values.

2.4 Evaluation Methods

Accuracy measures how well a diagnostic test identifies and excludes a condition. The standard deviation is a measure of the dispersion of a set of data from the average [19].

$$Accuracy = \frac{(TN + TP)}{(TN + TP + FN + FP)} \tag{5}$$

Information:
TP: True Positive is a positive image detected positively by the classification model
TN: True Negative is a positive image detected negative by the classification model
FN: False Negative is a negative image detected negative by the classification model
FP: False Positive is a negative image detected positively by the classification model.

A receiver operating characteristic curve (ROC) is a probability graph or curve that shows the performance of a classification model on all classification thresholds.

$$TPR = \frac{TP}{TP + FN} \tag{6}$$

$$FPR = \frac{FP}{FP + TN} \tag{7}$$

Area under curve (AUC) represents the degree or measure of separation. AUC tells how many models can distinguish between classes. The higher the AUC, the better the model in classifying data. AUC has a range of values from 0 to 1. Therefore, the AUC value that is close to number 1 means that the method can classify the data properly and the AUC value close to 0 means that the method cannot properly calcify the data.

3 Experiment and Results

The dataset contains X-ray images of children aged 1 to 5 years from "Guangzhou Women and Children's Medical Center" (GUANGZHOU WOMEN AND CHILDRENS MEDICAL CENTER, 2018) which can be accessed online on the Kaggle website. This dataset is divided into 2 classes, namely Normal and Pneumonia. There were 1,583 samples for the Normal class and 4,273 samples for the Pneumonia class. So that the total sample totaled 5,856 samples (Fig. 3).

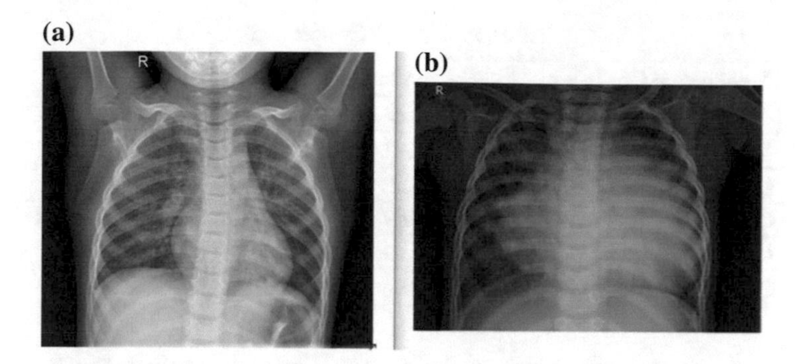

Fig. 3. a shows a normal X-ray chest image and b shows a chest image of X-ray Pneumonia

The following details on the dataset can be seen in Table 1.

Table 1. Dataset details

Data	Normal	Pneumonia
Data training	1.341 sampel	3.875 sampel
Data test	234 sampel	390 sampel
Data validasi	8 sampel	8 sampel

3.1 Result and Discussion

The first step is to resize all X-ray images into one size, 1024 × 1024. After all images are resized, the next step is extracting images with the HOG feature to produce 144 features and extracting the GLCM feature to produces 5 features. Five features extracted from GLCM features are contrast, homogeneity, energy, correlation and dissimilarity values. The next step is to use the CNN method to classify images. before that, it was tested with 37 different hidden layers with CNN architecture parameter configuration includes epoch – 10, kernel = 2, strides = 2, pool_size = 2, dropout = 0.5, dense = 1000, batch_size = 10.

After getting the five best hidden layers in Table 2, it was tested by repeating the training data model in 30 times and configuring the same parameters, epoch = 10, kernel = 2, strides = 2, pool_size = 2, dropout = 0.5, dense = 1000, batch_size = 10. From this experiment, the results of testing accuracy have a significant value compared to the previous results. The results of the CNN architectural model based on hidden layers with a range of values from 0 to 1 can be seen in Table 3.

Table 2. CNN model

Architecture	Hidden unit
CNN 1	[128, 16, 16, 16]
CNN 2	[128, 64, 16, 8]
CNN 3	[128, 64, 16, 16, 8, 8]
CNN 4	[128, 64, 16, 16, 8]
CNN 5	[128, 64, 64, 16, 16, 8]

Table 3. Best CNN accuracy

Architecture	Training		Validation		Testing	
	Accuracy	AUC	Accuracy	AUC	Accuracy	AUC
CNN 1	0.97 ± 0.01	0.97	0.91 ± 0.05	0.91	0.80 ± 0.02	0.74
CNN 2	0.91 ± 0.02	0.93	0.92 ± 0.07	0.94	0.83 ± 0.02	0.80
CNN 3	0.95 ± 0.02	0.96	0.97 ± 0.05	0.97	0.82 ± 0.02	0.78
CNN 4	0.92 ± 0.02	0.94	0.91 ± 0.09	0.95	0.84 ± 0.03	0.82
CNN 5	0.85 ± 0.05	0.90	0.94 ± 0.05	0.98	0.85 ± 0.02	0.85

The next experiment was to combine all the datasets and use the K-fold Cross Validation method to divide the training data, testing data and validation data, with k = 5 and configure the same CNN parameter epoch = 10, kernel = 2, strides = 2, pool_size = 2, dropout = 0.5, dense = 1000, batch_size = 10. In this experiment, the use of K-fold Cross Validation to divide data resulted in high accuracy values. So now it has been testing by trying 4 different classification methods to make a comparison with the CNN method which is seen from the values of accuracy, ROC and AUC. The method is Support Vector Machine (SVM), Logistic Regression (LR), Multi Layer Perceptron (MLP), and Random Forest (RF). The results can be seen in Tables 4, 5, and 6.

Table 4. Training data accuracy

Method	Fold 1	Fold 2	Fold 3	Fold 4	Fold 5	Mean accuracy
CNN 1	0.95 ± 0.01	0.96 ± 0.01	0.96 ± 0.02	0.97 ± 0.02	0.97 ± 0.02	0.96 ± 0.01
CNN 2	0.93 ± 0.0	0.94 ± 0.02	0.95 ± 0.02	0.95 ± 0.02	0.95 ± 0.02	0.94 ± 0.02
CNN 3	0.89 ± 0.04	0.85 ± 0.05	0.83 ± 0.06	0.82 ± 0.06	0.82 ± 0.07	0.84 ± 0.06
CNN 4	0.92 ± 0.01	0.92 ± 0.01	0.93 ± 0.02	0.93 ± 0.02	0.94 ± 0.02	0.93 ± 0.02
CNN 5	0.91 ± 0.02	0.93 ± 0.02	0.93 ± 0.02	0.94 ± 0.02	0.94 ± 0.02	0.93 ± 0.02
MLP	0.83 ± 0.08	0.82 ± 0.07	0.80 ± 0.07	0.80 ± 0.07	0.80 ± 0.07	0.81 ± 0.07
SVM	0.75 ± 0.00	0.75 ± 0.00	0.75 ± 0.00	0.75 ± 0.00	0.75 ± 0.00	0.75 ± 0.00
LR	0.72 ± 0.00	0.72 ± 0.00	0.72 ± 0.00	0.72 ± 0.00	0.72 ± 0.00	0.72 ± 0.00
RF	1.00 ± 0.00	1.00 ± 0.00	1.00 ± 0.00	1.00 ± 0.00	1.00 ± 0.00	1.00 ± 0.00

Table 5. Validation data accuracy

Method	Fold 1	Fold 2	Fold 3	Fold 4	Fold 5	Mean accuracy
CNN 1	0.94 ± 0.01	0.94 ± 0.01	0.94 ±0.01	0.95 ±0.01	0.95 ± 0.01	0.94 ± 0.01
CNN 2	0.92 ± 0.02	0.93 ± 0.02	0.94 ± 0.02	0.94 ±0.02	0.94 ± 0.02	0.93 ± 0.02
CNN 3	0.88 ± 0.05	0.93 ± 0.01	0.93 ± 0.01	0.93 ±0.01	0.93 ± 0.01	0.84 ± 0.06
CNN 4	0.90 ± 0.02	0.93 ± 0.01	0.93 ± 0.01	0.93 ±0.01	0.92 ± 0.02	0.93 ± 0.01
CNN 5	0.90 ± 0.02	0.92 ± 0.02	0.93 ± 0.02	0.93 ±0.02	0.93 ± 0.02	0.92 ± 0.02
MLP	0.82 ± 0.08	0.81 ± 0.07	0.79 ± 0.07	0.79 ±0.07	0.79 ± 0.07	0.80 ± 0.07
SVM	0.71 ± 0.01	0.71 ± 0.00	0.71 ± 0.00	0.71 ±0.01	0.71 ± 0.01	0.71 ± 0.00
LR	0.71 ± 0.00	0.71 ± 0.00	0.71 ± 0.00	0.71 ±0.01	0.71 ± 0.01	0.71 ± 0.00
RF	0.93 ± 0.00	0.94 ± 0.00	0.94 ± 0.01	0.94 ±0.01	0.94 ± 0.01	0.94 ± 0.01

Table 6. Testing data accuracy

Method	Fold 1	Fold 2	Fold 3	Fold 4	Fold 5	Mean accuracy
CNN 1	0.93 ± 0.01	0.94 ± 0.01	0.94 ± 0.01	0.95 ± 0.01	0.95 ± 0.02	0.94 ± 0.01
CNN 2	0.92 ± 0.02	0.93 ± 0.02	0.93 ± 0.02	0.94 ± 0.02	0.94 ± 0.02	0.93 ± 0.02
CNN 3	0.88 ± 0.05	0.86 ± 0.05	0.83 ± 0.06	0.82 ± 0.06	0.81 ± 0.07	0.84 ± 0.06
CNN 4	0.91 ± 0.02	0.92 ± 0.02	0.92 ± 0.02	0.93 ± 0.02	0.93 ± 0.02	0.93 ± 0.02
CNN 5	0.91 ± 0.02	0.91 ± 0.02	0.92 ± 0.02	0.93 ± 0.02	0.93 ± 0.02	0.92 ± 0.02
MLP	0.83 ± 0.07	0.83 ± 0.07	0.81 ± 0.07	0.80 ± 0.07	0.80 ± 0.07	0.81 ± 0.07
SVM	0.73 ± 0.00	0.73 ± 0.00	0.73 ± 0.00	0.73 ± 0.01	0.73 ± 0.01	0.73 ± 0.00
LR	0.75 ± 0.00	0.73 ± 0.02	0.73 ± 0.01	0.72 ± 0.02	0.72 ± 0.02	0.73 ± 0.01
RF	0.94 ± 0.00	0.94 ± 0.01	0.94± 0.00	0.94 ± 0.00	0.94 ± 0.01	0.94 ± 0.00

From the results above, I made average ROC and AUC on each fold of CNN architecture. Besides, I also made average ROC and AUC on each fold of the comparison method, SVM, MLP, LR, and RF. The following are ROC and AUC for each classification method (Figs. 4, 5, 6, 7, 8, 9, 10, 11 and 12).

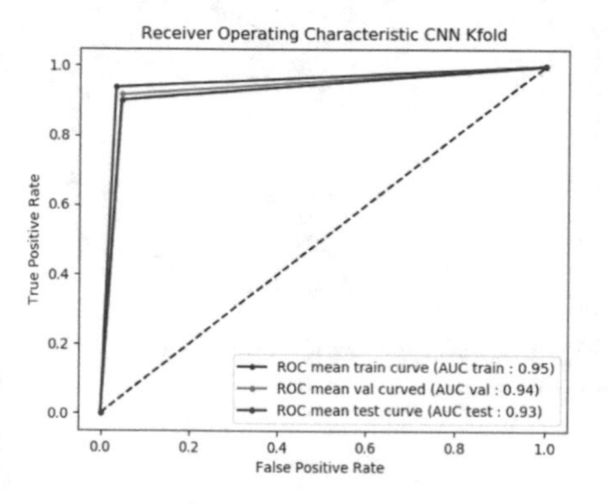

Fig. 4. Average ROC and AUC, CNN 1

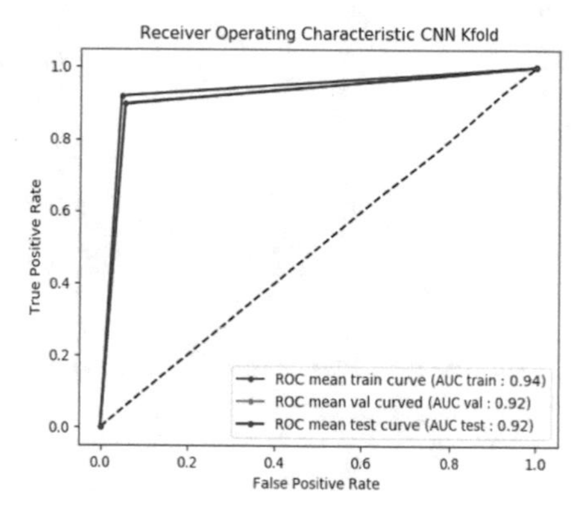

Fig. 5. Average ROC and AUC, CNN 2

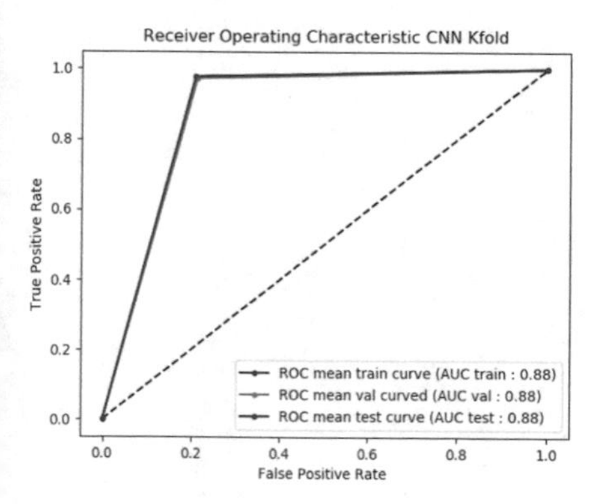

Fig. 6. Average ROC and AUC, CNN 3

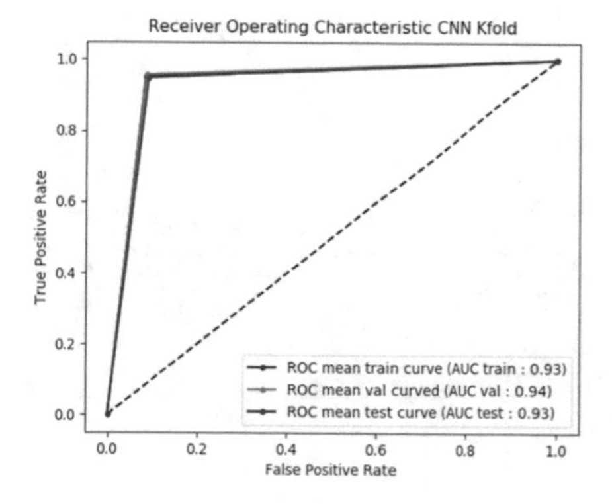

Fig. 7. Average ROC and AUC, CNN 4

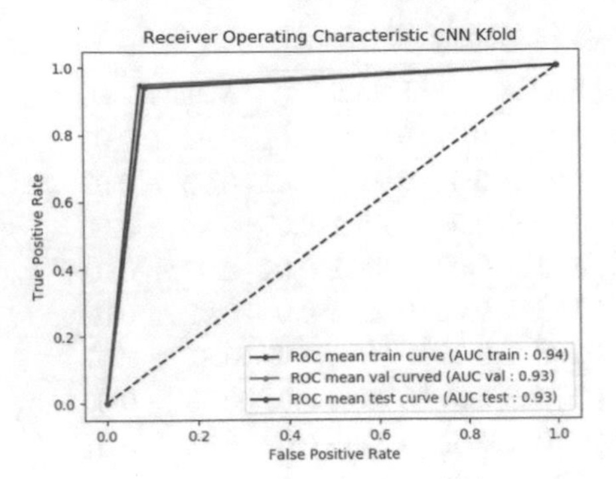

Fig. 8. Average ROC and AUC, CNN 5

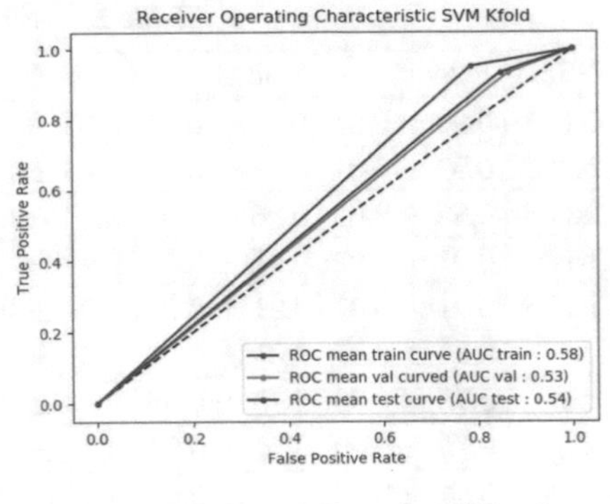

Fig. 9. Average ROC and AUC, Support Vector Machine

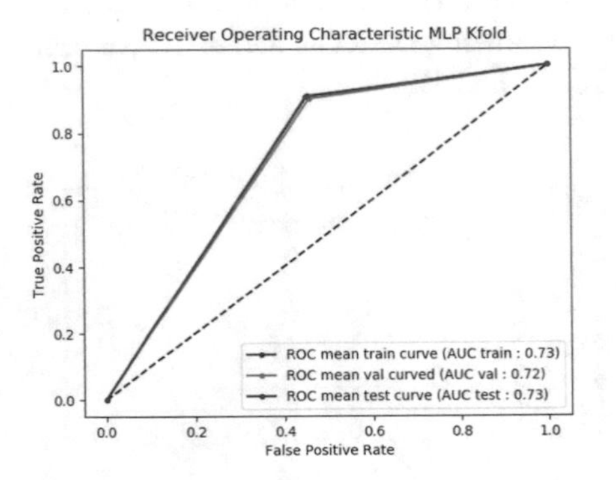

Fig. 10. Average ROC and AUC, Multi Layer Perceptron

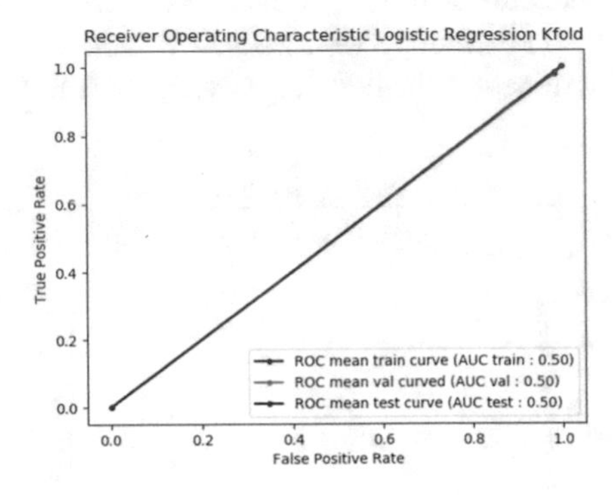

Fig. 11. Average ROC and AUC, Logistic Regression

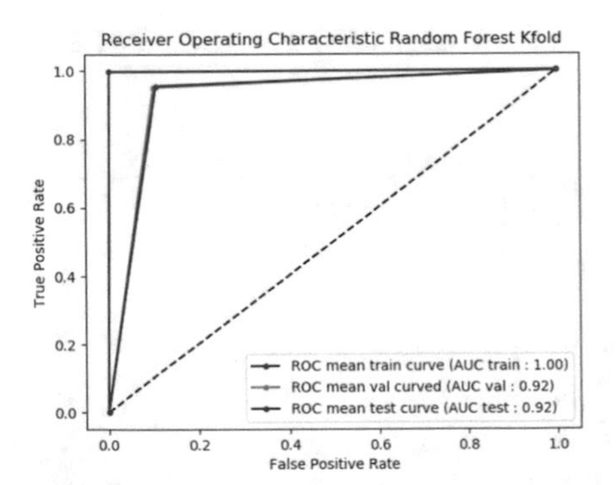

Fig. 12. Average ROC and AUC, Random Forest

4 Conclusion

In conclusion, CNN is successfully implemented for the classification of two classes of pneumonia based on chest X-ray images. This study evaluates the performance of CNN models with different parameter settings, in particular, the number of hidden layers and hidden units. Based on the experiment result, the number of hidden layers and hidden unit influences the classification accuracy. According to the experiment, CNN 1 has better performance followed by CNN 2, CNN 4, CNN 5 and CNN 3. After being evaluated on testing data, CNN 1 produces 0.94 in accuracy and 0.93 in AUC which is outperformed the other baseline methods such as SVM, MLP, and LR. However, it also produces a competitive result compared to that of RF. When compared with Saraiva's research. This research, using the extraction feature has accuracy 94%, while Saraiva's research using pixel of image produce accuracy 95.30%. However, this research produces a competitive result compared to that of Saraiva's research. In the future work, using more than 2 extraction feature or doing transfer learning maybe will be upgrading accuracy.

References

1. UNICEF, Pusat Pers | UNICEF Indonesia. UNICEF Indonesia, (2015). [Online]. Available: https://www.unicef.org/indonesia/id/pusat-pers. Accessed 30 Jun 2019
2. WHO, Pneumonia (2016). [Online]. Available: https://www.who.int/news-room/fact-sheets/detail/pneumonia. Accessed 30 Jun 2019
3. Ri KK (2010) Pneumonia Balita. Bul Jendela Epidemiol 3:43
4. Fimela, Cara Pencegahan Penyakit Mematikan Pneumonia Pada Anak-Anak—Parenting Fimela.com, (2016). [Online]. Available: https://www.fimela.com/parenting/read/3765370/cara-pencegahan-penyakit-mematikan-pneumonia-pada-anak-anak. Accessed 19 Feb 2019
5. Rachmad A (2008) Pengolahan Citra Digital Menggunakan Teknik Filtering Adaptive Noise Removal Pada Gambar Bernoise | Aeri Rachmad—Academia.edu. [Online]. Available: https://www.academia.edu/27291316/Pengolahan_Citra_Digital_Menggunakan_Teknik_Filtering_Adaptive_Noise_Removal_Pada_Gambar_Bernoise. Accessed 25 Feb 2019
6. Krizhevsky A, Sutskever I, Hinton GE, ImageNet classification with deep convolutional neural networks, pp 1–9
7. Bengio Y, Haffner P (1998) Gradient-based learning applied to document recognition. 86 (11)
8. Sun M, Song Z, Jiang X, Pang Y (2016) Author's accepted manuscript learning pooling for convolutional neural network. Neurocomputing
9. Abiyev RH, Ma'aitah MKS (2018) Deep convolutional neural networks for chest diseases detection. J Healthc Eng, 4168538, 2018
10. Bar Y, Diamant I, Wolf L, Lieberman S, Konen E, Greenspan H (2018) Chest pathology identification using deep feature selection with non-medical training. Comput Methods Biomech Biomed Eng Imaging Vis 6(3):259–263
11. Islam MT, Aowal MA, Minhaz AT, Ashraf K (2017) Abnormality detection and localization in chest X-rays using deep convolutional neural networks
12. Pattar S, Pavithra R (2015) Detection and classification of lung disease—pneumonia and lung cancer in chest radiology using artificial neural network. Int J Sci Res Publ, 5(1):2250–3153

13. Norris M (2018) Convolutional neural networks in radiology CS230-Fall 2018 course project
14. Saraiva A et al (2019) Classification of images of childhood pneumonia using convolutional neural networks
15. LeCun Y, Bengio Y, Hinton G (2015) Deep learning. Nature 521(7553):436–444
16. Ferreira A, Giraldi G (2017) Convolutional Neural Network approaches to granite tiles classification. Expert Syst Appl 84:1–11
17. Integrated Performance Primitives 2017 Developer Reference, "Histogram of Oriented Gradients (HOG) Descriptor."
18. Mohanaiah P, Sathyanarayana P, Gurukumar L (2013) Image texture feature extract approach. Int J Sci Res Publ 3(5):1–5
19. Zhu W, Zeng N, Wang N (2010) sensitivity_specificity_accuracy_CI.pdf, pp 1–9 2010

Semantic Information Retrieval Systems Costing in Big Data Environment

Khalid Mahmood[1(✉)], M. Rahmah[1], Md. Manjur Ahmed[2], and Muhammad Ahsan Raza[1]

[1] Faculty of Computing, College of Computing and Applied Science, Universiti Malaysia Pahang, Kuantan 26300, Malaysia
keeenkhan@gmail.com

[2] Department of Computer Science and Engineering, University of Barisal, 8200 Barisal, Bangladesh

Abstract. Nowadays, dealing with big data is a major challenge for application developers and researchers in several domains like storage, processing, indexing, integration, governance and semantic search. For decision-making and analysis purpose, semantic retrieval of information from big data is gaining more attention with the need of extracting accurate, meaningful and relevant results. Several semantic information retrieval techniques alternatively have been developed by researchers for retrieval of valuable information in big data environment. This article classifies literature and presents an analysis of five recent semantic information retrieval systems in terms of their methodologies, strengths and limitations. In addition, we evaluate these schemes on the basis of specific datasets and performance measures such as precision, recall and f-measure metrics. A comparative analysis of performance measures shows that IBRI-CASONTO achieves best f-measure value of 97.6 over other information retrieval systems.

Keywords: Information retrieval · Semantic search · Ontology · Clustering

1 Introduction

In the recent era of information technology, the systems and humans are interacting with the Web and other media by generating and extracting the data of large scale, termed as “big data”, is continuously growing day by day. Such data is measured in Exabytes (10^{18}) and Zettabytes (10^{21}) on web and other related media. It is predicted that, by 2025, the internet data storage size will exceed the brain capacity of everyone breathing in this world [1]. According to the National Security Agency report, approximately 1.8 petabytes of data is collected on the Internet each day [2]. This fast and continuously growing data size is only due to developments in digital age. It is reported by [1], that 2.5 billion gigabytes of data are being produced every day in the world, which consists of 90% unstructured data.

R. Ghazali et al. (Eds.): SCDM 2020, AISC 978, pp. 192–201, 2020.
https://doi.org/10.1007/978-3-030-36056-6_19

Searching precise and accurate information efficiently from data is an artistic technique, but it becomes more inventive when the amount of data is large as big data. Many optimal search schemes are introduced and implemented practically in various applications but extracting accurate information is still a big issue of searching such type of data in the present era. A few research scientists proposed semantic search schemes for extracting meaningful information from the data, but some issues are still there to acquire the desired and accurate results. The fulfillment of the semantic search is proposed in different ways by different authors. Clustering [4–7], Query processing [10], similarity measures [11], and use of annotation [12] are among the most semantic search approaches. However, most of the semantic approaches are based on ontologies [8, 9], through which they apply semantically to get desired results. This article provides an overlook of recent semantic information retrieval approaches, and a comparison of five such search schemes that provide better precision and recall rate. This manuscript also incorporates the analysis of different schemes through multiple aspects of semantic information retrieval, such as semantic applied, multiple language support, Natural Language Processing (NLP), weight computation, user interface and the key factor on which the retrieval is being performed. The purpose of this study is to demonstrate, how adopting a design attitude enables researchers and developers of information retrieval systems to enhance the quality of information retrieval process and potentially discovering effective outcomes of their work.

2 Literature Review

Finding results with semantic orientation is easy if searching from relational databases through SQL (Structured Query Language), but slightly different and challenging when searching from a large-scale data and distributed storage zones. Semantic Search is not a brand-new discipline in the world of researchers. Several researchers proposed their schemes to perform search efficiently and semantically, and efforts are continuing to make the search results more accurate. A classification and critical review of supporting and core schemes about the semantic search that are already proposed by different researchers are given below.

2.1 Cluster-Oriented Semantic Information Retrieval Systems

Various information retrieval schemes are proposed in which clustering is used as semantic technique to get desired results efficiently. We provide a brief review of some latest schemes in the following passages.

A ranking algorithm is introduced in [6], which represents the terms through the vector model known as K-dimensional. Every dimension of term of vector remains computed on the base of its rank distribution between relevant clusters. The authors in [14] introduced a clustering scheme based on proximity for grouping of data. This scheme precisely creates the clusters of large-scale data by using constraint reformulation. In this scheme, the authors search the presenter data of each cluster instead of searching all collection of data to answer the query. It is analyzed that this scheme is providing better results in terms of clustering as compared to the previous approach.

Another cluster oriented information retrieval approach [17] is designed in which a hybrid indexing methodology is adopted. Intelligent and efficient indexing is introduced in this scheme after the cluster formation of data. It is found that this scheme is fast in processing but provides poor results in terms of accuracy. In [18], after cluster formation, a Markov model of random field selection is presented for the assignment of grades to data clusters. The graph of data and queries is first created and then the probability of a relevant cluster to the given query is calculated. However, due to the complexity of the graph, the performance of this scheme is not good with multiple queries at a time. A personalized search engine based on clustering and a genetic algorithm is developed by the authors in [19]. This approach forms clusters of URLs first, clicked by the user and then applying a genetic algorithm on relevant clusters for ranking its URLs to answer the query.

In semantic data extraction techniques, clustering is typically used to implement semantics to find out more relevant and accurate results but greatly reduced query answering time as compared to traditional information retrieval schemes. However, if ranking algorithms are used along with clustering techniques then most probably it may retrieve less relevant and accurate results against given queries, this is the major disadvantage of cluster-based approaches.

2.2 Ontology-Oriented Semantic Search Systems

Using ontologies is a good idea in search engines because it is more intelligent and performs better as well. Ontological graphs are the reason to make ontologies more powerful and enable them to store accurate and more reliable data in repositories. Schemes based on ontologies enable users to extract information directly without involving into complexities. Various ontology-based search schemes have been introduced according to the researchers' interests involving single or multiple domains.

Several reference ontologies have been developed by the developers that focus on some specific communities such as UnivBench, HERO (Higher Education Reference Ontology), AIISO (Academic Institution Internal Structure Ontology) and university ontologies [21–23]. There are some search engines like Kngine (Knowledge Engine) and Wolfram Alpha [13, 24], comprise ontological semantic. A first question-answer based search system that supports 4 multiple languages is Google Kngine [24]. This search engine delivers accurate and comprehensive custom results against a given search query.

It is analyzed through various studies that ontologies are better to imply semantics in semantic search schemes because ontologies may also be utilized for domain knowledge specification of particular domain attention [23, 24]. Moreover, ontological theories are also suitable for the development and representation of conceptual data models in data mining [25].

2.3 Terms-Oriented Semantic Data Mining System

Terms mining or rules-based approaches are also playing an essential role in data retrieval discipline to find relevant information semantically in an efficient way. Recently, a cognitive approach for deep analysis of contents in the domain of semantic search systems is developed [26], which performs rule-based processing of natural language (NLP) together with the subject-regarding model and task-related intellectual structure in order to rank the data semantically. The major focus of this scheme is to increase the accuracy and the relevancy of the information search in a competent manner.

The major objective of the work presented in [27], is the integration of keywords, entities, and types of information. The findings are reported in terms of the performance of query processing as a ranked intersection for various indices and then select an appropriate index for semantic search. A critically major drawback is that the weights are computed at query time which increases the query processing delay.

Terms mining based information retrieval shows better performance in traditional systems but this scheme performs much better if it is used as a pre-processing phase for the development of cluster-based semantic information retrieval techniques. These schemes also can be used for ranking the clusters to select more relevant data clusters suitable for answering the query. An overall summary of some recent intelligent information retrieval schemes in Big Data is being provided in Table 1, which provides general strengths and limitations of each class or group of these schemes.

Table 1. Summary of semantic information retrieval schemes.

Class	Schemes	Strengths	Limitations
Cluster-oriented	K-dimensional [6], CAWP [14], Markov Model [18], hybrid indexing model [17], IC-GLS [20], Genetic Algorithm [19], KNNIR [5], ICIR [7], Wiser [15]	Greatly reduce query answering time and semantically enrich	Cluster ranking algorithms may cause of less relevant results
Ontology-oriented	UnivBench, HERO, AIISO [21, 22, 23], Google Kngine [24], Wolfram Alpha [13], IBRI-CASONTO [8], SOR [16]	Better to imply more semantics, specify domain knowledge and develop conceptual data models	Mapping and Transformation need special attention and require additional efficient algorithms
Terms-oriented	Cognitive-based [26], Efficient Indexing [27], UBFC [3], CSA [11], Wiser [15]	Best for traditional data mining, suitable cluster-based approaches	Low performance in terms of data retrieval time because examine whole collection of data

3 The Performance Evaluation

Based on literature review in Sect. 2, we evaluate the performance of five recent semantic information search approaches [7, 8, 11, 15, 16] with common measures. This critical analysis will be very helpful for future researchers and developers in utilization of these schemes as a part or full in their work. Firstly, we evaluate the architecture of these schemes, then we discuss different factors, after that, we provide performance comparison.

Now we have a brief look at each of these schemes in chronological order. The article selected to discuss first is IBRI-CASONTO [8], this approach is an ontology-based information search system for an Applied Sciences College situated in the Kingdome of Oman. This search engine is based on resource description framework (RDF) ontological graph, which supports two languages such as English and Arabic. The ontological structure of this scheme follows some common main steps including (1) design, (2) interference, (3) storage, (4) indexing, (5) searching, (6) query processing, and (7) interfacing. This search engine under discussion enables the searching of two types: (1) Simple or Traditional Search, (2) Intelligent or Cognitive Search. Keyword-based searching or classical searching computes keywords matching with the RDF dataset and ontological graph. The results are generated based on the highest matching scores. Entity-based search or semantic search aims to extract accurate results from the ontological graph and additionally to understand the context of the query and provide comprehensive results.

The second semantic information retrieval approach we need to discuss is SOR [16], which is an ontology-centric algorithm for the retrieval of heterogeneous type multimedia data from large-scale data collection. The major concerned issue that is being focused on by this scheme is the retrieval of required and semantic results from heterogeneous multimedia big data. In compliance to solve this issue major contributions of this approach are: (1) an algorithm based on semantic ontological matching is proposed by the author to search heterogeneous multimedia data, (2) the semantic information represented by the ontology and multimedia documents are stored together in the database, (3) An interface like traditional search engine based on the MapReduce framework, which stores and retrieve the data in parallel on low-end computers.

An Intelligent Cluster-based Information Retrieval is an effective scheme that uses frequent closed item sets, abbreviated as ICIR [7], another information searching algorithms which implies semantics through clusters. There are also threefold major contributions of this scheme: (1) it applies frequent closed item sets mining and clustering techniques to retrieve semantic information from a large-scale dataset, (2) it utilizes this extracted knowledge to search the query by calculating relevance score for each cluster and common trends or terms available in relevant bunch of data, (3) and performance evaluation of proposed approach on large-scale data collection. K-means algorithm is used in this approach by the authors for the cluster formation of sample datasets.

Relying on RDF knowledge-base is another technique, namely, Context Semantic Analysis (CSA) [11] that uses semantic analysis for inter-document similarity computation. This scheme comprises three stages: (1) the extraction of the contextual graph, which retrieves information from the knowledge graph, (2) the generation of Semantic Context Vector, which represents the actual context of generated data, (3) Context Similarity Evaluation compares context vectors of data. The vertices of the contextual weighted graph ranked to get the context of data and then the similarity is computed between two entities by using cosine similarity. Moreover, the scheme is scalable in cluster formation of large-scale data.

Wiser [15] is a semantic search engine based on entity linking that proposed for searching experts in an academic world. This is an unsupervised system, which implements together with the Wikipedia knowledge graph and text evidence-based classical language modeling techniques through entity linking. The graph edges labeled with weights represent semantic related to the entities. Each node has a relevance score calculated via random-walk over the graph. In query processing phase occurrences of the query-terms are exploited and relevant information is retrieved by using some data-centric ranking approaches that calculate the semantic connection. This system is tested on a huge standard dataset to examine its effectiveness.

We evaluated simply from the above discussion that each scheme is implementing different strategies through some common technologies and techniques. Another important evaluation is that the schemes are resulting in better performance in their application domain but still there is a need to optimize the algorithms to meet future challenges of Big Data which may relate to its different characteristics. The overall analysis of selected schemes for comparison is given in Table 2, which provides a brief overview based on different factors.

Table 2. Analysis of semantic search systems in terms of key features.

Scheme	Semantic source	Key search	Multiple language	NLP	Weights	User interface
IBRI-CASONTO	Ontological Graph	Entity based	Yes	No	Yes	Yes
SOR	Ontological matching	Multimedia Entity based	No	No	No	Yes
ICIR	Corpus cluster	Concept based	No	Yes	Yes	No
CSA	Knowledgebase	Entity based	No	No	Yes	No
Wiser	Knowledge graph	Entity based	No	Yes	Yes	Yes

While studying the schemes discussed above and selected to evaluate some common evaluation metrics found used by the authors to measure the performance of their schemes. Normally precision, recall, and f-measure are computed for evaluating the performance of an information retrieval scheme or search engine. These common metrics are also measured and represented the efficiency of results by representing computed values as comparison tables and/or charts in the articles regarding these schemes. The concept of these metrics is being briefly depicted in Fig. 1, which shows how input parameters are computed to calculate the values of these metrics. A common formula to calculate precision, see Eq. 1, is given in the following equation.

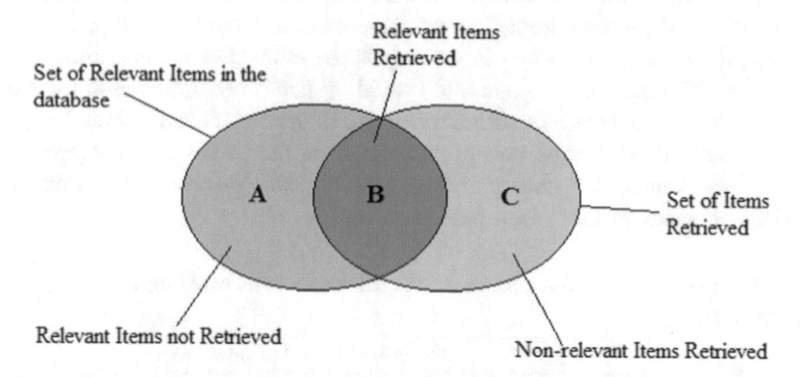

Fig. 1. Parameters description for evaluation of semantic information retrieval

Precision. It is the fraction retrieved relevant information to the total retrieved data against a search query. Consequently, the mathematical notation of the precision rate is given in the following equation.

$$Precision = \frac{|B|}{|B| + |C|} \times 100. \quad (1)$$

Here, $|B|$ denotes the relevant items retrieved, $|B| + |C|$ is the set of total items retrieved from the whole repository of data on which search query is performed. The following equation describes the computation of relevant items retrieved.

Recall. It is the retrieved relevant data proportion to all relevant data available in the data source. The Eq. 2 calculates the recall rate.

$$Recall = \frac{|B|}{|A| + |B|} \times 100. \quad (2)$$

Here, $|A| + |B|$ is the collection of total relevant data in the repository either that is retrieved or not retrieved against a search query.

F-measure. Traditionally known as F-score, is used to find out average (harmonic mean) by combining both precision and recall when they are almost close to each other. Following equation, see Eq. 3, is used to calculate F-scores for the evaluation purpose of selected schemes.

$$F - measure = 2 \times \frac{Precision \times Recall}{Precision + Recall}. \quad (3)$$

The above equations used as computation of evaluation metrics are general formulae to measure the accuracy and relevancy of results retrieved from the retrieval scheme against given queries from the repository of data or big data sources.

4 Results and Discussion

Our evaluation of semantic information retrieval approaches or semantic search engines alternatively based on the comparison of some recent schemes developed on different origins to achieve some common target that is the extraction of most relevant results semantically. All schemes are good and providing better performance in their relevant domains. In Table 3, there is a summary of the highest achieved values for precision and recall from different experiments performed on the specialized datasets for these schemes. The average measure of the harmonic mean (F-measure) is calculated for each of the schemes to cost mean performance.

Table 3. Comparison of semantic information retrieval systems based on Precision, Recall and F-measure values.

Scheme	Precision	Recall	F-measure
IBRI-CASONTO	97.9	97.3	97.6
SOR	81.7	78.2	79.9
ICIR	77.8	76.1	76.9
CSA	78.4	96.9	86.7
Wiser	41.5	43.5	42.5

A different dataset is selected from the relevant domains of each scheme to cost semantic information retrieval by calculating the values of previously discussed evaluation metrics on defined parameters. Some specific complex queries are tested on IBRI-CASONTO to search results from large-scale datasets (Arabic CAS_ Ontology and English CAS_ Ontology). A big collection of multimedia data consists of audio, video, image and text are used in SOR to test queries individually for each data type and then a mean value is selected for precision and recall.

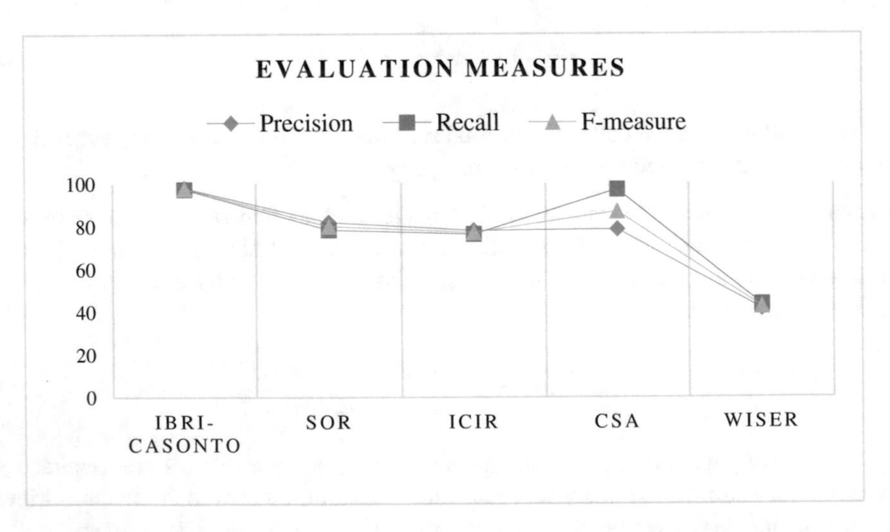

Fig. 2. Precision, recall and f-measure of five semantic search schemes

In ICIR, some specific datasets including CACM (Collection of ACM) and TREC (Text Retrieval Conference) selected for the costing or assessment purpose. The costing of second last scheme CSA is computed through applying queries on yovisto dataset targeting only text and entities. The semantics of Wiser is tested on topmost *k* result (different values of *k* and different queries) as a dataset of a large collection of documents. The chart given in Fig. 2 provides a compact description of costed values regarding each scheme selected for evaluation.

5 Conclusion

In this article, we present the evaluation of semantic information retrieval approaches through costing some common evaluation metrics. We first provide a review of some related approaches proposed by different authors. After review, we select some recent common originated schemes of this discipline and provide a brief overview of these schemes as an evaluation based on various factors. The assessment measures are settled for costing the performance outcomes of selected schemes. After this settlement, some experiments performed on specialized datasets to compute the values for these metrics. Our work provides a pathway to future researchers in finding new horizons in this area of study. In the future, we have a plan to introduce an intelligent model of big data recall engine for the improvement of semantics in the information retrieval domain.

Acknowledgments. The authors would like to thanks Universiti Malaysia Pahang for sponsoring this paper through RDU180362 grant. Special thanks also to Faculty of Computing, College of Computing and Applied Science, Universiti Malaysia Pahang.

References

1. Emani CK, Cullot N, Nicolle C (2015) Understandable big data: a survey. Comp Sci Rev 17:70–81
2. Zhang Q, Yang LT, Chen Z, Li P (2018) A survey on deep learning for big data. Inf F 42:146–157
3. Rajapoornima M, Tamilselvan L, Priyadarshini R (2016) Personalized semantic retrieval of information from large scale blog data. In: IEEE International conference on recent trends in electronics, information & communication technology (RTEICT), pp 1055–1059. IEEE Press
4. Arif MM, Peng S, Ata U, Khalid M, Abid M, Xiong L (2019) Logical tree based secure rekeying management for smart devices groups in IoT enabled WSN. IEEE Access 7:76699–76711
5. Lan M, Tan CL, Su J, Lu Y (2008) Supervised and traditional term weighting methods for automatic text categorization. IEEE Trans P Ann M Int 31:721–35
6. Cai X, Li W (2013) Ranking through clustering: an integrated approach to multi-document summarization. IEEE Trans Aud Sp Lang Proc 21:1424–33
7. Djenouri Y, Belhadi A, Fournier-Viger P, Lin JC (2018) Fast and effective cluster-based information retrieval using frequent closed itemsets. Info Sci 453:154–67
8. Sayed A, Al Muqrishi A (2017) IBRI-CASONTO: ontology-based semantic search engine. E Info J 18:181–192

9. Raza MA, Rahmah M, Ahmad N, Pasha M, Pasha U (2019) A taxonomy and survey of semantic approaches for query expansion. IEEE Access 7:17823–17833
10. Raza MA, Rahmah M, Noraziah A, Ashraf M (2018) Sensual semantic analysis for effective query expansion. Int J Adv C. S. App 9:55–60
11. Benedetti F, Beneventano D, Bergamaschi S, Simonini G (2019) Computing inter-document similarity with context semantic analysis. Inf S 80:136–147
12. Rani PS, Suresh RM, Sethukarasi R (2017) Multi-level semantic annotation and unified data integration using semantic web ontology in big data processing. C.C. 1–3
13. Hearst M (2009) Search user interfaces. C. Univ. Press
14. Mei JP, Chen L (2014) Proximity-based K-partitions clustering with ranking for document categorization and analysis. E Syst App 41:7095–7105
15. Cifariello P, Ferragina P, Ponza M (2019) Wiser: a semantic approach for expert finding in academia based on entity linking. Info S 82:1–6
16. Guo K, Liang Z, Tang Y, Chi T (2018) SOR: an optimized semantic ontology retrieval algorithm for heterogeneous multimedia big data. J Comp 28:455–465
17. Jin X, Agun D, Yang T, Wu Q, Shen Y, Zhao S (2016) Hybrid indexing for versioned document search with cluster-based retrieval. In: 25th ACM international conference on information and knowledge management, pp 377–386
18. Raiber F, Kurland O (2013) Ranking document clusters using markov random fields. In: 36th international ACM SIGIR conference on research and development in information retrieval, pp 333–342, ACM
19. Chawla S (2016) A novel approach of cluster based optimal ranking of clicked URLS Using genetic algorithm for effective personalized web search. App Soft Comput 46:90–103
20. Naini KD, Altingovde IS, Siberski W (2016) Scalable and efficient web search result diversification. ACM Trans Web (TWEB) 10(15)
21. Zemmouchi-Ghomari L, Ghomari AR (2013) Process of building reference ontology for higher education. In: Proceedings of the world congress on engineering, pp 1595–1600
22. Zemmouchi-Ghomar L, Ghomari AR (2013) Towards a reference ontology for higher education knowledge domain. Int R Comp S 2:474–88
23. Mesaric J, Dukic B (2007) An approach to creating domain ontologies for higher education in economics. In: 29th international conference on information technology interfaces, pp 75–80. IEEE Press
24. Ramachandran A, Sujatha R (2011) Semantic search engine: a survey. Int J C Tech Apps 2
25. Munir K, Anjum MS (2018) The use of ontologies for effective knowledge modelling and information retrieval. A Comp Info 14:116–126
26. Chen M, Décary M (2018) A cognitive-based semantic approach to deep content analysis in search engines. In: 12th IEEE international conference on semantic computing (ICSC), pp 131–139. IEEE Press
27. Lashkari F, Ensan F, Bagheri E, Ghorbani AA (2017) Efficient indexing for semantic search. E Sys App 73:92–114

The Comparison of Pooling Functions in Convolutional Neural Network for Sentiment Analysis Task

Nurul Ashikin Samat[1(✉)], Mohd Najib Mohd Salleh[2], and Haseeb Ali[1]

[1] Faculty of Computer Science and Information Technology, Universiti Tun Hussein Onn Malaysia (UTHM), 86400 Parit Raja, Johor, Malaysia
nurulashikinsamat@gmail.com,
chuadharyhaseebali@gmail.com

[2] Department of Software Engineering, Faculty of Computer Science and Information Technology, Universiti Tun Hussein Onn Malaysia (UTHM), Parit Raja, Johor, Malaysia
najib@uthm.edu.my

Abstract. Convolutional Neural Network (CNN) has gained considerable attention in many Natural Language Processing applications including sentiment analysis task. A typical CNN model usually is made up of several convolutional and pooling layers. In this paper, our aim is to acquire detailed understanding into different type of pooling function by directly differentiate them on a same architecture layers for sentiment analysis tasks. These insights should prove useful for future development of pooling function in CNN models for sentiment analysis task.

Keywords: Pooling function · Convolutional neural network · Natural language processing · Sentiment analysis

1 Introduction

In recent years, sentiment analysis has grown to be one of the most active researches in Natural Language Processing (NLP) as texts often carry rich semantic information that is useful for sentiment understandings [1]. It can also be called as opinion mining [1] or sentiment classification [2]. Sentiment analysis analyses people's sentiments, opinions, feedbacks, emotions and appraisals towards specific entities such as services, products, issues, events and individuals [3–5]. Generally, the text will be classified into 3 sentiments which are positive, negative or neutral sentiments [6]. However, there are some studies classify the text into sentiment ratings or scales such as 1 = very dissatisfied, 2 = dissatisfied, 3 = neutral, 4 = satisfied, and 5 = very satisfied. Sentiment analysis provides valuable insights to organizations, business opportunities, and marketing and promotions teams.

In addition with the rapid growth in various social medias such as Facebook and Twitter, and not to mention online review sites such as Amazon and Internet Movie

R. Ghazali et al. (Eds.): SCDM 2020, AISC 978, pp. 202–210, 2020.
https://doi.org/10.1007/978-3-030-36056-6_20

Database (IMDB) have draws great attentions from researchers and industries to gain valuable knowledge from those data [7, 8]. Although these data are valuable, they are typically unstructured. There are two common approaches applied in sentiment analysis which are (1) machine learning approach such as Decision Tree, Support Vector Machines (SVM), Neural Network and (2) lexicon-based approach.

With the recent developments in deep learning, the capacity of machine learning algorithms to analyse the text has improved greatly. Convolutional Neural Network (CNN) has been widely implemented as a powerful deep learning technique not only focused on visual application such as image processing [9, 10], video processing [11], and emotion computational models [12–14], but also in Natural Language Processing (NLP) tasks [15]. In this paper, CNN algorithm has been applied to perform the sentiment analysis. With great ability of feature extraction and self-learning, CNN can give higher accuracy in determining the sentiment. Lai et al. [16] mentioned that the CNN may capture the semantic of texts better compared to recursive or recurrent neural network. In CNN architecture, pooling layer is implemented in order to reduce the number of parameters during the trainings. To better understand of pooling layer, an experiment was conducted to compare the performance of different pooling functions.

The rest of the paper is organized as follows: Sect. 2 presents works related to sentiment analysis or polarity detection using CNN. Section 3 presents the architecture in CNN. Section 4 discussed the experimental results. Finally, conclusions and ideas for further works were stated in Sect. 5.

2 Related Works

The past decade has seen the rapid development of sentiment analysis in many research works. Most of the works were focused on English documents and texts. However, there have been increasing researches in Arabic, Chinese, and Spanish languages. In early study of sentiment analysis, Kim [17] has demonstrated sentence classification using CNN. Author achieved outstanding results with parameters tuning in simple CNN architecture. Kalchbrenner et al. [18] tested CNN in four experiments to perform sentiment analysis using Twitter dataset.

In recent years, several studies have focused and published on CNN for sentiment analysis. Dos Santos and Gatti [19] have proposed sentiment analysis focused on social media short text in Twitter such as single sentences and messages using deep convolutional neural network. Meanwhile, Xu et al. [20] have proposed a method for evaluating the sentiment in Chinese text based on the convolutional neural network in order to improve the analysis accuracy. The experiments performed better compared to traditional learning algorithms such as support vector machine and AdaBoost. Liao et al. [21] proposed an approach using CNN to perform sentiment analysis, in order to understand the situation in real world. Their analysis revealed that it is possible to forecast users' satisfaction, happiness, and miserable feelings with some products or situation. Li et al. [22] have proposed sentiment analysis towards Chinese micro blogging systems. Authors highlighted that CNN will automatically mine useful features and perform better sentiment analysis. Abid et al. [23] proposed joint architecture for the sentiment analysis on small, medium and large datasets using CNN.

There have been several studies in the literature reporting about pooling layer in different areas. Scherer et al. [24] carried out several experiments to test the performance of pooling functions for object recognitions. Wang et al. [25] compared five different pooling functions for sound event detection. Meanwhile, [26] presented their works on action recognition.

Although previous research works show that CNN have been applied for sentiment analysis and demonstrate good performance, there is still little discussion about the effects of different pooling layer on sentiment analysis tasks. Thus, in this paper, we apply different type of CNN architecture with different type of pooling function to predict the sentiment analysis. We aim to provide clear insights regarding pooling layer in CNN focusing on sentiment analysis task.

3 Architecture of CNN

In this paper, basic architecture of CNN was implemented for each dataset to test the effects of pooling layer. The architecture of CNN is typically consisted of three different types of layers which are (1) Convolutional Layer, (2) Pooling Layer, and (3) Fully Connected Layer. There are no definite rules to determine how many specific layers should be employed in the architecture. Nevertheless, CNN is typically consisted two parts which are (1) Feature Extraction Part using Convolutional Layer and Pooling Layer, and (2) Classification Part using Fully Connected Layer. Figure 1 shows one of the architectures used in this paper. The next subsections will give a brief overview of each layer in CNN.

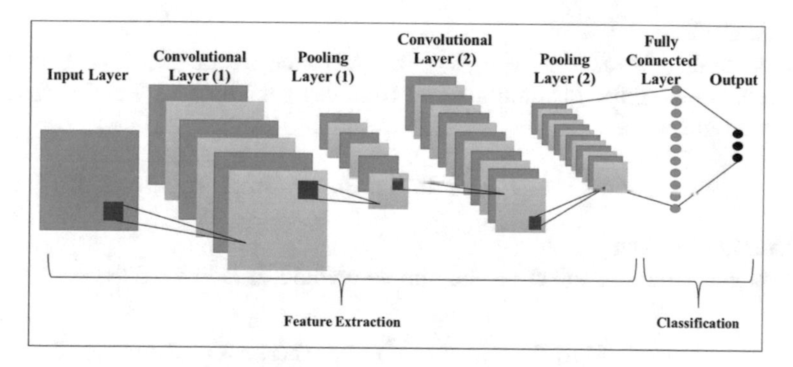

Fig. 1. Architecture of Convolutional Neural Network (CNN) consists of Input Layer, two Convolutional Layers, two Pooling Layers and Fully Connected Layer.

3.1 Convolutional Layer

Convolutional layer is the first layer to extract the features from an input. For NLP areas, the text input will be encoded into vectors form. Usually, these vectors are word embedding consist of pre trained word vectors such as word2vec or GloVe [17]. In this layer, filter is used and convolved (slide) across the input vectors, to identify the specific features. The dimensionality of the feature map generated by each filter will vary as a function of the sentence length and the filter region size. A pooling function is thus applied to each feature map to induce a fixed-length vector.

3.2 Pooling Layer

Pooling layer is typically implemented after the convolutional layers. Pooling layer will subsample the input from previous layer and lower the computational burden [27]. It is an important step in CNN. There are some common pooling functions have been implemented in CNN as shows in Fig. 2.

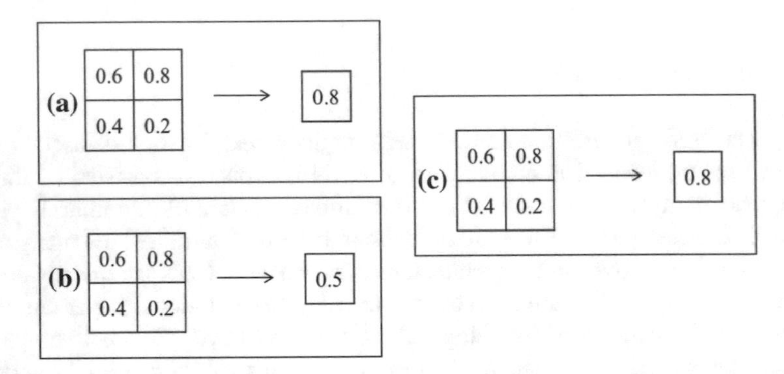

Fig. 2. Simple example for (a) Max Pooling, (b) Average Pooling and (c) Stochastic Pooling.

(1) **Max Pooling**
Max Pooling is used more widely in CNN architecture [28, 29]. Max pooling involves taking the maximum value from region. The Max Pooling function is expressed as:

$$\text{Let } x_i \in [0,1] \text{ then, } y = \max \; x_i \tag{1}$$

(2) **Average Pooling**
Average Pooling is effective and simple method. It is defined as:

$$\text{Let } x_i \in [0,1] \text{ then, } y = \frac{1}{n}\sum x_i \text{ where } n \text{ is total number of } x_i \tag{2}$$

(3) **Stochastic Pooling**
Stochastic Pooling [30] is improved pooling function from Max and Average poling shortcoming. It wills randomly sampling based on the probability values of each region. The probabilities, p is calculated as follow:

$$\text{Let}\, x_i \in [0,1],\ \text{then},\ p_{x_i} = \frac{x_i}{\frac{1}{n}\sum x_i} \tag{3}$$

Then, the pooled activation is selected from multinomial distribution based on p. The process is detailed as follow:

$$y = y_l \ \text{where}\ l \sim P(p_i, \ldots, p_n) \tag{4}$$

3.3 Fully Connected Layer

Fully connected layer accomplishes the training network based on the feature extracted in previous layers. It is similar to the fully connected network in the conventional Neural Network model. The output will pass through nonlinear activation function such as Rectified Linear Unit (ReLU) or Softmax approach.

4 Experimental Design and Results

In this paper, several experiments were conducted in order to evaluate the Pooling Layer in CNN. The purpose in these experiments is to answer whether different pooling function affects the accuracy of sentiment analysis.

4.1 Experimental Design

All the experiments were accomplished using Google Colaboratory, also known as Google Colab [31]. It is a cloud service based on Jupyter Notebooks. It provides free of charge access to a robust GPU. The performance of Google Colaboratory is equivalent to the performance of GPU hardware [32].

In this experiments, the benchmark dataset was collected from [33]. The dataset can be downloaded from Kaggle website [34]. The dataset consist 50,000 IMDB movie reviews for sentiment analysis. The sentiments score is divided into two labels which are positive and negative.

In this study, the data is partitioned into 75% for training and 25% for testing phase. Training set will build the CNN algorithm and then will be tested with remaining percentage for testing set.

Figure 3 presents the experimental design in this paper. In Stage 1, data cleansing and preprocessing were performed. Since the texts were obtained from IMDB website, it is necessary to remove the unwanted characters and tags. It is crucial to implement preprocessing in NLP applications in order to get a reliable knowledge. Then, sentiment analysis was performed using CNN.

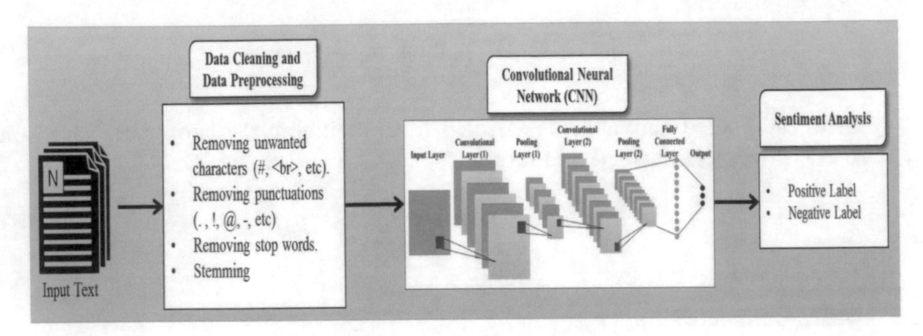

Fig. 3. Experimental design.

In order to compare the different pooling functions, we designed 3 different networks layer in CNN architecture. The summary of different architectures used was shown in Table 1.

Table 1. Summary of CNN architecture.

Type of architecture	Descriptions
Architecture 1	Input-Conv1-Pool1-Conv2-Pool2-FullConn-Output
Architecture 2	Input-Conv1-Pool1-Conv2-Pool2-Conv3-Pool3-FullConn-Output
Architecture 3	Input-Conv1-Pool1-Conv2-Pool2-Conv3-Pool3-Conv4-Pool4-FullConn-Output

Conv = Convolutional Layer, Pool = Pooling Layer, FullConn = Fully Connected Layer

4.2 Results

In this paper, CNN results are compared with 3 different pooling layers which are Max, Average and Stochastic Pooling. The performance will be evaluated with accuracy.

Table 2. Summary of performance comparison with 5 epochs.

	Architecture 1	Architecture 2	Architecture 3
(1) Max pooling	0.4976	0.9044	0.9722
(2) Average pooling	0.4979	0.9453	0.9021
(3) Stochastic pooling	0.4398	0.8535	0.9561

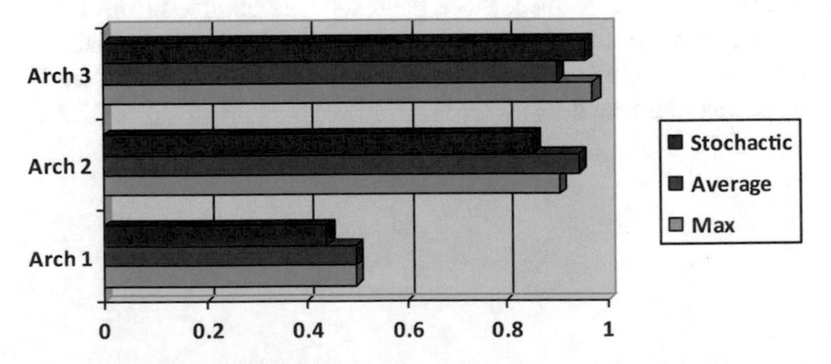

Fig. 4. Performance of accuracy comparison for 5 epochs.

Table 2 presents the results for 5 epochs CNN architectures. It shows that Max Pooling and Stochastic Pooling get better when we increase the number of convolutional layer and pooling layer. Afterwards, Fig. 3 illustrates the accuracy from the table.

Table 3. Summary of performance comparison with 10 epochs.

	Architecture 1	Architecture 2	Architecture 3
(1) Max pooling	0.5023	0. 9287	0. 9773
(2) Average pooling	0.4979	0.9073	0.9305
(3) Stochastic pooling	0.6047	0.9492	0.9558

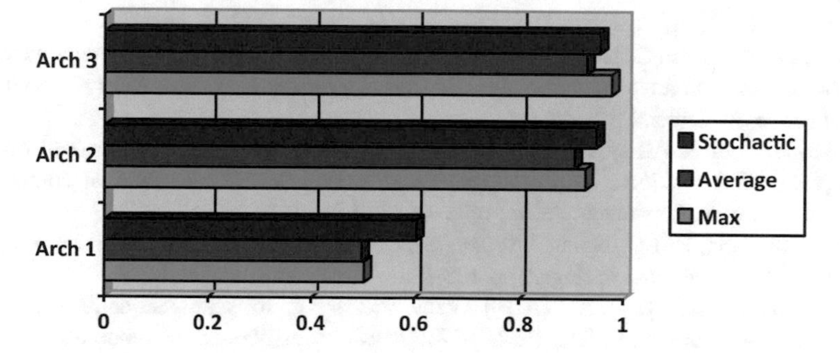

Fig. 5. Performance of accuracy comparison for 10 epochs.

Table 3 demonstrates the CNN accuracy for 10 epochs. It shows that Max, Average and Stochastic Pooling get better when we increase the number of convolutional layer and pooling layer. The Fig. 4 exhibits the accuracy from the table.

Figures 4 and 5 highlights the analysis of different pooling functions in CNN architecture. We implement three pooling functions: Max Pooling, Average Pooling and Stochastic Pooling. From our observations, Max and Average Pooling are simple and easy way to implement and consume less complexity. But both can lead to over fitting on training set and cause poor accuracy for sentiment analysis. Meanwhile, Stochastic Pooling is based on probability and success to overcome the issue in Max and Average Pooling. It also provides more constant accuracy for sentiment analysis. The main challenge is to create more efficient way to detect the relation in the sentences for sentiment analysis. In order to find the important relation in the sentences, Pooling Layer has been a great way to capture it. It is due to the ability of pooling layer to subsample the output without leaving any data. This allows for extracting most important values from the input sentences. Therefore, future works regarding pooling layer is important to increase the accuracy for CNN in sentiment analysis.

5 Conclusion

Pooling layer is necessary step in construction Convolutional Neural Network for sentiment analysis task. In this paper, we implement three different pooling layers in same architecture. We investigate the effects of pooling layer towards the accuracy of sentiment analysis. The results show that choosing the right pooling function can significantly improve the analysis accuracy on the dataset.

References

1. Sun S, Luo C, Chen J (2017) A review of natural language processing techniques for opinion mining systems. Inf Fusion 36:10–25
2. Wu Z, Dai X-Y, Yin C, Huang S, Chen J (2018) Improving review representations with user attention and product attention for sentiment classification. In: Thirty-second AAAI conference on artificial intelligence
3. Yaqub U, Chun S, Atluri V, Vaidya J (2017) Sentiment based analysis of tweets during the us presidential elections. In: Proceedings of the 18th annual international conference on digital government research. ACM, pp 1–10
4. Anandarajan M, Hill C, Nolan T (2019) Sentiment analysis of movie reviews using R. In: Practical text analytics. Springer, pp 193–220
5. Ramanathan V, Meyyappan T (2019) Twitter text mining for sentiment analysis on people's feedback about oman tourism. In: 2019 4th MEC international conference on big data and smart city (ICBDSC). IEEE, pp 1–5
6. Serrano-Guerrero J, Olivas JA, Romero FP, Herrera-Viedma E (2015) Sentiment analysis: a review and comparative analysis of web services. Inf Sci 311:18–38
7. Goswami S, Nandi S, Chatterjee S (2019) Sentiment analysis based potential customer base identification in social media. In: Contemporary advances in innovative and applicable information technology. Springer, pp 237–243
8. Rodríguez A, Argueta C, Chen Y-L (2019) Automatic detection of hate speech on facebook using sentiment and emotion analysis. In: 2019 international conference on artificial intelligence in information and communication (ICAIIC). IEEE, pp 169–174
9. Zhang Y-D, Dong Z, Chen X, Jia W, Du S, Muhammad K, Wang S-H (2019) Image based fruit category classification by 13-layer deep convolutional neural network and data augmentation. Multimed Tools Appl 78:3613–3632
10. Geetharamani G, Pandian A (2019) Identification of plant leaf diseases using a nine-layer deep convolutional neural network. Comput Electr Eng 76:323–338
11. Akiyama T, Kobayashi Y, Kishigami J, Muto K (2018) CNN-based boat detection model for alert system using surveillance video camera. In: 2018 IEEE 7th global conference on consumer electronics (GCCE). IEEE, pp 669–670
12. Kim H, Kim Y, Kim SJ, Lee I (2018) Building emotional machines: recognizing image emotions through deep neural networks. IEEE Trans Multimedia 20:2980–2992
13. Dhall A, Kaur A, Goecke R, Gedeon T (2018) Emotiw 2018: audio-video, student engagement and group-level affect prediction. In: Proceedings of the 2018 on international conference on multimodal interaction. ACM, pp 653–656
14. Ceruti C, Cuculo V, D'Amelio A, Grossi G, Lanzarotti R (2017) Taking the hidden route: deep mapping of affect via 3D neural networks. In: International conference on image analysis and processing. Springer, pp 189–196

15. Young T, Hazarika D, Poria S, Cambria E (2018) Recent trends in deep learning based natural language processing. ieee Comput Intell Mag 13:55–75
16. Lai S, Xu L, Liu K, Zhao J (2015) Recurrent convolutional neural networks for text classification. In: Twenty-ninth AAAI conference on artificial intelligence
17. Kim Y (2014) Convolutional neural networks for sentence classification. arXiv:1408.5882
18. Kalchbrenner N, Grefenstette E, Blunsom P (2014) A convolutional neural network for modelling sentences. arXiv:1404.2188
19. Dos Santos C, Gatti M (2014) Deep convolutional neural networks for sentiment analysis of short texts. In: Proceedings of COLING 2014, the 25th international conference on computational linguistics: technical papers, pp 69–78
20. Xu F, Zhang X, Xin Z, Yang A (2019) Investigation on the Chinese text sentiment analysis based on convolutional neural networks in deep learning. Comput Mater Contin 58(3):697–709
21. Liao S, Wang J, Yu R, Sato K, Cheng Z (2017) CNN for situations understanding based on sentiment analysis of twitter data. Procedia Comput Sci 111:376–381
22. Li Q, Jin Z, Wang C, Zeng DD (2016) Mining opinion summarizations using convolutional neural networks in Chinese microblogging systems. Knowl-Based Syst 107:289–300
23. Abid F, Alam M, Yasir M, Li C (2019) Sentiment analysis through recurrent variants latterly on convolutional neural network of Twitter. Futur Gener Comput Syst 95:292–308
24. Scherer D, Müller A, Behnke S (2010) Evaluation of pooling operations in convolutional architectures for object recognition. In: International conference on artificial neural networks. Springer, pp. 92–101
25. Wang Y, Li J, Metze F (2019) A comparison of five multiple instance learning pooling functions for sound event detection with weak labeling. In: ICASSP 2019-2019 IEEE international conference on acoustics, speech and signal processing (ICASSP). IEEE, pp 31–35
26. Wang X, Wang L, Qiao Y (2012) A comparative study of encoding, pooling and normalization methods for action recognition. In: Asian conference on computer vision. Springer, pp 572–585
27. Gu J, Wang Z, Kuen J, Ma L, Shahroudy A, Shuai B, Liu T, Wang X, Wang G, Cai J (2018) Recent advances in convolutional neural networks. Pattern Recogn 77:354–377
28. Severyn A, Moschitti A (2015) Learning to rank short text pairs with convolutional deep neural networks. In: Proceedings of the 38th international ACM SIGIR conference on research and development in information retrieval. ACM, pp 373–382
29. Williams T, Li R (2018) Wavelet pooling for convolutional neural networks
30. Zeiler MD, Fergus R (2013) Stochastic pooling for regularization of deep convolutional neural networks. arXiv:1301.3557
31. Google Colaboratory. https://colab.research.google.com/
32. Carneiro T, Da Nóbrega RVM, Nepomuceno T, Bian G-B, De Albuquerque VHC, Reboucas Filho PP (2018) Performance analysis of google colaboratory as a tool for accelerating deep learning applications. IEEE Access 6:61677–61685
33. Maas AL, Daly RE, Pham PT, Huang D, Ng AY, Potts C (2011) Learning word vectors for sentiment analysis. In: Proceedings of the 49th annual meeting of the association for computational linguistics: human language technologies, vol 1. Association for Computational Linguistics, pp 142–150
34. Kaggle. https://www.kaggle.com/c/word2vec-nlp-tutorial

Interval Type-2 Fuzzy Multi Criteria Decision Making Based on Intuitive Multiple Centroid

Ku Muhammad Naim Ku Khalif[1(✉)], Alexander Gegov[2], Ahmad Syafadhli Abu Bakar[3,4], and Noor Zuraidin Mohd Safar[5]

[1] Centre for Mathematical Sciences, Universiti Malaysia Pahang, Seberang Perai, Malaysia
kunaim@ump.edu.my

[2] School of Computing, University of Portsmouth, Portsmouth, UK
alexander.gegov@port.ac.uk

[3] Centre for Foundation Studies in Science, Universiti Malaya, Kuala Lumpur, Malaysia
ahmadsyafadhli@um.edu.my

[4] Centre of Research for Computational Sciences and Informatics in Biology, Bioindustry Environment, Agriculture and Healthcare (CRYSTAL), Universiti Malaya, Kuala Lumpur, Malaysia

[5] Faculty of Computer Science and Information Technology, Universiti Tun Hussien Onn Malaysia, Batu Pahat, Malaysia
zuraidin@uthm.edu.my

Abstract. This paper aims to introduce fuzzy multi criteria decision making model using consistent fuzzy preference relations and fuzzy technique for order performance by similarity to ideal solution sets that is incorporated with intuitive multiple centroid defuzzification in the context of interval type-2 fuzzy. The implementation of interval type-2 fuzzy sets is taken into consideration, where it has more authority to provide more degree of freedom in representing the uncertainty of human based decision making problems. It also highlights the combination of interval type-2 fuzzy sets with multi criteria decision making techniques allow the use of fuzzy linguistic by considering the need of human intuition in decision making problems under uncertain environment. Numerical example is included to illustrate the proposed model. The proposed model is importantly needed to validate using sensitivity analysis in order to analyse the quality and robustness of the model in giving the most promising alternative with respect to resources. The results show that it is highly practical to use the proposed model in decision making evaluation.

Keywords: Fuzzy multi criteria decision making · Consistent fuzzy preference relations · Fuzzy TOPSIS · Interval type-2 fuzzy sets · Intuitive multiple centroid · Uncertainty

R. Ghazali et al. (Eds.): SCDM 2020, AISC 978, pp. 211–221, 2020.
https://doi.org/10.1007/978-3-030-36056-6_21

1 Introduction

Uncertainty and fuzziness are well-known phenomena in many applications areas in science and engineering, where are often not crisp but there exist various degree of membership grade that practical automatically occurs in decision making problems. Type-2 fuzzy sets are appropriately tools for uncertainty or approximate reasoning modelling. It has more authority to provide more degree of freedom in representing the uncertainty of human based decision making problems. Klir and Yuan claim that the type-1 fuzzy sets inly describe imprecision not uncertainty [1]. On particular motivation for the further interest in type-2 fuzzy sets that its' provide a better scope for modelling uncertainty than type-1 fuzzy sets [2]. According to Karnik and Mendel, they claim that type-2 fuzzy sets can be characterised as fuzzy membership function where the membership value for type-2 fuzzy sets is in interval form [0,1], unlike type-1 fuzzy sets where the membership value is a crisp value in [0,1] [3].

Defuzzification plays important role in the performance of fuzzy systems' modelling techniques. Defuzzification process is guided by the output fuzzy subset that one value would be selected as a single crisp value as the system output. While much of the literature discuss there are variety of defuzzification methods have largely developed. Though, each of them have difference performance in difference applications and there is a general method can satisfy the performance in all conditions which is centroid method [4]. Centroid defuzzification methods of fuzzy numbers have been explored for the last decade that commonly used and have been applied in various discipline areas. The computational complexity of type-2 fuzzy sets is very difficult to handle into practical applications because of characterised by their footprint of uncertainty [5].

In literature, most of the hybrid MCDM model combined two techniques in order to tackle the evaluation of criteria and the evaluation of alternatives respectively. The evaluation process of criteria and alternatives play important role in MCDM techniques requirements. To identify the best decision to be made among the various alternatives with several criteria, the methodology has to study the preferences among the criteria to make sure the weights of criteria are reliable enough to be implemented in the selection of alternatives. In this paper, the hybrid of consistent fuzzy preference relations and fuzzy technique for order of preference by similarity to ideal solution (TOPSIS) using new centroid defuzzification for interval type-2 fuzzy sets is proposed in dealing with uncertainty events. The major weaknesses of classical TOPSIS are in not providing for weight elicitation, and consistency checking for judgments' evaluation. Hence, in this paper, the authors consider the fuzzy TOPSIS's employment that has been significantly restrained by the human capacity for information processing. Sensitivity analysis [6] is applied to validate the proposed model. It can effectively contributes to making accurate decisions by assuming that a set of weights for criteria or alternatives then obtained a new round of weights for them, so that the efficiency of alternatives has become equal or their order has changed.

The rest of this paper is organised as follows: Sect. 2 discusses the theoretical preliminaries of fuzzy set theory and generalised trapezoidal fuzzy numbers. This is then preceded to the proposed work of integrated fuzzy MCDM model that consist of consistent fuzzy preference relations and fuzzy TOPSIS using intuitive multiple centroid defuzzification in Sect. 3. Section 4 discusses the case study and results that illustrated the proposed model and validation processes using sensitivity analysis. Finally, Sect. 5 gives the conclusion.

2 Theoretical Preliminaries

In this section, the authors briefly review some definitions of interval type-2 fuzzy set that are illustrated as follows.

A. Interval Type-2 Fuzzy Set

Definition 1 [7] : A type-2 fuzzy set $\overset{\approx}{A}$ in the universe of discourse X represented by the type-2 membership function, μ. If all $\mu_{\overset{\approx}{A}}(x,u) = 1$, then $\overset{\approx}{A}$ is called an interval type-2 fuzzy sets. An interval type-2 fuzzy set can be considered as a special case of type-2 fuzzy sets, denoted as follows.

$$\overset{\approx}{A} = \int_{x \in X} \int_{u \in J_x} 1/(x,u), \text{ where } J_x \subseteq [0,1] \quad (1)$$

Definition 2 [7]: The upper and lower membership functions of an interval type-2 fuzzy set are type-1 fuzzy sets membership functions, respectively. A trapezoidal interval type-2 fuzzy set can be represented by, $\underset{i}{\overset{\approx}{A}} = (\tilde{A}_i^U, \tilde{A}_i^L) = ((a_{i1}^U, a_{i2}^U, a_{i3}^U, a_{i4}^U; H_1\ (\tilde{A}_i^U), \tilde{H}_2(\tilde{A}_i^L)), (a_{i1}^L, a_{i2}^L, a_{i3}^L, a_{i4}^L; H_1(\tilde{A}_i^L), H_2(\tilde{A}_i^L)))$ where can be depicted in Fig. 1 [8]. The $\tilde{A}_i^U$ and $\tilde{A}_i^L$ are type-1 fuzzy sets, $a_{i1}^U, a_{i2}^U, a_{i3}^U, a_{i4}^U, a_{i1}^L, a_{i2}^L, a_{i3}^L$ and a_{i4}^L are the reference points of the interval type-2 fuzzy sets $\overset{\approx}{A}$, $H_j(\tilde{A}_i^U)$ denote the membership value of the element $a_{i(j+1)}^U$ in the upper trapezoidal membership function $\tilde{A}_i^U$, $1 \le j \le 2$, $H_j(\tilde{A}_i^L)$ denotes the membership value of the element $a_{i(j+1)}^L$ in the lower trapezoidal membership function $\tilde{A}_i^L$, $1 \le j \le 2$, and for $H_1(\tilde{A}_i^U) \in [0,1]$, $H_2(\tilde{A}_i^U) \in [0,1]$, $H_1(\tilde{A}_i^L) \in [0,1]$, $H_2(\tilde{A}_i^L) \in [0,1]$ and $1 \le i \le n$, $H_2(\tilde{A}_i^U) \in [0,1]$.

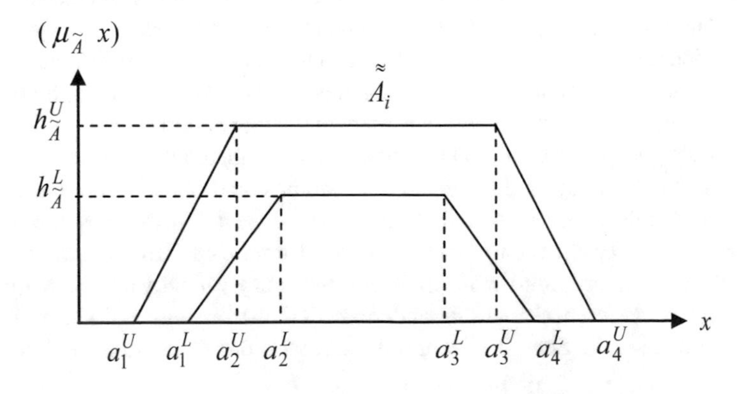

Fig. 1. The representation of interval type-2 fuzzy set.

3 Proposed Model

This section focuses on the development of fuzzy MCDM model that is incorporated with intuitive multiple centroid for interval type-2 fuzzy sets.

Step 1: Determine the weights of evaluation criteria.

The weighting of evaluation criteria are employed.

Step 2: Construct a pairwise comparison matrices.

The pairwise comparison matrices are constructed among all criteria in the dimension of the hierarchy systems based on the decision makers' preferences using following matrix:

$$A = \begin{bmatrix} 1 & \tilde{a}_{12} & \cdots & \tilde{a}_{1n} \\ \tilde{a}_{21} & 1 & \cdots & \tilde{a}_{2n} \\ \vdots & \vdots & \ddots & \vdots \\ \tilde{a}_{n1} & \tilde{a}_{n2} & \cdots & 1 \end{bmatrix} = \begin{bmatrix} 1 & \tilde{a}_{12} & \cdots & \tilde{a}_{1n} \\ 1/\tilde{a}_{12} & 1 & \cdots & \tilde{a}_{2n} \\ \vdots & \vdots & \ddots & \vdots \\ 1/\tilde{a}_{1n} & 1/\tilde{a}_{2n} & \cdots & 1 \end{bmatrix} \quad (2)$$

Step 3: Aggregate the decision makers' preferences.

The pairwise comparison matrices of decision makers' preferences are aggregated using equation below:

$$\tilde{a}_{ij} = (\tilde{a}_{ij}^{1} \times \tilde{a}_{ij}^{2} \times \ldots \times \tilde{a}_{ij}^{n})^{1/k} \quad (3)$$

where k is the number of decision makers and $i = 1,2,\ldots m$; $j = 1,2,\ldots n$.

Step 4: Defuzzify the fuzzy numbers of aggregation's result of decision makers' preferences.

The intuitive multiple centroid (IMC) defuzzification is extension from the classical vectorial centroid method for fuzzy numbers that proposed by [9, 10]. The concept is similar like the other centroid methods, to find the best centre point of fuzzy numbers that represent in crisp values or single values. Comparing to other centroid methods, IMC produces the appropriate way to get the output values that are more intelligent manner, easy to compute, more balance and consider all possible cases of fuzzy numbers. The IMC formula can be summarised as follows:

$$IMC_{\tilde{\tilde{A}}}(\tilde{x}_{\tilde{A}}, \tilde{y}_{\tilde{A}}) = \left(\frac{2(a_1^U + a_1^L + a_4^U + a_4^L) + 7(a_2^U + a_2^L + a_3^U + a_3^L)}{36}, \frac{7}{36}(h_{\tilde{A}}^U + h_{\tilde{A}}^L) \right) \quad (4)$$

Step 5: Compute the centroid index of intuitive multiple centroid of $(\tilde{x}_{\tilde{A}}, \tilde{y}_{\tilde{A}})$ ***with vertices*** $\alpha_{\overline{\alpha},\underline{\alpha}}$, $\beta_{\overline{\beta},\underline{\beta}}$ ***and*** $\gamma_{\overline{\gamma},\underline{\gamma}}$.

Centroid index of intuitive multiple centroid can be generated using Euclidean Distance by [11] as:

$$R(\tilde{A}) = \sqrt{\tilde{x}^2 + \tilde{y}^2} \tag{5}$$

Step 6: Compute the criteria values as weightage for alternatives' evaluation using consistent fuzzy preference relations.

Consistent fuzzy preference relations was proposed by [12] for constructing the decision matrices of pairwise comparisons based on additive transitivity property. Referring to [13], a fuzzy preference relation R on the set of the criteria or alternatives A is a fuzzy set stated on the Cartesian product set $A \times A$ with the membership function $\mu_R : A \times A \rightarrow [0, 1]$. The preference relation is denoted by $n \times n$ matrix $R = (r_{ij})$ where $r_{ij} = \mu_y(a_i, a_j)$ $\forall i, j \in \{1, \ldots, n\}$. The preference ratio, r_{ij} of the alternative a_i to a_j is determined by

$$r_{ij} = \begin{cases} 0.5 & a_i \textit{ is different to } a_j \\ (0.5, 1) & a_i \textit{ is preferred then } a_j \\ 1 & a_i \textit{ is absolutely preferred then } a_j \end{cases}$$

The preference matrix R is presumed to be additive reciprocal, $p_{ij} + p_{ji} = 1,\ \forall i,\ j \in \{1, \ldots, n\}$. Several propositions are associated to the consistent additive preference relations as follows:

Proposition 1 [14]: *Consider a set of criteria or alternatives, $X = \{x_1, \ldots, x_n\}$, and associated with a reciprocal multiplicative preference relation $A = (a_{ij})$ for $a_{ij} \in \left[\frac{1}{9}, 9\right]$. Then, the corresponding reciprocal fuzzy preference relation, $P = (p_{ij})$ with $p_{ij} \in [0, 1]$ associated with A is given by the equation*

$$p_{ij} = g(a_{ij}) = \frac{1}{2}(1 + \log_9 a_{ij}) \tag{6}$$

Generally, if $a_{ij} \in \left[\frac{1}{n}, n\right]$, then $\log_n a_{ij}$ is used, in particular, when $a_{ij} \in \left[\frac{1}{9}, 9\right]$; $\log_9 a_{ij}$ is considered as in the above proposition because a_{ij} is between $\frac{1}{9}$ *and 9. If a_{ij}* is between $\frac{1}{7}$ *and 7, then $\log_7 a_{ij}$ is used.*

Proposition 2 [14]: *For a reciprocal fuzzy preference relation $P = (p_{ij})$, the following statements are equivalent.*

$$(i)\quad p_{ij} + p_{jk} + p_{ki} = \frac{3}{2},\ \forall i, j, k \tag{7}$$

$$(ii)\quad p_{ij} + p_{jk} + p_{ki} = \frac{3}{2},\ \forall i < j < k \tag{8}$$

Proposition 3 [14]: *For a reciprocal fuzzy preference relation $P = (p_{ij})$, the following statements are equivalent*

$$(i)\quad p_{ij} + p_{jk} + p_{ki} = \frac{3}{2},\ \forall i < j < k$$

$$(ii) \quad p_{i(i+1)} + p_{(i+1)(i+2)} + \ldots + p_{(j-1)j} + p_{ji} = \frac{j-i+1}{2}, \ \forall i<j \tag{9}$$

Proposition 3 is crucial because it can be used to construct a consistent fuzzy preference relations form the set of $n-1$ values $\{p_{12}, p_{23}, \ldots, p_{n-1}\}$. A decision matrix with entries that are not in the interval $[0, 1]$, but in an interval $[-c, 1+c]$, $c > 0$, can be obtained by transforming the obtained values using a transformation function that preserves reciprocity and additive consistency with the function

$$f : [-c, 1+c] \rightarrow [0, 1], \ f(x) = \frac{(x+c)}{(1+2c)} \tag{10}$$

Step 7: Ranking evaluation of alternatives using fuzzy TOPSIS

Concept of TOPSIS method originally proposed by [15]. They claim that the alternative should not be chosen based on having the shortest distance from the positive ideal reference point (PIRT) only, but also have the longest distance from the negative ideal reference point (NIRP) in solving the MCDM problems. Here, the extension of fuzzy TOPSIS is illustrated differs from others in terms of the usage of defuzzification method, normalization process and ranking. The fuzzy decision matrix is constructed and the linguistic terms from fuzzy numbers are used to evaluate the alternatives with respect to criteria. Then, aggregate DMs' preferences:

$$\overline{DM} = \begin{array}{c} \\ A_1 \\ A_2 \\ \vdots \\ A_m \end{array} \begin{bmatrix} C_1 & C_2 & \cdots & C_n \\ \tilde{x}_{11} & \tilde{x}_{12} & \cdots & \tilde{x}_{1n} \\ \tilde{x}_{21} & \tilde{x}_{22} & \cdots & \tilde{a}_{2n} \\ \vdots & \vdots & \ddots & \vdots \\ \tilde{x}_{m1} & \tilde{x}_{m1} & \cdots & \tilde{x}_{mn} \end{bmatrix} \tag{11}$$

$$i = 1, 2, \ldots, m; \quad j = 1, 2, \ldots, n \quad , \tilde{x}_{ij} = \frac{1}{K}\left(\tilde{x}_{ij}^{1} \oplus, , , \oplus \tilde{x}_{ij}^{k} \oplus, , , \tilde{x}_{ij}^{K}\right)$$

where x_{ij} is the performance rating of alternatives, A_i with respect to criterion C_i evaluated by *kth* experts and $\tilde{x}_{ij} = \frac{1}{K}\left(\tilde{x}_{ij}^{1} \oplus, , , \oplus \tilde{x}_{ij}^{k} \oplus, , , \tilde{x}_{ij}^{K}\right)$. Fuzzy decision matrix is weighted and normalised. Then, defuzzify the standardised generalised fuzzy numbers into coordinate form, $(\tilde{x}, \tilde{y})$. The weighted fuzzy normalised decision matrix is denoted by $\tilde{V}$ as depicted below:

$$\tilde{V} = [\tilde{v}_{ij}]_{m \otimes n}; \quad i = 1, 2, \ldots, m; \quad j = 1, 2, \ldots, n \tag{12}$$

where

$$\tilde{v}_{ij} = \tilde{x}_{ij} \times \tilde{w}_j \tag{13}$$

Normalised each generalised trapezoidal fuzzy numbers into standardised generalised fuzzy numbers. The weights from consistent fuzzy preference relations are adopted here. Defuzzify the standardised generalised fuzzy numbers using intuitive multiple centroid, then translate them into the index point. Use the new point of $y_{\tilde{A}_i^*}$ to compute the index centroid point of standardised generalised trapezoidal fuzzy numbers using Euclidean distance equation: $R(\tilde{A}_i^*) = \sqrt{\tilde{x}_i^{*2} + \tilde{y}_i^{S2}}$. Determine the fuzzy positive-ideal solution (FPIS) and fuzzy negative-ideal solution (FNIS). Referring to normalise trapezoidal fuzzy weights, the FPIS, A^+ represents the compromise solution while FNIS, A^- represents the worst possible solution. The range belong to the closed interval [0,1]. The FPIS A^+ (aspiration levels) and FNIS A^- (worst levels) as following below:

$$A^+ = [1,1,1,1;1][1,1,1,1;0.9] \quad A^- = [-1,-1,-1,-1;1][-1,-1,-1,-1;0.9]$$

The FPIS, A^+ and FNIS, A^- can be obtained by centroid method for (x_{A^+}, y_{A^+}) and (x_{A^-}, y_{A^-}).

The distance $\tilde{d}_i^+$ and $\tilde{d}_i^-$ of each alternative from formulation A^+ and A^- can be calculated by the area of compensation method:

$$\bar{d}_i^+(\tilde{v}_{ij}, \tilde{v}_j^+) = \sqrt{(x_{\tilde{A}_i^*} - x_{A^+})^2 + (y_{\tilde{A}_i^*} - y_{A^+})^2} \tag{14}$$

$$\bar{d}_i^-(\tilde{v}_{ij}, \tilde{v}_j^-) = \sqrt{(x_{\tilde{A}_i^*} - x_{A^-})^2 + (y_{\tilde{A}_i^*} - y_{A^-})^2} \tag{15}$$

Find the closeness coefficient, CC_i and improve alternatives for achieving aspiration levels in each criteria. Notice that the highest CC_i value is used to determine the rank.

$$\overline{CC}_i = \frac{\bar{d}_i^-}{\bar{d}_i^+ + \bar{d}_i^-} = 1 - \frac{\bar{d}_i^+}{\bar{d}_i^+ + \bar{d}_i^-} \tag{16}$$

where, $\frac{\bar{d}_i^-}{\bar{d}_i^+ + \bar{d}_i^-}$ is satisfaction degree in *ith* alternative and $\frac{\bar{d}_i^+}{\bar{d}_i^+ + \bar{d}_i^-}$ is fuzzy gaps degree in *ith* alternative.

Fuzzy gap should be improvised for reaching aspiration levels and get the best mutually beneficial strategy from among a fuzzy set of feasible alternatives.

Step 8: Validation process using sensitivity analysis

The results of fuzzy MCDM models are importantly needed to validate using sensitivity analysis method to analyse the quality and how robustness of fuzzy MCDM model to reach a right decision under different conditions. In this paper, sensitivity analysis that proposed by [6] is utilised for validation purposes.

4 Case Study

This section illustrates a numerical example for proposed hybrid fuzzy MCDM methodology based on real case study for staff recruitment problem in MESSRS SAPRUDIN, IDRIS & CO firm in Malaysia. The legal company plan to hire the best candidate for executive post in several aspects which there are three decision makers (DMs) DM1, DM2, and DM3 of a firm and four alternatives or candidates *x1*, *x2*, *x3* and *x4*. Several criteria are considered which are: C1) Emotional steadiness, C2) Oration, C3) Personality, C4) Past experience and, C5) Self-confidence. This study simplifies the concept of attributes under fuzzy events. The values of attributes correspond to interval type-2 fuzzy sets. A comparative study was conducted to validate the results of the proposed model with established hybrid model which is fuzzy AHP – fuzzy TOPSIS proposed by [16].

Table 1. Ranking results of criteria for comparing study.

Fuzzy MCDM model	Criteria weight values					Ranking results
	ES	O	P	PE	S-C	
Fuzzy AHP – TOPSIS [16]	0.087	0.364	0.044	0.34	0.164	O > PE > S-C > ES > P
Proposed model	0.1172	0.2672	0.1190	0.2747	0.2219	PE > O > S-C > P > ES

Table 1 represents the criteria weight and ranking results between established fuzzy AHP – fuzzy TOPSIS [16] and proposed model. Based on decision makers' evaluation; Past experience, Oration and Self-confidence criteria play important aspects in recruiting new staff since the weight are greater than 0.2 respectively. These results of criteria's weights are implemented in following phase to evaluate for alternatives selection. The established fuzzy AHP – fuzzy TOPSIS model [16] produces different ranking results with rank Oration > Past Experience > Self-confidence > Emotion Steadiness > Personality.

Table 2. Ranking results of alternatives for comparing study

Fuzzy MCDM model	Alternatives ranking values, CCi				Ranking results
	Alt1	Alt2	Alt3	Alt4	
Fuzzy AHP – TOPSIS [16]	0.5497	0.5543	0.5616	0.5413	C3 > C2 > C1 > C4
Proposed model	0.7422	0.7823	0.83	0.6964	C3 > C2 > C1 > C4

Table 2 depicts the alternatives/candidates ranking results for CC_i values. The proposed model evaluates candidate 3 as the highest rank with 0.83 followed by candidate 2, candidate 1 and candidate 4 for the last rank. The results reveal that the candidate 3 is most suitable for this recruitment post. The established model [16] produces same ranking results for alternatives with the proposed model. Even, this model gives same ranking to proposed model, but the gaps of *CCi* values between each candidate are too small. Those ranking results by [16] will easily affected if the weightage of criteria are slightly changed. This can be evaluated by sensitivity analysis in studying how consistent and robust the model. In the context of sensitivity analysis evaluation, it presents that the proposed hybrid fuzzy MCDM model is definitely consistent even the weights of criteria are changed with several percentages. From the consistency results, the proposed hybrid fuzzy MCDM model is recommended to deal with bigger case study in real world phenomena in order to solve human based decision making problems under fuzzy environment.

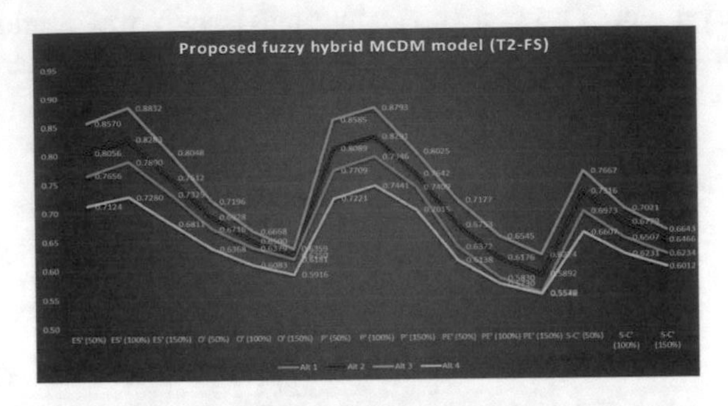

Fig. 2. Sensitivity analysis results by varying the weights of the criteria by proposed model.

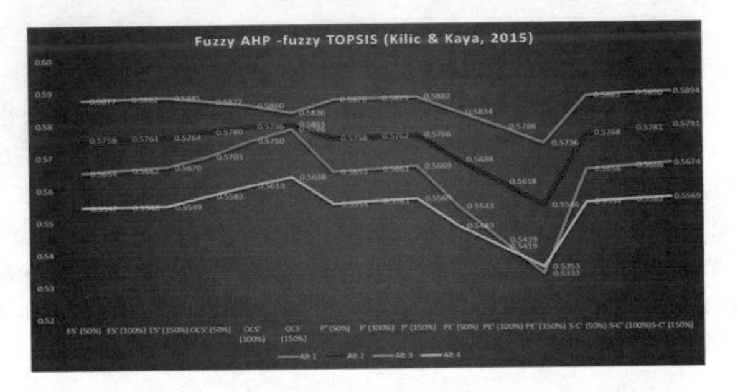

Fig. 3. Sensitivity analysis results by varying the weights of the criteria by [16].

Figure 2 illustrates the analysis results of changing the criteria weights for proposed model. It presents that when the weights of the criteria change, the values of the CC_i vary slightly. As can be seen from Fig. 3, the values and patterns of changes of CC_i are too small compare to the proposed model. The ranking values between alternative to other alternative are too small. That's mean that the gap are small to represent the assessment status of acceptance. This is depicted that the proposed model is good in robustness than established model [16].

5 Conclusion

This study has brought out the idea and concept regarding the fuzzy MCDM model that consist of consistent preference relations and fuzzy TOPSIS using intuitive multiple centroid (IMC) defuzzification method based on interval type-2 fuzzy sets. The development of IMC provides efficient computational defuzzification procedures for fuzzy sets. It presents in simple formulae that based on the perspective of analytic geometric principles. In developing an intuitionistic defuzzification, a novel manner of computing intuitive multiple centroid method has capability in dealing with all possible cases of interval type-2 fuzzy numbers. The development of fuzzy MCDM model provides better selection in human based decision making problems where at the same capable to deal with uncertainty in human judgment. Due to access information and availability of the uncertain data, it is hard to make right decision. In this sense, it is important to improvise the techniques or models form the classical one, adding intuitive reasoning and human subjectivity. As consequence, the proposed model is developed to design the robust and consistent methodology in order to give the most promising alternatives with respect to the resources. Therefore, this proposed model can be further proceeded in order to make some contributions by considering complicated case studies drawn for diverse fields crossing human based decision making problems.

References

1. Klir GJ, Clair US, Yuan B (1997) Fuzzy set theory: foundations and applications. Prentice Hall, Internatio
2. Wagner C, Hagras H (2010) Uncertainty and type-2 fuzzy sets and systems. Comput Intell (UKCI). 2010 UK Work
3. Karnik NN, Mendel JM (2001) Centroid of a type-2 fuzzy set. Informayion Sci 132:195–220
4. Mogharreban N, Dilalla LF (2006) Comparison of defuzzification techniques for analysis of non-interval data. Annu Conf North Am Fuzzy Inf Process Soc-NAFIPS (1):257–260
5. Karnik NN, Mendel JM (2001) Centroid of a type-2 fuzzy set. Inf Sci (Ny) 132(1–4):195–220
6. Amini A, Alinezhad A (2011) Sensitivity analysis of TOPSIS technique: the results of change in the weight of one attribute on the final ranking of alternatives. J Optim Ind Eng 7 (2011):23–28
7. Deng H (2014) Comparing and ranking fuzzy numbers using ideal solutions. Appl Math Model 38(5–6):1638–1646
8. Gong Y, Hu N, Zhang J, Liu G, Deng J (2015) Multi-attribute group decision making method based on geometric Bonferroni mean operator of trapezoidal interval type-2 fuzzy numbers. Comput Ind Eng 81:167–176
9. Ku Khalif KMN, Gegov A (2015) Generalised fuzzy bayesian network with adaptive vectorial centroid. In: 16th world congress of the international fuzzy systems association (IFSA) and the 9th conference of the european society for fuzzy logic and technology (EUSFLAT). pp 757–764
10. Ku Khalif KMN, Gegov A (2015) Bayesian logistic regression using vectorial centroid for interval type-2 fuzzy sets. In: 7th international joint conference on computational intelligence (IJCCI 2015). Pp 69–79
11. Cheng C-H (1998) A new approach for ranking fuzzy numbers by distance method. Fuzzy Sets Syst 95(3):307–317
12. Herrera-Viedma E, Herrera F, Chiclana F, Luque M (2004) Some issues on consistency of fuzzy preference relations. Eur J Oper Res 154(1):98–109

13. Kamis NH, Abdullah K, Mohamed H, Sudin S, Ishak WZAW (2011) Decision making models based on consistent fuzzy preference relations with different defuzzification methods. In: *2011* IEEE colloquium on humanities, science and engineering CHUSER 2011. no. Chuser, pp 845–850
14. Wang TC, Chen YH (2007) Applying consistent fuzzy preference relations to partnership selection. Omega 35(4):384–388
15. Hwang C-L, Yoon K (1981) Multiple attribute decision making. Springer, New York
16. Kiliç M, Kaya İ (2015) Investment project evaluation by a decision making methodology based on type-2 fuzzy sets. Appl Soft Comput 27(2015):399–410

A Spiking Neural Networks Model with Fuzzy-Weighted k-Nearest Neighbour Classifier for Real-World Flood Risk Assessment

Mohd Hafizul Afifi Abdullah[1,2](✉), Muhaini Othman[1], Shahreen Kasim[1,2], Shaznoor Shakira Saharuddin[1], and Siti Aisyah Mohamed[1]

[1] Faculty of Computer Science and Information Technology, Universiti Tun Husssein Onn Malaysia, 86400 Parit Raja,, Batu Pahat Johor, Malaysia
hafizul94@gmail.com, {muhaini,shahreen}@uthm.edu.my, shakirasahar@gmail.com, sitiaisyahmohamed@gmail.com

[2] Soft-Computing and Data Mining Center, Universiti Tun Husssein Onn Malaysia, 86400 Parit Raja,, Batu Pahat Johor, Malaysia

Abstract. Inspired by the brain working mechanism, the spiking neural networks has proven the capability of revealing significant association between different variables spike behavior during an event. The combination of the capability of SNN to produce personalised model has allowed high-precision for data classification. The exiting accuracy of weighted k-nearest neighbors classifier being used in the spiking neural networks architecture, noticeably can be further improved by implementing fuzzy-weights on the features, therefore allowing data to be classified more precisely to the high-impacting features. Simulation has been done by using three classifiers—Multi-layer Perceptron, weighted k-nearest neighbors, and Fuzzy-weighted k-nearest neighbors (FwkNN) using a real-world flood case study dataset and two benchmark dataset. Based on the result using the Kuala Krai Rainfall Dataset, FwkNN classifier has improved accuracy by 3.48% and 3.57% for 3-days earlier and 1-day earlier classification respectively. As compared to, FwkNN classifier has proven the capability to reduce misclassification and increase the accuracy of dataset classification.

Keywords: Personalised modelling · Spiking neural network · Spatio-temporal data · NeuCube · Fuzzy k-NN · Feature weighting · Classification · Flood data

1 Spiking Neural Network: A Brief Study

Spiking Neural Network, abbreviated as SNN is a branch of neural network models which has been built to imitate the brain working mechanism. SNN has proven its capability to recognise—and *utilise*—patterns. Similarly like other

R. Ghazali et al. (Eds.): SCDM 2020, AISC 978, pp. 222–230, 2020.
https://doi.org/10.1007/978-3-030-36056-6_22

neural networks, SNN are made up of the most basic unit called neurons (or *perceptrons*) which consists of a number of input channels, processing body, and an output channel. A neuron can be mathematically represented as Eq. 1 and complemented by Eq. 2:

$$a \circ f(x) \rightarrow y \tag{1}$$

$$y = \begin{cases} 1, & \text{if } a(x) > \tau, \\ 0, & \text{otherwise.} \end{cases} \tag{2}$$

where x is data vector fed to $f(x)$ and activated by activation function $a(x)$. If output of the activation function exceeds threshold τ, hence $y = 1$, or else 0.

SNN is however different from the other neural network models since the network explicitly consider the input timing—therefore the series of input signals (or *spikes*) keep changing over time. SNN has been applied in various domains consisting time-series data with multi-variables including ecological field [1,2], stroke-risk prediction [1,3,4], understanding fMRI data [5], flood risk assessment [6,7], aphids population prediction [1], and many others.

Clarified here—an investigation on natural events requires analysis of complex high dimension data in the form of spatio- and spectro-temporal data (SSTD), consisting of spatial and temporal components without losing inter-relationship connection between the components.

The brain deals extremely well as a spatio-temporal information processing, where when presented with information, complex spatio-temporal paths and patterns are formed across the brain [1,12]. This has motivated researchers to create spatio-temporal data machine (STDM) [1] for processing SSTD information based on the brain physiology. By imitating the behavior of the brain analysing data as close as possible, a data processing model can be constructed—which can be later on used to process new upcoming stream of data.

Therefore article presents an approach to analyse a real-world flood case study using evolving spiking neural network methods, an extended work for [7] where an attempt to produce a higher accuracy result using the same data set is presented.

The main approach for obtaining such result is by evaluation and justification of a more suitable classifier for the domain-specific task. The expected outcome is the architecture with the proposed classifier will be able to produce result with higher accuracy by reducing misclassification. The subject involves 5 years environmental-related temporal data (from 2012 until 2016) of Kuala Krai in Kelantan, Malaysia.

Therefore, the paper is organised as follows: Sect. 1 briefs regarding spiking neural networks and the predictability feature of natural occurrences, Sect. 2 explains the data modelling approaches suitable to be used to create an optimised data model for predicting flood risk. Within Sect. 3, the experimental design and procedure were explained. Section 4 defines and justifies the experimental result while Sect. 5 concludes the paper.

1.1 Time-Series Predictability

Flood occurrences, like any other natural disaster are predictable by using computational models such as neural networks. This article assumes less prior knowledge on spiking neural networks, therefore an extensive references on current and previous works are included.

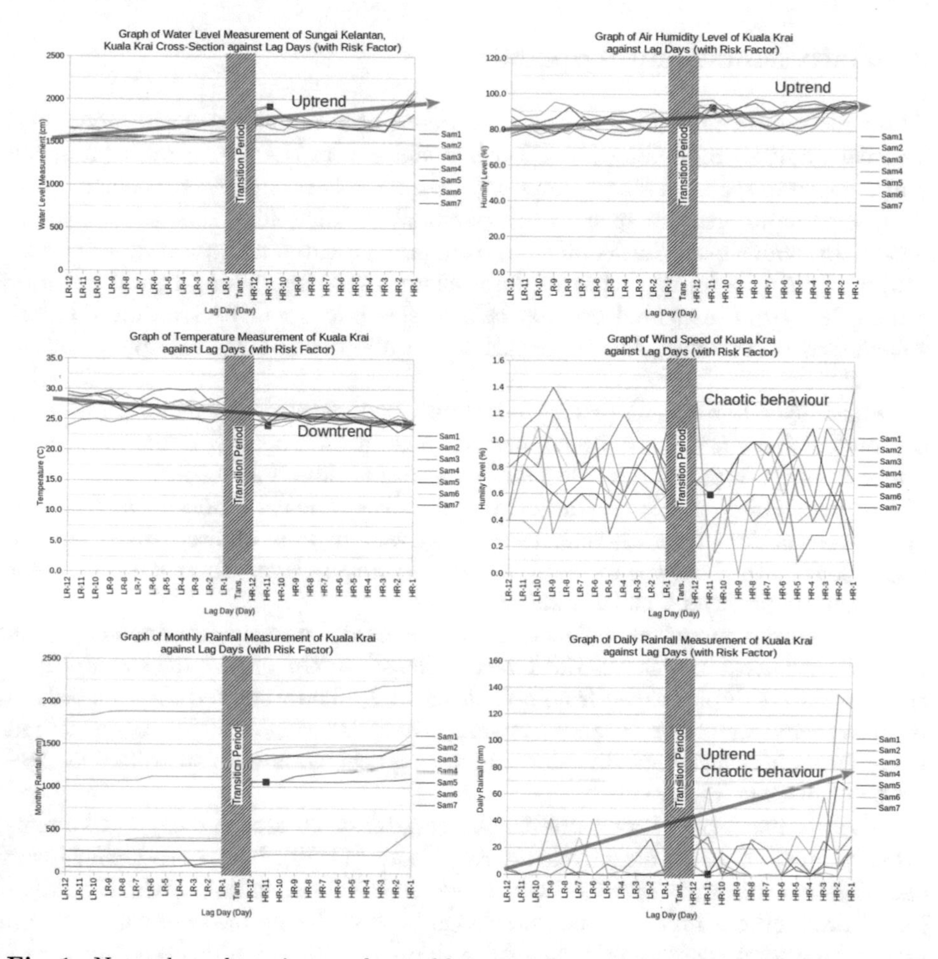

Fig. 1. Note that there is an observable trend for variables as flood day (day-0) approaches. The trend is distinctive depending on the variable. For instance, as flood day is approaching, water level increases while surrounding temperature decreases

Note that in Fig. 1, there are observable trends for each variables between the Low-Risk patterns (Left Hand Side) of the graphs and the High-Risk patterns (Right Hand Side) of the graphs. Understanding these spike pattern allows researchers to understand the significant association between different variables spike behavior during flood risk-events.

The existing architecture of personalised SNN approach which is based on weighted k-Nearest Neighbour (wkNN) classifier is noticibly prone to slight misclassification due to binary output produced by multi-class classifier. Therefore, the proposed fuzzy-integrated approach by using Fuzzy-weighted k-Nearest Neighbour (wkNN) classifier allows us to get non-binary output (similarly to the *confidence level* of the class).

2 Data Modelling Approaches

Three main data modelling approaches are—global modelling, local modelling, and personalised modelling [8]. Global model covers the whole problem space and is represented by a single function; local models represents sub-space (clusters) of a whole problem space; while personalised modelling represents a single data point which every data point has own model created using transductive reasoning [8]. An in-depth review of data clustering using spiking neural network models has been presented on [10]. Hence, this paper will be focusing on classifier based on two modelling approaches—localised modelling and personalised modelling.

Multi-layer Perceptron (MLP) is a supervised learning algorithm that learns a function $f(.) : R^m \rightarrow R^o$ by training on a dataset, where m is the number of dimensions for input and is the number of dimensions for output. Given a set of features $X = x_1, x_2, x_3, \ldots$ and a target y, it can learn a non-linear function approximator for either classification or regression. It is different from logistic regression, in that between the input and the output layer, there can be one or more non-linear layers, called hidden layers.

Personalised model classifiers are able to model data instantaneously ("*on the fly*") for every new input data vector, based on the closest data samples to the new data samples taken from a data set [8,9]. The weighted-kNN algorithm is an averaged weighted distance between k-neighbouring data points, measured using (usually) Euclidean distance and voting scheme is applied to define class for the new data vector.

FwkNN has been chosen due to the capability to assign fuzzy class membership function as in fuzzy-Nearest Neighbour [11,13], hence producing fuzzy classification rule and prevents new data vector misclassification. FwkNN is capable of producing a more reliabe knowledge by associating membership function rather than solely depend on k-Nearest Neighbours for assigning classes by defining class-specific weight and has been proven to be benificial. Algorithm 1 shows how the FwkNN training process is handled, while Fig. 2 visualises the comparison between wkNN algorithm and FwkNN algorithm classification process.

FwkNN is able to produce a classification outcome with a lower error rates by assigning membership values that serve as a confidence measure in the classification.

3 Experiment Design

Experiments were executed using Kuala Krai Rainfall Dataset consisting data from 2012 until 2016. The result is then compared with benchmark dataset to prove the credibility and integrity of the technique.

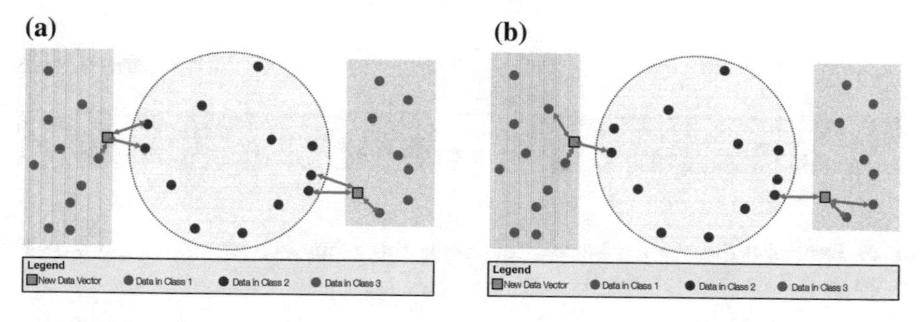

Fig. 2. Two classification algorithms with k-value $= 3$. **(a)** wkNN finds the 3 nearest neighbours to a new data vector and ranks the new data vector using majority voting rule causing misclassification. **(b)** FwkNN finds the 3 nearest neighbours based on their weighted-feature reducing the misclassification of new data vectors

---

Algorithm 1. Training of FwkNN classifier

for *every spike patterns from data* **do**
  Create new input neuron
  Calculate initial conn_weight using RO-learning rule
  Assign a fuzzy membership (weighted-distance) to the input neuron
  Adjust conn_weight for corresponding spikes usng STDP rule
  Calculate postsynaptic potential of the neuron
  Add neuron to neurons repository
end

3.1 Data Description

For the experiment, temporal data are-pre-processed to label the data with 2-classes (High-Risk, Low-Risk) for the training of the network. The Kuala Krai Rainfall Dataset is a sample of the temporal data which consists of 6 features reading measured once daily for 5 years based on Kuala Krai station; (i) water level in cm, (ii) daily rainfall in mm, (iii) monthly rainfall in mm, (iv) wind speed in ms^{-1}, (v) air humidity in percentage, and (vi) temperature in degree Celcius. The data is labelled and arranged to be fed backwards, as visualised in Fig. 3 Table 1.

Table 1. Training and testing samples of different timelength is used to feed the network with environmental data for an early prediction of flood risk

Early predict sample	Train:Test split	Train:All%
1-Day early	12:12	100.00%
3-Days early	10:12	83.30%

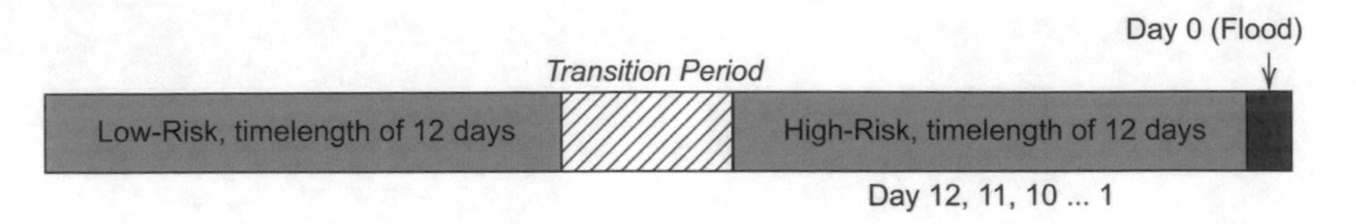

Fig. 3. Temporal data is fed backwards, beginning from day-12, day-11, and so forth approaching flood day on day-0

Visualised in Fig. 3, samples were created with 24-days timelength; 12 days of High-Risk, 12 days of Low-Risk, separated by a transition period. The daily features measurement of the 6-features patterns gradually change, approaching flood on day-0. Clearly, approaching to flood occurence, features measurement changes is expected such: water level, daily rainfall measurements increase, air humidity measurements increase; and temperature measurements decrease—as visualised in Fig. 1. Hidden patterns may be formed by monthly rainfall and wind speed reading. Therefore, spiking neural network models are used to analyse and reveal the hidden spike patterns from environment data, triggered during flood occurrence.

3.2 Experimental Procedure

Experiment is conducted in such a way the data pattern model is first constructed, then simulated in two phases; (i) model training and (ii) model testing.

Benchmark dataset used are obtained from UCL Machine Learning repository at https://archive.ics.uci.edu/ml and World Gold Council Website, GoldHub data source at https://www.gold.org/goldhub. Non-biased benchmark dataset are selected based on suitability to be used with the framework (Fig. 4).

An optimisation module has been executed using lower and upper limit values to obtain the optimised parameters which includes the AER threshold, small world radius, STDP rate, firing threshold, refractory time, time rounds, deSNN mod, and deSNN drift values. The lower and upper limit values were selected based on the standard configuration for SNN experiments. Table 2 summarizes the upper and lower limits of the parameters to be optimised and its optimal value used to construct the data model.

4 Results and Discussion

In this section, we will consider that the network model should correctly classify the result to 2 classes—Class 0 (Low-Risk for flood occurrences) and Class 1 (High-Risk of flood occurrences). Table below summarises the key results upon execution of the SNN models using three different classifiers—using Kuala Krai Rainfall Dataset versus benchmark datasets.

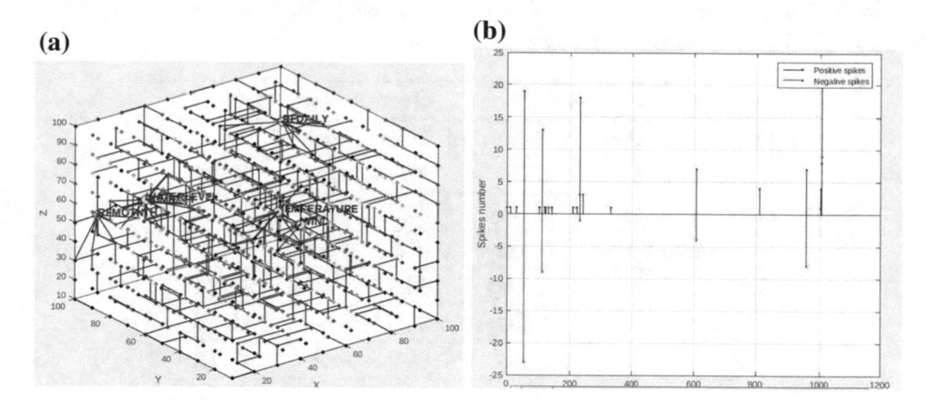

Fig. 4. **(a)** SNN reservoir with 1000 neurons showing parameters interrelationship connectivity for all 6 environmental-related parameters. Blue lines represents positive connections and red represent negative connection. Line thickness represent weight value and brighter neuron has stronger connection with other neurons. **(b)** Neuron spikes emitted

Table 2. Training and testing samples of different timelength is used to feed the network with environmental data for an early prediction of flood risk

Parameter	Upper Lim.	Lower Lim.	Optimal value
AER threshold	1.0	0.1	0.5
Small world radius	1.0	5.0	2.5
STDP rate	0.01	0.10	0.01
Firing threshold	0.1	1.0	0.47
Refractory time	1	10	2 or 7
Time rounds	1	10	4 or 5
deSNN mod	0.01	0.50	0.157
deSNN drift	0.01	0.50	0.329

The Dow Jones Index Data Set were used to predict if the index value drops below DJI 1,700 points while Gold Price Short & Long Term Dataset were used to predict new-high break prices (Table 3).

The proposed classifier algorithm has been compared against the conventional MLP and SNN architecture combined with wkNN classifier—and is proven to get higher overall accuracy compared to conventional approaches. On an instance, on DJI dataset, wkNN has performed slightly better than FwkNN possibly due to smaller training:testing portion for 3-days earlier prediction. Overall, FwkNN classifier in combination with SNN methods has scored most for overall accuracy.

Table 3. Comparison between results obtained using MLP and wkNN versus the proposed FwkNN classifier on 3 datasets

Early prediction level for		Overall accuracy (%)		
Dataset name	**Samples**	**MLP**	**wkNN**	**FwkNN**
Kuala Krai	3 Days	87.66	92.86	**96.34**
Rainfall Dataset	1 Day	88.24	91.25	**94.82**
Dow Jones Index	3 Days	95.53	**96.76**	96.42
Data Set	1 Day	96.22	97.60	**98.74**
Gold Price	3 Days	80.32	89.64	**91.45**
Short & Long Term	1 Day	84.74	92.88	**95.12**

5 Conclusion

The main goal of implementing FwkNN classifier in the SNN data modelling architecture is to improve the accuracy of classification result and reducing data misclassification. FwkNN is derived from the conventional wkNN classifier and is implemented by fuzzifying the weighted-distance between new data vector and existing data vector in the network. FwkNN has proven its capability to classify temporal data more accurately as compared to the conventional MLP and wkNN algorithms.

The combination if FwkNN algorithm as classifier for personalised SNN architecture is most welcomed in various domains—not only limited to ecological, financial, and economical fields. Given the right processing capacity, an optimised model, and parameters—the architecture in combination with FwkNN classifier is capable of producing a highly accurate predictive model for multiple domains which benefits the society.

Acknowledgments. The authors would like to thank Universiti Tun Hussein Onn Malaysia for supporting this paper publication.

References

1. Kasabov N, Scott NM, Tu E, Marks S, Sengupta N, Capecci E, Othman M, Doborjeh MG, Murli N, Hartono R, Espinosa-Ramos JI, Zhou L, Alvi FB, Wang G, Taylor D, Feigin V, Gulyaev S, Mahmoud M, Hou Z-G, Yang J (2016) Evolving spatio-temporal data machines based on the NeuCube neuromorphic framework: design methodology and selected applications. Neural Netw 78:1–14. https://doi.org/10.1016/j.neunet.2015.09.011
2. Tu E, Kasabov N, Othman M, Li Y, Worner S, Yang J, Jia Z (2014) NeuCube (ST) for spatio-temporal data predictive modelling with a case study on ecological data. In: 2014 international joint conference on neural networks (IJCNN), pp 638–645
3. Kasabov N, Feigin V, Hou Z-G, Chen Y, Liang L, Krishnamurthi R, Othman M, Parma P (2014) Evolving spiking neural networks for personalised modelling, classification and prediction of spatio-temporal patterns with a case study on stroke. Neurocomputing 134:269–279
4. Othman M, Kasabov N, Tu E, Feigin V, Krishnamurthi R, Hou ZG, Chen Y, Hu J (2014) Improved predictive personalized modelling with the use of Spiking Neural Network system and a case study on stroke occurrences data. In: 2014 international joint conference on neural networks (IJCNN), pp 3197–3204
5. Saharuddin SS, Murli N, Hasibuan MA (2018) Classification of spatio-temporal fMRI data in the spiking neural network. IJASEIT 8(6):2670–2676. https://doi.org/10.18517/ijaseit.8.6.5011
6. Othman M, Mohamed SA, Abdullah MHA, Yusof MM, Mohamed R (2018) A framework to cluster temporal data using personalised modelling approach. In: Recent advances on soft computing and data mining. Springer, International Publishing, pp 181–190. https://doi.org/10.1007/978-3-319-72550-5_18
7. Abdullah MHA, Othman M, Kasim S, Mohamed SA (2019) Evolving spiking neural networks methods for classification problem: a case study in flood events risk assessment. IJEECS 16(1):222–229. https://doi.org/10.11591/ijeecs.v16.i1.pp222-229
8. Kasabov N (2009) Soft computing methods for global, local and personalized modeling and applications in bioinformatics. In: Soft computing based modeling in intelligent systems. Springer, Berlin, pp. 1–18. https://doi.org/10.1007/978-3-642-00448-3_1
9. Kasabov N (2007) Global, local and personalised modeling and pattern discovery in bioinformatics: an integrated approach. Pattern Recognit Lett 28(6):673–685. https://doi.org/10.1016/j.patrec.2006.08.007
10. Mohamed SA, Othman M, Abdullah MHA (2019) A review on data clustering using spiking neural network (SNN) models. IJEECS 15(3):1392–1400. https://doi.org/10.11591/ijeecs.v15.i3.pp1392-1400
11. Keller JM, Gray MR, Givens JA (1985) A fuzzy k-nearest neighbor algorithm. IEEE Trans Syst Man Cybern 580–585. https://doi.org/10.1016/j.patrec.2006.08.007
12. Kasabov N (2014) NeuCube: a spiking neural network architecture for mapping, learning and understanding of spatio-temporal brain data. Neural Netw 52:62–76
13. Biswas N, Chakraborty S, Mullick SS, Das S (2018) A parameter independent fuzzy weighted k-Nearest neighbor classifier. Pattern Recognit Lett 101:80–87. https://doi.org/10.1016/j.patrec.2017.11.003

Linear Fuzzy Delay Differential Equation and Its Application in Biological Model with Fuzzy Stability Analysis

Animesh Mahata[1], Sankar Prasad Mondal[2], Ali Ahmadian[3(✉)], Shariful Alam[4], and Soheil Salahshour[5]

[1] Department of Mathematics, Netaji Subhash Engineering College, Panchpota Garia, Kolkata 700152, West Bengal, India
animeshmahata8@gmail.com

[2] Department of Applied Science, Maulana Abul Kalam Azad University of Technology, West Bengal, Nadia 741249, West Bengal, India
sankar.mondal02@gmail.com

[3] Institutes for Mathematical Research, Universiti Putra Malaysia, 43400 Serdang, Selangor, Malaysia
ahmadian.hosseini@gmail.com

[4] Department of Mathematics, Indian Institute of Engineering Science and Technology, Shibpur, Howrah 711103, West Bengal, India
salam50in@yahoo.co.in

[5] Young Research and Elite Club, Islamic Azad University, Mobarakeh branch, Mobarakeh, Iran
soheilsalahshour@yahoo.com

Abstract. In this present article solution and stability analysis of a fuzzy delay differential equation with application is presented. For the presence of the uncertainty the uncertainty parameter namely fuzzy number with the corresponding differential equation in time delay model becomes fuzzy delay differential equation (FDDE) model. Using generalized Hukuhara derivative technique the fuzzy delay differential equation transformed to system of two crisp delay differential equations. The fuzzy stability criterion is found for different cases. The results are followed by a real world problem delayed protein degradation model.

1 Introduction

Fuzzy set theory is an idea for measure the uncertainty. In 1965, Zadeh [1] proposed the fuzzy set theory. Fuzzy differential Equation (FDE) is very important for theatrical and applied study [2–4]. There are few ideas where researcher works on fuzzy differential equation with time delay [5–8]. A Differential equation will be Delay Differential Equation (DDE) if there is the state variable appears with delayed argument. It may several types like as constant Delay differential equation, variable Delays, State – dependent Delays etc. Fuzzy number is a reliable one for measure the uncertainty. Whenever we take any fuzzy parameter which may be co-efficient or initial condition or both of a DDE then Fuzzy Delay Differential Equation (FDDE) comes. As a result

R. Ghazali et al. (Eds.): SCDM 2020, AISC 978, pp. 231–240, 2020.
https://doi.org/10.1007/978-3-030-36056-6_23

the natural behavior of DDE is changed because the simple etiquettes of that fuzzy numbers are not same from a crisp number. Thus, fuzzy Delay Differential equation and its application is most important.

The main contributions of the work are as follows:

(i) Transfer the linear fuzzy delay differential equation to system of crisp delay differential equation using gH derivative approach.
(ii) The fuzzy stability analysis is done for every case.
(iii) The methods are followed by Application of Fuzzy Delay differential equation.

2 Preliminaries

Definition 2.1: Triangular fuzzy number: A fuzzy number is called a Triangular fuzzy number if its membership function is written as

$$\mu_{\tilde{A}}(y) = \begin{cases} 0, & y \leq y_{0,} \\ \frac{y - y_{0,}}{y_1 - y_{0,}}, & y_{0,} \leq y \leq y_1 \\ 1, & y = y_1 \\ \frac{y_2 - y}{y_2 - y_1}, & y \leq y \leq y_2 \\ 0, & y \geq y_2 \end{cases}$$

Where the notation is $\tilde{A} = (y_{0,}y_{1,}y_2)$.

Definition 2.2: α-cut of a fuzzy set: The α-cut of $\tilde{A} = (y_1, y_2, y_3)$ is given by

$$A_\alpha = [y_0 + \alpha(y_1 - y_0), y_2 - \alpha(y_3 - y_2)], \forall \alpha \in [0, 1]$$

Definition 2.3: Generalized Hukuhara derivative: The idea of generalized Hukuhara derivative of a given fuzzy function $p : (a, b) \to \Re_{\mathcal{F}}$ at s_0 is defined as

$$p'(s_0) = \lim_{h \to 0} \frac{p(s_0 + h) \ominus_g p(s_0)}{h}$$

If $p'(s_0) \in \Re_{\mathcal{F}}$ satisfying, we conclude that $p(s)$ is generalized Hukuhara differentiable at s_0.

Also we say that $g(s)$ is called (i)-gH differentiable at s_0 if

$$[p'(s_0)]_\alpha = [p'_1(s_0, \alpha), p'_2(s_0, \alpha)]$$

and $p(s)$ is called (ii)-gH differentiable at s_0 if

$$[p'(s_0)]_\alpha = [p'_2(s_0, \alpha), p'_1(s_0, \alpha)]$$

2.1 Generalized Characterization Theorem for FDDE Under Generalized Differentiability

Let us assume the FDDE $y'(s) = f(t, x(s), x(s-\sigma)$ for $s_0 \leq s \leq T$

Where $\sigma > 0$ is delay constant, $x \in R_F$ is a n-vector –valued fuzzy function and f is a continuous fuzzy function defined on a mapping $f : [s_0, T] \times R_F \times C_\sigma \to R_F$ with initial conditions

$$y(s_0) = y_0$$

$$y(s) = \varphi(s) \text{ for } s_0 - \sigma \leq s \leq s_0 \tag{1}$$

Where $y_0 \in R_F$, $\varphi : [s_0 - \sigma, s_0) \to R_F$ is a given continuous fuzzy function. The parametric forms are $[y'(s)]_\alpha = [y_1'(s,\alpha), y_2'(s,\alpha)]$, $[y(s)]_\alpha = [y_1(s,\alpha), y_2(s,\alpha)]$

$$[y(s-\sigma)]_\alpha = [y_1(s-\sigma,\alpha), y_2(s-\sigma,\alpha)] \text{ for } s \in [s_0, T], \alpha \in [0,1] \tag{2}$$

3 Elucidation of Fuzzy Delay Differential Equation Using Generalized Differentiability Concept

Let us study the fuzzy delay differential equation of Type I i.e.,

$$\begin{gathered} \tfrac{d\tilde{y}(s)}{ds} = f(s, \tilde{y}(s), \tilde{y}(s-r)), x \in I = [s_0, X] \\ y(s_0) = \tilde{y}_0, s \in [s_0 - r, s_0] \end{gathered}$$

Now two different cases arise

3.1 When $\tilde{y}(x)$ is (i)-gH Differentiable

The above fuzzy delay differential equation transform to

$$\begin{aligned} \frac{dy_1(s,\alpha)}{ds} &= f_1(s, y_1(s,\alpha), y_2(s,\alpha), y_1(s-r,\alpha), y_2(s-r,\alpha)) \\ \frac{dy_2(s,\alpha)}{dx} &= f_2(s, y_1(s,\alpha), y_2(s,\alpha), y_1(s-r,\alpha), y_2(s-r,\alpha)) \end{aligned}$$

Where,

$$\begin{aligned} f_1 = \inf\{f(s, y(s), y(s-r)) | y(s) \in [y_1(s,\alpha), y_2(s,\alpha)], y(s-r) \\ \in [y_1(s-r,\alpha), y_2(s-r,\alpha)]\} \end{aligned}$$

and

$$f_1 = \sup\ \{f(s, y(s), y(s-r)) | y(s) \in [y_1(s,\alpha), y_2(s,\alpha)], y(s-r) \in [y_1(s-r,\alpha), y_2(s-r,\alpha)]\}$$

3.2 When $\tilde{y}(x)$ is (ii)-gH Differentiable

The above fuzzy delay differential equation transform to

$$\frac{dy_1(s,\alpha)}{ds} = f_2(s, y_1(s,\alpha), y_2(s,\alpha), y_1(s-r,\alpha), y_2(s-r,\alpha))$$
$$\frac{dy_2(s,\alpha)}{dx} = f_1(s, y_1(s,\alpha), y_2(s,\alpha), y_1(s-r,\alpha), y_2(s-r,\alpha))$$

Where,

$$f_1 = \inf\{f(s, y(s), y(s-r)) | y(s) \in [y_1(s,\alpha), y_2(s,\alpha)], y(s-r) \in [y_1(s-r,\alpha), y_2(s-r,\alpha)]\}$$

and

$$f_1 = \sup\ \{f(s, y(s), y(s-r)) | y(s) \in [y_1(s,\alpha), y_2(s,\alpha)], y(s-r) \in [y_1(s-r,\alpha), y_2(s-r,\alpha)]\}$$

4 Stability Analysis of Fuzzy Delay Differential Equation

Let us consider the first order linear delay differential equation of the form

$$\frac{dy(s)}{ds} = Ay(s) + By(s-r)$$

With fuzzy initial condition $y(0) = y_0$
Four different cases arise for the above differential equation
Case 1: $A > 0, B > 0$
Sub case 4.1: When $y(s)$ is (i)-gH differentiable
The upstairs differential equation adapted to

$$\frac{dy_1(s,\alpha)}{ds} = Ay_1(s,\alpha) + By_1(s-r,\alpha)$$

$$\frac{dy_2(s,\alpha)}{ds} = Ay_2(s,\alpha) + By_2(s-r,\alpha) \tag{3}$$

The steady state solution of (4.1) is $(y_1^*, y_2^*) = (0, 0)$

The characteristic equation of the variational matrix corresponding the system of (4.1) at $(0,0)$ taking λ being eigen value is given by

$$det\begin{bmatrix} A+Be^{-\lambda r}-\lambda & 0 \\ 0 & A+Be^{-\lambda r}-\lambda \end{bmatrix}=0$$

Or, $A+Be^{-\lambda r}=\lambda$

Since $A>0, B>0$ together imply $A+B>0$ the steady state (y_1^*, y_2^*) is unstable.

Sub case 4.2: When $y(s)$ is (ii)-gH differentiable the transferred system is

$$\frac{dy_1(s,\alpha)}{dx}=Ay_2(s,\alpha)+By_2(s-r,\alpha)$$

$$\frac{dy_2(s,\alpha)}{ds}=Ay_1(s,\alpha)+By_1(s-r,\alpha) \tag{4}$$

The steady state solution of (4.2) is $(y_1^*, y_2^*)=(0,0)$

The characteristic equation of the variational matrix corresponding the system of (4.2) at $(0,0)$ taking λ being eigen value is given by

$$det\begin{bmatrix} -\lambda & A+Be^{-\lambda r} \\ A+Be^{-\lambda r} & -\lambda \end{bmatrix}=0$$

Characteristic equation of V_2 is $\lambda^2-(A+Be^{-\lambda t})^2=0$

If put delay constant $r=0$ then the characteristic equation becomes $\lambda^2-(A+B)^2=0$ or,$\lambda=\pm(A+B)$. Clearly the given system is unstable. Since generally the non-delay system is unstable then delay system goes to unstable.

Case 2: $A>0, B<0$

Sub case 4.3: When $y(s)$ is (i)-gH differentiable the transferred system is

$$\frac{dy_1(s,\alpha)}{ds}=Ay_1(s,\alpha)+By_2(s-r,\alpha)$$

$$\frac{dy_2(s,\alpha)}{ds}=Ay_2(s,\alpha)+By_1(s-r,\alpha) \tag{5}$$

Sub case 4.4: When $y(s)$ is (ii)-gH differentiable the transferred system is

$$\frac{dy_1(s,\alpha)}{ds}=Ay_2(s,\alpha)+By_1(s-r,\alpha)$$

$$\frac{dy_2(s,\alpha)}{ds}=Ay_1(s,\alpha)+By_2(s-r,\alpha) \tag{6}$$

Case 3: $A<0, B>0$

Sub case 4.5: When $y(s)$ is (i)-gH differentiable the transferred system is

$$\frac{dy_1(s,\alpha)}{ds} = Ay_2(s,\alpha) + By_1(s-r,\alpha)$$

$$\frac{dy_2(s,\alpha)}{ds} = Ay_1(s,\alpha) + By_2(s-r,\alpha) \tag{7}$$

Sub case 4.6: When $y(s)$ is (ii)-gH differentiable the transferred system is

$$\frac{dy_1(s,\alpha)}{ds} = Ay_1(s,\alpha) + By_2(s-r,\alpha)$$

$$\frac{dy_2(s,\alpha)}{ds} = Ay_2(s,\alpha) + By_1(s-r,\alpha) \tag{8}$$

Case 4: $A<0, B<0$

Sub case 4.7: When $y(s)$ is (i)-gH differentiable the transferred system is

$$\frac{dy_1(s,\alpha)}{ds} = Ay_2(s,\alpha) + By_2(s-r,\alpha)$$

$$\frac{dy_2(s,\alpha)}{ds} = Ay_1(s,\alpha) + By_1(s-r,\alpha) \tag{9}$$

Sub case 4.8: When $y(s)$ is (ii)-gH differentiable the transferred system is

$$\frac{dy_1(s,\alpha)}{ds} = Ay_1(s,\alpha) + By_1(s-r,\alpha)$$

$$\frac{dy_2(s,\alpha)}{ds} = Ay_2(s,\alpha) + By_2(s-r,\alpha) \tag{10}$$

Remarks: The main concept is that the fuzzy delay differentia equation converted to system of two crisp delay differential equation. We need to know the stability analysis of crisp system of delay differentia equation.

5 Application in Delayed Protein Degradation Model

Considered the delayed Protein Degradation model is given by

$$\frac{dY(t)}{dt} = m - nY(t) - pY(t-\tau) \tag{11}$$

The entire variables and parameters castoff in mathematical models are described as:

Variables and parameters	Explanation of variables and constraints
Y(t)	The concentration of proteins at any time t in a system.
m	The constant rate of protein production.
n	The rate of non-delayed protein degradation.
P	The rate of delayed protein degradation.
τ	The protein degradation machine degrades the protein after a time τ (the discrete time delay) after initiation.

Now find the amount of protein degradation after certain times. If the initial condition is $Y(t=0) = (0.2, 0.3, 0.45)$ with $m = 40, n = 0.3, p = 0.1$ and $\tau = 20$. Find the amount of drug after 50 s.

5.1 Solution

The solution when $Y(t)$ is (i)-gH differentiable then

$$\frac{dY_1(t,\alpha)}{dt} = 40 - 0.3Y_2(t,\alpha) - 0.1Y_2(t-20,\alpha)$$

$$\frac{dY_2(t,\alpha)}{dt} = 40 - 0.3Y_1(t,\alpha) - 0.1Y_1(t-20,\alpha) \tag{12}$$

Where $Y_1(t=0,\alpha) = 0.2 + 0.1\alpha$ and $Y_2(t=0,\alpha) = 0.45 - 0.15\alpha$

5.1.1 Equilibrium Points

The equilibrium point of (5.1) is E $(Y_1^*, Y_2^*) = (200, 200)$

The system characteristic equation of the variational matrix corresponding to the system (5.2) is given by

$$det\begin{bmatrix} -\lambda & -0.3 - 0.1e^{-\tau\lambda} \\ -0.3 - 0.1e^{-\tau t} & -\lambda \end{bmatrix} = 0 \text{ for } \tau = 20$$

Characteristic equation of is $\lambda^2 - (-0.3 - 0.1e^{-\tau\lambda})^2 = 0$ (5.3)

If we take $\tau = 0$, the Eigen values of (5.3) $\lambda = \pm 0.2$, so the steady state E_1 $(Y_1^*, Y_2^*) = (200, 200)$ of (5.3) is unstable.

Remark 5.1.2: Here in Figs. 1, 2, we see that $Y_1(t,\alpha) \leq Y_2(t,\alpha)$ and $Y_1(t,\alpha)$ is increasing and $Y_2(t,\alpha)$ is decreasing as time goes. Hence both the cases the solution is strong fuzzy solution of system (5.2)

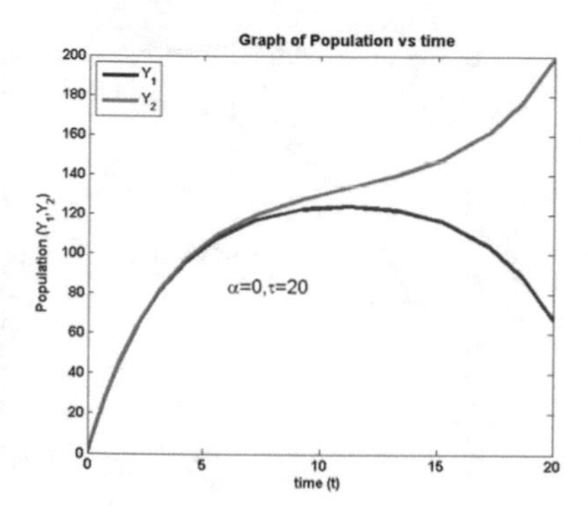

Fig. 1. For $\alpha = 0, \tau = 20$, the solution of the converted Delay model (5.2) when $t \in [0, 20]$.

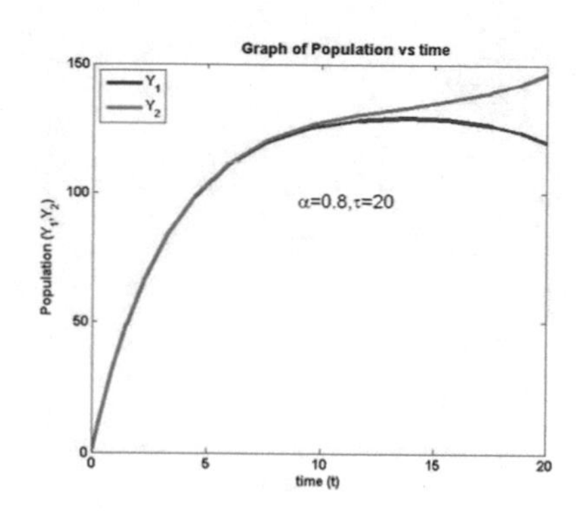

Fig. 2. For $\alpha = 0.8, \tau = 20$, the solution of the converted Delay model (5.2) when $t \in [0, 20]$.

The solution when $Y(t)$ is (ii)-gH differentiable then

$$\frac{dY_1(t,\alpha)}{dt} = 40 - 0.3Y_1(t,\alpha) - 0.1Y_1(t-20,\alpha) \tag{13}$$

$$\frac{dY_2(t,\alpha)}{dt} = 40 - 0.3Y_2(t,\alpha) - 0.1Y_2(t-20,\alpha) \tag{14}$$

Where $Y_1(t=0,\alpha) = 140 + 10\alpha$ and $Y_2(t=0,\alpha) = 165 - 15\alpha$. The system characteristic equation of variational matrix corresponding to the system (5.2) taking λ being eigen value is given by

$$det\begin{bmatrix} -0.3 - 0.1e^{-\tau\lambda} - \lambda & 0 \\ 0 & -0.3 - 0.1e^{-\tau t} - \lambda \end{bmatrix} = 0 \text{ for } \tau = 20 \tag{15}$$

$\lambda = -0.3 - 0.1e^{-\tau\lambda}$

Comparing with $\lambda = M + Ne^{-\tau\lambda}$ where $M = -0.3, N = -0.1$

$M + N = -0.3 + -0.1 = -0.4 < 0$ and $-0.1 \geq -0.4$,therefore by Theorem 3.1.1 the steady state equilibrium E_2 $(Y_1^*, Y_2^*) = (200, 200)$ of (5.4) is asymptotically stable.

Remarks: Here in Figs. 3 and 4 we see that $Y_1(t,\alpha) = Y_2(t,\alpha)$ that means in this case we get a crisp solution although we take fuzzy initial condition. We strongly recommend for taking this solution.

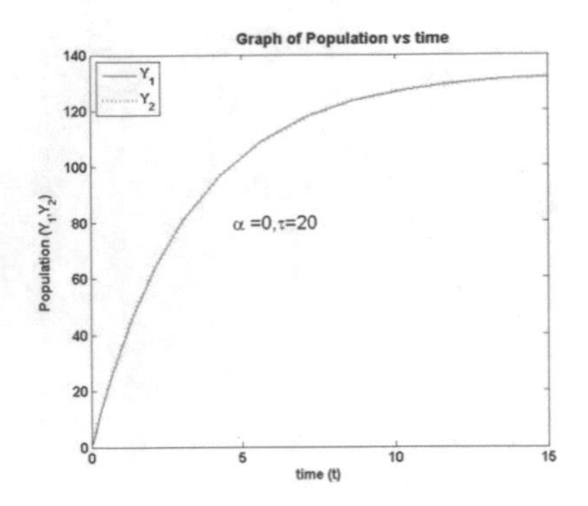

Fig. 3. For $\alpha = 0, \tau = 20$, the solution of the converted delay model (5.3) when $t \in [0, 15]$.

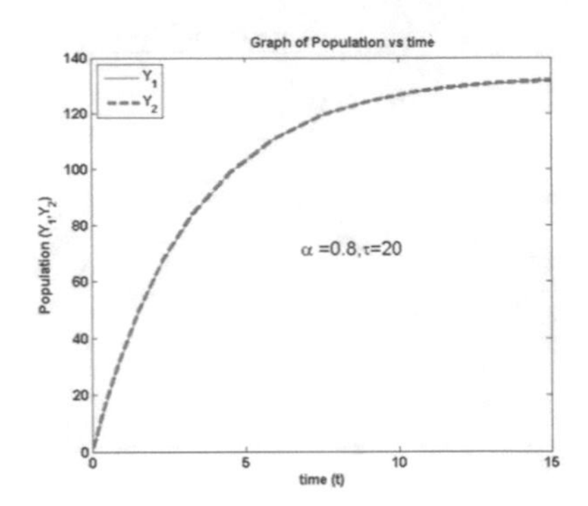

Fig. 4. For $\alpha = 0.8, \tau = 20$, the solution of the converted Delay model (5.3) when $t \in [0, 15]$.

6 Conclusion

Mathematical modeling using either different type of differential equation (ordinary, partial, delay, stochastic etc.) is an area of tremendous development and research. Delay Differential equation and biology can an important topic for modeling now days. Now if there is an uncertainty with differentia equation then we can practice the conception of fuzzy differential equation. For the paper purpose we concentrate in the delay differential equation and its application in fuzzy environment. In this paper we consider the transformation idea for fuzzy delay differential equation to crisp system of delay differentia equation using the concept of generalized Hukuhara derivative technique. Using MATLAB software numerical result found. The discussion is also done with some idea of dynamic behavior of delayed protein degradation model in fuzzy environment.

Acknowledgement. This research was financially supported by the Ministry of Education, Malaysia under FRGS Grant (Project No.: 01-01-18-2031FR).

References

1. Zadeh LA (1965) Fuzzy Sets Inf Control 8:338–353
2. Dubois D, Prade H (1982) Towards fuzzy differential calculus part 3 differentiations. Fuzzy Sets Syst 8:225–233
3. Kaleva O (1987) Fuzzy differential equations. Fuzzy Sets Syst 24:301–317
4. Wu C, Song S (1996) Approximate solutions, existence, and uniqueness of the cauchy problem of fuzzy differential equations. J Math Anal Appl 202:629–644
5. Min C, Huang N-J, Zhang LH (2014) Existence of local and global solutions of fuzzy delay differential inclusions. Adv Differ Equ 108:1–14

6. Lupulescu V, Abbas U (2012) Fuzzy delay differential equations. Fuzzy Optim Decis Mak 11:99–111
7. Barzinji K, Maan N, Aris N (2014) Linear fuzzy delay differential system: analysis on stability of steady state. MATEMATIKA 30(1a):1–7
8. Narayanamoorthy S, Yookesh TL (2015) Approximate method for solving the linear fuzzy delay differential equations. Discret Dyn Nat Soc 2015, Article ID 273830:1–9
9. Jafelice RM, Barros LC, Bassanezi RC (2009) A fuzzy delay differential equation model for HIV dynamics. In: IFSA-EUSFLAT, pp 265–270

Rainfall Intensity Forecast Using Ensemble Artificial Neural Network and Data Fusion for Tropical Climate

Noor Zuraidin Mohd Safar[1,2(✉)], David Ndzi[3], Hairulnizam Mahdin[1,2], and Ku Muhammad Naim Ku Khalif[4]

[1] Sensor and Internet of Things (SioT) Faculty of Computer Science and Information Technology, Universiti Tun Hussein Onn Malaysia, Batu Pahat, Johor, Malaysia
zuraidin@uthm.edu.my
[2] Faculty of Computer Science and Information Technology, Universiti Tun Hussein Onn Malaysia, Batu Pahat, Johor, Malaysia
[3] School of Engineering and Computing, University of the West of Scotland, Glasgow, UK
[4] Faculty of Industrial Sciences and Technology, Universiti Malaysia Pahang, Pekan, Pahang, Malaysia

Abstract. This paper proposes an ensemble method based on neural network architecture and stacking generalization. The objective is to develop a novel ensemble of Artificial Neural Network models with back propagation network and dynamic Recurrent Neural Network to improve prediction accuracy. Historical meteorological parameters and rainfall intensity have been used for predicting the rainfall intensity forecast. Hourly predicted rainfall intensity forecast are compared and analyzed for all models. The result shows that for 1 h of prediction, the neural network ensemble forecast model returns 94% of precision value. The study achieves that the ensemble neural network model shows significant improvement in prediction performance as compared to the individual neural network model.

Keywords: Rainfall forecasting · Artificial Neural Network · Recurrent neural network · Expert system · Ensemble learning · Tropical climate

1 Introduction

Rainfall is an environmental process in natural water cycle. Developments in weather forecast models have been substantial interest in meteorological study. The emerging of new technologies and the extensive of natural phenomena have enabled the used of computational prediction methods. Rainfall intensity forecasting continues one of the most challenging in hydrological process due to spatial and temporal deviations in rainfall dissemination. The objective is to determine the feasibility of using a data collected from a single source point as input to neural networks in order to achieve predictions of regression and rain classification from different environmental

R. Ghazali et al. (Eds.): SCDM 2020, AISC 978, pp. 241–250, 2020.
https://doi.org/10.1007/978-3-030-36056-6_24

parameters. In the proposed model, however it uses a real time data that has smaller significant delay. This prediction model use a fusion of meteorological data within the same location of rain event occurred. In general, meteorological processes are extremely non-linear and complicated to predict at high spatial, they are non-linear and follows a very irregular trend. An ensemble of Artificial Neural Network (ANN) has the capability to bring out the structural relationship between the environmental parameters for better forecasting output [1]. The ANN with network back propagation (BPN), Recurrent Neural Network (RNN) of nonlinear autoregressive network with exogenous inputs (NARX) and Ensemble Neural Network (ENN) approach have been used to make the forecasting of 1, 3 and 6 h rainfall intensity prediction. Data used to train the Neural Network corresponds to historical time series of environmental parameters and rainfall rate prediction. The fusion of environmental parameters are used in ANN architecture and training development. The performance estimation is determined using Mean Square Error (MSE), Mean Absolute Error (MAE) and the measurement correlation coefficient between actual and forecast value.

The environmental parameters and rainfall data from Malaysia is used in this work. The goal of this model is to use ensemble of neural network with stack generalization of the environmental parameters in predicting rainfall intensity. An ensemble technique based upon the stacking generalization of second level technique has been proposed to aggregate the output predictions from ANN-BPN and NARX models. The results are computed and analyzed. The comparison between the individual ANN-BPN, NARX and ENN models with the ensemble model has been carried out and outcome is analyzed. The results indicate that ENN model achieved 94% prediction accuracy for 1 h rain intensity prediction. This paper has been organized into five sections with the starting section is an introduction of the study. Section 2 presents the information of study area and meteorological data, Sect. 3 presents the methodology, Sect. 4 presents result and discussion of the finding and finally, Sect. 5 discusses the conclusion and future works.

2 Study Area and Meteorological Data

The study area is a small town in North-West of Malaysia (Fig. 1) that has a tropical type of climate with long hours of sunny day, high temperature and humidity. The average temperature is 27.5 °C

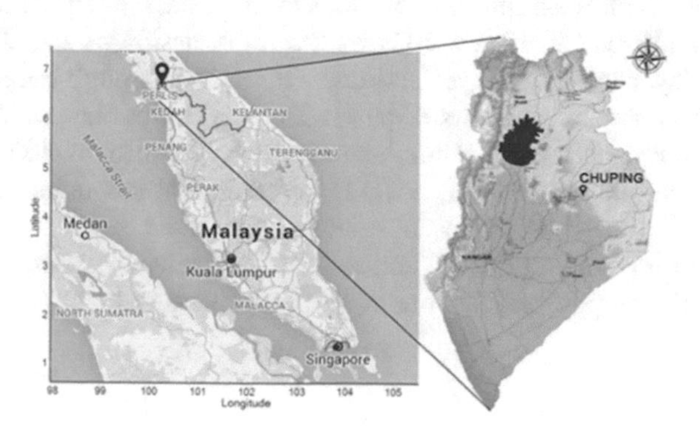

Fig. 1. Map of Perlis and study area.

The average annual rain and its rainfall rate are 3.09 mm and 2.94 mm s^{-1} respectively. Meteorological information and weather forecasting is managed by Malaysia Meteorological Department. Chuping weather station has a meteorological data for temperature, pressure, dew point, wind speed, humidity and rainfall intensity. Three years of data (hourly recorded) from 2012 to 2014 that consist of 26304.

3 Methodology

3.1 Data Preparation

Hourly meteorological data is provided by Malaysian Meteorological Department. The rainfall stations selected is Chuping. Basic analysis of the parameter pattern such as minimum, maximum, and mean values is been calculated and identified. This will give quick analysis to inspect the relationship between parameters themselves and relationship towards the rain condition.

Table 1. Environmental parameter analysis for Chuping weather station

Parameters	Unit	Max	Mean	Min
Pressure	HPa	1017.0	1009.6	1001.2
Temperature	°C	38.1	27.5	16.1
Dew point	°C	29.4	24.5	15.3
Humidity	%	100	83.7	29.0
Wind speed	ms^{-1}	5.1	1.1	0.0

3.2 Artificial Neural Network

ANN's concept is to mimic the procedures of human neurons. ANN is an interconnected group of linked artificial neurons. Feed forward neural network is implemented in the early ANN model and is known as perceptron. This model uses a single layer of perceptron, fed straight to the output input. The drawback of this approach is that many classes of pattern cannot be recognize by perceptron. Neural network will learn from the input signal training data set and its required target output. Network training is the iterative method. With fresh information from training data, every iteration coefficient of each node is altered [2]. Figure 2 shows the transmission of signal that propagate from the input layer to hidden layer and lastly the target through each neuron.

Supposed L is amount of layers (input, hidden and output layer). Whereas l is the input and hidden layer that carry N number of nodes in the form of $N(l)$, where $l = (0,1, ...,L; l = 0)$ is the input) and $i = 1,..,N(l)$ is the node that has output from preceding layer and it will be the input for the subsequent layer, $y_{l,i}$ is the output that is dependable from inbound signals $x_{l,i}$ and α, β, Υ parameters. As a result, the equation that generalize the output from each node can be expressed in:

$$y_{l,i} = f_{l,i}(x_{l-1,1}w_{l-1,1}, \ldots, x_{l-1,N(l-1)}, \alpha, \beta, \gamma, \ldots) \tag{1}$$

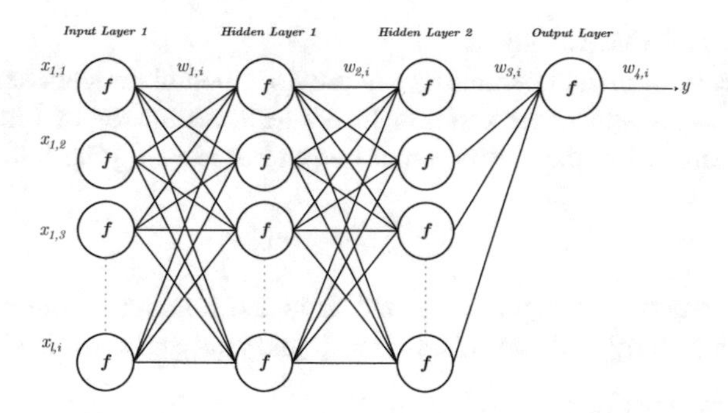

Fig. 2. ANN layer representation

The next step is to compare the output signal from the node with the required output from the training data (z) after the signal propagation is completed. The different is known as an error signal $\delta_{l,i}$.

$$\delta_{l,i} = z - y_{l,i} \tag{2}$$

Supposing training data has P items, the error measured for the Pth item of the training data is the sum of the squared error:

$$E_P = \sum_{k=1}^{N(L)} (z - y_{l,i})^2 \tag{3}$$

where k is the quantity of component z (actual) and y_{li} (prediction) The weights for a specific node are adapted by propagating back to all neurons in direct proportion to the error in the units to which it is attached. Gradient Descent algorithm is use to implement this. Gradient descent approach is technically to find weight that will deliver the minimize error. Step by step algorithm to minimize the error are measured by Acquiring gradient vector and Calculating signal error of $\varepsilon_{l,i}$ as the derivative of the error measure in E_p with the respect to the output of node i in layer l for both direct and indirect path. The derivation order can be express in:

$$\varepsilon_{l,i} = \frac{\partial^+ E_p}{\partial y_{l,i}} \tag{4}$$

ANN initial execution.
Data for both ANN-BPN and NARX are separated into training, validating and testing set. A small proportion of the training data is used for cross validation to prevent overfitting and ascertain for a better generalization. In training process, monitoring the errors in training and validation is necessity. Three years data with hourly parameters

reading together with the rainfall records were selected to train ANN models, the randomly selected data is divided in which 70% of the data was set as a training set, 15% for testing and 15% of the data was set aside to use for validation.

Input, Hidden and Output layer.
Initial input is environmental parameters. In previous neural network studies there is no solid evidence about the finite solution to define the number of hidden layers and hidden layer neurons but there is an input-output based using the following form:

$$[2n^{1/2}, 2n + m] \tag{5}$$

where n is the amount of input value and m is the number of output in the neural network [3]. This study will use this approach and trial and error basis.

Neural Network architecture.
The proposed ANN prediction model is based on MLP with back propagation architecture. Various different neural network structures have been tried with trial and error basis and the best performance were selected for forecast results. In general the input parameters P_{input_t} and output parameters P_{output_t} for the neural network architecture can be describe as follow:

$$P_{input_t} = [R_t, R_{t-1}, \ldots, R_{t-n,} P_t, T_t, DP_t, H_t, WS_t] \tag{6}$$

$$P_{output_t} = [R_{t+1}] \tag{7}$$

where R is the rainfall intensity detected in time t, n is the hour(s) before time of the prediction.

3.3 Back Propagation Network

ANN have been integrated to predict multiple meteorological parameters including rainfall, temperature, wind speed and humidity [4]. A related study was organized in the Philippines by utilizing day-to-day meteorological data during training. The experimentation showed that forecast results achieved 90% when the combination of ANN and Bayesian network were used [5]. Most research on adapting ANN for rainfall forecasting has been done with feed forward neural networks with BPN. A typical BPN contains of input-output layer and at least a single hidden layer [6]. Number of neurons at each layer and number of hidden layers determine the network's ability on predicting accurate output for a given data set. For example, in this study, 5 input parameters and 20 nodes (in first hidden layer) are used. The output from each node will feed for the next hidden layer that has 10 nodes as depicted in Fig. 3.

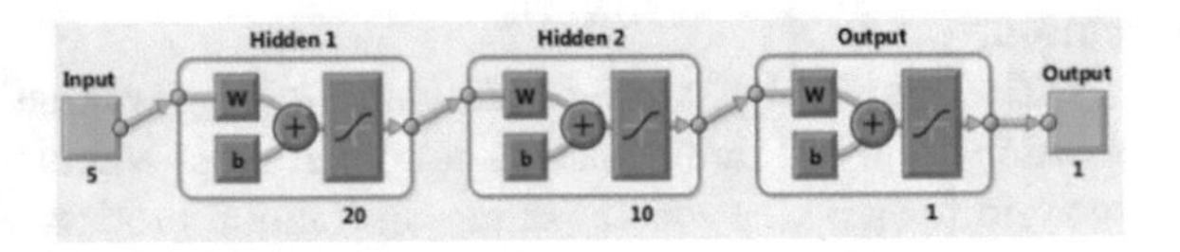

Fig. 3. ANN-BPN architecture

3.4 Recurrent Neural Network with NARX

Study on recurrent neural network of rainfall forecasting in Indonesia gave an output of 75% accuracy by using El-Nino Southern Oscillation (ENSO) as neural network input [7]. The NARX model is based on Autoregressive with Exogenous linear model, which is commonly used in time-series modeling [8]. NARX address the vanishing gradient problem by imposing delays in its network [9]. The general form of NARX can be expressed in the following equation:

$$y_t = f(y_{t-1}, y_{t-2}, \ldots, x_t, x_{t-1}, \ldots, \theta) \tag{8}$$

where y_t is a new state in terms of the previous state y_{t-1}, the current input x_t, and some parameters θ that are shared over time. An overview of NARX configuration and its structure for this work is depicted in Fig. 4.

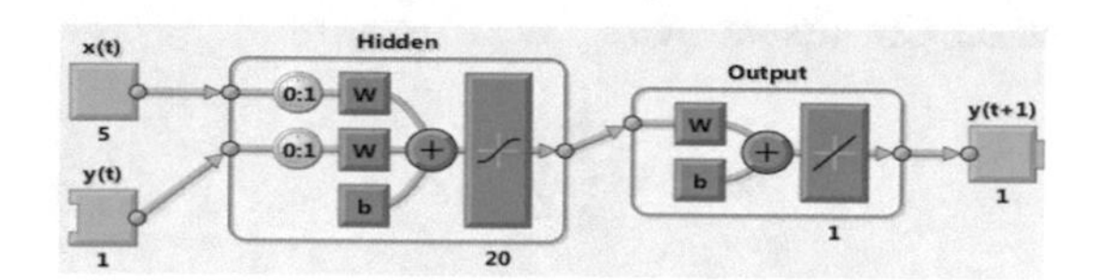

Fig. 4. NARX configuration and its structure

3.5 Ensemble Neural Network with Stacking Generalization

The basic idea of ensemble learning is to combine multiple models to improve prediction performance [2, 10]. This model based upon multiple learners have been shown to perform better than models with single learners, especially when dealing with complex dataset [5]. They are considered meta-algorithms designed to work on top of existing learning algorithms. The generalization capability of a neural network system can be significantly improved with respect to the best performing single neural network through a neural network ensemble [11]. Depending on the type of architecture selected, the precision is accomplished by appropriate training algorithms and parameters [12]. The variety of the ensemble members can be accomplished by distinct techniques and most commonly used by controlling the set of original random weights, varying topologies, varying the training algorithm and manipulating the training, such as mixing trained networks with different sample information [13]. These can be used on additional layers, or the process can stop with a final result. In this work, two level of stacking were applied, the schematic diagram of the implementation is depicted in Fig. 5

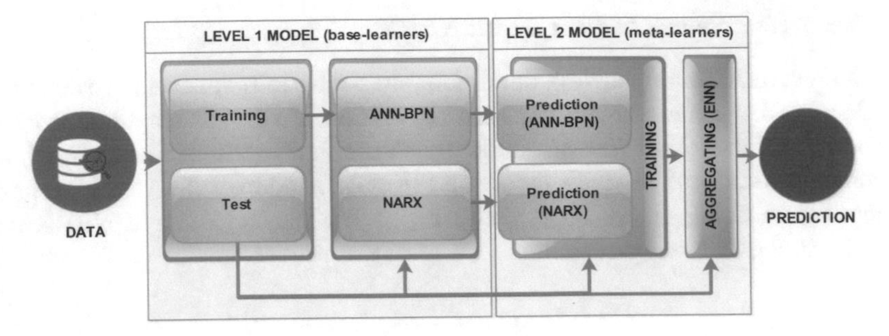

Fig. 5. Stacking generalization

3.6 Evaluation of the Forecasting Result

The performance was evaluated using Mean Absolute Error (*MAE*), Mean Square Error (*MSE*) and Correlation Coefficient (*R*).

$$MAE = \frac{1}{n}\sum_{t=1}^{n} |y_t - \hat{y}_t| \tag{9}$$

$$MSE = \frac{1}{n}\sum_{t=1}^{n} (y_t - \hat{y}_t)^2 \tag{10}$$

$$R = \frac{n\sum y_t\hat{y}_t - (\sum y_t)(\sum \hat{y}_t)}{\sqrt{n(\sum y_t^2) - (\sum y_t)^2} - \sqrt{n(\sum \hat{y}_t^2) - (\sum \hat{y}_t)^2}} \tag{11}$$

where y_t is the real value, $\hat{y}_t$ is forecast value and n is the observation number and the R value indicates the relationship between forecast and real value.

4 Results and Discussion

The experimental setup will find the finest outcome for hourly prediction; evaluating the magnitude of error (MAE, MSE) and the regression (R) of predictive output against the desired output. Rain intensity prediction is difficult for a single neural network prediction methodology. This study's main goal is to define potential improvements in forecast efficiency using network ensemble. A neural network ensemble is a technique in which the outputs of a collection of distinctly trained networks are combined to create an embedded forecast. The ensemble neural network is achieved by using diverse network architectures. In this study ANN-BPN and NARX are used as a base-learners and the ensemble is created using stacking generalization methods. Using an appropriate training algorithms, choosing number of hidden layers and quantity of neurons in hidden layer by trial error, the precision of individual networks is accomplished. An ensemble technique based upon the stacking generalization of second level technique has been proposed to aggregate the output predictions from ANN-BPN and

NARX models. The results are measured and analyzed. The comparison between ANN-BPN, NARX and ENN has been carried out and outcome is analyzed.

Table 2. Prediction results for ANN-BPN, NARX and ENN model

Rain intensity prediction	Performance matrices	NARX	ANN-BPN	ENN
1 h	*MAE*	0.305	0.152	0.128
	MSE	1.768	0.932	0.929
	R	0.853	0.926	0.935
3 h	*MAE*	0.412	0.394	0.401
	MSE	3.344	3.566	3.455
	R	0.696	0.671	0.716
6 h	*MAE*	0.448	0.419	0.385
	MSE	4.327	4.452	4.383
	R	0.578	0.561	0.595

The obtained results as shown in Table 3 indicate that, an ensemble of neural networks increases forecast accuracy compared to stand alone ANN-BPN and NARX. ENN reached 0.935, 0.716 and 0.595 for 1, 3 and 6 h prediction respectively for correlation coefficient accuracy evaluation. Figure 6 shows an ENN regression result for 1 and 3 h prediction. The accuracy of prediction degraded after the time windows prediction increases (comparison between actual against predicted value are depicted in Figs. 7 and 8). The magnitude of error matrices for MAE and MSE also show ENN produced significant improvement compared to ANN-BPN and NARX. The enhancement of accuracy was possible based on the various type of neural networks that reduce the errors at the same point of targeted output. Then, using the ensemble the errors can be compensated for better forecast accuracy.

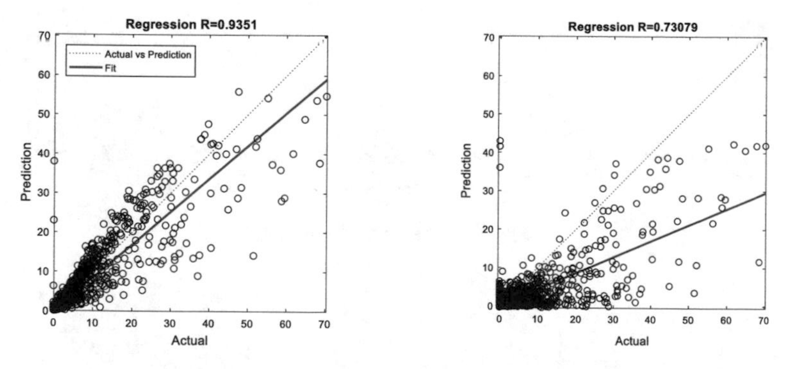

Fig. 6. ENN Regression for 1 h and 3 h rainfall intensity prediction

The comparison of actual against predicted rain intensity are presented in Figs. 7 and 8. Three neural network architectures (ANN-BPN, NARX and ENN) were compared. In Fig. 7, 1 h actual rain intensity is show the same pattern to the predicted intensity for ENN. The benefit of the neural network ensemble is that it can compensate for the network residual errors resulting from Level 1 (base-learners), by integrating their outputs for training features for Level 2 (meta-layer) ensemble network. Consequently, the prediction is better than single network.

5 Conclusion

An ensemble model based on the neural network stacking generalization of multilayer perceptron is developed to forecast and analyze hourly prediction of rainfall intensity for tropical climate. Weather data between 1st January 2012 and 31st December 2014 are collected to fit the model. ENN had an additional input for training from the predicted outcome of ANN-BPN and NARX model. The performance of the proposed ensemble model (ENN) is evaluated and compared to different single network prediction models of ANN-BPN and NARX. The evaluation demonstrates that an ensemble model performs 5% better in determining rainfall intensity values together with magnitude error of 0.13 (MAE) and 0.92 (MSE) for 1 h prediction. ENN reached 0.94, 0.72 and 0.60 for 1, 3 and 6 h prediction respectively for correlation coefficient accuracy evaluation. These results demonstrate the interpretability and reliability of the stacked ensemble model and a need to improve rainfall intensity forecast methods in future works. As this study shows a better result with an ensemble of neural network technique, other models need to be contrived for further improvements in prediction. In addition to further improvements in model performance, larger historical data spanning years of recorded data should be adopted towards addressing the important of localized of the short period of rain forecasting. Furthermore, the availability of continuous hourly forecast results can be used in user friendly tools and an interactive platform that will truly benefits the target users.

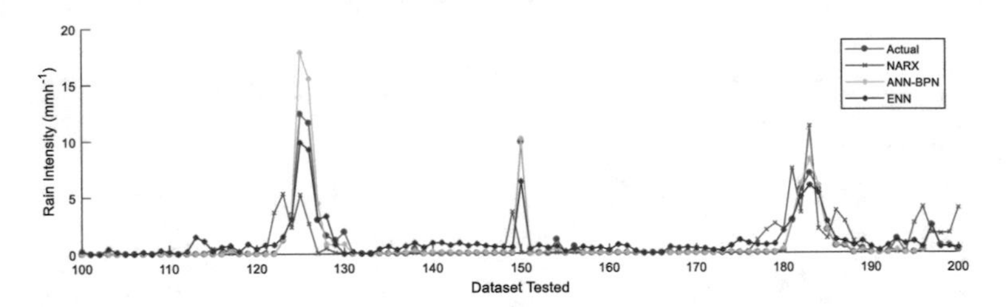

Fig. 7. One hour prediction

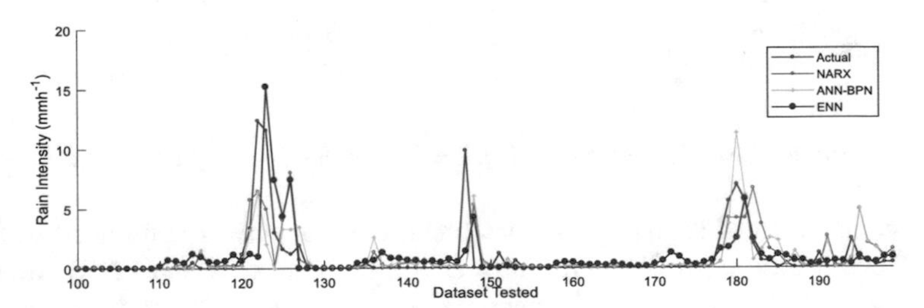

Fig. 8. Three hours prediction

Acknowledgments. This work is sponsored by University Tun Hussein Onn Malaysia.

References

1. Tyagi V, Mishra A (2014) A survey on ensemble combination schemes of neural network. Int J Comput Appl 95(16):18–21
2. Haidar A, Verma B, Sinha T (2018) A novel approach for optimizing ensemble components in rainfall prediction. In: 2018 Proceeding of IEEE congress on evolutionary computation (CEC), no 978, pp 1–8
3. Goss DFE (1993) Forecasting with neural networks: an application using bankruptcy data. Inf Manag 24(3):159–167
4. Kashiwao T, Nakayama K, Ando S, Ikeda K, Lee M, Bahadori A (2017) A neural network-based local rainfall prediction system using meteorological data on the Internet: a case study using data from the Japan Meteorological Agency. Appl Soft Comput 56:317–330
5. Klent Gomez Abistado CNA, Maravillas EA (2014) Weather forecasting using artificial neural network and Bayesian network. J Adv Comput Intell Intell Inf. 18(5):812–817
6. Lawrence S, Giles CL, Tsoi AC (1996) What size neural network gives optimal generalization ? Convergence properties of backpropagation, Networks, no. UMIACS-TR-96-22 and CS-TR-3617, pp 1–37
7. Kanigoro B, Salman AG (2016) Recurrent gradient descent adaptive learning rate and momentum neural network for rainfall forecasting. In: 2016 international seminar on application for technology of information and communication (ISemantic), pp 23–26
8. Mohd R, Butt MA, Zaman Baba M (2019) SALM-NARX: Self adaptive LM-based NARX model for the prediction of rainfall. In: Proceedings of International Conference on I-SMAC (IoT Soc. Mobile, Anal. Cloud), I-SMAC 2018, pp 580–585
9. Noor HM, Ndzi D, Yang G, Safar NZM (2017) Rainfall-based river flow prediction using NARX in Malaysia. In: Proceedings of 2017 IEEE 13th international colloquium on signal processing and its applications, CSPA 2017
10. Razak IAWA, Abidin IZ, Siah YK, Abidin AAZ, Rahman TKA (2017) Ensemble of ANN and ANFIS for water quality prediction and analysis—a data driven approach. J Telecommun Electron Comput Eng 9(2–9):117–122
11. Ma Z, Wang P, Gao Z, Wang R, Khalighi K (2018) Ensemble of machine learning algorithms using the stacked generalization approach to estimate the warfarin dose. PLoS ONE 13(10):1–12
12. Velasco LCP, Granados ARB, Ortega JMA, Pagtalunan KVD (2018) Performance analysis of artificial neural networks training algorithms and transfer functions for medium-term water consumption forecasting. Int J Adv Comput Sci Appl (IJACSA) 9(4):109–116
13. Souto YM, Porto F, Moura AM, Bezerra EA (2018) A Spatiotemporal ensemble approach to rainfall forecasting. In: Proceedings of international joint conference on neural networks, July 2018, pp 1–8

Applied Artificial Intelligence

Recommending Domain Specific Keywords for Twitter

Muhammad Adeel Abid, Muhammad Faheem Mushtaq(✉), Urooj Akram, Bushra Mughal, Maqsood Ahmad, and Muhammad Imran

Faculty of Computer Science and Information Technology, Khwaja Fareed University of Engineering and Information Technology, 64200 Rahim Yar Khan, Pakistan
faheem.mushtaq, bushra.mughal@kfueit.edu.pk, Creativemind.adeel@gmail.com, uroojakram.cs, maqsoodzee, mimran047@gmail.com

Abstract. Twitter has become the most popular social media in today's world. More than 284 million users are online monthly, and 80% user accesses their twitter account through mobile. A tweet is limited to 140 characters, so it contains concise information about particulars. Due to its popularity and usage, near about 500 million tweets are sent per day that relates to different domains. This work focuses on recommending domain specific keywords for twitter. For this purpose, 10 domains are chosen as a sample. Then we apply Term Frequency-Inverse Document Frequency (TF-IDF) and Log likelihood methods and compared the keywords extracted from both against each domain to make our result much valuable. Furthermore, the categorization of keywords is made as noun and verb, and also finds out the sentiment words. At the end, a relevancy test is performed from five users. These keywords can be great value in clustering tweets data and can be used for identifying a user's interest in any specific domain. Furthermore, these keywords are of the great asset for advertisement purpose.

Keywords: Twitter · Social media · Domain-specific keywords · Clustering tweets data · Advertisement

1 Introduction

Twitter is a widely used information network that connects you with the latest news, ideas, opinions and researches of your interest. Organizations, groups, individuals, school, colleges, universities and other groups start a twitter account to promote their status, views, ideas, events, news and to inform others about their personal network. Twitter is used by people belonging to different categories i.e. bands, showbiz personalities, students, doctors, singers, researchers, sportsmen, politicians, etc. Twitter users belong to all age groups who use different sources to access Twitter [1]. Most of the users use a Twitter account on frequent basis in a day through computer/mobile and a Twitter official told that mostly more than 50% users access their Twitter accounts on mobile. The usage of Twitter is estimated from the fact that in August 2019

R. Ghazali et al. (Eds.): SCDM 2020, AISC 978, pp. 253–263, 2020.
https://doi.org/10.1007/978-3-030-36056-6_25

approximately 500 million tweets are sent in a day [2]. Twitter is now widely used by the business organization to show and highlight their information. This is a key point through which they interact with their customers, partners and influencers to build a strong relationship [3]. Twitter is also a platform between audience/viewers and media/TV organizations. Interesting facts related to Twitter are given below in 2019 [2].

- 335 million monthly active users
- 500 million Tweets are sent per day
- 80% of Twitter active users are on a mobile
- 79% of accounts are outside the U.S.
- Twitter supports 35 + languages
- US, UK and Japan are the top countries by users.

When a large population of world keeps posting information in the form the messages on Twitter on daily basis then we have large collection [4] of important information such as latest news, user's views, and interesting ideas about everything in world; we have bands of favorite music, foods, what is happening in sports and also have bird's eye view of trending topics and so on. Persons post tweets that relate to the different domain i.e. Travel, news, politics, business, health, charity, technology, music, sports, science etc. Tweets about a particular domain are identified on the basis of common words found in tweets [5]. In this research, we especially focus on finding out domain specific keywords [6], sentiment words [7] in each domain and categorize them [8]. It helps in clustering a large number of Tweets data. Keywords related to different domains can be used for advertisement purpose.

This remaining part of this paper is organized as follows: Sect. 2 describes some related work which includes the techniques related to the extraction of keywords. Section 3 explains the collection of data from different domains and way to extract the keywords based on TF-IDF and LogLikelihood method. Section 4 discusses the results and discussion of proposed method. Section 5 presents the conclusion and future work of this research.

2 Related Work

There are numerous techniques that can be used to focus on word/keyword extraction [8] and can be related to our work.

2.1 Word Frequency Analysis

In the early time, researches are on finding the frequency of a term usage but much more concerned in finding the keywords form a corpus. In 1972 the extraction of keywords against a document with the respect to another document is common [9]. This technique is commonly known as TF-IDF. It works with specifying a term within a document corpus. The following equation is used for the computation of word frequency analysis,

$$TFIDF(t,d,n,N) = \left(\sum_{word \in d} {}^{1 \text{ if word=t}}_{0 \text{ else}} \right) \times \log\left(\frac{N-n}{n}\right) \tag{1}$$

2.2 Using a Document Corpus

Another technique used for extraction information uses a Markov Chain that evaluates every word that is found in the document corpus [10]. This technique mainly stress on defining a Markov Chain for the term t and document d followed by the two states (C, T), terms are refer to be as related if two terms arrived at the same state with similar regularity.

2.3 Frequency-Based Single Document Keyword Extraction

This Technique was developed by Matsuo and Ishizuka [11] that illustrates the extraction of keywords from a given document that utilize a technique named word co-occurrence to format a co-occurrence matrix illustrated in Table 1.

Table 1. The co-occurrence matrix

	a	B	c	D
a		5	13	7
b	5		42	3
c	13	42		25
d	7	3	25	

If there is only one occurrence of a word in the document, the authors uses another technique named Pearson's chi-squared to test for each word in the document.

Let n = Total number of words
Let O = frequency observed
Let E = Frequency expected

$$x^2 = \sum_{i=0}^{n} \left(\frac{(O_i - F_i)^2}{E_i} \right) \tag{2}$$

This test allows the test of the frequency distribution of each word with the expected distribution.

2.4 Content-Sensitive Single Document Keyword Extraction

There is another method used for extraction of keyword that was introduced by Ohsawa, et al. [12], solves the problem in a different way. This method, called Key Graph constructs a graph for a document, with the representation of terms as nodes, and edges to connect nodes and shows association. Connected groups are identified as words Clusters within a document graph. Searching and locating nodes and find ranking by probability for each cluster is identified as candidate keywords. Tests that are applied on Key Graph illustrates that it is unable to match, and exceeded, Term Frequency-Inverse Document Frequency (TF-IDF) in a series of tests [12].

2.5 Keyphrase Extraction Using Bayes Classifier

An alteration of TF-IDF was used in conjunction with a naive Bayes classifier by Frank et al. [13] to locate key phrases in a document within the corpus. This method works by running the TF-IDF variation in equation over every phrase in the document.

$$TFIDF(p, d) = \Pr[\mathit{phrase} \text{ in d is p}] \times -\log \Pr[\text{p appears in any document}] \quad (3)$$

where p stand for the phrase, and d shown the document. Bayes theorem is used to find keyphrase in the document:

$$\Pr[key|T, D] = \frac{\Pr[T|key] \times \Pr[D|key] \times \Pr[key]}{\Pr[T, D]} \quad (4)$$

3 Materials and Methods

Twitter APIs are used to collect tweets data for each of the 10-domain using different keywords. Then we extract terms of tweets for identifying the keywords against each domain and further operations are performed.

3.1 Data Collection

Twitter APIs and downloaded tweets are utilized from different domains. A java program is used to download tweets and php language is used for performing procedure. Firstly, the 7 sample of keywords from each domain to collects 21000 tweets against each domain (Travel, Music, News, Sports, Health, Science, Charity, Technology, Politics, and Business). Tweets dataset having an attribute “language” which is beneficial in order to collect only English tweets. The quantitative detail of collected data shown in Table 1.

Table 2. Quantitative details of Twitter Data

Total number of tweets	Total size of the data in MB	Total number of English tweets	Total number of English tweets selected	Total number of non-English tweets
691604	288 (approximate)	353783	210000 (21000 for each of 10 domains)	337821

The existing methods having problems in computation and their result regarding keywords are not up to the mark and used in rare cases. TF-IDF and Loglikelihood method are well recognized methods regarding keywords extraction and dominant on all existing techniques.

TF-IDF Method: TF-IDF is used to rank the keywords according to the occurrence/frequency. For the computation of TF-IDF [14] from TF_{ij}, the following Eq. 5 is used to illustrated as the frequency of the term i in j documents.

$$TF_{ij} = \frac{n_{ij}}{\sum_k nkj} \tag{5}$$

And the IDF_i (Inverse document frequency) is calculated as

$$IDF_i = \log\left(\frac{|D|}{|\{d : t_i \in d\}|}\right) \tag{6}$$

Where |D| is the set of all domains and t_i is the term.

The final TF-IDF is calculated by multiply term frequency with relevant Inverse document frequency.

$$TF - IDF_i = TF_{ij} \times IDF_i \tag{7}$$

LogLikelihood Method: This method is used to determine keywords in the corpora which differentiate one from another [15]. The main formula is

$$G = 2 \times \left[\begin{array}{l}\left[freq_{domain} \times \log\left[\frac{freq_{domain}}{freq\_Expected_{domain}}\right]\right] + \\ \left[freq_{general} \log\left[\frac{freq_{general}}{freq\_Expected_{general}}\right]\right]\end{array}\right] \tag{8}$$

where $freq_{domain}$ and $freq_{general}$ are actual frequencies in the domain corpus and reference corpus. The $freq\_Expected_{domain}$ and $freq\_Expected_{general}$ are expected frequencies that are calculated according to the following formulae:

$$freq\_Expected_{domain} = size_{domain} \times \frac{freq_{domain} + freq_{general}}{size_{domain} + size_{general}} \tag{9}$$

$$freq\_Expected_{general} = size_{general} \times \frac{freq_{domain} + freq_{general}}{size_{domain} + size_{general}} \tag{10}$$

Furthermore, the Eq. 8 cannot distinguish between the corpus of two domains. Formula 1 worked as symmetrical for both corpuses. We should correct this situation with another Eq. 11, i.e. their relative frequency is bigger in the domain corpus than in the reference corpus. If the condition is false, then we will discard the word as a possible term: in our case, we multiply its weight by −1.

$$\frac{freq_{domain}}{size_{domain}} > \frac{freq_{general}}{size_{general}} \quad (11)$$

4 Results and Discussion

TF-IDF and loglikelihood method are applied for each domain and find out 100 keywords for each domain. To more purify this research, we further find common keywords in TF-IDF and loglikelihood method. Following graph illustrates the statistics.

The keywords that are common between TF-IDF and Loglikelihood method is shown in Fig. 1. The keywords extracted for each domain are as follows:

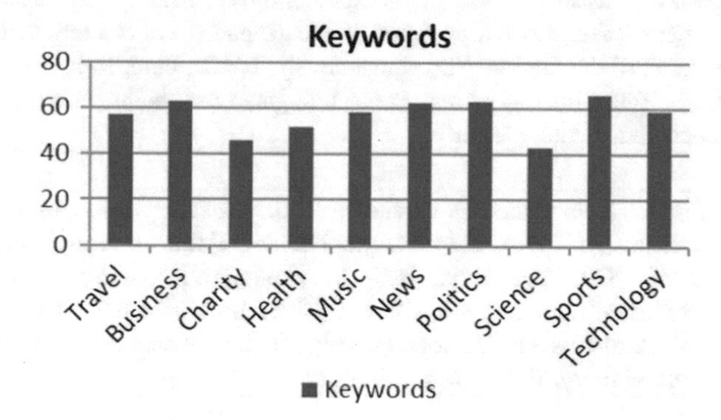

Fig. 1. Graph showing keywords that are common between TF-IDF and Loglikelihood method

Travel
Travel, lake, tour, tourist, Travelling, Transport, tourism, France, de, Public, Durban, Zone, Chill, gear, hitsbiebx, Yorkshire, vacation, justinbieber, VISIT, Yorkshires, followed, Attractions, loves, Pages, files, front, brother, warped, hotel, amazing, missing, Airport, bus, beach, Top, essential, Trip, City, le, fireworks, East, holiday, abuse, summer, road, California, July, Child, home, London, Morning, Meet, Park, weekend, fun, 4th, Time
Business
Credit, Stock, debit, Financial, Profit, loss, card, exchange, business, pips, Weight, Counseling, Closed, Consumer, stocks, Total, cards, lots, pipsforex, Call, York, sell, Exchange-traded, AA, posting, motoring, EURUSD, Largest, mt4forexrobot, Market, trading, Fat, Dakota, CASH, forex, Insurance, pricing, Foods|, Buy, Consumer, Bank, London, Lost, Analyst, Nonprofit, twitter, comes, Burning, North, followers, model, Hearing, healthy, Aid, Online, money, OFFER, Job, Success, Youtube, Services, share, company, REUTERS
Charity
Help, needy, poor, orphan, donation, Charity, Fund, raising, piety, Donate, Black, Fundraising, gofundme, Im, money, Click, please, Allah, Event, circulation, ease, leg, Thank, Ramadan, dark, people, Support, film, rich, Campaign, consider, amount, feed, Children, ministry, raise, girl, watching, Watch, SEASON, thanks, God, chocolate, Building, month, helps

Category	Keywords
Music	Music, song, songs, lyrics, album, band, artist, video, love, Listening, listen, Posted, NowPlaying, favorite, Format, MP3, iTunes, Facebook, 5sos, photos, SING, YouTube, Rock, live, movie, lol, bands, ass, Download, Radio, hip-hop, single, Luke5SOS, Michael5SOS, singing, Follow, Calum5SOS, Ashton5SOS, art, playlist, pizza, ft, Festival, check, playing, hear, release, You've, dance, Brothers, liked, Added, hot, tweet, official, ur, guys
News	Killed, Kill, Rape, gun, MURDER, Affairs, KNICKS, Felton, news, Pull, Trigger, Jail, trusting, Olivia, trial, teen, Raymond, guard, Robber, shoots, voicemail, Veterans, Women, handing, avoids, drivers, letting, Police, reaches, Daytime, jokes, plea, worst, robbed, Emmy, HitTheFloor, accused, carpet, guilty, woman, heart, goes, hosts, held, VA, NY, Gonna, Avoid, Foreign, yourself, Bitch, race, head, won't, deal, giving, Red, death, DONT, York, nigga, women
Health	Stress, blood, pressure, disease, antibiotics, Medicine, Syrup, health, Sagittarius, Currently, attention, issues, maple, feel, Lower, pay, cough, Heart, pancakes, Natural, Chocolate, Healthcare, reduce, PAIN, infection, Superbugs, levels, Foods, cure, low, sleep, Risk, fitness, cream, Drink, Care, Taste, doctor, taking, Prevent, sugar, Lose, helps, Food, sick, weight, mental, Body, Diet, Eat, Healthy, Don't
Politics	Nawaz, Sharif, Obama, Democratic, Government, PMLN, PTI, Politics, Pakistan, Pml, PM, Party, Govt, President, Iraq, Qadri, TuQ, MaryamNSharif, Political, Army, Operation, PDP, OurFuture, leader, ul, GOP, PPP, prime, Tahir, Iran, Immigration, Independence, Barack, Lahore, Peoples, minister, Supreme, mICHELLE, Khan, ImranKhanPTI, gov, leaders, Iraqi, Putin, Imran, Modi, Border, Nisar, Democracy, Pak, Punjab, Obamas, tcot, IK, Threat, Nation, tells, officials, Court, country, Children, rules, heard
Sports	Tennis, Table, Match, Hockey, football, Rugby, sports, Cup, World, SOCCER, play, game, team, Ghana, Conversion, players, vs, victory, sport, England, playing, Sourav, Saha, overshadowed, BengalhttptcoeEAGnQh0t4, prodigy, Player, win, pong, Brazil, spurs, mens, United, Ping, Argentina, Brooks, Mayor, season, ball, basketball, NBA, world cup, Wales, fans, played, league, west, Kings, watch, Queens, Final, USA, title, Tournament, won, leads, games, wins, watching, Baseball, Williams, Goal, club, Women's, John, Field
Science	Chemical, chemicals, invention, Discovery, Science, NASA, Space, Romance, CassiniSaturn, weapons, bloodstream, Toxic, rocket, Harmful, Loral, broccoli, Forests, burn, Backtrack, NASA, moon, telescope, fundamental, BANG, researchers, planet, human, Brain, reaction, Original, earth, Channel, Mistakes, mars, AIR, Products, Syria, study, picture, Launch, claims, water
Technology	Research, Engineering, Android, electronics, circuit, Technology, Space, Google, AndroidGames, domain, game insight, service, Station, name, testing, collected, Jobs, ive, featured, Engineer, coins, tape, Unveils, androidandroidgamesgameinsight, Social, media, You're, registration, Market, dope, job, GOLD, Tech, Tribez, LG, plans, MSFT, Boyd, dana, Scientific, harvested, Free, ISSResearch, Samsung, Silverstone, Design, supporting, Apple, creation, iOS, Indian, Amazon, Digital, University, app, Manager, download, NASA, Thanks

Then we find out the nouns and verbs in keywords. The statistics of keywords are shown in following graph.

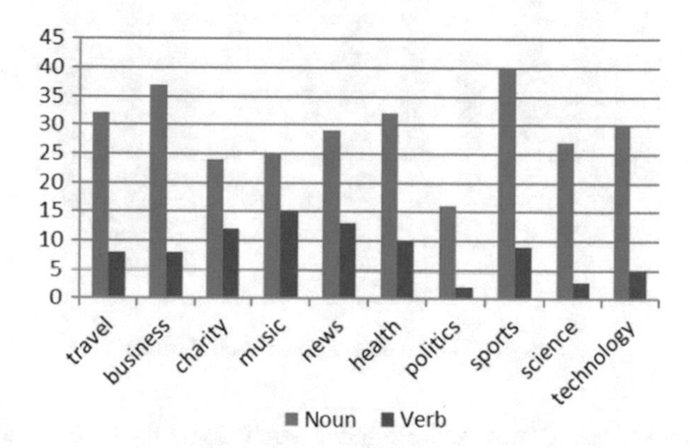

Fig. 2. Nouns and Verb in keywords of each domain.

Figure 2 shows the number of nouns and verb found in keywords against each domain. We also find sentiment words [7] in each domain from sentiwordnet.

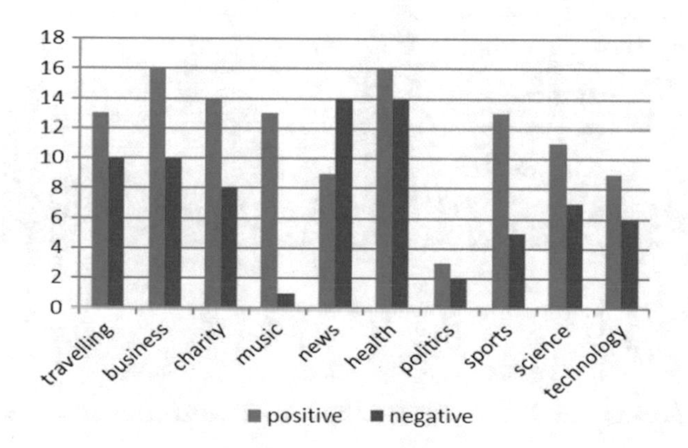

Fig. 3. keywords as positive or negative of each domain

Figure 3 shows the keywords as positive and negative of each domain. We also perform TTest against each domain.

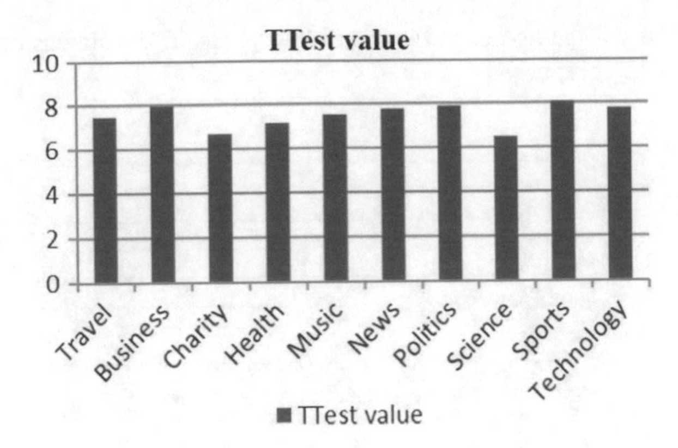

Fig. 4. TTest value against each domain

Figure 4 shows the TTtest value against each domain. At the end we perform relevancy test from 5 users and ask each user to check each domain keyword and categorize them as relevant, related and not relevant.

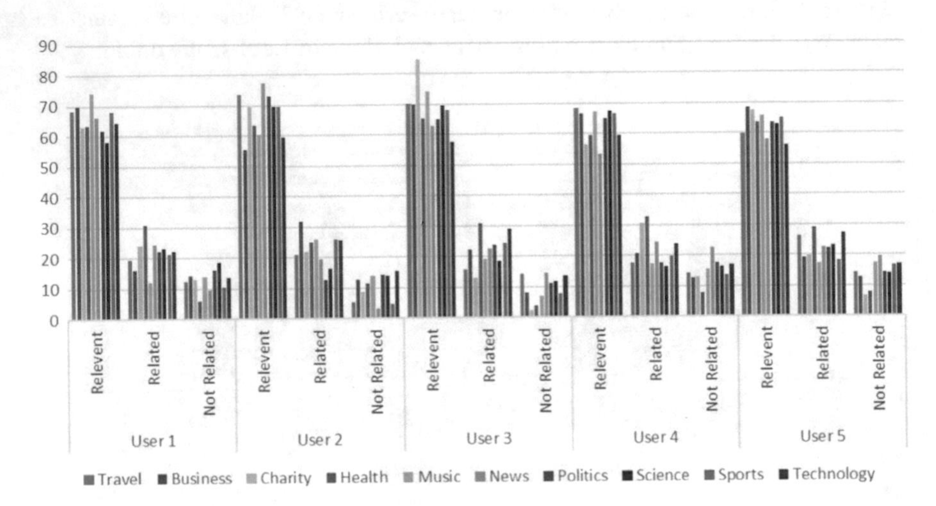

Fig. 5. Details about relevancy test percentage wise

Figure 5 shows the graph detail about relevancy test percentage wise. From this graph we can see that we obtain relevant results from 58 to 70% and near about 10–25% results related after combining relevant and related results we obtain near about 80% results relevant/related with the respect to each domain.

5 Conclusion and Future Work

Twitter becomes the most growing and popular social media and microblogging service in today's world. Due to its popularity, a large number of tweets are available. This work focuses on finding Domain Specific keywords for twitter. It is found that keywords are used to specifying a domain in twitter because all keywords are different for each domain. Keywords are also used to show trends in twitter. Moreover, the keywords are utilized for the advertisement and clustering purpose in tweets data. This research can be further analyzed in the future to some other method for specifying domain specific keywords would be used which would enhance the relevancy percentage of keywords against a domain. Furthermore, the alpha and beta test can be applied to rectify the correctness of keywords, and the keywords extraction techniques can be applied to HTML pages for the advertisement purpose.

References

1. Asghar M, Mushtaq MF, Asmat H, Saad Missen MM, Khan TA, Ullah S (2014) Finding correlation between content based features and the popularity of a celebrity on Twitter. Int J Comput Sci 11:177–181
2. Twitter Statistics. https://techjury.net/stats-about/twitter
3. Twitter. https://about.twitter.com/en_gb.html
4. Cai Y, Chen Y (2009) Mining Influential bloggers: from general to domain specific. In: Proceedings of 13th international conference on knowledge-based and intelligent information and engineering systems, pp 1–8
5. Fan W, Bifet A (2014) Mining big data: current status, and forecast to the future. SIGKDD Explor. 14:1–5
6. Prasad AVK, Saibaba CM (2016) Mining big data: current status, and forecast to the future for telecom data. Int J Priv Cloud Comput Environ Manag 1–10
7. Agarwal A, Xie B, Vovsha I, Rambow O, Passonneau R (2011) Sentiment analysis of Twitter data. In: Proceedings of the workshop on language in social media, pp 30–38
8. Lott B (2012) Survey of keyword extraction techniques, Semantic Scholar, 1–10
9. Jones KS (2004) A statistical interpretation of term specificity and its application in retrieval. J Doc 28:11–21
10. Wartena C, Brussee R, Slakhorst W (2010) Keyword extraction using word co-occurrence. In: Proceedings of 21st international workshops on database and expert systems applications DEXA, pp 54–58
11. Matsuo MIY (2004) Keyword extraction from a single document using word co-occurrence statistical information. Int J Artif Intell Tools 13:157–169
12. Ohsawa Y, Benson NE, Yachida M, Science H (1998) KeyGraph: indexing by a co-occurrence graph based on building construction metaphor. In: Proceedings IEEE international forum on research and technology advances in digital libraries, ADL'98
13. Wu YB, Li Q, Bot RS, Chen X (2006) Domain-specific keyphrase extraction. In: Proceedings of the 14th ACM international conference on information and knowledge management

14. Kim SN, Baldwin T, Kan M (2009) An unsupervised approach to domain-specific term extraction. In: Proceedings of Australasian language technology association workshop, pp 94–98
15. Gelbukh A, Sidorov G, Lavin-Villa E, Chanona-Hernandez L (2010) Automatic term extraction using log-likelihood based comparison with general reference corpus. In: International conference on application of natural language to information systems. Lecture notes in computer science, vol 6177, pp 248–255

A Review on Smart IOT Based Parking System

Adil Ali Saleem[1(✉)], Hafeez Ur Rehman Siddiqui[2], Rahman Shafique[1], Abdullah Haider[1], and Muhammad Ali[1]

[1] Faculty of Computer Science and Information Technology, Khwaja Fareed University of Engineering and Information Technology, Rahim Yar Khan, Pakistan
Adilalisaleem@gmail.com, projectbs50@gmail.com, fromabdullah@gmail.com, ab417044@gmail.com
[2] School of Engineering and Design, London South Bank University, London, UK
siddiqov@gmail.com

Abstract. With the rapid proliferation in population of the big cities, numbers of vehicles are also growing with proportionate pace. People prefer to use private transport rather than public transport which results an increase in the number of vehicles. Thus, finding vacant parking place becomes cumbersome, resulting in several issues like increase in consumption of fuel, time and energy. Thanks to the Internet of things (IOT) that has advanced so much and opened a new spectrum of technologies that were previously considered as science fiction. Almost every field is inclined to get benefits from IOT. Big cities or metropolitan cities are also on the track of smart cities by exploiting IOT and addressing issues such as traffic congestion, limited car parking facilities and road safety. In this paper, we have reviewed some research papers from the recent work carried out in the year 2018 to 2019 related to IOT based smart parking. In their investigations different models augmented with sensors, integrating cloud and mobile application are presented, which subsequently yields in a smart parking system which saves time, energy, fuel and subsequently carbon footprint. It is noticed that disabled people are ignored to be addressed in their systems and there is no single global parking solution exists so far. The rapid development in miniaturization and robustness of processors, sensor and coupling with machine learning added to the potential of smart and single global solution feasible for future researchers to address parking issues exists in both indoor and outdoor parking lots.

Keywords: IOT · Smart parking · Sensors · Clouds · Microcontroller

1 Introduction

During recent years with the increase in population in metropolitan cities the number of vehicles also increased with the same pace. This leads to an issue for drivers to park their cars in parking. Extra Time, efforts and fuel consumption to find a parking space and they end up parking their cars on the streets which then leads to road blocks during rush hours. Sometimes during festive seasons or peak hours people cannot find a space

R. Ghazali et al. (Eds.): SCDM 2020, AISC 978, pp. 264–273, 2020.
https://doi.org/10.1007/978-3-030-36056-6_26

for parking. INRIX (a Washington based company that provides data to automaker and parking services) during a research shows that about 3.6 billion hours of time, 70 billion dollars and 1.7 billion gallons of fuel were wasted to find a parking space in a year in USA [1]. In Germany, one of the studies conducted by INRIX, average driver pays 98 euros (112$) for parking that makes Germans waste an average of 896 euros to find parking in a year. Another issue linked to this problem is air pollution longer the time it takes to find parking space more fuel will be consumed and resulting in the release of pollutant gases.

IOT is a system of interlinked computation devices, objects or people with unique ID's. Transferring of data in IOT doesn't require human interaction. By using IOT system, devices can connect and communicate with each other. IOT makes physical devices able to think, sense, and share information with each other and make decisions [2].

To resolve these issues many advanced countries are choosing smart IOT based parking systems and thus enhance the driver safety. Smart parking systems helps provide the information about the vacant spot by using diverse sensors deployed in the parking site. Smart parking systems are capable to provide necessary information and make an easy access to drivers and subsequently it makes a best use of the space in parking lot. It also helps drivers to find a place for parking in a specific area which yields in the reduction of time used to find the parking slot. This helps in reduction of carbon foot print.

In this paper recent diverse solution on IOT based smart parking systems are explored and presented. Aiming at compactly review the recent progress in the said domain to make an understanding of IOT based smart parking for the readers.

2 Literature Review

Author [1] proposed a model to save the time, effort and obviously fuel using IOT while searching for parking. The system consists of hardware components like master Raspberry-pi (RPi), infrared (IR) sensor, and sixteen RPi each equipped with a camera and a display board for each lane [1]. A master RPi is placed at the entrance of each lane connected to an ultrasonic sensor. The sensor's frequency of sensing is kept 2 s. When a car approaches, the ultrasonic sensor attached to the master RPi detects the incoming car and signals to the array of 16 RPi's to take an image of the number plate displayed at rear view. A cloud database is used for data management. Google cloud vision API is used to extract license number from image previously captured by sixteen RPi cameras. Sixteen RPi units were used because the number plate mounted on each vehicle is not fixed at specific place at front or rear. With this arrangement, it is observed that, the set of RPi's is capable to capture 97% to 99% the number plate of any vehicle. Once the google API for vision extracted the number plate, it is sent back to the specific RPi unit. Subsequently the number and the index of the specific RPi unit is transmitted to the master RPi. In this system a small pager is handed over to each car as it enters parking lot. The master RPi updates the database with the information, license number and the specific RPi's index, obtained previously from RPi unit. The driver obtained a direction message about parking slot at the pager device via

SMTP. When the vehicle parked on the allocated parking slot, the status of that parking slot is updated as occupied. The same procedure in reversed is carried out and master RPi updates the database and sets the slot as vacant.

Another approach to manage indoor car parking is proposed by [3]. It helps the user by providing the nearest vacant parking slot at the users set destination. The system comprises an IR, RPi and Arduino board and Wi-Fi module for data transmission and a gateway. The setup informed and guide to current vacant slot in the parking. The infrared transmitter and receiver detect the occupancy and non-occupancy of the parking slot. The transmitter is attached to the footpath besides the parking slot. The Wi-Fi module is attached to the light bulb at each parking slot. These Wi-Fi modules gather data from the IR receiver. Every parking slot has a light bulb and a Wi-Fi module. These Wi-Fi modules pass data to the gateway. The gateway sends data to cloud, where it is stored for analysis and later retrieval. The driver must register oneself to reserve a parking slot by using a mobile phone application. The mobile application provides global positioning system (GPS) to direct the driver to a parking slot already reserved for him/her. Moreover, the application also provides fare and modes of payment option and register driver's feedback after leaving the parking.

An end to end cloud-based solution for parking problems based on IOT has been proposed by [4]. Using low cost wireless sensor network (WSN) incorporates with cloud help to reduce the overall cost, time and efforts that yields in environment improvement by reducing carbon footprint. Proposed system presented by [4] detects the vacant parking slot by IR sensors and then directs the drivers to locate and navigate to the best parking slot which is easy to access for them.

The deployed sensors transmit the parking slot status (occupied or unoccupied) in real time to an RPi (gateway), the RPi in return relays the data to a cloud. A web-based application uses cloud data to facilitate the users with slot availability information, advance booking and automatic payment through a wallet system.

Both online and on-field booking are available. In online booking user book parking slot prior to his/her arrival using web-based application and a quick response (QR) code is generated. As the driver enters the parking lot the QR code, previously generated at time of reservation is scanned which helps to keep check in, checkout and fare records. In On-field booking, as the driver arrives at parking, an automatic toll machine registers the vehicle number and the user cell number and generates a QR code. The QR code is scanned and grants an access to the parking and the timer for that vehicle is set. On exit the QR code is scanned again and the bill is prepared followed by stopping the timer. At the end, the bill is deducted via card payment at exit gate.

Another approach to IOT based smart parking is presented by [5]. The solution comprises of Micro-controller (Arduino) with ZigBee are used for real time notifications regarding parking slot occupancy or non-occupancy and navigation to target is provided to driver via android based application.

The ZigBee standard is developed on top of the IEEE 802.15.4 standard and defines the network and application layers [6]. The network layer provides networking functionalities for different network topologies, and the application layer provides a framework for distributed application development and communication [6]. The two protocol stacks can be combined to support short range low data rate wireless communication with battery-powered wireless devices [6].

Data Transmission is different in Zigbee from Bluetooth and Wi-Fi module. Because it is connected to the devices having 16-bit address. It also uses unique ID for communication. ZigBee's self-routing is its unique quality. This model consists of two main components Arduino and Zigbee. At one end Arduino interacts with different electronic components, here it interacts with IR sensor, Servo Motor, LED's and LCD. On the other end another Arduino acts as a gateway and also uses global system for mobiles communication (GSM) module in general packet radio service (GPRS) mode. The gateway (Arduino) is deployed in base station to update driver on fare and car location if he/she lost his/her car. Arduino microcontroller uses IR sensor to sense the presence of vehicle in parking slot. The LED turns red in case if it is occupied otherwise green. An LCD screen at parking slot entrance is also used to display the status of parking slot availability. A barrier controlled by a servo motor is also controlled. The barrier is set opened if parking slot is available otherwise closed to indicate the parking slot occupancy. Wi-Fi and Bluetooth module are out performed by the Zigbee. It uses 50% of less power than other two said modules.

Author [7] uses RPi board mounted with a camera for facial and character recognition for two-way security. To use this system driver must get oneself registered on an android based application. The vehicle registration number is stored at the time of reservation. A fifteen minutes long window is set to avail the reserved parking lot otherwise the reservation will be cancelled. Upon arrival at parking lot, RPi camera reads the number plate and verified it with the one stored at the time of reservation. RPi also captures the driver facial picture to prevent car lifting. Once registration number verification and facial image process has been accomplished, the vehicle enters the nearest vacant parking slot and parking time sets and stops upon leaving the parking lot. The fare receipt is sent to the driver on mobile application. Subsequently he/she pays the bill using E-Wallet. Driver face detection is also carried out on the time of exit, if it matches with already stored picture taken at time of arrival, the vehicle is allowed to exit.

An arrangement of IR sensor, RPi board and ultrasonic sensor is proposed by [8] for IOT based smart parking. Information are sent wirelessly to the cloud for storage. User can view this information using an android application. An ultrasonic sensor senses the parking slot's occupancy or non-occupancy based on the length of time of flight (TOF) of sound waves. Obviously, a non-occupied parking slot will have longer TOF of sound waves. This information is shared with RPi, which in turns transfers it to the cloud. The data from the cloud is accessed by the user with the help of android based application to see which slot is vacant in the nearest parking lot.

Another model is proposed by [9] aims at managing parking efficiently and smartly. Anisotropic Magneto-Resistive (AMR) is used to find out the occupancy of the parking slot. AMR sensor is made up of Wheatstone bridge which converts the magnetic field intensity to small values of electric voltage. AMR sensor senses the occupancy by detecting a change in the magnetic field of earth. The frequency of sensing is kept five minutes to save power. If the sensor senses the change in magnetic field which is greater than the set threshold, it sends the status of the slot as occupied to the gateway which is used as a bridge between Wireless Sensor Network (WSN) and the IP based internet protocols. On IP side message queuing telemetry transport (MQTT) protocol is used due to its lower bandwidth and power efficiency. Sensors sense the occupancy and send a short message to MQTT server using Zigbee network. The message contains the location, slot number and the time when vehicle is entered in the slot. Forward or backward parking is identified by sensing the magnetic field because the change in magnetic field on the front and middle is greater than the rear side.

Author [10] suggested a system in which an ultrasonic sensor is used to sense the occupancy of slot. For ultrasonic sensor a specific distance value is set as threshold for non-occupied status. When the slot is occupied, the distance calculated by sound waves will always be less than the set threshold. All distance values below the set threshold classifies the slot as occupied. The Parking slot status is sent to Arduino board and RPi serially and stored in the cloud using MQTT generated instance. An Android based application for user is developed to access, view and reserve the parking slot data stored in the cloud. Upon reservation a QR code is generated which is scanned at the entrance of parking. This app shows the slot's availability in real time.

An android based mobile application integrating it with IOT module is presented by [11] manage parking efficiently. In this system wireless sensor network (WSN), near field communication (NFC), Cloud and other IOT technologies are used. This paper proposed a micro controller-based and GSM parking lots is used for their monitoring. A mobile app is developed for viewing and reservation of parking lot. For this purpose, PIC16F87XA micro controller is used on the parking lot side to check and control the occupancy of the parking lot. On the user side the user must get register by providing required credentials. Once the registration is completed, they can search for available parking slots. A one-time password (OTP) is generated, on the time of reservation of available slot. This OTP is used at the time of Entrance and Exit at the entry gate of parking. Total amount of receipt is sent to the user on the android application which then can be charged online.

Author [12] anticipated a solution that uses ultrasonic sensors with long range wireless data telemetry (LoRa) transmitter and receiver to make parking system smart. Ultrasonic sensor is connected to TTGO ESP32 LoRa board. These sensors are deployed in each parking slot to sense the occupancy of the parking slot. A Threshold of 25 cm is set for ultrasonic sensor. If the slot is vacant flag is set to 0 and 1 in the contrary case. LoRa Transmitter sends flag value with the slot name to the LoRa receiver that in turns transfers this data to the cloud. Android app is developed to view slot availability.

Author [13] recommended a system for smart parking system using Bluetooth Low Energy (BLE) communication. For this purpose, an Arduino board, RPi which works on Bluetooth communication, IR sensor and ultra-sonic sensor controlled by Arduino is used. For vehicle's drivers an android application has been developed. The IR sensor is used to sense the occupancy for outdoor parking while ultrasonic sensor is used to sense the occupancy for indoor parking.

Considering low power consumption, Arduino send the data every two seconds to central devices such as RPi. In between the data transmission the sensors are kept in sleeping mode. Subsequently data is stored in SQLite (a small database used in mobiles) database within RPi, which then synced with back end data base after every Twenty-five seconds. Bluetooth communication is used to send data from Arduino to RPi. User can view the available parking slot on a mobile application.

A solution for the road side parking has been proposed based on Bluetooth low energy beacon technology is proposed by [14]. The proposed solution consists of beacon reader, beacon transmitter, gateway and a server. In this model author assumes that a beacon transmitter is attached to the right-side mirror in right hand driving countries, which emits beacon packets regularly. The beacon transmitter has a specific ID which is used to identify the vehicle. These packets are read by the beacon reader which is placed at every parking slot. A gateway device connected to each beacon reader using Bluetooth is used to read the values from them. Low-power wireless personal area network (6LoWPAN) technology is used in the Bluetooth network. RPi is used as a gateway that gets the information from beacon reader and sends them to the server for analysis regarding parking slot occupancy.

3 Analysis and Discussion

An IOT based smart parking system is presented by [1] doesn't provide any kind of information regarding available slots remotely and lack of parking slot reservation. The purpose of the two extra cameras, other than sixteen cameras were used to take the vehicle's front and rear picture, has not been mentioned.

The system [3] provides not only reservation via android application but also directed to parking slot by GPS assisted navigation. Albeit it covers the deficiency presented in [1], it doesn't explain the system response against a false reservation i.e. slot is reserved but didn't avail. A similar work is presented by [4] with considering power consumption as additional feature by setting scanning frequency. Upon reservation a QR code is generated to be scanned at the entrance on arrival. The payment mode (card payment) is also taken into account. The system is specific for indoor parking. An online payment for parking fee and navigation to parking slot should also be considered. Moreover, no mechanism presents to counter false reservation.

Zigbee has been used in between edge layer and gateway layer in the system presented by [5]. An apparent downside is the use of GSM that adds additional operating cost and the use of LED lights in the parking area are of no use. Moreover, the purpose of automatic barrier is not clear especially, how the barrier will be opened when a parked car is about to leave. Navigation to parking slots is not taken into account [7]. Provides an additional security using facial recognition besides IR and ultrasonic sensors. The offered system is equipped with online reservation system and payment methods. A fifteen minutes long reservation window is set to counter the false reservation. Ultrasonic and IR sensors are used in [8] to find the availability of parking slot. Android based application is used to fetch data from cloud at one end and present it to user on the other end [10]. Has the same system as [8] except it is using only ultrasonic sensor for parking availability and it also provides parking slot reservation [12]. Uses a long-range wireless transmission between edge layer and gateway i.e. LoRa. Ultrasonic sensor is used to detect parking slot occupancy [13]. Differentiate their system from [10] by introducing Bluetooth communication technology in between edge layer and gateway at cost of short range but benefitting from power harnessing. No reservation and payment method are explained in [8, 12, 13]. Moreover, those systems only targets indoor parking lot.

Author [9] suggested an innovative technique for occupancy detection using change in earth magnetic field. Zigbee protocol is used for communication between edge and gateway layer while MQTT protocol is used at user end (application layer). This system has same lack of features described in [8]. Author in [11] implemented an indoor smart parking system using Ultrasonic sensor with the micro controller PIC16F87XA. However, in this model OTP is provided at the entrance or parking lot exit which results in wastage of time. Lack of navigation is another issue noticed in this system. A slightly different approach is adopted by [14] using BLE beacon technology for smart parking system. The system only covers one end i.e. parking administration to manage the parking lot by obtaining information regarding occupied and vacant places.

Table 1. The comparison between different reviewed investigations

References	Technologies used	Pros	Cons
[1]	IR Sensor, Ultrasonic Sensor, RPI camera and Board, Pager Device	Navigation text message to the vacant parking slot is via pager device	No mobile application, no reservation, no GPS navigation, no facility for disabled
[3]	IR Sensor, RPI, Arduino Board, Wi-Fi Module for data transmission	Equipped with dedicated mobile application, GPS navigation and Reservation, different modes for online payments	No explanation of system response against false reservations, no facility for disabled
[4]	WSN, IR Sensor, RPI	Web application, Navigation to vacant parking slot, online and at spot reservation facility	No mechanism to counter false reservations, no online payment method, lack of navigation, no facility for disabled
[5]	ZigBee, Arduino, IR Sensor, Servo Motor, LED, LCD, GSM	LCD at entrance to show vacant parking slots, Reservation system	GSM adds extra cost, lack of navigation, no facility for disabled
[7]	RPI board Mounted with a camera, IR Sensor	Reservations via mobile application, counter mechanism for false reservation, Payment through E-Wallet	Lack of navigation, no facility for disabled
[8]	IR Sensor, Ultra Sonic Senor, RPI	Mobile Application to view vacant parking slots	No reservation, no online payment method, targets indoor parking, lack of navigation, no facility for disabled

(*continued*)

Table 1. (*continued*)

References	Technologies used	Pros	Cons
[9]	AMR Sensor, RPI	MQTT is used at Application Layer	No reservation, no online payment method, targets indoor parking, lack of navigation, no facility for disabled
[10]	Ultra Sonic Sensor, Arduino, RPI	Mobile Application to view the information about slot occupancy	No reservation, no online payment method, targets indoor parking, lack of navigation, no facility for disabled
[11]	PIC16F87XA micro controller, IR Sensor	Mobile Application, Reservation, Online payment	Use of OTP at entrance and exit is wastage of time, lack of navigation, no facility for disabled
[12]	TTGO ESP32 LoRa board, Ultra sonic Sensor	Android App to view slot availability	No reservation, no online payment method, targets indoor parking, lack of navigation, no facility for disabled
[13]	Arduino, RPI, Ultrasonic Sensor, IR Sensor	Mobile Application is used to view information about slot availability	No reservation, no online payment method, targets indoor parking, lack of navigation, no facility for disabled
[14]	Beacon reader and sender, RPI	Information about the occupancy of parking slots is stored in database, which then used by managers to manage the Parking	It covers only one end only for managers, no reservation, no online payment method, lack of navigation, no facility for disabled

It doesn't provide any information regarding slot availability, reservations or payment method to the drivers.

These reviewed systems don't provide any kind of facility for disabled people. Moreover, no single global parking solution exists. Some approaches addressed indoor while other addressed outdoor parking. The rapid development in miniaturization and robustness of cost effective processors and sensor coupling with machine learning will open smart and effective ways for future researchers to address current challenges in smart parking. Subsequently it will pave the way towards a global IoT based parking solution equally for both outdoor and indoor parking system. Automated Vehicles (AV's) and electric vehicles will affect the parking system design in the near future. The AV's will be kept on charging while parked in parking lot. So, while developing a smart parking system the issues previously found and the issues that will arise in the near future should be addressed. Following table shows the comparison between different reviewed investigations (Table 1).

4 Conclusion

Parking issues related to time, efforts and fuel consumption in finding empty slots, reservations facility and its procedure, are major challenges in big cities. Different approaches have been applied by researcher to address the parking issues. This paper reviewed approaches adopted by many researchers alongside technologies implementation and analysis are presented to address those prime issues as well as secondary issues such as traffic congestions and wastage of resources. The aim is to summaries most recent techniques and technologies to assist researchers to materialize a global solution of an IOT based parking.

References

1. Das S (2019) A novel parking management system, for smart cities, to save fuel, time, and money. In: 2019 IEEE 9th annual computing and communication workshop and conference (CCWC), pp 0950–0954
2. Al-Fuqaha A, Guizani M, Mohammadi M, Aledhari M, Ayyash M (2015) Internet of things: a survey on enabling technologies, protocols, and applications. IEEE Commun Surv Tutor 17(4):2347–2376
3. Pavan Kumar G, Rajeev Kumar C, Manikanta VJ, Azhagiri M (2018) IoT based sensor enabled smart parking system. Int Res J Eng Technol 10(5):1174–1176
4. Poornimakkani S, Senthilkumar S, Daniel SF (2018) A cloud based end-to-end smart parking solution powered by IoT. Int Res J Eng Technol 3(5):3559–3565
5. Qadir Z, Al-Turjman F, Khan MA, Nesimoglu T (2018) ZIGBEE based time and energy efficient smart parking system using IOT. In: 2018 18th mediterranean microwave symposium (MMS), pp 295–298
6. Kumar A, Sharma A, Grewal K (2014) Resolving the paradox between IEEE 802.15.4 and Zigbee. In: 2014 international conference on reliability optimization and information technology (ICROIT). IEEE, pp 484–486
7. Cassin Thangam E, Mohan M, Ganesh J, Sukesh CV (2018) Internet of things (IoT) based smart parking reservation system using Raspberry-pi. Int J Appl Eng Res 8(13):5759–5765
8. Balhwan S, Gupta D, Reddy SRN (2019) Smart parking—a wireless sensor networks application using IoT. In: Proceedings of 2nd international conference on communication, computing and networking. Springer, Singapore, pp 217–230
9. Dhaou IB, Kondoro A, Alsabhawi AH, Guedhami O, Tenhunen H (2018) A smart parking management system using IoT technology. In: Proceedings of the 22st conference of open innovations association FRUCT, p 43
10. Gupta R, Pradhan S, Haridas A, Karia DC (2018) Cloud based smart parking system. In: 2018 second international conference on inventive communication and computational technologies (ICICCT). IEEE, pp 341–345
11. Kamble P, Chandgude S, Deshpande K, Kumari C, Gaikwad KM (2018) Smart parking system. Int J Adv Res Dev 3(4):183–186

12. Kodali RK, Borra KY, Gn SS, Domma HJ (2018) An IoT based smart parking system using LoRa. In: 2018 international conference on cyber-enabled distributed computing and knowledge discovery (CyberC). IEEE, pp 151–1513
13. Marso K, Macko D (2019) A new parking-space detection system using prototyping devices and bluetooth low energy communication. Int J Eng Technol 9(2):108–118
14. Chen HT, Lin PY, Lin CY (2019) A smart roadside parking system using bluetooth low energy beacons. In: Workshops of the international conference on advanced information networking and applications. Springer, pp 471–480

Mental Health App Reviews Analyzer (MHARA) Using Logistic Regression and Tri-Gram

Maqsood Ahmad[1,2(✉)], Noorhaniza Wahid[1], Arif Mehmood[2], Gyu Sang Choi[3], Rahayu A. Hamid[1], Muhammad Faheem Mushtaq[1], and Shaznoor Shakira Saharuddin[1]

[1] Faculty of Computer Science and Information Technology, Universiti Tun Hussein Onn Malaysia, 86400 Parit Raja, Batu Pahat, Johor, Malaysia
maqsoodzee@gmail.com, shakirasahar@gmail.com, {nhaniza, rahayu}@uthm.edu.my, faheem.mushtaq@kfueit.edu.pk

[2] Faculty of Computer Science and Information Technology, Khwaja Fareed University of Engineering and Information Technology, 64200 Rahim Yar Khan, Pakistan
maqsoodzee@gmail.com, arifnhmp@gmail.com

[3] Department of Information and Communication Engineering, Yeungnam University, Gyeongsan 38542, Korea
castchoi@ynu.ac.kr

Abstract. This study is related to assist for finding the mental health Apps on Google Play Store using text reviews and rating. The apps are related to Anxiety, Bipolar disorder, Epilepsy, Migraine and Mental retarded. A prediction system has been proposed using Tri-Gram features and Logistic Regression. The system is trained using reviews as features and rating as labels/classes. The model has been evaluated using accuracy, precision, recall and f1-score. It is also compared with other state-of-the-art classifiers, but the proposed approach provides the good accuracy. The accuracy, precision, recall and f1-score of the proposed model respectively are 0.69, 0.67, 0.69 and 0.66. The results exhibited that the Logistic Regression out-performs as compare to other state-of-the-art algorithms.

Keywords: Data mining · Text mining · Tri-Gram · Logistic regression · Prediction system

1 Introduction

The advancement of technology is helping in many fields of life, especially in smart phone. Now a day most of the users are using Apps on smart phone for getting all kind of information. They not only are getting the information but use the Apps as an assistant in the treatment of diseases. Therefore, for selection of Apps is an important part. Mostly, they study the review before using the App. The reviews on the Mental Health Apps (MHA) uploaded on Google Play Store by the inference using text mining

R. Ghazali et al. (Eds.): SCDM 2020, AISC 978, pp. 274–282, 2020.
https://doi.org/10.1007/978-3-030-36056-6_27

techniques are a challenging task. These reviews assist the people to develop a priority list to choose most proliferating and best MHA app for them. The best method is to involve the user in the process of development, shaping and discovering the application. The repositories of the reviews of user feedback are developed by the world top companies that used to analysis for the improvement purposes.

The MHA applications are developed to help the people who are suffering with mental diseases and also contribute in the academia. There is thousands of applications related to mental health in the Google Play Store e.g. PsyTests, Depression Treatment and this number significantly increasing from developers that uploaded on daily basis. On the other hand, the number of users also increases to download and use the MHA applications to get quick assistance.

The U.S. National Institute of Mental Health (NIMH) and the UK's National Health Service (NHS) are the public health organizations pointed the MHA applications as scalable and cost-effective solutions to address the shortage of mental health provider, thus, the effectiveness of the MHA applications remains challenged [1]. These applications help the patients to self-manage their mental health condition and prescribed the supplement psychiatric treatment that proved the effective treatment to the patient.

The users of MHA application share their own experience, views about the specific application with feedback comments and rating, that makes such application robust and unique sources of information [2]. The selection of Apps using review is hard and not possible to read all the reviews about the App. In this research, a model has been proposed to predict best app and applied on a dataset of thousands of user's reviews from different MHA under five categories (Anxiety, Bipolar Disorder, Epilepsy, Mental Retarded, and Migraine) of mental disease. The developed model called Mental Health App Reviews Analyzer (MHARA) is a state-of-the-art to make a knowledge base system. This knowledge base is used for textual information extraction that focusing on the power of text mining and Natural Language Processing (NLP) techniques which can be utilized in the MHARA. The results produced are used to evaluate the users' reviews based on the accuracy, precision, recall, and f1-score. The NLP techniques have been identified as an important area of growth within the artificial intelligence in medicine community [3]. In NLP, the classification of documents and strings into different categories are the essential part of the process. It is difficult in text classification to automatic tagging of customer queries, classification of blogs in different categories, and having little training dataset. The text classification problem is too difficult to generalize by the learners. For example, most of the patients satisfied with the news filtering services required a hundred day are worth of training data.

In terms of MHA Google Play Store categories, different machine learning (ML) algorithms and text mining classifications techniques for android application reviews had been applied [4]. We have extracted 5,309 reviews from five different categories related to mental health. The statistical information has been evaluated based on the text features and ML algorithms like Logistic Regression (LR), Random Forest (RF), and Multinomial Naïve Bayes (MNB) classifiers. The performance of each algorithm has been evaluated using different parameters. The results exhibited that the LR outperformed as compared to other state-of-the-art algorithms.

The remaining sections of the article are organized as follows: Sect. 2 is describing the related work whereas Sect. 3 discusses about the proposed model. Section 4 explains the experimental results and discussion while Sect. 5 presents the conclusion and future work.

2 Literature Review

In recent years, mHealth is the rapidly expanding practice that used the mobile technology to improve and track health outcomes [5]. ML provides a way that exploits the correlations between the high volume datasets with different kind of information such as text and image to be classified in such a way that it could be easily readable for human. The main utilization of ML can be categorized into clustering, classification, association, regression and optimization tasks in all type of data [6].

We know that in ML, classification problems are classified according to class labels that indicative of the features or instances of input. In [7], the authors defined the Multi-Label Classification (MLC) as a supervised ML problem and classifications algorithm learns from the features and have multi class labels, where each class has two values. Moreover, the feature may have different set of labels. Single Label Classification (SLC) and MLC both are different with respect to the data features instances.

The previous studies [8] showed that the effectiveness of the classifiers can be improved by using label dependencies. Moreover, in [9] patient dataset with a modification of a multi-label validated the proposed method for prediction. If we compare the studies of [10, 11], both are not label dependencies, while in [12] used the label dependencies. Sometimes, simple features also have significance in classification predictions [7, 13] that used simple Bag of Words (BoW). Further, [14] also used BoW to improve the performance on medical dataset.

3 Proposed Methodology

In this section, the MHARA is introduced as a new approach for classification and prediction of users' reviews on MHA of five different categories i.e Anxiety, Bipolar Disorder, Epilepsy, Mental Retarded, and Migraine. Figure 1 summarizes a high-level architecture of the proposed framework for MHARA. The framework consists of four activities: data acquisition, pre-processing, feature engineering, and classification. Detail explanations for every activity are discussed in the following subsections.

3.1 Data Acquisition

The experiment is performed on the dataset that collected from Google Play Store. It is based on Android apps that are rich in review after installing the apps. Moreover, it is very popular in Android Market Place. This online marketplaces provide free and paid access for mobile users to over a million mobile applications [15]. To limit our scope with MHA, 250 apps are analyzed from five different categories. Next, top 27 apps that mostly have maximum download and rating are selected. The list of AppID of the 27

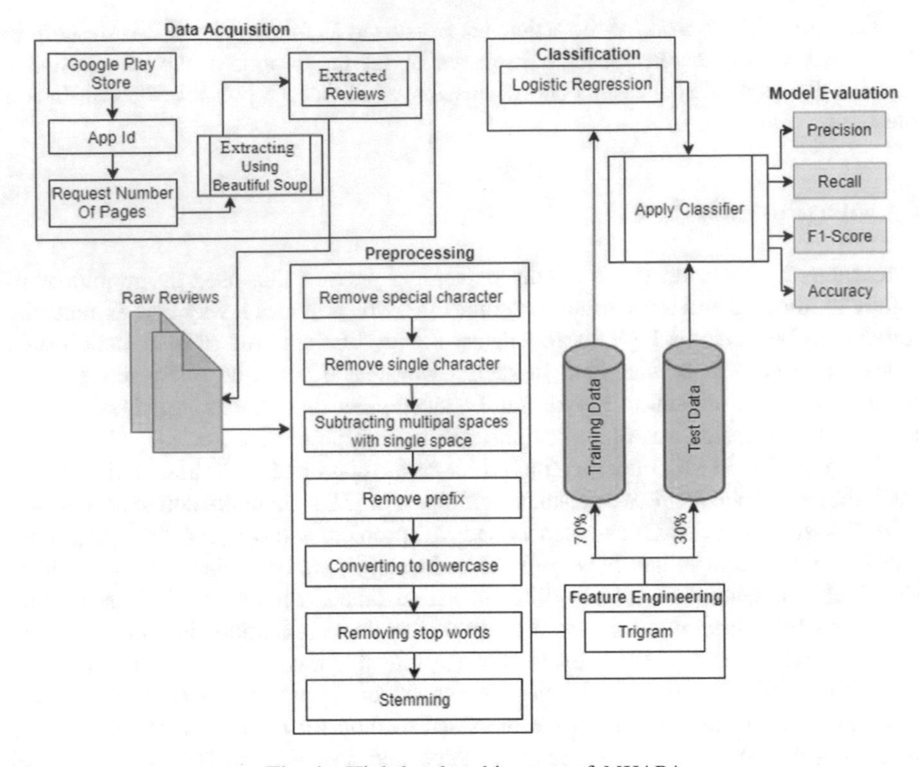

Fig. 1. High level architecture of *MHARA*

apps are used as an input to extract their associated user reviews (max 401 reviews per app) and rating. The initial dataset contained of 5309 users' reviews associated to the 27 apps within five app categories as shown in Table 1.

Table 1. Detail dataset overview under various categories

Sr. no	Category of application	Total number of reviews
1	Anxiety	3909
2	Bipolar disorder	551
3	Epilepsy	199
4	Mental retarded	342
5	Migraine	308
Total		5309

Furthermore, the Beautiful Soup Library of Python is used to extract data from the HTML and XML files [6]. This library works very quickly and saves the programmer's time. The sample dataset can be viewed in Fig. 2.

rating	review	app_id	app_name	category
4	This is a great app that I find helpful for anxiety or for a restless mind. I wa	ie.armour.dare2	Dare - Break Free From	Anxiety
2	Â£8.99 a month, a bit expensive really, are Dare here to help People or just	ie.armour.dare2	Dare - Break Free From	Anxiety
5	helpful but cost to much monthly I think I will buy the book and use the fre	ie.armour.dare2	Dare - Break Free From	Anxiety
5	Dont look.any further. This program works wonders helping you achieve y	ie.armour.dare2	Dare - Break Free From	Anxiety
5	this app has helped me massively, Barry's voice is calming and clear and the	ie.armour.dare2	Dare - Break Free From	Anxiety
1	Almost$10/month for what, poorly drawn cartoon characters that magical	ie.armour.dare2	Dare - Break Free From	Anxiety
5	really enjoy the extra support you need durning a panic attack. great layou	ie.armour.dare2	Dare - Break Free From	Anxiety
5	So I left a review yesterday saying i paid for the rapid relief and can no long	ie.armour.dare2	Dare - Break Free From	Anxiety
5	excellent tool to hell through anxiety	ie.armour.dare2	Dare - Break Free From	Anxiety
4	I loved the older version and am sorry I \"upgraded\". I understand that it c	ie.armour.dare2	Dare - Break Free From	Anxiety
5	Barry, you are a hero	ie.armour.dare2	Dare - Break Free From	Anxiety
5	Amazing app! Helps me when I need it the most	ie.armour.dare2	Dare - Break Free From	Anxiety
5	This app is truly so amazing. I suffer from Depression \u0026amp; Social an	ie.armour.dare2	Dare - Break Free From	Anxiety
5	Wow! I have spent years in therapy and have read countless self help book	ie.armour.dare2	Dare - Break Free From	Anxiety
5	The deep release meditation, which is free, takes me deeper into relaxatio	ie.armour.dare2	Dare - Break Free From	Anxiety
5	This is one of the best tools I have for my panic attacks right now. The voic	ie.armour.dare2	Dare - Break Free From	Anxiety
5	I was a bit skeptical at first but had been trying a whole bunch of different	ie.armour.dare2	Dare - Break Free From	Anxiety
5	This has really helped me with my panic/anxiety attacks both during and th	ie.armour.dare2	Dare - Break Free From	Anxiety
5	Whenever I'm going through something that makes me anxious I listen to o	ie.armour.dare2	Dare - Break Free From	Anxiety
5	I've been fighting bouts of depression since my teenage years. I wish I coul	ie.armour.dare2	Dare - Break Free From	Anxiety
4	I have tried multiple applications for stress relief, Dare has a \nuser-friendl	ie.armour.dare2	Dare - Break Free From	Anxiety
5	If you suffer from anxiety on a daily basis this app will help immensely! I \n	ie.armour.dare2	Dare - Break Free From	Anxiety
5	Thank you for making this app. It made me cry out my pain which I was hidi	ie.armour.dare2	Dare - Break Free From	Anxiety
5	Very helpful. I'm going through very bad anxiety these days, this app is kee	ie.armour.dare2	Dare - Break Free From	Anxiety
5	This app helps me to feel calm and focused throughout the day. I am addin	ie.armour.dare2	Dare - Break Free From	Anxiety
5	I felt like a friend who knows me was talking. Now I know it's not me it's ar	ie.armour.dare2	Dare - Break Free From	Anxiety

Fig. 2. Sample review screen short with description extracted from Google play store

3.2 Pre-processing

Data cleaning processes have been applied after extracting the raw reviews related to MHA from Google play store. These reviews are contained high level of noise, sparsity, miss spell words and poor grammatical sentences. Therefore, removing the noises requires a complete pre-processing on dataset for classification task. In pre-processing, various standard techniques are used and took significant steps to improve the quality of dataset for classification [16]. These steps includes tokenization [1], removing a punctuation marks and white spaces, eliminate a single character, converting data into lowercase, spell correction, removing stop words, lemmatization and then stemming [15].

The pre-processing steps produce high quality data and experimental findings, showed that aforementioned text normalization techniques improve the classification accuracy. It also overcome the issue of dimensionality in supervise learning approaches [15]. After refining through pre-processing, a corpus is developed which contained thousands of terms occurrences.

3.3 Feature Engineering

Feature engineering is a key step in text classification field [17]. The feature engineering is a combination of three sub techniques that are feature extraction, feature value representation, and feature selection. Mostly, two approaches are used for feature extraction: expert-driven and fully automated feature extraction [17]. In this study, automated feature extraction is used which reduce machine learning processing time by a factor of 10, compared to manual feature extraction while delivering better modeling performance. Different features are used for experiment purpose on four techniques: Tri-Grams, TF/IDF, Uni-Gram and Bi-Gram. Tri-Gram was found as an appropriate feature extraction technique in this model. It is important to mention here that Tri-Gram can capture the semantic information of review more accurately [18]. In addition, it also improves the prediction as compared to Term Frequency/Inverse Document Frequency (TF/IDF), Uni-Gram and Bi-Gram. Tri-Grams are more likely to generate "new" but still somewhat recognizable text.

3.4 Classification

Selection of classifier is very important in the process of experiment. The main hurdles are continuous experiments and foundation of accurate parameters for classifiers. We have tested different classifiers and the LR providing the more accurate results as compared to other classifiers. Since, there are five classes involved therefore; five classifiers are implemented including LR, Stochastic Gradient Descent (SGD), Multinomial NB, RF and K-neighbors.

4 Results and Discussion

The proposed MHARA model is evaluated and tuned the parameters which are exhibited in Table 2. These parameters are performing the best prediction in our model.

Table 2. LR tuned parameters with their values

Parameter name	Value	Parameter name	Value
penalty	12	dual	False
tol	0.0001	C	1.0
fit_intercept	True	intercept_scaling	1
class_weight	None	solver	warn
max_iter	100	multi_class	Warn
verbose	0	warm_start	False
n_jobs	None	11_ratio	None

The Dataset is also split into training and testing data (70% for training and 30% for testing). Several classification algorithms were exploited using scikit-learn. The performances of each classifier were evaluated based on accuracy, precision, recall and f1-score. The values of these metrics are recorded with respect to classifier which shows that how well the MHARA results correspond to the annotated results.

The results of different classifier depicted in Fig. 3 clearly show that LR performed well as compared to other classifiers. It is also noted that LR is performed well in three evaluation metrics: recall, f1-score and accuracy.

However, precision is slightly less as compared to LR and RF has a high precision. Since we are developing a model that related to prediction that required high accuracy,

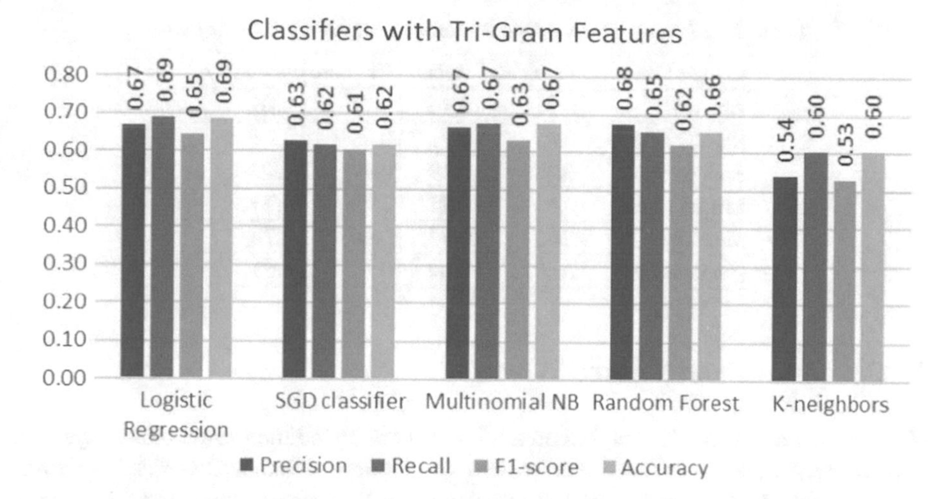

Fig. 3. Five classifiers performance comparison using Tri-Gram features

therefore, low precision i.e. 0.68 of RF has no impact as compared to LR of 0.67 precision. The study not only confined for Tri-Gram features but also used other features: Uni-Gram, Bi-Gram and TF/IDF.

Meanwhile, Fig. 4 depicts the performance of LR on four features. It is clearly shows that Tri-Gram provides more efficient feature for LR classifier with higher performance in all evaluation metrics. Thus, it shows that the proposed approach has a good accuracy in all categories (Table 3) as compared to other classifier. Therefore, the proposed model is a very good for those whom are struggling to find the perfect app using the correlation of reviews and rating of apps.

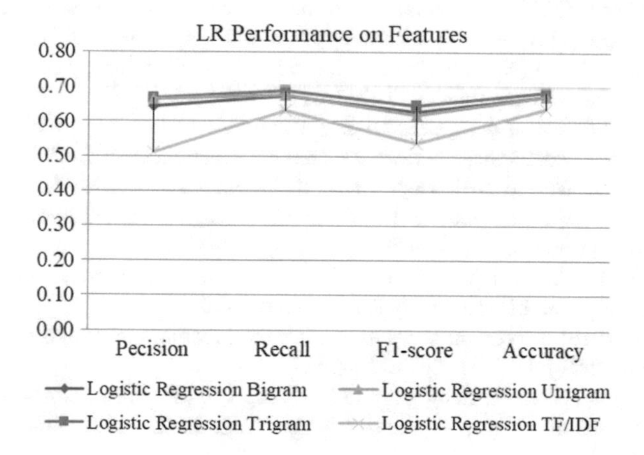

Fig. 4. LR performance comparison on four features

Table 3. LR with Tri-Gram performance comparison category wise.

Categories	Precision	Recall	F1-score	Accuracy
Anxiety	0.71	0.76	0.72	0.76
Bipolar disorder	0.65	0.71	0.66	0.71
Epilepsy	0.48	0.49	0.47	0.49
Mental retarde	0.81	0.74	0.70	0.74
Migraine	0.69	0.74	0.68	0.72
Average	0.67	0.69	0.65	0.69

5 Conclusion and Future Work

We proposed a new model that has an ability to recommend the mental related apps on Google Play Store. The apps are related to Anxiety, Bipolar disorder, Epilepsy, Migraine and Mental retarded. The model using the Tri-Gram features and the features are trained through LR for multi-class classification. The model has some limits i.e. it required much data to train. This training is time consuming and have limit of accuracy. The accuracy can be improved through hybrid the various classifiers and merging the text features. Moreover, in future work this study can be extended using deep learning.

Acknowledgments. This study has been supported in part of the Fundamental Research Grant Scheme (FRGS) Vot K006 under the Malaysia Ministry of Higher Education (MOHE) and Universiti Tun Hussein Onn Malaysia (UTHM).

References

1. Chandrashekar P (2018) Do mental health mobile apps work: evidence and recommendations for designing high-efficacy mental health mobile apps. mHealth 4(6):1–6
2. Sarker A, Gonzalez G (2015) Portable automatic text classification for adverse drug reaction detection via multi-corpus training. J Biomed Inform 53:196–207
3. Peek N, Combi C, Marin R, Bellazzi R (2015) Thirty years of artificial intelligence in medicine (AIME) conferences: a review of research themes. Artif Intell Med 65(1):61–73
4. Boudreaux ED, Waring ME, Hayes RB, Sadasivam RS, Mullen S, Pagoto S (2014) Evaluating and selecting mobile health apps: strategies for healthcare providers and healthcare organizations. Transl Behav Med 4(4):363–371
5. Istepanian RSH, Al-Anzi T (2018) m-Health 2.0: new perspectives on mobile health, machine learning and big data analytics. Methods 1–16
6. Kadhim AI (2019) Survey on supervised machine learning techniques for automatic text classification. Artif Intell Rev 52(1):273–292
7. Wang S, Chang XJ, Li X, Long G, Yao L, Sheng QZ (2016) Diagnosis code assignment using sparsity-based disease correlation embedding. IEEE Trans Knowl Data Eng 28 (12):3191–3202
8. Dembczy K (2010) Bayes optimal multilabel classification via probabilistic classifier Chains. In: 27th international conference on machine learning. Omnipress, USA, pp 279–286

9. Qu G, Zhang H, Hartrick CT (2011) Multi-label classification with Bayes' Theorem. In: 4th international conference on biomedical engineering and informatics. IEEE Press, New York, pp 2281–2285
10. Tsoumakas G, Katakis I, Vlahavas I (2005) Data mining and knowledge discovery handbook. In: Data mining and knowledge discovery handbook. Springer, Boston, MA
11. Zhang ML, Zhou ZH (2007) ML-KNN: a lazy learning approach to multi-label learning. Pattern Recogn 40(7):2038–2048
12. Qu G, Wu H, Hartrick CT, Niu J (2012) Local analgesia adverse effects prediction using multi-label classification. Neurocomputing 92:18–27
13. Bromuri S, Zufferey D, Hennebert J, Schumacher M (2014) Multi-label classification of chronically ill patients with bag of words and supervised dimensionality reduction algorithms. J Biomed Inform 51:165–175
14. Zufferey D, Hofer T, Hennebert J, Schumacher M, Ingold R, Bromuri S (2015) Performance comparison of multi-label learning algorithms on clinical data for chronic diseases. Comput Biol Med 65:34–43
15. Mujtaba G, Shuib L, Idris N, Hoo WL, Raj RG, Khowaja K, Shaikh K, Nweke HF (2019) Clinical text classification research trends: systematic literature review and open issues. Expert Syst Appl 116:494–520
16. Hatamian M, Serna J, Rannenberg K (2019) Revealing the unrevealed: mining smartphone users privacy perception on app markets. Comput Secur 83:332–353
17. Heer J, Washington U, Hellerstein JM, Berkeley UC, Kandel S (2015) Predictive interaction for data transformation. In: 7th biennial conference on innovative data system research, CIDR conference
18. Saumya S, Singh JP, Dwivedi YK (2019) Predicting the helpfulness score of online reviews using convolutional neural network. Soft Comput 1–17

Source Camera Identification for Online Social Network Images Using Texture Feature

Nordiana Rahim(✉) and Cik Feresa Mohd Foozy

Faculty of Computer Science and Information Technology, University of Tun Hussein Onn Malaysia, 86400 Batu Pahat, Johor, Malaysia
{nordiana, feresa}@uthm.edu.my

Abstract. Nowadays, digital images are easily shared through social media and it common amongst internet users. It was a convenient way to share a moment and communicate with people all over the worlds through social media on the internet. However, this has caused the increasing number of crimes involving digital images in social media. It is well known that each digital image that passes through online social networks (OSNs) is explicitly modified by Web 2.0 tools. Thus, it is challenging for authorities to probe further, including identifying the source of the digital images. Considering this limitation, an alternative method to identify source camera based on the texture feature for OSNs images is proposed. This technique uses texture feature characteristics, namely, Gray Level Co-Occurrence Matrix (GLCM) and Gray Level Run-Length Matrix (GLRLM). Original and OSNs images were tested to determine whether the proposed method is robust for both image types and gives higher accuracy than previous methods. Four types of camera models were used in this research. The results prove that the method tested in this study is accurate with an average accuracy of 97.00% and 99.59% for original and OSNs images, respectively, and is capable to read up to 600 images.

Keywords: Digital image · Source camera identification
Online social network · Texture feature

1 Introduction

Nowadays, online social networks (OSNs) are becoming more popular as the technological advancement of digital devices is increasing rapidly. This development allows users to easily create their network and enable them to upload their multimedia content such as images, videos, and audios via various OSNs platforms through the internet. Unfortunately, this improvement has caused the public posting of personal and sensitive information including date of birth, phone number, relationship status and home addresses [1, 2].

Moreover, these sophisticated technologies have caused an increase in digital crimes [3]. Digital images are used as clues for detecting certain crimes committed online. One main issue faced by digital image forensic personnel is image source identification. Source identification is a branch of digital image forensics that is widely discussed among academicians and industrial community members. There are a lot of

R. Ghazali et al. (Eds.): SCDM 2020, AISC 978, pp. 283–296, 2020.
https://doi.org/10.1007/978-3-030-36056-6_28

studies and techniques been proposed to determine the source of the image via the identification of the device [4–6]. Thus, this paper focuses on source camera identification and texture feature, as described in Sect. 2. Sections 3 and 4 present the experimental settings and analysis of the technique used for source camera identification, particularly for OSN images. Section 5 discusses and concludes the findings in this paper.

2 OSN and Source Camera Identification

Ellison [7] has defined a social network as a web-based service that creates connections between users. Social network users are required to create a user profile using their details. According to Alexa [8], excessive usage of social network has been recorded and is continuously growing. This has increased digital crime [9, 10] including privacy risk [11], identity theft [12, 13] and the emergence of fake profiles [14, 15]. This situation has triggered a concern among academicians to discuss security and privacy issues, especially on OSN users. OSNs holds an abundance of valuable user information (e.g. digital photos, videos, a user-friend list with the identifier (ID), and table of login and IP data [16]). Users tend to share their photos as a channel to express their feelings and as a starting point to start conversations. Over 300 million OSN users (for instance, Facebook) upload photos and videos that are widely accessible all over the internet. This allows criminals to easily access user information, current status and posted photos and videos.

Typically, criminals tend to download photos from OSN applications and use the data for illegal activities. Hence, taking into consideration the underlying issues, an improved approach of digital image source identification is necessary to be in line with the fast-growing OSN applications. Digital image forensic practitioners are working towards developing the best approach to solve this issue. Each digital image contains a random feature and pattern noise that depends on the sensor noise used to shoot the image. This paper aims to identify the camera used to generate a digital picture by examining the images texture feature. Based on the literature review, there are three main techniques proposed to solve source camera identification problems. These techniques are identification based on sensor pattern noise (SPN), color filter array CFA interpolation, and image statistical feature. Details of the techniques are explained in the next subsection.

2.1 Image Source Identification Based on SPN

During the process of image generation, internal defects such as pixel defect and pattern noise are created. These defects can be extracted and used to match with reference patterns of digital cameras. Subsequently, the source camera of an image can be identified. Lukas et al. [18–20] have suggested techniques based on pixel non-uniformity (PNU) which is a good source to retrieve noise patterns. However, these techniques do not work with cropped and different sizes of images [21].

Gharibi et al. [22] proposed a method that uses local information of the inherent pattern such as photo-response non-uniformity (PRNU). This pattern is used as the signature feature of the digital camera. In the proposed algorithm, fuzzy-based classification is used to select the best region according to the camera local information to extract the noise pattern. Next, the noise pattern is evaluated the correlation between the image pattern and camera pattern, thus it determined the source of camera identification.

A technique based on cross-correlation between PRNU and image residual noise has been proposed by Xie et al. [23]. This technique identifies the source of the image by the sharp position of the correlation value. It is based on detecting digital camera output reference PRNU in the noise residual of images. Thus, the results show that the average of the recognition accuracy rate is 96.3%.

2.2 Image Source Identification Based on CFA Interpolation

Demosaicing algorithm is a crucial part for correcting high spatial rendering in images, using a single imaging sensor. This process highly impacts the edge and color quality of an image. The forming of interpolation presents specific types of interdependencies between color values of an image's pixels. The specific forms of dependencies can be extracted from images as fingerprints. Bayram et al. [24] proposed the Expectation-Maximization (EM) algorithm to estimate weighting (interpolation) coefficients, which elect the amount of contribution from each pixel in the CFA interpolation kernel. A set of weighting coefficients was then used to design a classifier to distinguish the source of the camera. Unfortunately, this technique does not give a satisfactory result for heavily compressed images. Therefore, the technique was improvised by separately capturing the periodicity in second-order derivatives on smooth and non-smooth parts of images [24]. Long et al. [25] has proposed a quadratic pixel correlation model by obtaining the coefficient matrix for each color channel. The principal components were extracted and fed to a feed-forward back propagation neural network for source camera identification. On the other hand, Ho et al. [26] have developed a method using the variance of different colored planes to identify the model of source camera. Unfortunately, his method has been improvised by Hu et al. [27] which they extracted two variance maps by estimating the variance of every component in a green-to-red and green-to-blue spectrum. Then, the shape and texture feature was obtained for camera model identification.

2.3 Image Source Identification Based on Image Statistical Feature

Kharrazi et al. [28] proposed a method to identify source camera based on image statistical feature. They focused on 34-dimensional features such as average pixel value, RGB pair energy ratio, RGB pair correlation of neighbour distribution centre of mass, wavelet domain statistics, and image quality measures. These features are fed to a Support Vector Machine (SVM) classifier to identify camera source. Furthermore, another effective method proposed by Wang et al. [29] which presented 216-dimensional higher-order wavelet features and 135-dimensional wavelet coefficient co-occurrence features.

This feature was extracted from images and applied the Sequential Forward Feature Selection (SFFS) method to reduce redundancy and correlation. Additionally, they utilized a multi-class SVM classifier to identify camera source. Results show that this method identified 98% of camera source.

Gao et al. [30] approach, detection accuracy reached up to 99.27% for seven different cameras based on multi-step transition matrices. Meanwhile, Xu et al. [11] recommended a method based on Local Binary Pattern (LBP) involving three groups of 59-dimensional LBP features. These features were extracted from the spatial domain of red and green colored channels, corresponding to prediction-error arrays and 1-level diagonal wavelet sub-bands of the image, respectively. This approach gave an average accuracy that reached up to 98.08%. The technique that based on GLCM feature was proposed by Kulkarni et al. [31] which uses a gradient-based and Laplacian operator to smooth the noise and detect the edge in the image. These two operators obtained a third image present with edge and noise. Edges were removed by applying the threshold, which allows noise to be present in the image. This noisy image was then subjected to an extraction module that consisted of GLCM to extract various features based on its properties. GLCM feature properties consist of homogeneity, contrast, correlation, and entropy. As a result, the detection accuracy for this method was shown reach up to 99.59%.

3 Proposed Technique

Most existing methods focus on identifying the camera based on the provided image, which is known as the original image. Nevertheless, not many studies focused on determining source camera based on the given image, particularly those downloaded from OSNs. Thus, this research aims to conduct an experiment to detect the source camera of images from OSNs. An algorithm is proposed using image texture features extracted from noisy images. These noisy images were then used to extract GLCM and GLRLM. These two features are important to provide a piece of pattern information or arrangement on the structure of the image [34]. Texture feature is utilized in this experiment because it does not rely on color or intensity and reflects the intrinsic phenomenon of images. Both GLCM and GLRLM properties were combined and fed to several classifiers. The proposed method consists of three phases, namely, detection and extraction of noise and edges, feature extraction, and classification process.

3.1 Detection and Extraction of Noise Image

The extraction process is a low-level processing step. It is used to identify image regions, points, and lines/curves that highlight significant clues in a digital image [35]. A variety of applications is available for this feature, such as object recognition, visual tracking [32], content-based image retrieval (CBIR), and wide baseline matching [33, 34]. Moreover, in computer vision, feature detection can be classified into several methods, which consist of edge, contour/boundary, corner, and region detection. (see Fig. 1)

In image processing, an edge refers to pixels that can change unexpectedly. On the other hand, contour or boundary is an intersection line or curve of different segmented regions. Corner denotes the points at which two different edges of direction occur in the

Fig. 1. The classification of detection features

local neighborhood. It is an intersection between two connected contour lines. Besides that, region represents a closed set of connected points. Nearby and similar pixels are grouped together to form the interest region. This research particularly focuses on edge detection method consisting of Canny edge detection and Laplacian of Gaussian filter.

Canny detection [35] is a technique used to extract useful information from different vision objects and reduce amount of data that needs to be processed. Canny algorithm aims to satisfy three main criteria, namely good detection, satisfactory localization, and able to perform single response per edge. The Canny operator works in multi-stage processes. First stage involves smoothing the image using Gaussian convolution. Next, a simple 2-D first derivative operator [36] is applied to the smoothed image in order to highlight the image region with high spatial derivatives. Algorithm is tracked along the top of ridges and set to zero. This process is known as non-maximal suppression. Moreover, the tracking process indicates hysteresis controlled by two thresholds that are Threshold 1 (T1) and Threshold 2 (T2); T1 > T2. The process can only begin at a point on the ridges, which is higher than T1. Tracking is continued in both directions, from T1 until height of ridges falls below T2. This helps to ensure noisy edges are not broken into multiple edge fragments.

Laplacian filter is a derivative filter used to find rapidly changing areas (edges) in images. Derivative filters are very sensitive to noise. Therefore, a Gaussian filter was used to smooth the image before applying it to a Laplacian derivative filter. This two-step process is called the Laplacian of Gaussian (LoG) operation and is given as follows:

$$L(x,y) = \nabla 2f(x,y) = \frac{\delta^2 f(x,y)}{\delta x^2} + \frac{\delta^2 f(x,y)}{\delta y^2} \quad (1)$$

Meanwhile the 2-D Laplacian of Gaussian is derived as:

$$Log(\mathrm{x,y}) = -\frac{1}{4}\left[1 - \frac{x^2+y^2}{2\sigma^2}\right] e^{-\frac{x^2+y^2}{2\sigma^2}} \quad (2)$$

where σ^2 (standard deviation) control amount of the process.

3.2 Texture Feature Extraction

Feature extraction is a process that defines a set of features or image characteristics that represent a group of important information for analysis and classification process. Texture feature is an arrangement of structures found in an image. It can be used as a pattern of information that is valuable for digital investigations [37, 38]. Furthermore, texture feature has been used in various industries such as bio-medical, remote-sensing, agriculture, and more. Texture analysis can be grouped into four categories of method that are model-based, statistical-based, structural-based and transform-based. The model-based method revolves around a mathematical model used to predict the value of the pixel. On the other hand, statistical-based method indirectly represents texture via non-deterministic properties, which govern distributions and relationships between gray levels of an image. Meanwhile, for the structural-based method, it works by understanding the hierarchical structure of an image. In the transform-based method, generally a modification is performed and a new response from images are obtained, which are then analyzed as a representative of the original image [39]. This research implemented the statistical-based method texture analysis to extract GLCM and GLRLM features.

Gray Level Co-Occurrence Matrix (GLCM). Texture features were computed based on statistical distribution of pixel intensity at a given position relative to others in a matrix of pixels that rep- resents the image. Several order statistics were involved, namely, first-order, second-order and higher-order. Feature extraction based on GLCM used to analyze image as a texture comes under second order statistics. GLCM is tabulation of frequencies or how often a combination of pixel brightness values occurs in an image [40]. Figure 2(c) shows formation of GLCM from a four-level gray image at distance d = 1 and direction θ = 0on the other hand, Fig. 2(a) is an example of a pixel intensity matrix representing an image with four levels of gray. Intensity levels 0 and 1 are marked with a thin box. The thin box represents pixel intensity 0 with pixel intensity 1 as its neighbor (in the horizontal direction or direction of 0). There are two occurrences of such pixels. Therefore, GLCM matrix is formed (see Fig. 2b) with value 2 in row 0, column 1

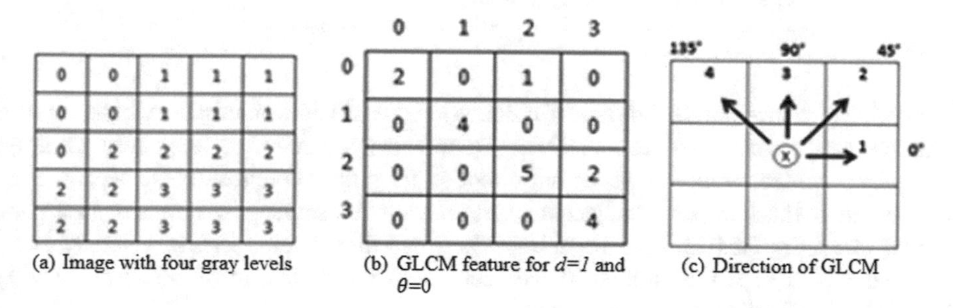

(a) Image with four gray levels (b) GLCM feature for $d=1$ and $\theta=0$ (c) Direction of GLCM

Fig. 2. Formation of GLCM feature

Similarly, row 0, column 0 also gives a value of 2 because there are two occurrences in which 0 value pixels have 0 value pixels as its neighbor (horizontal direction). Hence, the pixel matrix as illustrated in Fig. 2(a). It can be transformed into a GLCM as shown in Fig. 2(b). In addition to horizontal direction (0), GLCM can also be formed in 45°, 90° and 135° directions, as presented in Fig. 2(c). Haralick et al. [40] extracted 14 features from the co-occurrence matrix, although only 8 features are widely used in many applications such as energy, entropy, max probability, inverse different moment, contrast, homogeneity, inertia, and correlation. Although co-occurrence matrixes can capture texture properties, they have not been directly used as a tool for analysis. A matrixes data has to be extracted to get the properties numbers that used for classification of textures. Table 1 gives the lists of properties of GLCM features.

Table 1. GLCM features set

Moment	Equation
Energy	$f_1 = \sum_i \sum_j p(i,j)^2$
Contrast	$f_2 = \sum_{n=0}^{N_g-1} n^2 \left\{ \sum_{i=1}^{N_g} \sum_{j=1}^{N_g} p(i,j) \| \|i-j\| = n \right\}$
Correlation	$f_3 = \frac{\sum_i \sum_j (ij)p(i,j) - \mu_x\mu_y}{\sigma_x\sigma_y}$
Homogeneity	$f_4 = \sum_i \sum_j \frac{1}{1+(i-j)^2} p(i,j)$
Entropy	$f_5 = \sum_i \sum_j p(i,j)\log(p(i,j))$
Autocorrelation	$f_6 = \sum_i \sum_j (ij)(p(i,j))$
Dissimilarity	$f_7 = \sum_i \sum_j \|i-j\|, p(i,j)$
Cluster shade	$f_8 = \sum_i \sum_j (i, i+j-\mu_x-\mu_y)^3 p(i,j)$
Cluster prominence	$f_9 = \sum_i \sum_j (i, i+j-\mu_x-\mu_y)^4 p(i,j)$
Maximum probability	$f_{10} = \underset{i,j}{MAX}\, p(i,j)$

Gray Level Run-Length Matrix (GLRLM). GLRLM is a matrix in which texture features can be extracted for analysis. Texture is recognized as a pattern of gray intensity pixels, moving in a particular direction from the reference pixels. Meanwhile, run length is the number of adjacent pixels having the same gray intensity in a particular direction. GLRLM can be defined as a two-dimensional matrix whereby each element p (i,j | θ) is the number of element j with intensity i, in the direction θ = 0. Figure 3, GLRM feature are presented.

Tang et al. [41] proposed another five texture features, namely, short run emphasis (SRE), long run emphasis (LRE), gray-level non-uniformity (GLN), run length non-uniformity (RLN) and run percentage (RP). Besides that, Chu et al. [46] added two

1	2	3	4
1	3	4	4
3	2	2	2
4	1	4	1

(a) Matrix of 4x4 pixel image

Gray Level	Run Length (j)			
i	1	2	3	4
1	4	0	0	0
2	1	0	1	0
3	3	0	0	0
4	3	1	0	0

(b) GLRLM feature

Fig. 3. Formation of GLRM feature

extra features called low gray-level run emphasis (LGRE) and high gray- level run emphasis (HGRE). These features use the gray level of pixels in sequence and is intended to distinguish textures that have the same value of SRE and LRE. Nonetheless, it displays some differences in distribution of gray levels. Dasarathy and Holder [42] added additional four features extracted from GLRLM, that are short-run low gray-level emphasis (SRLGE), short-run high gray-level emphasis (SRHGE), long-run low gray-level emphasis (LRLGE), and long-run high gray-level emphasis (LRHGE). Table 2 lists GLRLM features that are commonly used.

Table 2. GLRLM features set

Moment	Equation
Short run emphasis (SRE)	$\frac{1}{n}\sum_{i,j}\frac{p(i,j)}{j^2}$
Long run emphasis (LRE)	$\sum_{i,j} j^2 p(i,j)$
Gray level non-uniformity (RLN)	$\frac{1}{n}\sum_{i}\left(\sum_{j} p(i,j)\right)^2$
Run percentage (RP)	$\sum_{i,j}\frac{n}{p(i,j)j}$
Low gray level run emphasis (LGLE)	$\frac{1}{n}\sum_{i,j}\frac{p(i,j)}{i^2}$
High gray level run emphasis (HGRE)	$\frac{1}{n}\sum_{i,j} i^2 p(i,j)$

3.3 Classification Process

LibSVM was one of the classifiers used in this experiment. It is a state-of-the-art classification tool developed based on statistical learning theory. LibSVM has been successfully implemented in image recognition, text classification, and biological information processing. In this experiment, other classifiers like nave Bayesian and neural network (multilayer perceptron) are also employed and implemented in WEKA, data mining tool [43]. The total training and testing for the three classifiers are selected using a 10-fold-cross validation. Both training and testing datasets for each classifier contained an equal number of data instances from each class.

4 Experimental Setup, Result and Analysis

In this section, details of the work done are explained. First, image databases used in this experiment are briefly described. Next, test results and analysis of extraction features for source camera identification, particularly of OSN images, are presented. This section is concluded with descriptions of the experimental process and results. Performance of the proposed method was tested via three scenarios:

- Scenario 1: Tested using the original image.
- Scenario 2: Tested using the image from OSN.
- Scenario 3: Evaluated using other state-of-the-art methods [28, 31].

4.1 Image Database

To verify the performance of the proposed method, 600 digital images from four different types of camera models taken from *Dresden Image Dataset [44] are tested. All images were uploaded and then downloaded from *Facebook in JPEG format. Resolution of downloaded images is slightly different from original images since OSNs have an auto-resizing function along with the ability to auto-change the resolution of images. Lists of a digital camera used in this experiment are listed in Table 3.

Table 3. List of digital cameras used in this experiment

Camera id	Camera model
AGDC504	Agfa_DC-504
AGDC733	Agfa_DC-733 s
AGDC830	Agfa_DC-830i
AGS505	Agfa_Sensor505-x

The experiment was run and tested using MATLAB 2014b, installed on a computer with Inter(R) Core (TM) i7-4700MQ CPU 8.00 GB RAM.

4.2 The Proposed Method on Original Images

All original images were subjected to edge detection and noise extraction through Laplacian and Canny operators. This process created a new image that contains edges and noise. Then, GLCM and GLRLM features and combination of GLCM and GLRLM features were extracted from the new images. All four types of cameras were evaluated by three classifiers (SVM, naive Bayers, and neural network Multi-Layer Perceptron [MLP]) through 10-fold cross-validation. Result of the accuracy detection is recorded in Fig. 4.

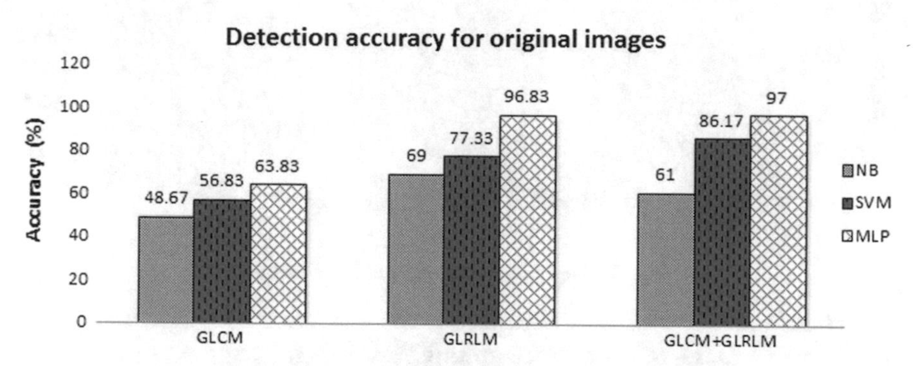

Fig. 4. Detection accuracy for original image

Based on the result, the average performance of the method is recorded 96.83% with GLRLM features and 63.83% with GLCM features using the neural network (MLP) classifier. Combination of GLCM and GLRLM features produced the best result with 97% using the same classifier. This has shown that the proposed method can detect and identify the image source and the experiment is continued with the OSN images.

4.3 The Proposed Method on OSN Images

The second experiment was conducted to demonstrate that the proposed method can detect and identify the camera source of OSN images. The procedure for this experiment was similar as conducted in the first experiment.

It is important to note that the downloaded images have been auto-resized and the resolution auto-modified. Detection accuracy results are shown in Fig. 5. Based on Fig. 5, GLRLM produced the best result using the same classifier from the previous experiment. Besides that, accuracy was slightly reduced to 99.5% for the combination of GLCM and GLRLM features.

4.4 Performance of OSN Images Towards Proposed Method Against State-of-the-Art Methods

This experiment was designed to compare the proposed technique with state-of-the-art methods using OSN images. The experiment started with original images and findings show that the state-of-the-art methods could identify the images camera source, as illustrated in Fig. 6.

As seen in Fig. 6, the experimental results show that the proposed technique able to identify the image source with higher accuracy detection compared to the others state-of-art techniques. Two types of images have been tested; original and OSN images.

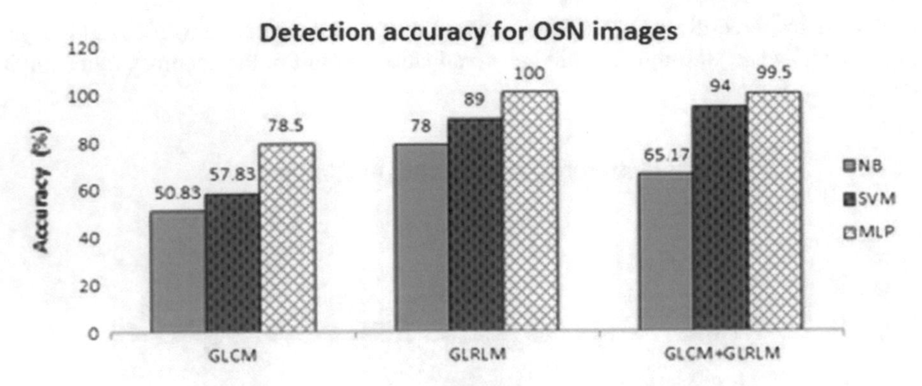

Fig. 5. Detection accuracy for OSN images

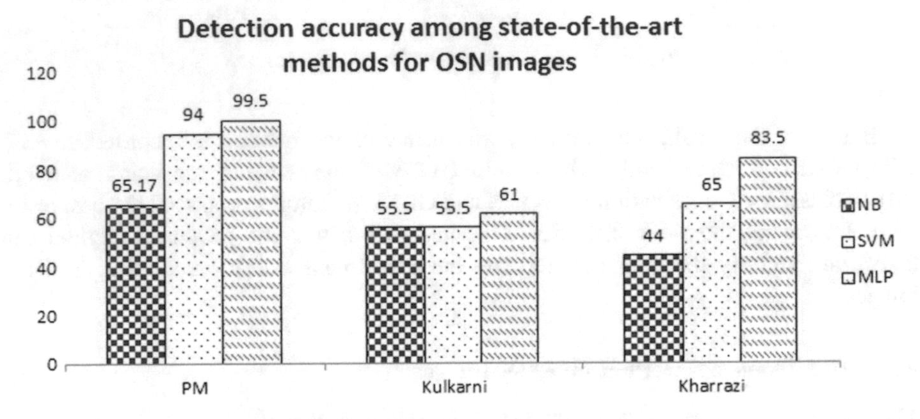

Fig. 6. Detection accuracy for OSN images

The experiments have shown that, the proposed method is capable in identifying the camera source for both images with detection accuracies of 97.0% and 99.5%. Furthermore, this method gave the highest accuracy result for OSN images compared to original images. This is due to usage of texture feature in the algorithm. Texture can be defined as a kind of visual characteristic that does not rely on color. Moreover, it reflects the intrinsic surface properties of images. The second order statistics of GLCM and GLRLM gave statistical values for each property, making it possible to identify source of images.

5 Conclusion

In this research, the proposed methods are effective in identifying the source of the camera on original and OSN images. It has shown that the proposed method is capable to identify the camera source for both images with average detection is 97.0% and 99.5%. Furthermore, this method gave the highest accuracy result for OSN images compared to original images. This is due to the usage of the texture feature in the algorithm. Texture can be defined as a kind of visual characteristic that does not rely on color. Moreover, it reflects the intrinsic surface properties of images. The second-order statistics of GLCM and GLRLM gave statistical values for each property, making it possible to identify the source of images. Another factor is pre-processing of uploaded pictures affects the detection accuracy of OSN images [17].

Three classifiers have been used in this study; nave Bayes, LibSVM, and MLP. SVM and MLP classifiers gave the best performance for original and OSN images. With an average detection accuracy of 99.5%, this technique presented the best performance among other state-of-the-art techniques.

References

1. Yang Z, Wilson C, Wang X, Gao T, Zhao BY, Dai Y (2014) Uncovering social network sybils in the wild. ACM Trans Knowl Discov Data (TKDD) 8(1):2
2. Jin L, Chen Y, Wang T, Hui P, Vasilakos AV (2013) Understanding user behavior in online social networks: a survey. IEEE Commun Mag 51(9):144–150
3. Abdalla A, Yayilgan SY (2014) A review of using online social networks for investigative activities. In: International conference on social computing and social media. Springer, pp 3–12
4. Redi JA, Taktak W, Dugelay J-L (2011) Digital image forensics: a booklet for beginners. Multimed Tools Appl 51(1):133–162
5. Garfinkel S (2010) Digital forensics research: the next 10 years. Digit Investig 7:S64–S73
6. Piva A (2013) An overview on image forensics. ISRN Signal Process
7. Ellison NB (2007) Social network sites: definition, history, and scholarship. J Comput Mediat Commun 13(1):210–230
8. Blog V (2012) World map of social networks
9. Zainudin NM, Merabti M, Llewellyn-Jones D (2010) A digital forensic investigation model for online social networking. In: Proceedings of the 11th annual conference on the convergence of telecommunications, networking & broadcasting, Liverpool, pp. 21–22
10. Huber M, Mulazzani M, Leithner M, Schrittwieser S, Wondracek G, Weippl E (2011) Social snapshots: digital forensics for online social networks. In: Proceedings of the 27th annual computer security applications conference. ACM, pp 113–122
11. Xu B, Wang X, Zhou X, Xi J, Wang S Source camera identification from image texture features. Neurocomputing 207:131–140
12. Caviglione L, Coccoli M, Merlo A (2014) A taxonomy-based model of security and privacy in online social networks. Int J Comput Sci Eng 9(4):325–338
13. Rizi FS, Khayyambashi MR (2013) Profile cloning in online social networks. Int J Comput Sci Inf Secur 11(8):82
14. Adikari S, Dutta K (2014) Identifying fake profiles in linkedin. In: Pasific Asia conference on information systems, p. 278
15. Xiao C, Freeman DM, Hwa T (2015) Detecting clusters of fake accounts in online social networks. In: Proceedings of the 8th ACM workshop on artificial intelligence and security. ACM, pp 91–101
16. Murphy JP, Fontecilla A (2013) Social media evidence in government investigations and criminal proceedings: a frontier of new legal issues. Rich J Law Technol 19:11–14

17. Castiglione A, Cattaneo G, Cembalo M, Petrillo UF (2013) Experimentations with source camera identification and online social networks. J Ambient Intell Humaniz Comput 4 (2):265–274
18. Lukas J, Fridrich J, Goljan M (2005) Determining digital image origin using sensor imperfections. In: Electronic imaging, international society for optics and photonics, pp 249–260
19. Lukas J, Fridrich J, Goljan M (2006) Digital camera identification from sensor pattern noise. IEEE Trans Inf Forensics Secur 1(2):205–214
20. Luks J, Fridrich J, Goljan M (2005) Digital bullet scratches for images. In: IEEE international conference on image processing, 2005. ICIP 2005, vol 3, IEEE, pp III–65–8
21. Van Lanh T, Chong K-S, Emmanuel S, Kankanhalli MS (2007) A survey on digital camera image forensic methods, In: 2007 IEEE international conference on multimedia and expo. IEEE, pp 16–19
22. Gharibi F, Akhlaghian F, RavanJamjah J, ZahirAzami B (2010) Using the local information of image to identify the source camera. In: The 10th IEEE international symposium on signal processing and information technology. IEEE, pp 515–519
23. Xie Y-J, Bao Y, Tong S-F, Yang Y-H (2013) Source digital image identification based on cross-correlation. In: Proceedings of the 2nd international conference on computer science and electronics engineering. Atlantis Press
24. Bayram S, Sencar H, Memon N, Avcibas I (2005) Source camera identification based on cfa interpolation. In: IEEE international conference on image processing, 2005. ICIP 2005, vol 3. IEEE, pp III–69–72
25. Long Y, Huang Y (2006) Image based source camera identification using de- mosaicking. In: IEEE workshop on multimedia signal processing, pp 419–424
26. Ho JS, Au OC, Zhou J, Guo Y (2010) Inter-channel demosaicking traces for digital image forensics. In: 2010 IEEE international conference on multimedia and expo (ICME). IEEE, pp 1475–1480
27. Hu Y, Li C-T, Lin X, Liu B-B (2012) An improved algorithm for camera model identification using inter-channel demosaicking traces. In: 2012 eighth international conference on intelligent information hiding and multimedia signal processing (IIH-MSP). IEEE, pp 325–330
28. Kharrazi M, Sencar HT, Memon N (2004) Blind source camera identification. In: 2004 International conference on image processing, 2004. ICIP'04, vol 1, IEEE, pp 709–712
29. Wang B, Guo Y, Kong X, Meng F (2009) Source camera identification forensics based on wavelet features. In: Fifth international conference on intelligent information hiding and multimedia signal processing, 2009. IIH-MSP'09. IEEE, pp 702–705
30. Gao S, Hu R-M, Tian G (2012) Using multi-step transition matrices for camera model identification. Int J Hybrid Inf Technol 5(2):275–288
31. Kulkarni N, Mane V (2015) Source camera identification using glcm. In: Advance computing conference (IACC), 2015 IEEE international. IEEE, pp 1242–1246
32. Gauglitz S, Hllerer T, Turk M (2011) Evaluation of interest point detectors and feature descriptors for visual tracking. Int J Comput Vis 94(3):335
33. Tuytelaars T, Van Gool L (2004) Matching widely separated views based on affine invariant regions. Int J Comput Vis 59(1):61–85
34. Aanes H, Lindbjerg-Dahl A, Steenstrup-Pedersen K (2012) Interesting interest points-a comparative study of interest point performance on a unique data set. Int J Comput Vis 97:18–35
35. Canny J (1986) A computational approach to edge detection. IEEE Trans Pattern Anal Mach Intell (6):679–698
36. Roberts LG (1963) Machine perception of three-dimensional soups, Ph.D. thesis

37. Conners RW, Harlow CA (1980) A theoretical comparison of texture algorithms. IEEE Trans Pattern Anal Mach Intell (3):204–222
38. Mohanty AK, Senapati MR, Beberta S, Lenka SK (2013) Texture-based features for classification of mammograms using decision tree. Neural Comput Appl 23(3–4):1011–1017
39. Mohamed SS, Salama MM (2005) Computer-aided diagnosis for prostate cancer using support vector machine. In: Medical imaging, international society for optics and photonics, pp 898–906
40. Haralick RM, Shanmugam K, Dinstein IH (1973) Textural features for image classification. IEEE Trans Syst Man Cybern (6):610–621
41. Tang X (1998) Texture information in run-length matrices. IEEE Trans Image Process 7 (11):1602–1609
42. Dasarathy BV, Holder EB (1991) Image characterizations based on joint gray level run length distributions. Pattern Recogn Lett 12(8):497–502
43. Hall M, Frank E, Holmes G, Pfahringer B, Reutemann P, Wit-ten IH (2009) The weka data mining software: an update. ACM SIGKDD explor Newsl 11(1):10–18
44. Gloe T, Bhme R (2010) The dresden image database for benchmarking digital image forensics. J Digit Forensic Pract 3(2–4):150–159
45. Bayram S, Sencar HT, Memon N, Avcibas I (2006) Improvements on source camera-model identification based on CFA interpolation. Proc WG 11:24–27
46. Chu A, Sehgal CM, Greenleaf JF (1990) Use of gray value distribution of run lengths for texture analysis. Pattern Recogn Lett 11(6):415–419

Automatic Patents Classification Using Supervised Machine Learning

Muhammad Shahid[1], Adeel Ahmed[2],
Muhammad Faheem Mushtaq[3(✉)], Saleem Ullah[3], Matiullah[3],
and Urooj Akram[3]

[1] Department of Physics, Government Sadiq Egerton (S.E) College, Bahawalpur, Pakistan
mathok@yahoo.com

[2] Department of Computer Science, Quaid-I-Azam University, Islamabad, Pakistan
adeel.ahmed@cs.qau.edu.pk

[3] Faculty of Computer Science and Information Technology, Khwaja Fareed University of Engineering and Information Technology, Rahim Yar Khan, Pakistan
{faheem.mushtaq, saleem.ullah}@kfueit.edu.pk,
mutiullah@cs.qau.edu.pk, uroojakram.cs@gmail.com

Abstract. Every year, approximately one million patent documents are issued with unique patent number or symbol. In order to find the relevant patent document, several users query the IPC documents using IPC symbols. So, there is a need of automatic classification and ranking of patent documents w.r.t. user query. Automatic classification is only possible through supervised machine learning techniques. In this paper, we classified patent documents using common classifiers. We collected 1625 $_{\text{patent}}$ documents related to eight different classes taken from IPC website using web crawler in an unstructured text. We considered 90% of training and 10% of test samples of the total patents. We built a feature matrix using tf-idf, smart notations and BM25 weighting schemes. This feature matrix is given to each classifier as input and output of each classifier consists of correctly classified and incorrectly classified instances. Finally, we evaluated the accuracy of each classifier using precision, recall and F-measure. We performed comparative analysis of classifiers and observed that by adding more features to each classifier, accuracy of classifier can be improved.

Keywords: Classification · Supervised learning · Unigram · BM25 · TF-IDF · SMART notations

1 Introduction

With the advancement of technology, most of the researchers taking an interest in the categorization of structured and unstructured text. So, an automatic classification of digital objects gains serious attention of many researchers due to huge amount of data. International patent documentation is a database that contains collection of documents

R. Ghazali et al. (Eds.): SCDM 2020, AISC 978, pp. 297–307, 2020.
https://doi.org/10.1007/978-3-030-36056-6_29

related to international patents. This database is maintained by European Patent Office (EPO). It contains patent families (A, B, C, D, E, F, G, H) and legal status information. This database is funded by the World Intellectual Property Organization (WIPO) and the government of Austria under an agreement on May 2, 1972. In 1991, it was integrated into the European Patent Office at Vienna, Austria.

According to Hogenboom et al., the number of digital objects in the form of text exceeds the number of documents so there is a need to redefine the business process area unit [1]. Thus, automatic document classification is an important issue for managers. There have been lot of research available to classify the patent documents in which some methods are not so effective and some of these are un-implemented. Patent documents classification involves the extraction of unique terms from all available classes. Then these terms are indexed in such a way that the terms present in a particular document of a class are correlated with terms present in other documents of a class. Patents analysis includes patent citation analysis, patent statistical analysis, patent biblometric data analysis, automated patent classification and cluster using the techniques of machine learning. The analysis of Patent biblometric is proposed using bibliographic knowledge like inventors, patent numbers, assignees, title, date of application, country and international patent classifications to discover citation contributions [2].

In 1997, Karki et al. described that patent citation analysis provides information about primarily cited patents, impact index and the technological strength of patents [3]. Henry et al. described the intellectual properties of applied mathematics as a live of invention [4]. Malerba et al. suggested that application knowledge can be helpful as the live of a firm's technological diversification and advancement [5]. Moreover, Giovanna et al. utilized patent statistics as a sign of technical specialization in a country [6]. Abbas et al. proposed a patent group system for prime technology analysis [7]. The patent document has a lot of information for analysis. Thia information can be divided into two types of groups; structured and unstructured. In structural group, items include patent number, filing date, or assignees; and in unstructured group, data is available in the form of free text it means there is no fixed location for any piece of information because of various contents, such as abstracts, claims, title, and summary. Some patent documents contain visualized items called patent graphs which belong to both structural and unstructured data. Patent documents provided by World Intellectual Property Organization (WIPO) that has hierarchal structure. Stein et al. performed working on analysis of hierarchical text classification with modern machine learning algorithm [8].

In this study, we focused on the patent documents classification and their accuracy using different classifiers. We identified features by assigning a score to each term present in a patent document based on different weighting schemes. Then this feature matrix is given to classifier as input and observed the classifier accuracy.

The rest of the paper is as follows: Sect. 2 describes the related work; Sect. 3 describes about proposed methodology for classifying patents. Section 4 presents experimental setup and results. Section 5 represents evaluation. Section 6 presents the conclusion of this research.

2 Related Work

A method of assigning a document into one pre-defined class is called document classification. Each document cluster-based on document features extracted are compared with all document's classes and then associate it with nearest available appropriate document category. Document clumps divide a collection of documents into teams while not discrimination pre-defined categories. Zhao et al. and Alba et al., assign an application to amazon in 2018 and international business corporation in 2019 respectively, that is capable to map the input phrase to segment of related documents [9–11]. Due to various classifiers is unable to provide reliable and acceptable output. So it gives the opportunity for automatic document categorization has become popular. The many researchers are trying to find out the way of automatically categorization of documents. In this regards, there are various approaches such as k-nearest-neighbor approach, naıve mathematician analysis, artificial neural networks, and genetic algorithm, used for classification.

The vector space model has been widely employed in ancient information retrieval and for automatic document categorization [12]. Govindarajan et al. proposed a novel preprocessed framework for topic analysis and visualization called Excessive Topic Generation (ETG). In vector space model, terms are extracted from the documents, then the weight of the indexed terms is computed in the form of term frequency, inverse document frequency that enhances the document retrieval accuracy, then the documents relevancy is calculated using cosine similarity measure [13, 14]. Here, X = (x_1, x_2, . . ., x_i) represents the vector space model for document U, where variable u_i represents the ith feature of document U. Likewise, V = (v_1, v_2, . . ., v_i) represents the vector space model of document V, then the similarity between U and V is calculated as

$$COS(U, V) = \frac{\sum_{i=1}^{n} u_i v_i}{\sqrt{\sum_{i=1}^{n} u_i^2 u \sum_{i=1}^{n} v_i^2}} \tag{1}$$

McCallum et al. compared two naive Bayes models (multi-variant Bernoulli model and multinomial model) based on different vocabulary sizes [15]. In 1999, Nigam et al. used naive Bayes to handle tagged and untagged documents [16]. An assumption behind naïve Bayes is referred to as category conditional independence. P(c_i) is the probability of a given document and P(d'/c_i) is the conditional probability of the document d' belonging to class c_i. The probability of document d' belonging to class c_i is explained in Eq. (3),

$$P(c_i/d') = \frac{P(d'/c_i).p(c_i)}{\sum_{cj \in c} P(d'/c_j).p(c_j)} \tag{2}$$

$$p(d'/c_i) = \prod_{j=1}^{d'} p(w_j/c_i) \tag{3}$$

Agrawal et al. proposed a bayesian patent system that classified the patents automatically [17]. Larkey et al. developed a K nearest neighbor-based (KNN) model for U.S. patent document searching and classification [18]. KNN is used to train the documents for classify test documents by using the following similarity measure.

$$Z(x, c_i) = \frac{\sum_{d\_(j\in)KNN)} sim(x, d_j) Y(d_j, c_i)}{\sum_{d\_(j\in)KNN)} sim(x, d_j)} \tag{4}$$

Here, sim(x, d_j) is the similarity between tested document 'x' and trained document d_j that computes geometric distance or the circular function cost between two document vectors. And y(d_j, c_i) = 1 if d_j belong to c_i.

Artificial Neural Network (ANN) is an interconnected process of nodes computationally attached to unravel issues. ANN is often used for overlap classes and document classification also and it trained by large datasets to fit the model for better results. Farkas et al. provided a synonym locator and neural network to come up with document clusters [12]. Massey et al. incorporated Adaptive Resonance Theory, for the classification of documents [19]. Selamat et al. proposed a web-page organization that used neural network with inputs gained by principal element analysis [20]. Lam et al. described the feature extraction technique that used linguistics and co-occurrence analysis in neural networks for document classification [21]. Mostafa et al. proposed a multi-layer feed-forward neural network to classify medical documents [22].

3 Proposed Approach for Patents Classification

The methodology adopted for patents classification consists of the following subsections:

3.1 IPC Hierarchy of Patents Classification

IPC (International Patent Classification) is a hierarchical classification scheme developed by the World Intellectual Property Organization (WIPO) and consists of hierarchical class-layers, such as sections, classes, subclasses, and groups, respectively. We have collected about 1625 patent documents related to eight different IPC classes. Further, we have taken the eight subclasses and collected the documents. Table 1 shows the statistics related to each class considered for training data and test data.

3.2 Collecting Patent Documents

We collected about 1625 patent documents from the IPC website. These documents are extracted in the form of unstructured text. Since the data were in HTML format so we used web crawler for extracting contents from patent documents.

3.3 Text Segmentation

The dataset obtained from the United States Patent and Trademark Office's main website (USPTO) [24]. All documents are given is in HTML format. Each document has some fixed features which are helpful for the segmentation of various contents of documentation like title of patent, abstract, claims, and description of that cable. Moreover some document has more portions or subsections with a specific tag such as summary of the invention, field of the invention, background, and detailed description of the preferred embodiment. Although, some of the patents may have more or fewer sub-sections most of the patents follow this format.

3.4 Terms Extraction and Indexing

We performed tokenization and stop-words removal on each document and extracted the words. Since stemming increases recall so we used porter stemmer [23] to perform stemming of each word or terms so that the affix part can be removed, and stem part can be separated. Then an inverted index is built that represents the list of terms (vocabulary), their frequencies and postings.

Table 1. Statistics related to each class

IPC class	Subclass	Total no. of documents in each class	Number of documents taken for training data (90%)	Number of documents taken for test data (10%)
A	A61	592	532	60
B	B60	147	132	15
C	C07	89	54	9
D	D01	8	7	1
E	E21	63	56	7
F	F16	153	137	16
G	G06	396	356	40
H	H04	177	159	18
Total		1625	1433	166

3.5 Computing Weight Related to Terms Using Different Weighting Schemes

We have collected the keywords regarding each subclass, A61, B60, C07, D01, E21, F16, G06, H04, given at the IPC hierarchy. We considered these keywords given under each subclass as a query. We computed weight of terms between documents and a query (keywords) of a particular class.

3.5.1 TF-IDF Weighting

We collected terms related to each patent document. The TF-IDF weight is computed using the following equation

$$W_{t,d} = \log(1 + tf_{t,d}) \times log_{10}\left(\frac{N}{df_t}\right) \quad (5)$$

Here 'tf' is the term frequency and 'df' is the document frequency of each term, N is the total number of documents in a class and $W_{t\in q\cap d}$.

3.5.2 BM25 Transformation

BM25 transformation of a term can be computed as

$$W_{t\in q\cap d} = C(t,q)\frac{(k+1)c(t,d)}{c(t,d)+k}log\frac{M+1}{df(t)}, k \geq 0 \quad (6)$$

where 'w' is the weight of specific term 't' that belongs to query 'q' and document, 'd'. c(t, q) is the count of occurrence of term 't' in a query 'q'. c(t, d) is the count of occurrence of term 't' in a document. The 'df' is a document frequency of a term that is number of documents that contain the term 't' and 'M' is the total number of documents present in a class.

3.5.3 SMART Notations

Most of the search engines allow for different weightings for documents vs queries. SMART notation denotes the combination in use with the notation *ddd.qqq for* both documents and a query. A very standard weighting scheme used is lnc.ltc. Here for; *Document*: tf (l) is used for logarithmic, (n) is used for no-idf and (c) is used for cosine normalization query is given is below.

Query: logarithmic tf (l), idf (t), cosine normalization (c). Acronyms (n) are computed using the weighting schemes as shown in Table 2. We have used the SMART notations as shown in Table 3, for computing weight (score) of each term.

Table 2. Acronyms description and their weighting

Acronyms	Weighting
n (natural)	$tf_{t,d}$
l (logarithm)	1 + logarithm $tf_{t,d}$
t (idf)	idf = log N/df_t
c (cosine)	$\frac{1}{\sqrt{w_1^2+w_1^2+\ldots+w_1^2}}$

Table 3. SMART notations

SMART notations (ddd.qqq)	Document (ddd)	Query (qqq)
ntc.ntc	$tf_{t,d}$ (n), idf (t), cosine normalization (c)	$tf_{t,d}$ (n), idf (t), cosine normalization (c)
ntn.ntn	$tf_{t,d}$ (n), idf (t), no normalization (n)	$tf_{t,d}$ (n), idf (t), cosine normalization (c)

(continued)

Table 3. (*continued*)

SMART notations (ddd.qqq)	Document (ddd)	Query (qqq)
ntc.ntn	$tf_{t,d}$ (n), idf (t), cosine normalization (c)	$tf_{t,d}$ (n), idf (t), no normalization (n)
ntn.ntc	$tf_{t,d}$ (n), idf (t), no normalization (n)	$tf_{t,d}$ (n), idf (t), cosine normalization (c)
ltc.ltc	logarithmic $tf_{t,d}$ (l), idf (t), cosine normalization (c)	logarithmic $tf_{t,d}$ (l), idf (t), cosine normalization (c)
ltn.ltn	logarithmic $tf_{t,d}$ (l), idf (t), no normalization (n)	logarithmic $tf_{t,d}$ (l), idf (t), no normalization (n)
ltc.ltn	logarithmic $tf_{t,d}$ (l), idf (t), cosine normalization (c)	logarithmic $tf_{t,d}$ (l), idf (t), no normalization (n)
ltn.ltc	logarithmic $tf_{t,d}$ (l), idf (t), no normalization (n)	logarithmic $tf_{t,d}$ (l), idf (t), cosine normalization (c)

3.6 Classification

We classified the patent documents using Naïve Bayes [10], Bayes Net [12], k-nearest neighbor (KNN) [13], Artificial Neural Network (ANN) [14], Forest tree classifier, support vector machine (SVM) and Ada BoostM1 classifier. The probabilistic unigram model is used for classifying the patent documents. This model classifies the sample 'x' to class 'y' with minimum risk that is with the highest probability. We used 90% of each class as training samples and 10% of documents of each class as test samples. Terms $\{O_i\}$ in 10% of documents are test samples and terms $\{W_i\}$ in 90% of documents are training samples. Each instance (term) W_i, O_i can be classified to 'class' C_i in both training and test samples.

4 Experimental Setup and Results

We took 1625 documents related to eight IPC subclasses as shown in Table 1. For training, we chose about 90% samples and for testing, we took about 10% samples. Table 4 gives the statistics of samples related to each class. Table 5 shows the correct and incorrect samples classified by each classifier with tf-idf weighting as features obtained by using 10-fold cross-validation on data. We observed that ANN classified the instances more accurately as compared to other classifiers. Table 6 shows the correct and incorrect samples classified by each classifier with the BM25 transformation weighting scheme by using 10-fold cross-validation on data. Table 7 shows the correct and incorrect samples classified by each classifier with SMART notations as features obtained by using 10-fold cross-validation. Finally, we utilized all weighting schemes and find the score of each term belongs to both training and test samples and then input the weight of each terms to the classifiers as features. Table 8 shows the correct and incorrect samples classified by each classifier obtained by using combined (tf-idf, BM25, SMART notations) weighting scheme. In comparison, from

Tables 5, 6, 7, we observed that those classifiers in which a ranking function 'BM25' used as a feature, has given better results than those classifiers in which TF-IDF and smart notations are considered.

Table 4. Statistics related to training and test samples

Class	Training sample	Test sample
A61	301248	100015
B60	178089	35340
C07	2212536	59630
D01	20126	3246
E21	512637	55632
F16	251236	69235
G06	633214	200004
H04	1120364	150012

Table 5. Classifier results with TF-IDF weighting Scheme (10 fold)

Classifier	% Correctly classified	% Incorrect classified
Random Forest	59.74	40.25
kNN	59.68	40.31
ANN	60.3	39.69
Bayes Net	60.29	39.7
Ada BoostM1	60.29	39.7
SVM	60.29	39.7
Naïve Bayes	59.69	40.2

Table 6. Classifier results using BM25 weighting (10 fold)

Classifier	% Correctly classified	% Incorrect classified
Random Forest	80.15	19.84
kNN	80.08	19.91
ANN	72.63	27.36
Bayes Net	79.49	20.5
Ada BoostM1	72.63	27.36
SVM	60.29	39.7
Naïve Bayes	41.28	58.71

Table 7. Classifier results using SMART notations (10 fold)

Classifier	% Correctly classified	% Incorrect classified
Random Forest	70.64	29.35
kNN	69.62	30.37
ANN	65.37	34.62
Bayes Net	58.4	41.59
Ada BoostM1	60.29	39.7
SVM	61.59	38.4
Naïve Bayes	60.99	39

Table 8. Classifier results using combined weighting scheme (10 fold)

Classifier	% Correctly classified	% Incorrect classified
Random Forest	91.47	8.5
kNN	89.76	10.23
ANN	76.12	23.87
Bayes Net	76.6	23.34
Ada BoostM1	72.63	27.36
SVM	62.47	37.52

To combined all of these weighting schemes and used as features to input to each classifier, then we obtained an effective result, as shown in Table 8.

5 Evaluation

We evaluated each classifier using precision, recall and balanced F-measure. The effectiveness of the proposed methodology is evaluated with 1625 patents were imported to derive a classifier model. We defined the precision and recall as

$$Precision = \frac{TruePositive\,(Ai)}{TruePositive\,(Ai) + FalsePositive\,(Bi)} \tag{7}$$

$$Recall = \frac{TruePositive\,(Ai)}{TruePositive\,(Ai) + FalseNegative\,(Ci)} \tag{8}$$

Here A_i is the number of instances or samples that are classified to class i, correctly, and B_i, C_i is the number of instances that are classified to class i, incorrectly. F-measure can be defined as

$$F = \frac{2 * Precision\ * \ Recall}{Precision\ + \ Recall} \tag{9}$$

Figure 1 shows the evaluated result of each classifier with tf-idf weighting scheme in terms of evaluation parameters such as precision, recall and F-measure. Figure 2 shows the evaluated result of each classifier with BM25 weighting scheme in terms of evaluation parameters. Figure 3 shows the evaluated result of each classifier with SMART notations weighting scheme, in terms of evaluation parameters, and Fig. 4 shows the best-evaluated result of each classifier with combined weighting scheme in terms of these parameters.

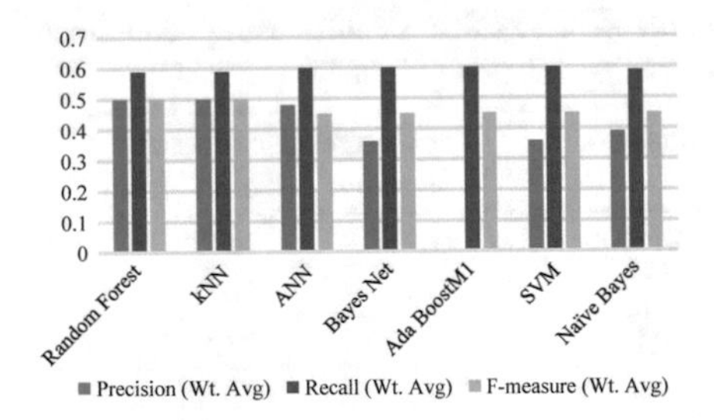

Fig. 1. Evaluation of classifiers with TF-IDF

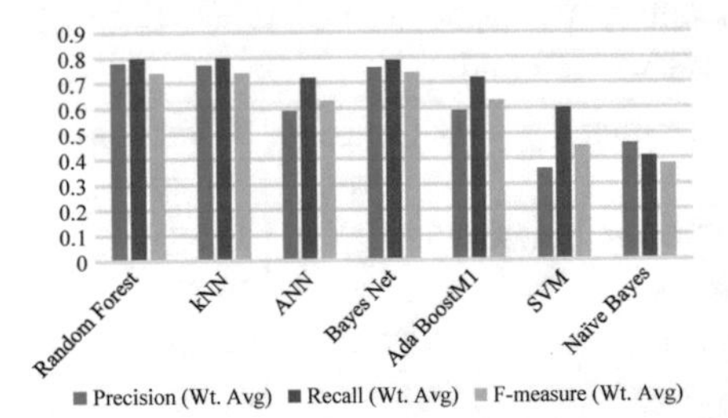

Fig. 2. Evaluation of classifiers with BM25

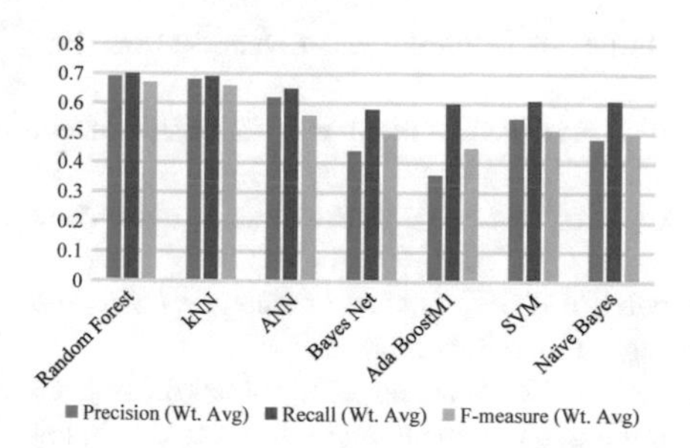

Fig. 3. Evaluation with smart notations

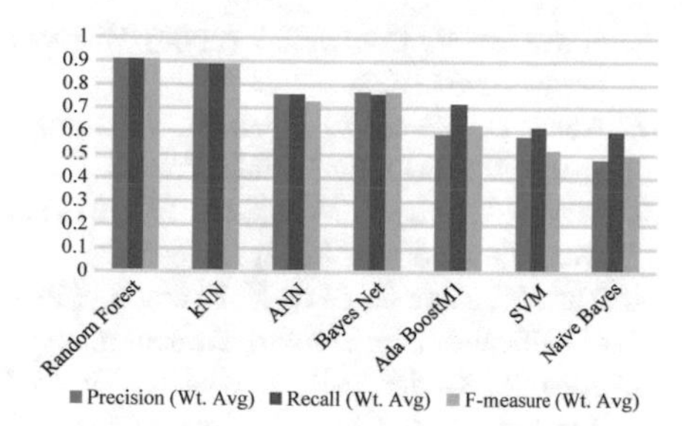

Fig. 4. Evaluation with combined weighting

6 Conclusion

Automatic classification of digital objects taking the serious attention of many researchers due to huge amount of data. International patent documentation is a database that contains international patent collection of documents. This database contains patent families (A, B, C, D, E, F, G, H) in the form of International Patent classes (IPC). To obtain relevant patents, users query the IPC documents using IPC symbols. So, there is a need for automatic classification w.r.t. user query. The purpose of this paper is to classify the patent documents by using different weighting schemes as features. We classified patients by commonly used classifiers to observe the accuracy of each classifier. We gathered about 1625 patent documents of eight subclasses from IPC website in which 90% documents are taken as training samples and 10% documents are taken as test samples. We evaluated the results of classifier model by using precision, recall, and F-measure and performed a comparative study. It is observed that those classifiers in which a ranking function 'BM25' used as a feature, has given better results than those classifiers in which TF-IDF and smart notations are used. But when we combined all of these weighting schemes and used as features to input to each classifier, then we obtained considerable results. We also observed that by increasing features, performance of some classifiers can be improved in terms of time and accuracy. Thus, increasing features, decrease the classification error.

References

1. Hogenboom F, Frasincar F, Kaymak U, Jong F, De Caron E (2016) A survey of event extraction methods from the text for decision support systems. Decis Support Syst 85:12–22
2. Narin F (1994) Patent bibliometrics. Scientometics 1(30):147–155
3. Karki M (1997) Patent citation analysis: a policy analysis tool. World Patent Inf 19(4):269–272
4. Henry C, Stiglitz JE (2010) Intellectual property, dissemination of innovation and sustainable development. Global Policy 1(3):237–251
5. Malerba F, Breschi S, Lissoni F (2003) Knowledge-relatedness in firm technological diversification. Res Policy 32:69–87

6. Giovanna V, Cantwell J (2004) Historical evolution of technological diversification. Res Policy 33:511–529
7. Abbas A, Zhang L, Khan SU (2014) A literature review on the state-of-the-art in patent analysis. World Patent Inf 37:3–13
8. Stein RA, Jaques PA, Valiati JF (2019) An analysis of hierarchical text classification using word embeddings. Inf Sci 216–232
9. Zhao M, Zhou H, Long X, Zhang X, Lin J, Yacoub SM (2018) Document analysis for region classification. In: Book document analysis for region classification
10. Alba A, Coden AR, Drews C, Gruhl DF, Lewis NR, Mendes PN, Ramakrishnan C, Terdiman JF (eds) (2019) Segmenting and interpreting a document, and relocating document fragments to corresponding sections
11. Antonie M-L, Zaiane OR (2002) Text document categorization by term association. In: Proceedings of IEEE international conference on data mining, pp 19–26
12. Khatavkar V, Kulkarni P (2019) Trends in document analysis. In: Data management, analytics and innovation. Springer, Singapore, pp 249–262
13. Govindarajan UH, Trappey AJ, Trappey CV (2019) Intelligent collaborative patent mining using excessive topic generation. Adv Eng Inf 42:100955
14. Raghavan VV, Wong SKM (1986) A critical analysis of vector space model for information retrieval. J Am Soc Inf Sci 37(5):279–287
15. McCallum A, Nigam K (1998) A comparison of event models for naive Bayes text classification. In: Proceeding of the workshop on learning for text categorization, AAAI'98, Madison, WI, pp 41–48
16. Nigam K, McCallum AK, Thrun S, Mitchell T (1999) Text classification from labeled and unlabeled documents using EM. Mach Learn J 39(2):103–134
17. Agrawal R, Chakrabarti S, Dom B, Raghavan P (1998) Scalable feature selection, classification and signature generation for organizing large text databases into hierarchical topic taxonomies. Int J Very Large Data Bases 7(3):163–178
18. Larkey LS (1999) A patent search and classification system. In: Proceedings of the fourth ACM conference on digital libraries, pp 179–183
19. Massey L (2003) On the quality ART1 text clustering. Neural Netw 16:771–778
20. Selamat A, Omatu S (2004) Web page feature selection and classification using neural networks. Inf Sci 158:69–88
21. Lam W, Han Y (2003) Automatic textual document categorization based on generalized instance sets and a metamodel. IEEE Trans Pattern Anal Mach Intell 25:628–633
22. Mostafa J, Lam W (2000) Automatic classification using supervised learning in a medical document filtering application. Inf Process Manage 36(3):415–444
23. Van Rijsbergen CJ, Robertson SE, Porter MF (1980) New models in probabilistic information retrieval. British Library, London. (British Library Research and Development Report, No 5587
24. United States Patent and Trademark Office (USPTO), https://www.uspto.gov/patents-application-process/patent-search/classification-standards-and-development

A Fuzzy Logic Technique for Optimizing Follicular Units Measurement of Hair Transplantation

Salama A. Mostafa[1(✉)], Abdullah S. Alsobiae[2], Azizul Azhar Ramli[1], Aida Mustapha[1], and Rabei Raad Ali[1]

[1] Faculty of Computer Science and Information Technology, Universiti Tun Hussein Onn Malaysia, Batu Pahat, Johor, Malaysia
{salama, azizulr, aidam}@uthm.edu.my, rabei.aljawary@gmail.com

[2] College of Computing and Informatics, Universiti Tenaga Nasional, 43000 Kajang, Selangor, Malaysia
abooody_333@hotmail.com

Abstract. Hair transplantation medical procedure is one of the main methods that are at present utilized in the treatment of balding of the scalp. It is essentially a procedure of extricating or taking a particular number of Follicular Units (FUs) from the back of the head which serves as the contributor or donor region and transplanting them in the region of the scalp that is going bald. A FU comprises one to five normally occurring human skin hairs. The most mainstream techniques designed for hair transplantation dependent on the FUs idea is the Follicular Units Extraction (FUE). Past endeavors to calculate the needed number of FUs for the FUE failed to put into consideration various metrics or indices (parameters) associated with the determination procedure. This paper expounds a Fuzzy Logic Follicular Units Measurement (FL-FUM) strategy for hair transplantation of the FUT and FUE techniques. The FL-FUM technique gives a progressively exact estimation of the needed FUs number by envisaging about three fuzzy metrics of Age, Race and Donor Area Density (DAD). Its objective is to help hair reclamation people who utilize the FUT and FUE techniques in assessing the needed number of necessary grafts that fulfill a patient's baldness state. The FL-FUM strategy employs a Fuzzy Logic system on the three metrics (fuzzy sets) to defuzzify the assessment of the FUs dependent on Visualized Male Pattern Baldness Schema. The strategy is tried and assessed by contrasting its outcomes and the comparable existing strategies and is observed to be highly productive for real estimation cases.

Keywords: Hair transplantation · Follicular units transplantation · Follicular units extraction · Follicular units measurement · Fuzzy logic

R. Ghazali et al. (Eds.): SCDM 2020, AISC 978, pp. 308–319, 2020.
https://doi.org/10.1007/978-3-030-36056-6_30

1 Introduction

Male pattern baldness or hair loss conditions are in a way inevitable. It influences an individual irrespective of age and race. Baldness is today in almost 35 million individuals all over the USA alone. Hair transplant medical procedure is progressively turning into an extremely well-known fix for normal hair gains [1]. It is intriguing looking at the available data in the hair transplant domain concerning the real number of new patients appearing in comparison with the measures that are being carried out every year [2, 3]. Nonetheless, every member of this domain concurred that hair transplant is more to workmanship than to medical practice where the aptitudes and the encounters play a lot on the eventual outcome quality. In addition, the necessities and the desire of the current male pattern baldness patients are expanding [4]. Thus, candidates' disappointment is increasing, despite the monumental advancement in the hair transplant outcomes. A couple of instruments and strategies have been created to aid hair transplanting.

The Follicular Unit (FU) transplant is the fundamental operative principle or model for estimating hair transplants [2]. Various suggested methods in performing hair transplantation medical procedure dependent on the FU idea such as the Follicular Units Extraction (FUE) and Follicular Units Transplantation (FUT) are available. FUE method i.e. small grafts extraction has given the chance to make practically imperceptible new hair inclusion. Accordingly, there is much advancement in the hair transplant outcomes for a couple of years now [5]. Presently, the specialists including patients realize this innovation and are as well mindful of its fundamental ideas. Notwithstanding, most advanced hair transplants are yet an objective that numerous specialists need to accomplish.

One of the fundamental instruments/techniques in hair transplant is utilized for Follicular Units Measurement (FUM). Doctors utilize this strategy for carrying out straightforward counts for FUE and FUT operative intakes. In any case, the computation procedure generally pursues the standard guideline and a fundamental digital calculator, which produces sketchy estimations. As of late, patients and specialists utilize a few available online apparatuses for giving direction and cost estimation to the medical procedure as the medical procedure cost is for the most part identified with the needed FUs to cover the hair loss territory. In this manner, this study suggests an improved Fuzzy Logic Follicular Units Measurement (FL-FUM) strategy that can deliver better estimates about the needed number of hair FUs or grafts for the cases in question. The FL-FUM strategy coordinates a fuzzy logic or rationale method to give increasingly exact FUM products. The fuzzy logic takes into consideration the patient's age, race and contributor region density in the FUM procedure.

2 Research Method

2.1 Hair Transplantation Techniques

In 1998 a gathering of hair reclamation specialists in Dermatologic Surgery coined the name follicular unit transplantation. FUT is basically the transplantation of FUs using any hair transplantation methods [6]. The fundamental methods proposed in actualizing hair transplantation medical procedure depending on the FU idea are the Follicular Units Extraction (FUE) and Follicular Units Transplantation (FUT) [7]. The primary objective of these two methods is to obtain the most elevated regular or normal outcomes of hair reclamation. There are two different approaches in FUT of stereomicroscopic dismemberment (dissection) and Limmer single-strip reaping (harvesting), as a major aspect of the procedure of activity. In spite of the recognition differences among patients and experts, FUE is progressively getting to be prominent procedures in restoring a patient's hair as a result of transcendently normally existing individual FUs. In FUE, individual follicular units (FUs) are split into little entities while in the FUT they are consolidated into huge collections resulting in a scar on the back of the head. Since FUs can be taken forthwith from the giver region without the need of a strip laceration (i.e., utilizing FUE), the inceptive meaning of FUT has turned out to be out of date. Nevertheless, FUE strategy does not generally collect unblemished units and requires a semi-grouping procedure to bare a portion of the connective tissue support structures [6].

In this manner, there are two main interests for transplanting individual FUs. Firstly, utilizing individual FUs will deliver the most normal outcomes. Secondly, it enables the specialist to transplant the greatest measure of hair into the littlest conceivable injury. Overall, patients with light-shaded or fine hair, a few professionals consolidate two FUs to make greater thickness in the forelock region. Conversely, patients who have FUs with a limit of three hairs or have coarse hair, a few experts are encouraged to transplant grafts in satisfactory thicknesses or densities. On one hand, a fix is to partition the hair of one FU into little grafts to make smaller-scale grafts transplantation [7]. On the other hand, however, a few doctors attempt to accumulate one-hair FUs into 'twos' to limit the number of beneficiary locales required.

2.2 Hair Follicular Units Calculations

The absolute quantity of unions or grafts for complete operative hair rebuilding is changing from patient to patient as a result of the differences in their hair attributes. Table 1 provides a sketchy estimation to the total number of hair grafts required to achieve a total reclamation as indicated by Norwood Class of hair loss which is inferred from [8, 9]. Experts created various instruments to help doctors in hair transplant medical procedure. This segment outlines a portion of the related works in hair transplant medical procedure on hair follicular units' counts.

Different tools have been developed to assist physicians in hair transplant surgery. Some of the related works in hair transplant surgery to the hair follicular units' calculations are summarized in this section. The research that is proposed by [10] results in a Hair Calculator Test tool that can be used in hair follicular units' calculation. In this tool, the physicians may choose the desired density by scrolling the bar where the natural-looking for the hair starts usually from 50%. Once the physicians determine the right density of grafts per cm^2, then he can select which areas he needs for hair restoration. Once the user did that by hitting the Continue button the total number of grafts that needed for the hair restoration is calculated. Another tool is proposed by [11] that allows the user to choose the desired density and the required areas in order to

Table 1. The Norwood Classes with FUs range modified from [8]

Norwood class	Visual class	Follicular units	Number of sessions
I		400–600	1
II		800–1000	1
III		1400–1800	1
IV		2000–2400	1 or 2
V		2800–3200	2
VI		3600–4000	2
VII		4400–5000	2

calculate the recommended follicular units. This tool only gives a rough estimation and should not be considered as a definite estimator.

The physiological properties of every individual patient have to be considered in order to come up with satisfactory measurement results. Graft calculator of [12] is a more advanced tool that has different types of attractive features. It considers many factors regarding hair follicular units' calculation. The user can easily use this tool to determine the need of hair grafts by selecting a style that described the hair loss case and then selects the area by clicking on its corresponding grid location and then by dragging the slider to adjust hair growth. Once done, the user can simply proceed to the final stage of the measurement results. This tool can give a better estimation as it covers many considerations in its calculations.

2.3 Fuzzy Logic Technique

The fuzzy logic or rationale method gives an efficient device to coordinate people groups' understanding and encounters of a specific field in a framework. It carries out estimated reasoning dependent on four fundamental ideas: language factors, fuzzy sets, probability disseminations and fuzzy inferencing. A fuzzy set is utilized to describe etymological or language factors. Their qualities can be depicted subjectively utilizing an etymological articulation and quantitatively utilizing a membership function [13]. Etymological articulations are valuable for interfacing ideas and learning. Membership functions are important in preparing inputs of numerical information type. At the point when a fuzzy set is allocated to a language variable or factor, it forces a versatile requirement called a probability distribution to the quality of the factors. The surmised reasoning is a direct formalism for deciphering human information or rule of thumb in a numerical structure. It incorporates fuzzy inferencing (defuzzifier) for approximating self-assertively inputs to an increasingly strong yield [14, 15]. The least difficult and the most generally utilized technique for the defuzzification is the focal point of gravity or the centroid which is embraced in this study. The general architectural design of fuzzy logic for inexact reasoning procedure is as in Fig. 1.

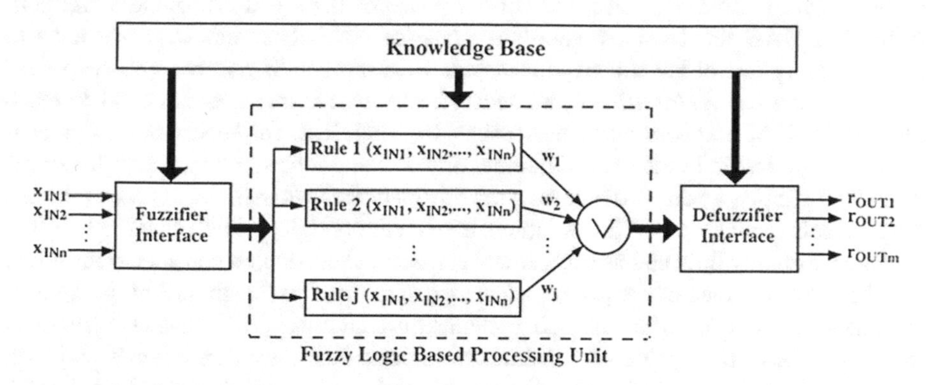

Fig. 1. Fuzzy logic architecture for approximate reasoning [15]

3 Modelling and Implementation

Proper planning is one of the vital issues in hair restoration surgery. The main rule for successful planning is to anticipate the worst-case scenario for a patient. Even though hair balding is unpredictable, still, there are many variables that can assist in an educated judgment about the patient hair loss future in the short-term and the long-term as well. Hence, an accurate estimation of the number of grafts is achievable. This section gives an illustration of the variables that are selected in determining the number of FUs for a patient. how the Fuzzy Logic (FL) manipulates the variables to provide a more accurate estimation. Subsequently, it illustrates the modelling of the Fuzzy Logic Follicular Units Measurement (FL-FUM) method that is proposed for hair FUs calculations.

Appropriate planning is one of the fundamental problems in hair rebuilding medical procedure. The primary standard for the fruitful plan is to envision the direst outcome imaginable for a patient. Despite that, hair loss is flighty; still, there are numerous factors that can aid in an informed judgment about the patient male pattern baldness future for the time being and in the long run too. Thus, a precise estimation of the number of grafts is attainable. This segment gives an outline of the chosen factors in deciding the quantity of FUs for a patient and how the Fuzzy Logic (FL) controls the factors to give an increasingly exact estimation. In this manner, it presents the demonstration of the suggested technique for hair FUs quantity counts in the form of Fuzzy Logic Follicular Units Measurement (FL-FUM).

3.1 The Fuzzy Sets

The estimations of the quantity of the required grafts are abstract and not exact (e.g., ranges from 500 to 1000 FUs) which is unquestionably not a precise technique. In light of our investigation to the real factors that affect the estimation of the required quantity of hair FUs, patients age, race, the current phase of male pattern baldness and contributor or DAD are the most significant metrics. To bolster this case, we have to evaluate every one of the chosen four factors. Beginning with variable age, the patient of age 20 and has a *classIII* (as outlined in Table 1), for instance, is bound to reach *classVI* or *VI* of hair loss when he reaches 40. Therefore, for such case, a low estimation is applicable to the needed number of grafts and more grafts are kept for future use. Conflictingly, a patient of age 40 and with a hair loss *classIII*, the impact of the age factor is substantially less. The variations in his hair conditions towards more hair loss are more probably little and as such, fewer grafts are needed to be considered for future use. More so, the race of the patient affects the donor region thickness and the figment thickness of the scalp after the hair reclamation technique [16]. Besides, Caucasian individuals have the upside of constructive impact, then Asian individuals and then Africans. Thusly, the quantity of grafts that are evaluated to Africans is to be stricter to low, Asians medium and Caucasian high. Lastly, the donor density or thickness greatly affects ascertaining the number of grafts where the patient with higher donor region thickness can get a greater quantity of grafts and the other way around.

3.2 The Fuzzy Logic Follicular Units Measurement

The Fuzzy Logic Follicular Units Measurement (FL-FUM) method utilizes the fuzzy logic technique. The FL-FUM architecture comprises four primary parts, which are the Fuzzifier, the Fuzzy Inference Engine (FIE), the Fuzzy Rule Base (FRB) and the Defuzzifier as in Fig. 2. The fuzzy sets factors utilized in the FUs count are Age, Race and DAD while Baldness class is a crisp set. The fuzzy logic utilizes the suggested factors of fuzzy sets and the crisp set as contributions to compute the FUs quantity yields.

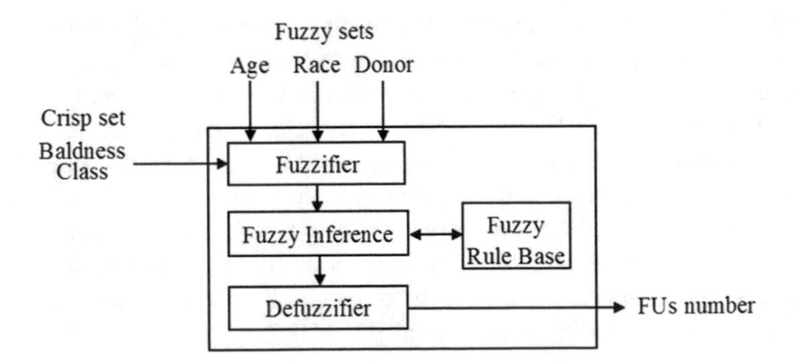

Fig. 2. The architecture of the FL-FUM method

The major steps that the FL-FUM method carries out to estimate the required FUs of a specific class are explained in the following algorithm.

Algorithm 1: The FL-FUM methods

step1: fuzzify the inputs by assigning the degree of the membership through the membership functions to determine the belonging for each of the fuzzy sets;
step2: adjust the fuzzy rules by applying fuzzy logical operators to evaluate the strength of the rules for the given case;
step3: select the rules that determine the fuzzy variable output boundaries to find solutions;
step4: aggregate the outputs of the fuzzy set by combining the outputs of each triggered rule;
step5: defuzzify the aggregated output fuzzy set to obtain a single crisp (non-fuzzy) output variable (i.e. FUs)
step6: display the final output;

3.3 Memberships

The fuzzy sets for the age input variable have the following names: *Very Young*, *Young*, *Intermediate, Old* and *Very Old*. Age range division might not reflect real ageing sequence by years but it is specified to suit the system needs. Figure 3(a) shows the assignment of range and degree of membership for the Age variable. Age range classification probably will not show genuine maturing grouping by years however, it is indicated to meet the framework requirements. The fuzzy sets for the Race variable have the following names: *African*, *Asian* and *Caucasian*. Figure 3(b) shows the assignment of range and degree of membership for the Race variable. The fuzzy sets for the DAD variable have the following names: *Low*, *Medium*, *High* and *Very High*. Figure 3(c) shows the assignment of range and degree of membership for the DAD variable. The FUs variable represents a different number of classes that represent baldness types: *classI*, *classII*, *classIII*, *classIV*, *classV*, *classVI* and *classVII* in which the classes' selection by the user is represented by crisp set. Each class has different fuzzy sets for the FUs output variable: *Very Low*, *Low*, *Medium, High* and *Very High*. Figure 3(d) shows the crisp sets of the baldness classes along with the rang of each corresponding class.

The rules of the FL-FUM method take the form of IF (condition 1) [AND] (condition 2) [AND] (condition 3) THEN (conclusion). Some examples of the utilized rules in the method are as follows:

RULE 1.1: IF **Age** is *Very Young* AND **Race** is *African* AND **DAD** is *Low* THEN ***classI*-FUs** is *Law*

RULE 1.2: IF **Age** is *Young* AND **Race** is *Asian* AND **DAD** is *Low* THEN ***classI*-FUs** is *Law*

After applying the corresponding rules for the boldness class according to the membership, the fuzzy boundaries of the fuzzy variables are discovered. The defuzzifier defuzzifies the aggregated output fuzzy set using the centre of gravity or the centroid technique. The following equation is used to determine the centroid of composite fuzzy regions.

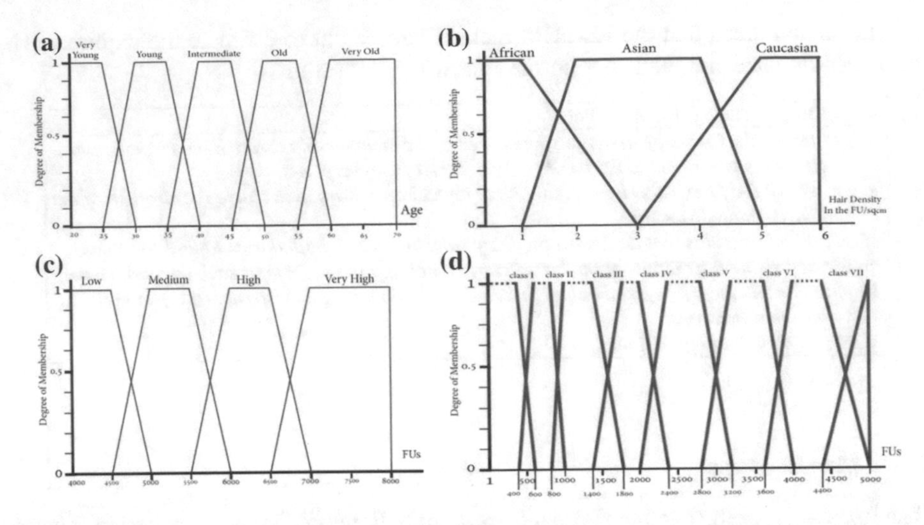

Fig. 3. The memberships of the fuzzy variables, (a) Age membership, (b) Race membership, (c) DAD membership and (d) baldness classes

$$centroid = \frac{\sum_{i=1}^{n} x1 + [(\sum_{i=1}^{n} x2 - \sum_{i=1}^{n} x1) * f(u_i)}{n} \quad (1)$$

where $x1$and $x2$ are points that respectively represent the begin and the end of the fuzzified area and n is the number of the applicable rules as in some cases more than on rule constraints are fulfilled. As a result, some rules with conflict resolutions need to have extended fuzzy area to be defuzzified.

3.4 Implementation

This study comes up with the Fuzzy Logic Follicular Units Measurement (FL-FUM) strategy for hair FUs computation. The FL-FUM is executed in Java programming language. Figure 4 illustrates the user interface of the FL-FUM method. The FL-FUM cycle begins with the client inputting the patient's data, for example, the patient's name, age, race, case number and the date. At that point, the client checks and determines the hair loss level of the patient and by this figure out the patient's baldness class. The technique standardizes the gathered information into the needed or specification format. The standardized information is transformed into fuzzy and crisp sets utilizing a particular scientific model to pre-process the information and empower the fuzzy logic to recognize and examine the information. From the given arrangements of information, the fuzzy logic assesses the needed number of hair FUs. Based on the chosen class, the technique evaluates the hair FUs number to meet with the balding regions as necessary.

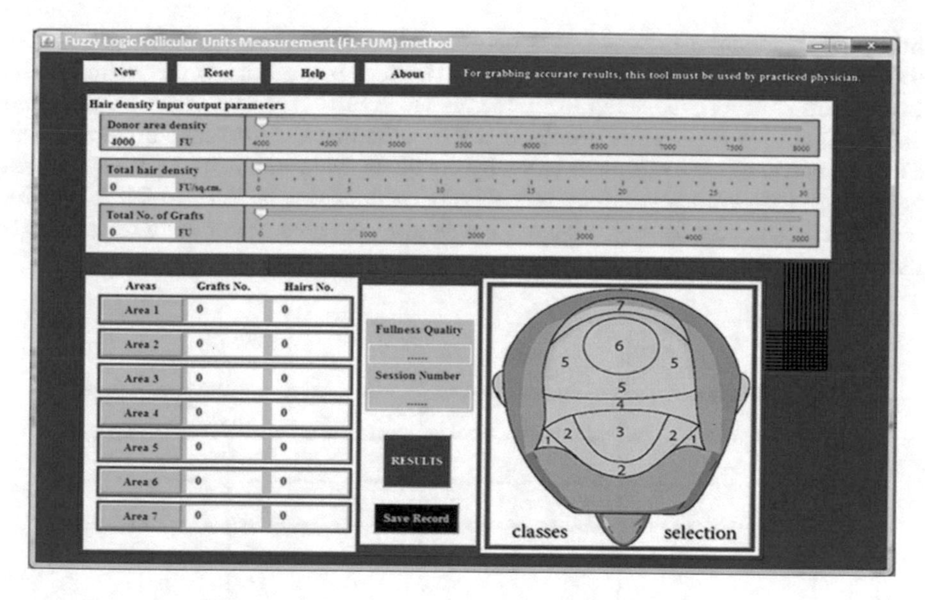

Fig. 4. The implementation of the FL-FUM method

3.5 Results and Discussion

The FU-FUM technique exhibits its operation and exactness using three test situations of a *classIV* hair loss. Utilizing various classes will not reflect a clear-cut correlation as there exists a noteworthy distinction between the metrics of these classes. In addition, this class serves as the most intricate case and can demonstrate the significance and basic proportions of the technique. For example, according to Norwood's male hair pattern grouping, the patient that has *classVII* will nearly expend his entire accessible DAD because it is the last class of the hair loss wherein there is no more hair loss later on. Table 2 depicts the three test situations of the *classIV* hair loss.

Table 2. The parameters of the three test scenarios

Scenario	Age	Race	DAD	FUs No.	Density
1	*Very young*	*African*	*Low*	2045	21%
2	*Intermediate*	*Asian*	*High*	2216	23%
3	*Very old*	*Caucasian*	*Very high*	2379	25%

The outcomes of the *classIV* baldness demonstrate that the framework evaluates the quantity of FUs between 2000–2400, which is the limits of *classIV* (as in Table 1). In the initial situation, the provided fuzzy factors are Age = *Very Young*, Race = *African* and DAD = *Low*, the result of the surmising principle is *Low* FUs and the assessed estimation of the *Low* dependent on the provided fuzzy sets is 2045 and with a thickness of 21% which is taken as precise thickness. In the subsequent situation, the fuzzy factors are Age = *Intermediate*, Race = *Asian* and DAD = *High*. This situation brings about Medium FUs, which is between 2100–2300. The evaluated outcome of the FUs is 2216 and with hair thickness of 23% which is additionally precise thickness. In the third situation, the setup fuzzy factors are Age = *Very Old*, Race = *Caucasian* and DAD = *Very High* which results in a *High* FUs that has the scope of 2250–2400. The evaluated outcome of this situation by the FL-FUM technique is 2379 FUs, which is inside the *High* range, and the hair density or thickness is 25%, which in consideration is as an exact density. Figure 5 condenses the outcomes of the three *classIV*'s hair loss situations.

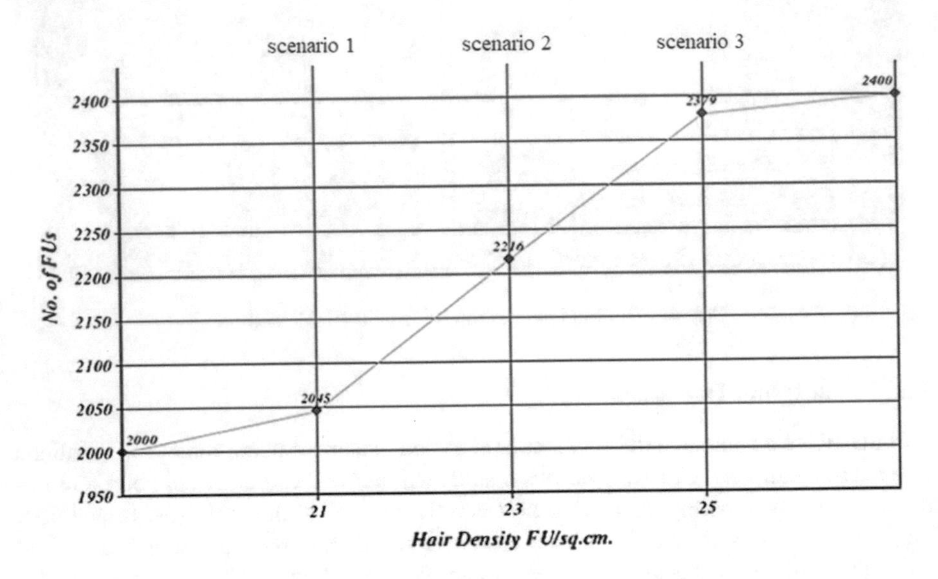

Fig. 5. The three experiments' results graph

The total Density of the hair in the last section to the right side of Table 2 is in percentage and 25% in situation 3, for example, it typifies 25% in comparison with the first hair density. Indeed, in hair transplantation, delivering more than 40% of the initial hair thickness is impossible for those with hair loss level of *classIII* or more. That is a direct result of the donor region that has a set number of FUs. Subsequently, the hair fullness appearance for *classIII* or more are aesthetic and to some degree appears to be artificial. The three situations confirm or substantiate that there is a distinction in the estimations of the needed hair FUs depending on the patients' situation. Numerous potential situations request techniques, for example, the FU-FUM to carry out such confounded and comprehensive computations. The confinement or constraint of this work lies in lack or inadequate size consideration of the donor and baldness regions in the estimations and calculation procedure as they are thought to be relative or commensurate.

4 Conclusion

In the last 40 years, numerous researches concentrated on human hair reclamations evolution and advancement. Follicular Units (FUs) are scalp groupings out of which human hair grows. Additionally, these researches explored the two most outstanding methodologies in hair transplantation, which are Follicular Unit Transplantation (FUT) and Follicular Unit Extraction (FUE). The FUT and FUE methodologies acknowledge the FUs as the essential component of tissue taken from the donor region to the trans-planted scalp zone. This study recommends a strategy that helps hair transplantation specialists in evaluating the quantity of FUs appropriate to fulfil a patient state of baldness. The Fuzzy Logic Follicular Units Measurement (FL-FUM) technique's goal is to aid an exceptional hair transplant outcome by giving (i) a fitting or suitable planning procedure and (ii) an exact evaluation of the hair loss condition that leads to less operative sessions and higher quality outcomes or yields. The precise estimations show superior mitigating the restricted transplantation assets of FUs and in turn conserve time and money. The future endeavour looks at examining the probability of incorporating different metrics or factors in the FUs estimation. In addition, there is a requirement for a procedure that handles and deals with the estimation of numerous transplantation sessions, for example, Case-based Reasoning (CBR) method.

Acknowledgements. This work is sponsored by Universiti Tun Hussein Onn Malaysia under TIER1 FASA 1/2007, UTHM Research Grant (VOT U896) and Gates IT Sdn. Bhd.

References

1. Zingaretti G, Petrovic O (2019) Systems and methods for creating hair transplantation procedure sites. U.S. Patent Application No. 10/327,801 (2019)
2. Umar S (2012) The transplanted hairline: leg room for improvement. Archiv Dermatol 148 (2):239
3. Unger W, Solish N, Giguere D, Bertucci V, Coleman W, Loukas M, Unger R (1994) Delineating the "Safe" donor area for hair transplanting. Am J Cosmet Surg 11(4):239–243
4. Zingaretti G, Canales MG, McCollum JW (2018) Systems and methods for hair loss management. U.S. Patent Application 15/748,119, filed 2 August 2018
5. Gho CG, Martino Neumann HA.: Donor hair follicle preservation by partial follicular unit extraction. A method to optimize hair transplantation. J Dermatol Treat 21(6):337–349 (2010)
6. Barrera A, Uebel C (eds) (2013) Hair transplantation: the art of micrografting and minigrafting. CRC Press
7. Sadick NS, White MP (2007) Basic hair transplantation: 2007. Dermatol Ther 20(6):436–447
8. Rawnsley JD (2008) Hair restoration. Facial Plast Surg Clin N Am 16(3):289–297
9. Jimenez F, Izeta A, Poblet E (2011) Morphometric analysis of the human scalp hair follicle: practical implications for the hair transplant surgeon and hair regeneration studies. Dermatol Sur 37(1):58–64
10. Ilter D, Korkmaz K (2018) FAQ about hair transplant and FUE. ILTER CLINIC. Accessed 13 Sept 2018

11. The Specialist of Hair Transplant Aboard (2010) Hair Loss Learning Center Hair Restoration Network. chirurgie esthétique de A à Z Paris. Accessed 6 Sept 2018
12. Choudhuri (2012) Graft calculator Hair Perfect Clinic Limited. Accessed 6 Sept 2018
13. Das TK, Acharjya DP (2014) A decision making model using soft set and rough set on fuzzy approximation spaces. Int J Intell Syst Technol Appl 13(3):170–186
14. Mostafa SA, Mustapha A, Mohammed MA, Ahmad MS, Mahmoud MA (2018) A fuzzy logic control in adjustable autonomy of a multi-agent system for an automated elderly movement monitoring application. Int J Med Inf 112:173–184
15. Muroga S (1982) VLSI system design. Research supported by the national science foundation. New York, Wiley-Interscience
16. Mostafa SA, Darman R, Khaleefah SH, Mustapha A, Abdullah N, Hafit H (2018) A general framework for formulating adjustable autonomy of multi-agent systems by fuzzy logic. In: KES international symposium on agent and multi-agent systems: technologies and applications. Springer, Cham, pp 23–33

Soft Computing and Data Mining in Economics and Engineering

Divisible Load Framework and Close Form for Scheduling in Fog Computing Systems

Mojtaba Kazemi[1], Shamsollah Ghanbari[2(✉)], and Manochehr Kazemi[2]

[1] Department of Computer Engineering, Qom Branch, Islamic Azad University, Qom, Iran
mojkaz@gmail.com

[2] Department of Computer Science, Ashtian Branch, Islamic Azad University, Ashtian, Iran
myrshg@gmail.com, univer_ka@yahoo.com

Abstract. Fog computing is a possible way to reduce the latency of requests which have been sent to the cloud centres. It means the jobs can be scheduled to fog systems before being sent to the cloud centres. There is an extensive literature concerning to scheduling in fog computing systems. This paper mainly proposes a divisible load framework for scheduling in fog computing system. The divisible load theory is a suitable method for scheduling of data intensive jobs. This paper illustrates that the divisible load scheduling(DLS) method can be performed in the area of fog computing. This paper proposes a three-layer architecture for load scheduling in fog computing using divisible load theory. We formulate a close form for the proposed model. Finally we solve the close form.

Keywords: Fog computing · Divisible load scheduling · Latency · Close form

1 Introduction

Fog computing, is a structure that applies edge devices to accomplish a substantial amount of communication, computation, and storage regionally and routed over the internet backbone. Actually, the main goal of using fog computing is to decrease the latency of cloud computing [1]. There are a considerable number of publication concerning to the scheduling of fog computing, which are listed in Sect. 2. We have a load scheduling approach to the problem. For this purpose a divisible load scheduling method has been used. The divisible load theory acts based on the reality that the load can be distributed on a large number of independent processors. The processors should stop processing simultaneously,

R. Ghazali et al. (Eds.): SCDM 2020, AISC 978, pp. 323–333, 2020.
https://doi.org/10.1007/978-3-030-36056-6_31

otherwise; it fails to achieve it optimal performance [2]. The first articles in relation to the divisible load theory were published in 1988 [3,4]. Essentially, the fact regarding to the DLS or DLT is that the computation and communication can be distributed independently into some parts of arbitrary sizes, and each part can be executed separately in parallel. In relation to the DLS it is supposed that, preliminary amount V of load is maintained by the root processor denoted by p_0. In fact, the root processor dispatches the load to the worker processors denoted by p_1, p_2, ..., p_m. Assume that the ith processor $(0 \leq i \leq m)$ obtains α_i fraction of load. The condition for optimal dispatching is that all processor finish the processing at the same time. Under other circumstances, the load could be sent from busy to idle processors [5]. The goal is to calculate α_0, α_1, ..., α_m in the DLS timing equation. In the last three decades, regarding to DLS there have been a broad range of applications around parallel and distributed computing, consists of image and vision processing [6,7], data grid applications [8], and linear algebra [9]. Additionally, the DLS was applied to diverse topologies, including single-level tree, multi-level tree, bus and daisy chain, three-dimensional meshes, k-dimensional meshes, hypercubes and arbitrary graphs. To address these applications see [2]. Furthermore, the DLS has been accomplished for scheduling in homogeneous and heterogeneous platforms, grid-based environment [10,11] and cloud-based applications [12]. Lately, some other aspect of the DLS, such as multi-layer divisible computations [13], hierarchical memory [14], multi-criteria DLS [15], multi-objective DLS [16] and priority-based DLS [17] have been broadly investigated. Additionally, there are some comprehensive reviews regarding to DLS respecting the strategies, applications, and also open issues [2,18,19].

In this investigation we approach to the divisible load theory as a tools to reduce the complexity and finish time in transferring and processing of data in fog environment. This paper proposes a three-layer architecture for load scheduling in fog computing using divisible load theory. We formulate a close form for the proposed model as well. The objectives of this paper are to

- investigate the suitability of divisible load scheduling theory for applying in the fog environments
- proposing a divisible load framework and a close form for fog computing.
- computing the close form of divisible load scheduling model of fog computing.

Then we will go on organizing the other sections as follow. Section 2 explores the existing studies related to divisible load scheduling as well as fog computing. Section 3 illustrates the notation and parameters used in the paper. Section 4 explains a three-layer divisible load scheduling architecture for fog computing systems. Section 5 formulates the proposed model as a divisible load scheduling timing equation. In Sect. 6 we solve the proposed timing equation. A general algorithm has been presented in Sect. 7. Finally, we provide a conclusion in Sect. 8.

2 Fog Computing

Fog computing is a conception that was presented primitively by CISCO in 2012 [20]. It is the expansion of cloud computing that was transferred from core to edge of network. The mentioned reason causes computing to be done in the edge of network, where it is closer to end users and IOT/END devices. In fact, Fog computing is a powerful complement for cloud computing, but it is not a substitute of that. Although, Fog computing enables processing at the edge, but it still suggests possibility of interact with the cloud [21]. Therefore, Fog computing proposed for extension of cloud computing to produce services such as computation, storage and network services in the network edge [22]. In other word, Fog is "a cloud that is closer to ground" as it is really in nature [21]. Based on CISCO prediction, there would be about 50 billion IOT devices in 2020. This exponential increase is due to proliferation of smart sensors which serve various vertical markets (e.g. autonomous and driverless vehicles and transportation, industrial controls, smart cities and smart wearables, etc.), mobile instruments (e.g. tablets, mobile phones and GPS devices for disabled person), wireless sensors and actuators networks (WSAN) [23]. Therefore, new technologies and new concepts are needed for managing the growing fleet of IoT devices [23]. The increasing trend of IOT devices also leads us to direct and transfer a huge amount of data to cloud for processing. Cisco predicted annual IP traffic of data centers will meet 15.3 ZB (ZB=10^{21} Byte). This causes appearing (emerging) of many difficulties on cloud computing. Rising high delay and congestion of network, is consequence of using current cloud computing paradigm for handling of enormous amount of devices and data [23,24]. Considering the many types of IoT devices which need delay-sensitive and location aware applications, cannot fully conform with cloud computing. Based on these issues, fog computing was proposed by CISCO in 2012 [20]. Fog computing is a layered model. It enables ubiquitous access to many shared computing resources which facilitate deployment of distributed and latency-aware applications and services [23]. Fog computing plays an effective role in decreasing latency, energy consumption, traffic network, content distribution network, radio access network (RAN), vehicle network and so on [24]. Fog computing not only minimizes the request-response time from/to supported applications, but also provides local computing resources and network connectivity to centralized services (clouds)[23]. Fog computing consist of fog nodes which they are physical (servers, routers, gateways, set top boxes, proxy servers, bas stations) or virtual (cloudlets, virtual switches, virtual machines) and resides between cloud services and IOT/end devices. The fog nodes are context aware and can have a data management and communication system [23]. Fog nodes are distributed across the network. All of them do not occupy resources and also all of them are not so rich and powerful from resource aspect. In this case, development of large scales applications in one node is impossible. Their organization can be clusters (vertically or horizontally to support isolation and federation respectively) and peer to peer and master slave. Fog nodes are core of fog computing architecture. They are connected tightly coupled with smart end devices and network access and provide computational resources for them [23,24]. They have one HOP distance to end devices.

3 Notations and Definitions

In this research, we consider that m parallel processors which are embedded in the m fog nodes, denoted by N_1, N_2, ..., N_m are interconnected to the root fog node (F_0) equipped with a processor that are called root processors. We also assumed that n adjacent fog computing environment as micro data centers, are interconnected to the root fog node (F_0) which are denoted by F_1, F_2, ..., F_n and equipped with processors, and finally one link from cloud computing environment, is connected to the root fog node (F_0).

The subsequent parameters are applied throughout the paper.

- w_j: This item is the inverted computing speed (computation rate) of the jth processor which is located in the root fog.
- z: This item is the reverse transmission speed of the jth processor which is located in the root fog. It is assumed that the nodes of root fog are tightly coupled connections.
- w_c: This item is the inverted computing speed of the cloudlet.
- z_c: It is the reverse transmission speed of the 0th fog node to the cloudlet.
- w_{F_i}: It is the converse computing speed of the ith fog node.
- z_{F_i}: It is the reverse transmission speed of the ith fog node.
- T_{cp}: It is the time extended by a standard processor to compute a unit load.
- T_{cm}: It is the time taken to transmit a unit load on a standard link.
- α_j: It is the fraction of load assigned for the jth processor which is located in the root fog.
- β_i: It is the fraction of load assigned for the ith fog node.
- γ: It is the fraction of load assigned for the cloudlet.

4 Proposed Method

As we mentioned in the previous sections, divisible load scheduling method is a suitable way for using in fog computing environment. A three-layer architecture for the proposed method has been depicted in Fig. 1. As the figure shows, the proposed architecture consists of three different layers including IOT/end-device layer, fog layer, and cloud layer.

The first layer involves the data gathered by sensors and IOT devices. The second is a fog layer. The third layer called cloud layer which indicates that the fogs are connected to a cloud center. We propose a divisible load scheduling solution for managing the distribution of loads among the fog nodes, adjacent fogs, and cloud center. The proposed method consists of the following stages.

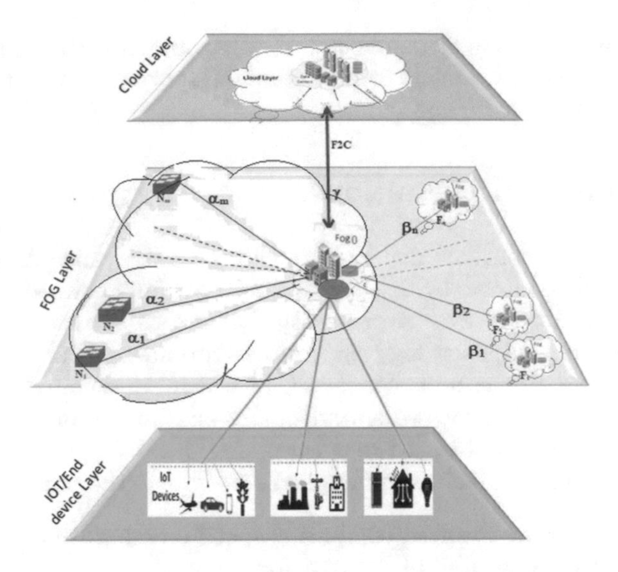

Fig. 1. A three-layer architecture for the proposed method

4.1 Initializing

In this stage the gathered data is sent to the nearest fog which is denoted by f_0. The f_0 consists of a cluster of fog nodes. The first node which receives the input data, is called root node (denoted by N_0).

4.2 Feasibility Checking

In this stage the two conditions including divisibility and intensiveness of load should be investigated. If the input data satisfy the two mentioned conditions then we perform the next stage. There are various methods for investigating the two mentioned conditions e.g, **CloneCloud**, **ThinkAir**) [26,27].

CloneCloud is an approach that automatically identifies the communication and computation costs for migrating the workload from local to edge and cloud [26]. ThinkAir is an approach that aims to parallel task execution by dynamically allocating cloud resources instead of statically analyzing partitions of applications [27]. There are a huge number of sensors and IOT devices in a network which create many data that may have same format and must be sent to processors for processing. This data which are produced by various sensors and end devices, essentially and inherently, have divisible property. This inherent property, pertains to structure and function of sensors. They are electronic instruments that according to their functionality, meters temperature, humidity, motion, traffic, movement, velocity, status, ... and according to their scheduling, or an event, transfer their recorded data to a data collector. Gateways and switches or a network of such devices, acts as a data collector. This data are independent records and will save in storage devices. In fact, some applications, due to multitude of sensors and end devices, and continual metering creates huge amount of (massive) data that can be called as Big data [25]. On the other hand, in some applications such as scheduling for smart cities, or connected vehicles [23], fire detection and fire fighting [24], smart grid [23] and health care systems have many data computing, and therefore they are data intensive. The above mentioned applications, have a heavy data computation and arbitrary divisible data, hence they are suitable for using DLS.

4.3 Divisible Load Modeling

In this stage the input data distribute using divisible load scheduling method. As Fig. 1 shows, the data can be classified in three different groups. The first group of data distribute on the fog nodes which are tightly connected to the root node. In fact α_0 fraction of load is distributed on F_0. Furthermore, α_0, α_1, ... ,α_m fractions of load are distributed on fog node N_0, N_1, ..., N_m respectively. The second group of data must be distributed on the other adjacent fogs. In this case β_1, β_2, ..., β_n are the fractions of load that are distributed on F_0, F_1, ..., F_m respectively. Finally, the third group of data must be sent to the cloud center. The fraction of assigned data to the cloud center is γ.

5 Close Form for the Proposed Divisible Load Model

A close form for the proposed method can be formed by the following equations.

$$(\alpha_0 T_{cp})w_0 = (\alpha_1 T_{cm})z_1 + (\alpha_1 T_{cp})w_1, \tag{5.1}$$

$$(\alpha_1 T_{cp})w_1 = (\alpha_2 T_{cm})z_2 + (\alpha_2 T_{cp})w_2,$$

$$(\alpha_2 T_{cp})w_2 = (\alpha_3 T_{cm})z_3 + (\alpha_3 T_{cp})w_3$$

$$\vdots$$

$$(\alpha_{m-1} T_{cp})w_{m-1} = (\alpha_m T_{cm})z_m + (\alpha_m T_{cp})w_m, \tag{5.2}$$

$$(\alpha_0 T_{cp})w_0 = (\gamma T_{cm})z_c + (\gamma T_{cp})w_c, \tag{5.3}$$

$$(\alpha_0 T_{cp})w_0 = (\beta_1 T_{cm})z_{F_1} + (\beta_1 T_{cp})w_{F_1}, \tag{5.4}$$

$$(\alpha_0 T_{cp})w_0 = (\beta_2 T_{cm})z_{F_2} + (\beta_2 T_{cp})w_{F_2}, \tag{5.5}$$

$$\vdots$$

$$(\alpha_0 T_{cp})w_0 = (\beta_n T_{cm})z_{F_n} + (\beta_n T_{cp})w_{F_n}, \tag{5.6}$$

$$\alpha_0 + \alpha_1 + \alpha_2 + ... + \alpha_m + \gamma + \beta_1 + \beta_2 + ... + \beta_n = 1 \tag{5.7}$$

$$z_1 = z_2 = z_3 = \cdots = z_m = z \tag{5.8}$$

6 Solving the Proposed Close Form

From Eq. (5.1), we have

$$\alpha_o = \frac{w_1 T_{cp} + z T_{cm}}{w_0 T_{cp}} \alpha_1 \tag{6.1}$$

From Eq.(5.1–5.2), we have

$$\alpha_i = \frac{w_{i-1}T_{cp} + zT_{cm}}{w_i T_{cp}} \alpha_{i-1}, i = 2, ..., m \tag{6.2}$$

Let $\kappa_i = \frac{w_{i-1}T_{cp}+zT_{cm}}{w_i T_{cp}}$, then from Eq. (6.2) we have

$$\alpha_i = \kappa_i \alpha_{i-1} \tag{6.3}$$

and

$$\alpha_i = (\Pi_{l=2}^{i} \kappa_l)\alpha_1 \tag{6.4}$$

From Eq.(5.3), we have

$$\gamma = \frac{w_0 T_{cp}}{T_{cm} z_c + T_{cp} w_c} \alpha_0 \tag{6.5}$$

and by 6.1 we have

$$\gamma = \frac{w_0 T_{cp}}{T_{cm} z_c + T_{cp} w_c} \alpha_0 = \frac{w_1 T_{cp} + zT_{cm}}{T_{cm} z_c + T_{cp} w_c} \alpha_1 \tag{6.6}$$

From Eq. (5.6), we have

$$\beta_n = \frac{w_0 T_{cp}}{T_{cm} z_{F_n} + T_{cp} w_{F_n}} \alpha_0 \tag{6.7}$$

Combining Eq.(6.1) and Eq.(6.7) we have

$$\beta_n = \frac{w_1 T_{cp} + zT_{cm}}{T_{cm} z_{F_n} + T_{cp} w_{F_n}} \alpha_1 \tag{6.8}$$

Based on Eq. (5.7), it is clear that:

$$\alpha_0 + \alpha_1 + \sum_{i=2}^{m} \alpha_i + \gamma + \sum_{j=1}^{n} \beta_j = 1 \tag{6.9}$$

By using Eq.(6.4) and Eq.(6.1) we have

$$\left(\frac{w_1 T_{cp} + zT_{cm}}{w_0 T_{cp}} + 1 + \sum_{i=2}^{m} \frac{w_1 T_{cp} + zT_{cm}}{w_i T_{cp} + zT_{cm}} + \frac{w_1 T_{cp} + zT_{cm}}{T_{cm} z_c + T_{cp} w_c} + \sum_{j=1}^{n} \frac{w_1 T_{cp} + zT_{cm}}{T_{cm} z_{F_j} + T_{cp} w_{F_j}} \right) \alpha_1 = 1 \tag{6.10}$$

Consequently, we obtain:

$$\alpha_1 = \frac{1}{\sum_{l=0}^{m} \frac{w_1 T_{cp}+zT_{cm}}{w_l T_{cp}+z_l T_{cm}} + \frac{w_1 T_{cp}+zT_{cm}}{T_{cm} z_c+T_{cp} w_c} + \sum_{j=1}^{n} \frac{w_1 T_{cp}+zT_{cm}}{T_{cm} z_{F_j}+T_{cp} w_{F_j}}} \tag{6.11}$$

From Eq.(6.1)

$$\alpha_o = \frac{w_1 T_{cp} + zT_{cm}}{w_0 T_{cp}} \cdot \frac{1}{\sum_{l=0}^{m} \frac{w_1 T_{cp}+zT_{cm}}{w_l T_{cp}+z_l T_{cm}} + \frac{w_1 T_{cp}+zT_{cm}}{T_{cm} z_c+T_{cp} w_c} + \sum_{j=1}^{n} \frac{w_1 T_{cp}+zT_{cm}}{T_{cm} z_{F_j}+T_{cp} w_{F_j}}} \tag{6.12}$$

Also, we obtain

$$\alpha_o = \frac{1}{\sum_{l=0}^{m} \frac{w_0 T_{cp}}{w_l T_{cp}+zT_{cm}} + \frac{w_0 T_{cp}}{T_{cm} z_c+T_{cp} w_c} + \sum_{j=1}^{n} \frac{w_0 T_{cp}}{(w_1 T_{cp}+zT_{cm})(T_{cm} z_{F_j}+T_{cp} w_{F_j})}} \tag{6.13}$$

Based on Eq.(6.4) and Eq.(6.11), the optimal fractions of load α_i can be calculated as the following equations.

$$\alpha_i = \frac{1}{\sum_{l=0}^{m} \frac{w_i T_{cp}+zT_{cm}}{w_l T_{cp}+z_l T_{cm}} + \frac{w_i T_{cp}+zT_{cm}}{T_{cm} z_c+T_{cp} w_c} + \sum_{j=1}^{n} \frac{w_i T_{cp}+zT_{cm}}{T_{cm} z_{F_j}+T_{cp} w_{F_j}}} \tag{6.14}$$

and the total finish time is calculated by the following equation.

$$T_{f,m} = \alpha_0 T_{cp} w_0 \tag{6.15}$$

$$= \frac{1}{\sum_{l=0}^{m} \frac{w_0 T_{cp}}{w_l T_{cp}+zT_{cm}} + \frac{w_0 T_{cp}}{T_{cm} z_c+T_{cp} w_c} + \sum_{j=1}^{n} \frac{w_0 T_{cp}}{(w_1 T_{cp}+zT_{cm})(T_{cm} z_{F_j}+T_{cp} w_{F_j})}} T_{cp} w_0 \tag{6.16}$$

7 Load Distribution

By using Eq. (6.14) we can calculate α_0,α_1, ..., α_m. Similarly, β_1,β_1, ...,β_n can be calculated using Eq. (6.8). In the same way γ can be calculated using Eq. (6.6). After that the load is distributed on current fog nodes, and adjacent fogs and cloud. The load can be allocated using Algorithm 1.

Algorithm 1 Allocating fraction of load

Input: N_1, N_2, ..., N_m are fog nodes with processor F_1, F_2,..., F_n are adjacent fogs (micro data centers) and T_{cp},T_{cm} are time of processing and communicating of processor nodes and W_i,Z_i are working and linking parameters respectively.
Output: α_i for internal fog nodes fractions, β_j for adjacent fogs fractions and, γ for cloud fraction.

1: $i \leftarrow 1$
2: **while** $(i \leq m)$ **do**
3: $\kappa_i = \frac{w_{i-1}T_{cp}+z_{i-1}T_{cm}}{w_iT_{cp}+z_iT_{cm}}$
4: **end while**
5: $\alpha_o \leftarrow \frac{1}{\sum_{l=0}^{m} \frac{w_0T_{cp}}{w_lT_{cp}+z_lT_{cm}} + \frac{w_0T_{cp}}{T_{cm}z_c+T_{cp}w_c} + \sum_{j=1}^{n} \frac{w_0T_{cp}}{(w_1T_{cp}+zT_{cm})(T_{cm}z_{F_j}+T_{cp}w_{F_j})}}$
6: $i \leftarrow 1$
7: **while** $(i \leq m)$ **do**
8: $\alpha_i \leftarrow \kappa_i\alpha_{i-1}$
9: **end while**
10: $j \leftarrow 1$
11: **while** $(j \leq n)$ **do**
12: $\beta_j \leftarrow \frac{w_0T_{cp}}{T_{cm}z_{F_n}+T_{cp}w_{F_n}}\alpha_0$
13: **end while**
14: $\gamma \leftarrow \frac{w_0T_{cp}}{T_{cm}z_c+T_{cp}w_c}\alpha_0$

8 Conclusion

This paper presented a divisible load method for scheduling the resources in fog computing environments. We proposed a divisible load framework for the scheduling of fog models. We also formulated a close form for the proposed method. Finally the formulated close form of divisible load method has been solved in order in order to find an optimal fraction of load for allocating data to fog nodes.

References

1. Bonomi F, Milito R, Zhu J, Addepalli S (2012) Fog computing and its role in the internet of things. In: Proceedings of the first edition of the MCC workshop on Mobile cloud computing, pp 13–16. ACM
2. Shamsollah G, Mohamed O (2014) Comprehensive review on divisible load theory: concepts, strategies, and approaches. Hindawi Publishing Corporation, Math Problems Eng. https://doi.org/10.1155/2014/460354
3. Cheng Y-C, Robertazzi Thomas G (1988) Distributed computation with communication delay. IEEE Trans Aerosp Electr Syst 24(6):700–712
4. Agrawal R, Jagadish HV (1988) Partitioning techniques for large grained parallelism. IEEE Trans Comput 37(12):1627–1634
5. Sohnn J, Robertazzi TG (1993) Optimal load sharing for a divisible job on a bus network. In: Proceedings of the 1993 conference on information sciences and systems

6. Bharadwaj V, Li X, Ko CC (2000) Efficient partitioning and scheduling of computer vision and image processing data on bus networks using divisible load analysis. Image Vis Comput 18(11):919–938
7. Lee C-K, Hamdi M (1995) Parallel image processing applications on a network of workstations. Parallel Comput 21(1):137–160
8. Korkhov VV, Moscicki JT, Krzhizhanovskaya VV (2009) The user-level scheduling of divisible load parallel applications with resource selection and adaptive workload balancing on the grid. Syst J IEEE 3(1):121–130
9. Chan SK, Veeravalli B, Ghose D (2001) Large matrix vector products on distributed bus networks with communication delays using the divisible load paradigm: performance analysis and simulation. Math Comput Simul 58(1):71–92
10. Krijn R Van Der, Yang Y, Henri C (2005) Practical divisible load scheduling on grid platforms with APST-DV. In: The 19th IEEE international parallel and distributed processing symposium
11. Yu D, Robertazzi TG (2003) Divisible load scheduling for grid computing. In: Fifteenth IASTED international conference on parallel and distributed computing and systems l(1):1–6
12. Suresh S, Huang H, Kim HJ (2015) Scheduling in compute cloud with multiple data banks using divisible load paradigm. IEEE Trans Aerosp Electr Syst 51(2):1288–1297
13. Berliska J, Maciej D (2015) Scheduling multilayer divisible computations. RAIRO-Oper Res 49(2):339–368
14. Marszakowski Jdrzej M, Drozdowski M, Marszakowski J (2016) Time and energy performance of parallel systems with hierarchical memory. J Grid Comput 14(1):153–170
15. Ghanbari S, Othman M, Leong WJ, Abu Bakar MR (2014) Multi-criteria based algorithm for scheduling divisible load. Lect Notes Electr Eng 285:547–554
16. Shamsollah G, Mohamed O, Leong WJ, Abu Bakar MR (2016) Multi-objective method for divisible load scheduling in multi-level tree network. Future Gener Comput Syst 54:132–143
17. Shamsollah G, Mohamed O, Leong WJ, Abu Bakar MR (2015) Priority-based divisible load scheduling using analytical hierarchy process. Appl Math Inf Sci 9(5):2541–2552
18. Bharadwaj V, Ghose D, Mani V, Robertazzi TG (1996) Sheduling divisible loads in parallel and distributed systems, vol 8. Wiley
19. Robertazzi Thomas G (2003) Ten reasons to use divisible load theory. Computer 36(5):63–68
20. Networking, Cisco Visual. Cisco global cloud index: Forecast and methodology 2015–2020. White paper (2016)
21. Carla M, Diala N, Sami Y, Glitho Roch H, Morrow Monique J, Polakos Paul A (2017) A comprehensive survey on fog computing: state-of-the-art and research challenges. IEEE Commun Surv Tutor 20(1):416–464
22. Luxiu Y, Juan L, Haibo L (2018) Tasks scheduling and resource allocation in fog computing based on containers for smart manufacturing. IEEE Trans Ind Inf 14(10):4712–4721
23. Iorga M, Feldman L, Barton R, Martin MJ, Goren NS, Mahmoudi C (2018) Fog computing conceptual model. No. Special Publication (NIST SP)-500-325
24. Mahmud R, Kotagiri R, Buyya R (2018) Fog computing: a taxonomy, survey and future directions. Internet of everything. Springer, Singapore, pp 103–130

25. Nazmudeen N, Saleem Haja M, Wan AT, Buhari SM (2016) Improved throughput for power line communication (plc) for smart meters using fog computing based data aggregation approach. In: 2016 IEEE international smart cities conference (ISC2), pp 1–4. IEEE, 2016
26. Chun BG, Ihm S, Maniatis P, Naik M, Patti A (2011) Clonecloud: elastic execution between mobile device and cloud. In: Proceedings of the sixth conference on Computer systems (pp. 301–314). ACM
27. Kosta S, Aucinas A, Hui P, Mortier R, Zhang X (2012) ThinkAir: dynamic resource allocation and parallel execution in the cloud for mobile code offloading. In: 2012 proceedings IEEE Infocom (pp 945–953). IEEE

An Enhanced Evolutionary Algorithm for Detecting Complexes in Protein Interaction Networks with Heuristic Biological Operator

Dhuha Abdulhadi Abduljabbar[1,2](✉), Siti Zaiton Mohd Hashim[1], and Roselina Sallehuddin[1]

[1] School of Computing, Faculty of Engineering, Universiti Teknologi Malaysia, 81310 Johor Bahru, Johor, Malaysia
dhuhaftsm@gmail.com, {sitizaiton, roselina}@utm.my
[2] Department of Computer Science, Baghdad University, Baghdad, Iraq

Abstract. Detecting complexes in protein interaction networks is one of the most important topics of current computational biology research due to its prominent role in predicting functions of yet uncharacterized proteins and in diseases diagnosis. Evolutionary Algorithms (EAs) have been adopted recently to identify significant protein complexes. Conductance, expansion, normalized cut, modularity, and internal density are some well-known examples of complex detection models. In spite of the improvements and the robustness of predictive functions introduced by complex detection models based on EA and regardless of the general topological properties of protein interaction networks, inherent biological data of protein complexes has not, or rarely exploited and incorporated inside the methods as a specific heuristic operator. The aim of this operator is to guide the search process towards discovering hyper-connected and biologically related complexes by allowing a more effective exploration of the state space of possible solutions. Thus, the main contribution of this study is to develop a heuristic biological operator based on Gene Ontology (GO) annotations where it can serve as a local-common optimization approach. In the experiments, the performance of eight EA-based complex detection models has analyzed when applied on the yeast protein networks that are publicly available. The results give a clear argument for the positive effect of the proposed heuristic biological operator to considerably enhance the reliability of the current state-of-the-art optimization models.

Keywords: Evolutionary algorithm · Complex detection models · Heuristic biological operator · Gene ontology · Protein complexes

1 Introduction

Every living cell consists of proteins that continuously interact with each other to execute and regulate most biological functions. In the last few years, with the advent of bioinformatics and associated fields of study, several empirical data on protein interactions became available through the development of advanced high-throughput technologies [1, 2]. The interactions between proteins can form protein-protein

R. Ghazali et al. (Eds.): SCDM 2020, AISC 978, pp. 334–345, 2020.
https://doi.org/10.1007/978-3-030-36056-6_32

interaction (PPI) networks, where nodes correspond to proteins and edges correspond to physical or functional interactions between them.

Owing to the fact that proteins distribution is not uniform in the PPI networks, some proteins have formed well-connected sub-networks. Brohée and van Helden [3] furthermore disclosed that protein complexes and functional modules form hyper-connected components in the PPI networks. The detection of such natural complexes in the PPI networks is an important and ongoing challenge because many biological network problems like aligning, clustering, and the search for relationships are equivalent to the sub-graph isomorphism problem which is known to be Non-deterministic Polynomial-time hard (NP-hard) that recently acquired great attention [4, 5]. Protein complexes detection can aid in the description of the evolutionary orthology signal [6], prediction of the biological functions of yet uncharacterized proteins, cancer detection, as well as, in therapeutic purposes [7].

Many scholars have proposed a lot of methods to identify protein complexes. Generally, complex detection techniques in PPI networks are categorized into two classes: the topology-free clustering methods as proposed in [8, 9] and the graph-based clustering methods that have been proposed by several researchers [2, 10–13]. In the literature, the evolutionary algorithms (EAs)-based methods have been regarded as competitive, and powerful computational techniques to address complex detection problem in PPI networks. Pizzuti and Rombo [2, 12] addressed the issue by optimizing single objective models, where the authors proposed various quality functions (as fitness functions) to extract the hidden protein complexes in PPI networks. These include Modularity (Q), Conductance (CO), Expansion (EX), Cut Ratio (CR), Normalized Cut (NC), Internal Density (ID), Community Score (CS), and Scaled Cost Function (SCF). The experimental results revealed that genetic algorithms have the competitive ability, and practicable computational procedure to solve the issue more accurately than other traditional methods. Bandyopadhyay et al. [13] proposed a multi-objective formulation which aims to optimize three criteria for identifying protein complexes of human PPI networks, besides discovering their disease associations. Accordingly, the predicted complexes showed consistently better results in the context of some performance metrics. Ramadan et al. [14] constructed a new clustering algorithm for predicting protein complexes through the use of genetic algorithms. They used three objective functions min-max-cut, ratio cut, normalized cut [15], and their proposed density cut function which is outperformed other competing approaches and proved it's capable of efficiently detecting densely and sparsely protein complexes. Recently, Bara'a and Abdullah [16] introduced a new multiobjective optimization model to formulate complex detection problem. The study also introduced protein complex attraction and repulsion, a heuristic operator, for the improvement of the quality of a given solution. The results showed that the introduction of the proposed heuristic operator had a positive influence on improving the quality of complexes detected by the current EA models.

In spite of the improvements and the robustness of predictive functions introduced by complex detection models based on EA and regardless of the general topological properties of PPI networks, inherent biological data of protein complexes has not, or rarely exploited and incorporated inside the methods as a specific heuristic operator. The purpose of this operator is to guide the search process towards discovering hyper-connected and biologically related complexes by avoiding premature convergence of the evolutionary algorithm and enhancing its prediction power. Where, biological data that are available, including genes and topology, could be used and incorporated inside the method as specific variation operators instead of random operators, letting a more effective exploration of the state space of possible solutions. To this end, the contribution of this study is to develop heuristic biological operator based on Gene Ontology (GO) annotations which can serve as a local-common optimization approach helps in reinforcing prediction ability of the complex detection models based on EA.

2 Methods

Modelling of a protein interaction network $\mathcal{N}$ is done as an undirected graph G, in which the pair (V, E) signifies the whole pairwise connections between various proteins in the network . In $\mathcal{N}$, the set of v proteins is regarded as a set of nodes of graph, i.e. $G : V(G) = \{v_1, v_2, \ldots v_n\}$ while representation of the mutual connection between any pair of proteins in $\mathcal{N}$ is corresponding to the edges (v_i, v_j) in E, viz., $E(G) \subseteq V(G) \times V(G) = \{(v_i, v_j) | 1 \leq i, j \leq n \wedge i \neq j\}$. The problem of detecting protein complex has been modelled as an optimization issue, wherein developing various optimization functions have done by employing different kinds of quality indices. Here, we put forward utilizing eight known quality indices as fitness functions. Furthermore, the new heuristic biological operator has been developed with an aim to enhance the identification of complexes in PPI networks.

2.1 Fitness Functions

Let us assume a graph $G = (V, E)$ that has been employed to model a PPI network $\mathcal{N}$ with cardinality $n = |V|$ and volume $m = |E|$. Suppose $\{C_1, \ldots, C_K\}$ be a partition of $\mathcal{N}$ in k clusters, let us assume a cluster C_i of $\mathcal{N}$ that possesses $n(C_i)$ nodes and $m(C_i)$ edges. For any of the node $v \in C_i$, the number of connections incident to node v as a degree of v can be defined, which can be formally noted as $m(v)$, and can be divided into two concepts: inter-edges pertaining to node v expressed as $\bar{m}(v, C_i) = |\{(v, w) \in E \wedge w \notin C_i\}|$ and intra-edges pertaining to node v expressed as $m(v, C_i) = |\{(v, w) \in E \wedge w \in C_i\}|$ [17, 18]. The following metrics, derived from [2, 12, 19] and designed to accommodate the concept of the clustering quality, can be defined as:

Community Score (CS): Measures the edge density of each cluster with respect to its size.

$$max\, CS(C) = \sum_{i=1}^{k} \left[\frac{2m(C_i)}{n(C_i)} \right]^2 \tag{1}$$

Conductance (*CO*): Measures the fraction of edges that point outwards of the clustering.

$$min\, CO(C) = \sum_{i=1}^{k} \frac{\sum_{v \in C_i} \bar{m}(v, C_i)}{2m(C_i) + \sum_{v \in C_i} \bar{m}(v, C_i)} \tag{2}$$

Cut Ratio (*CR*): Measures the fraction of all possible edges leaving the clustering.

$$min\, CR\ (\mathcal{C}) = \sum_{i=1}^{K} \frac{\sum_{v \in C_i} \bar{m}(v, C_i)}{n(C_i)(n - n(C_i))} \tag{3}$$

Expansion (*EX*): Measures the number of edges per nodes which pointing outwards of the clustering.

$$min\, EX(\mathcal{C}) = \sum_{i=1}^{K} \frac{\sum_{v \in C_i} \bar{m}(v, C_i)}{n(C_i)} \tag{4}$$

Internal Density (*ID*): Measures the intra-edge density of a clustering.

$$min\, ID(\mathcal{C}) = \sum_{i=1}^{K} 1 - \frac{m(C_i)}{n(C_i)(n(C_i) - 1)/2} \tag{5}$$

Normalized Cut (*NC*): Measures the fraction of total edge connections to all the nodes in the graph.

$$min\, NC(C) = \sum_{i=1}^{k} \frac{\sum_{v \in C_i} \bar{m}(v, C_i)}{2m(C_i) + \sum_{v \in C_i} \bar{m}(v, C_i)} + \frac{\sum_{v \in C_i} \bar{m}(v, C_i)}{2(m - m(C_i)) + \sum_{v \in C_i} \bar{m}(v, C_i)} \tag{6}$$

Modularity (*Q*): Measures the expected number of edges between the nodes of a cluster C_i in a random graph with the same degree sequence.

$$max\, Q(\mathcal{C}) = \sum_{i=1}^{K} \left[\frac{2m(C_i)}{m} - \left(\frac{\sum_{v \in C_i} m(v)}{2m}\right)^2 \right] \tag{7}$$

Scaled Cost Function (*SCF*): Measures the total number of bad connection incidents along with node v, i.e. connections between v and a node that do not fall under the same cluster of v or connections not existing between v and any other node pertaining to the same cluster of v, in terms of the area size impacted by v in the clustering.

$$min\,SCF = \frac{n-1}{3}\sum_{v\in V}\frac{(\overline{m}(v, C_i) + \overline{l}_{c_i}(v))}{|nn(v) \cup \{u \in C_i\}|} \tag{8}$$

Here, $\overline{l}_{c_i}(v) = |\{u \in C_i|(v,u) \notin E\}|$ denotes the number of nodes present in C_i and are not linked to v, while $nn(v)$ can be defined as the set of neighbour nodes of v.

2.2 Method Representation and Evolutionary Operators

While applying the evolutionary algorithm to detect protein complex in PPI networks, selecting the representation of the encoding scheme is an important and influential step on the algorithm performance. Park and Song [20] put forward the locus-based adjacency representation, which was adopted in this study. The decoding step is crucial for each individual I in the population to identify the structure of the complexes pertaining to a network. Regarding the evolutionary operators, the type of adopted crossover operator in this study can be defined as a uniform crossover, which is applied with a respective probability p_c. As for the mutation operator, we employed a neighbour mutation operator with p_m probability [12].

Unlike the traditional neighbour mutation operator, where there is no exploited to any biological knowledge about the problem including genes and topology, we propose a heuristic biological operator to move nodes from their complexes to more appropriate complexes in an effort to provide more reliable solutions comprising of hyper-connected and biologically related complexes. In the next section, the proposed heuristic biological operator is discussed in details.

2.3 Heuristic Biological Operator

The main contribution of our work involves incorporating heuristic biological operator inside the EA framework, to enhance the obtained solution, instead of random mutation operator. Based on Gene Ontology Consistency (GOC) which is a Jaccard index on sets of GO annotations [5], defined formally as:

$$GOC(I_i, I_j) = \frac{|GO(I_i) \cap GO(I_j)|}{|GO(I_i) \cup GO(I_j)|} \tag{9}$$

where I_i and I_j are two proteins (or nodes) in the network, and $GO(i)$ is the set of GO annotations of a protein (or node) i. Then, we can define the GO score (GOS) of node $I_i \in C$ as:

$$GOS(I_i, C) = \sum_{I_j\in C;\, I_j\in N_{I_i}} GOC(I_i, I_j) \tag{10}$$

where N_{I_i} represents the set of neighbor nodes of node I_i. Under the probability of the heuristic biological operator p_{GO}, and for an individual I, a node I_i from cluster C will change its cluster belongingness to another new cluster $\acute{C}$, where GOS of node I_i with $\acute{C}$ is equal or larger than its GOS with C. In other words, if and only if GOS of node I_i, that belongs to cluster C, is smaller than its GOS when comprised in another cluster (i.e. $\acute{C} \cup I_i$), then, there will be a movement of node I_i by the heuristic biological operator to another cluster in order to achieve highest GOS value with the new cluster's nodes. On the other hand, if GOS of node I_i, that belongs to cluster C, is equal to its GOS when comprised in another cluster (i.e. $\acute{C} \cup I_i$), then, the heuristic operator will either move node I_i to another cluster that reports with its nodes an equal GOS value, or leave the node I_i inside its cluster. Algorithm 1 highlights the major steps concerning to this heuristic biological operator.

Algorithm 1: Heuristic Biological Operator

Input : I: individual, n: number of nodes in the network, p_{GO}: probability of the heuristic biological operator

Output: I': individual after applying heuristic biological operator

```
1   set C = {C_1, C_2, ..., C_k} ← δ(I) // decode I into its complex structure
2   for i ← 1 to n do
3       if( rand ≤ p_GO)
4           set C_i ← Cluster_ID(I_i)
5           set K ← max(Cluster_ID(I))
6           set k_I_i_GO ← GOS(I_i, C_i)
7           set k_I_i_GO'_{Ć_j∈C∧i≠j} ← GOS(I_i, (Ć_j ∪ I_i))
8           if ( k_I_i_GO < k_I_i_GO' )
9               set C' ← argmax_{C'_j∈C, C'_j≠C_i}(GOS(I_i, (Ć_j ∪ I_i)))
10              set Cluster_ID(I_i) ← C'
11          else if ( k_I_i_GO = k_I_i_GO' )
12              if (rand ≤ p_GO )
13                  set C' ← argmax_{C'_j∈C, C'_j≠C_i}(GOS(I_i, (Ć_j ∪ I_i)))
14                  set Cluster_ID(I_i) ← C'
15              end if
16          end if
17      end if
18  end for
19  Return (I')
```

3 Experimental Setting

3.1 Dataset

To analyze the performance of the detection models, this study will use two yeast PPI networks (PPI_D1, and PPI_D2), where their generic complexes are known and publicly available at MIPS databases [21]. The PPI_D1 network has initially been generated by a study of Gavin et al. [22] and later filtered by a study of Zaki et al. [23]. The second network was also filtered by Zaki et al. [23] where they presented a filtered version of PPI_D2 network, which is a combined protein interaction dataset containing yeast protein interactions generated by six experiments. Table 1 reports the graph characteristics of the tested PPI networks.

Table 1. Graph characteristics of the tested PPI networks ($\mathcal{N}$). S^* is the generic known complexes, and, K_{S^*} and $n(S^*)$ are, respectively, the number and the scope size of S^*

$\mathcal{N}$	n	m	S^*	K_{S^*}	$n(S^*)$
PPI_D1	990	4687	*Cmplx_PPI_D1*	81	6-to-38
PPI_D2	1443	6993	*Cmplx_PPI_D2*	162	4-to-266

3.2 Evaluation Measures

Regarding validity indices, we used the most prevalent measures, which are a recall, precision, F-measure [3]. The predicted cluster $C_i \in \{C_1, C_2, \ldots, C_{K_C}\}$ is considered to match the known complex $S_j \in \{S_1, S_2, \ldots, S_{K_{S^*}}\}$ if their overlapping score (OS) defined in Eq. 11 is equal to or larger than a specific threshold σ_{OS} [3].

$$OS(C_i, S_j) = \frac{|C_i \cap S_j|^2}{|C_i||S_j|} \tag{11}$$

$$match(C_i, S_j) = \begin{cases} 1 \; if \; OS(C_i, S_j) \geq \sigma_{OS} \\ 0 \quad\quad\quad\quad otherwise \end{cases} \tag{12}$$

The notions of *recall*, *precision*, and cumulative *F*-measure are defined based on the matching represented in Eq. 12, as following:

$$recall = \frac{|S_i | S_i \in S^* \wedge \exists C_j \in C \rightarrow match(S_i, C_j)|}{K_S} \tag{13}$$

$$precision = \frac{|C_i | C_i \in C \wedge \exists S_j \in S^* \rightarrow match(C_i, S_j)|}{K_C} \tag{14}$$

$$F\_measure = \frac{2 * recall * precision}{recall + precision} \tag{15}$$

4 Experimental Results and Discussions

The functioning of eight state-of-the-art evolutionary-based complex detection models (which are $CS, CO, CR, EX, ID, NC, Q$ *and* SCF) is studied and tested. For fair comparisons, the distinctive components of the EA technique are adjusted to their commonly used settings present in the literature. Population size $pop = 100$, number of generations $t = 100$, probability of traditional neighbour mutation $p_m = 0.2$, and probability of crossover $p_c = 0.8$. Moreover, the outcomes discussed the effect of the proposed heuristic biological operator on the EA models' ultimate performance with the probability of $p_{GO} = 0.5$, where each of the EA models will be tested with two versions. First, when the function of EA-based complex detection models applied with traditional neighbour mutation (noted hereinafter as p_m version); second, when the function of EA-based complex detection models applied with the proposed heuristic biological operator (noted hereinafter as p_{GO} version).

The outcomes of the EA models testing on PPI networks (*PPI_D*1 *and PPI_D*2) are presented in terms of the average of best values found in the archive, over the 10 independent runs in both two versions given above, while varying the threshold of overlapping score *OS* from 0.1 to 0.5 in a step size equal to 0.1. Since maximizing both metrics (recall and precision) is often difficult, therefore, the *F*-measure reflects the tradeoff between the two and measures the model performance by taking into account both sensitivity (recall) and specificity (precision). Figures 1 and 2 present the *F*-measure values of the EA-based complex detection models for the two PPI networks with different overlapping scores of each version. As shown in Figs. 1 and 2, the application of the heuristic biological operator helped to improve the prediction power of all EA models, except few cases with CS model. High values of F-measure means that both recall and precision are sufficiently high. As for CS model when the overlapping score $OS = 0.2$ with PPI_D1 or $OS \leq 0.2$ with PPI_D2, the model performance at p_m version is slightly higher than its performance at p_{GO} version.

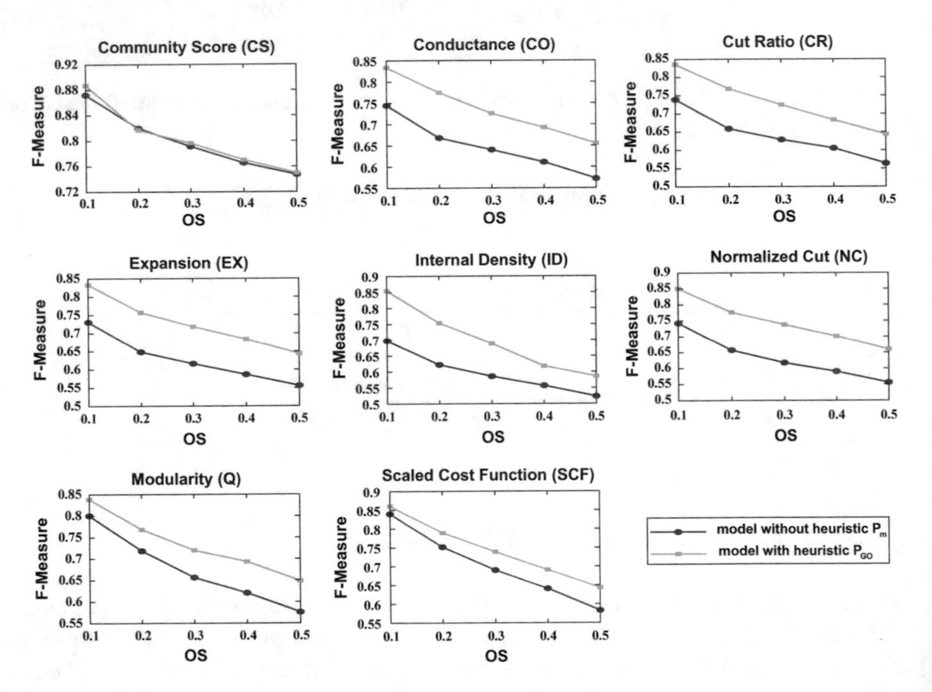

Fig. 1. Quality of complex detection in *PPI_D*1 network in terms of *F_measure* vs. overlapping score (*OS*) for all EA models in both versions (p_m *and* p_{GO}).

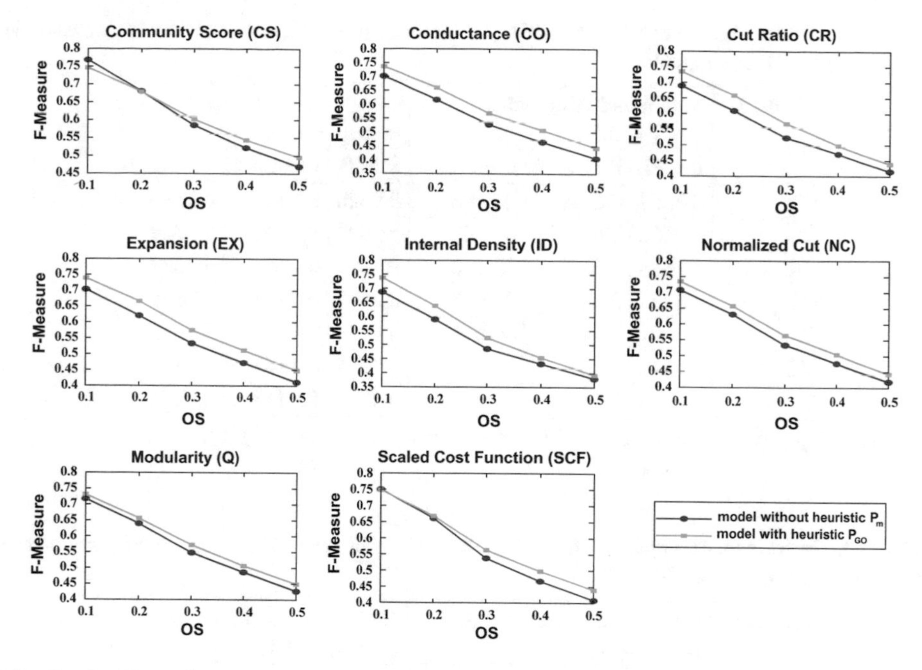

Fig. 2. Quality of complex detection in *PPI_D2* network in terms of *F_measure* vs. overlapping score (*OS*) for all EA models in both versions ($p_m \, and \, p_{GO}$).

Since CS model with the heuristic operator resulted of partition the network into a large number of small clusters, compared with partition the network at p_m version. Generally, the results revealed and proved that the proposed heuristic biological operator is beneficial to the state-of-the-art complex detection models in terms of F-measure in all overlapping scores.

Next, the research results at the p_{GO} version were compared with the latest published result, for a study conducted by Bara'a and Abdullah [16] at the overlapping score (OS) equal to 0.2 for *PPI_D1* and *PPI_D2* networks. As shown in Tables (2 and 3) our research results proved its superiority at all EA models in terms of *precision* and *F_measure*, and in most models in terms of *recall* measure. In Table 2, only the models EX, ID, and Q obtained low recall values compared with the results of [16], this means that these models along with p_h predicted a higher number of complexes out of all the standard complexes. But with p_{GO}, the models predicted complexes that composed of a high percentage of proteins belonging to the standard complex, allowed them to get higher precision values. The best outcome of each model in the tables is shown in bold.

Table 2. Performance comparison for *PPI _D*1 (in terms of *recall*, *precision*, *and F_measure* at $OS = 0.2$) for all EA models.

EA model	Bara'a and Abdullah [16] (at $p_h = 0.5$)			Our proposed solution (at $p_{GO} = 0.5$)		
	Recall	Precision	F-measure	Recall	Precision	F-measure
CS	0.8564	0.6196	0.7188	**0.9026**	**0.746**	**0.8168**
CO	0.7462	0.6793	0.7108	**0.7923**	**0.7583**	**0.7744**
CR	0.7538	0.6742	0.7110	**0.7821**	**0.7567**	**0.7687**
EX	**0.8474**	0.6073	0.7073	0.7577	**0.7567**	**0.7569**
ID	**0.8282**	0.6139	0.7048	0.8256	**0.694**	**0.7533**
NC	0.7513	0.6839	0.7157	**0.7885**	**0.7643**	**0.776**
Q	**0.8103**	0.6453	0.7181	0.791	**0.7473**	**0.7682**

* p_h: represents the probability of the heuristic operator of [16].

Table 3. Performance comparison for *PPI _D*2 (in terms of *recall*, *precision*, *and F_measure* at $OS = 0.2$) for all EA models.

EA model	Bara'a and Abdullah [16] (at $p_h = 0.5$)			Our proposed solution (at $p_{GO} = 0.5$)		
	Recall	Precision	F-measure	Recall	Precision	F-measure
CS	0.7687	0.4339	0.5543	**0.882**	**0.5524**	**0.6791**
CO	0.6773	0.4534	0.5428	**0.8147**	**0.5549**	**0.6599**
CR	0.6893	0.4502	0.5446	**0.81**	**0.5553**	**0.6587**
EX	0.7513	0.4201	0.5389	**0.8167**	**0.5625**	**0.6661**
ID	0.7227	0.4228	0.5333	**0.8067**	**0.5282**	**0.6379**
NC	0.6740	0.4388	0.5313	**0.8033**	**0.5561**	**0.6572**
Q	0.7013	0.4414	0.5417	**0.8093**	**0.5524**	**0.6565**

* p_h: represents the probability of the heuristic operator of [16].

5 Conclusion

The main contribution of this study is to exploit the biological data, including genes and topology, of the protein complexes in designing new-local heuristic biological operator. Experiments presented a comparison of eight state-of-the-art EA models in two different versions (with and without heuristic operator), the results revealed the benefits of the proposed operator to all *EA_* based complex detection models. Moreover, the research result, at the heuristic version, proved its superiority over the latest competitive study in terms of *recall*, *precision*, and *F_measure*. As future work, it is interesting to consider extending the proposed method for identifying overlapping communities in PPI networks, since the communities in most biological networks are overlapped.

References

1. Krogan NJ et al (2006) Global landscape of protein complexes in the yeast Saccharomyces cerevisiae. Nature 440(7084):637
2. Pizzuti C, Rombo SE (2014) Algorithms and tools for protein–protein interaction networks clustering, with a special focus on population-based stochastic methods. Bioinformatics 30 (10):1343–1352
3. Brohée S, van Helden J (2006) Evaluation of clustering algorithms for protein-protein interaction networks. BMC Bioinform 7(1):488
4. Bui TN, Jones C (1992) Finding good approximate vertex and edge partitions is NP-hard. Inf Process Lett 42(3):153–159
5. Elmsallati A, Clark C, Kalita J (2016) Global alignment of protein-protein interaction networks: A survey. IEEE/ACM Trans Comput Biol Bioinform 13(4):689–705
6. Jancura P, Mavridou E, Carrillo-de Santa Pau E, Marchiori E (2012, June) A methodology for detecting the orthology signal in a PPI network at a functional complex level. In: BMC bioinformatics, vol 13, No 10, p S18. (BioMed Central)
7. Pizzuti C (2014) Computational intelligence for community detection in complex networks and bio-medical applications. [Sl: Sn]
8. Arnau V, Mars S, Marín I (2005) Iterative cluster analysis of protein interaction data. Bioinformatics 21(3):364–378
9. Pei P, Zhang A (2005) A two-step approach for clustering proteins based on protein interaction profile. In: Proceedings of IEEE computational systems bioinformatics conference, CSB, vol 2005, no 1544467, p 201. NIH Public Access
10. Cho YR, Hwang W, Ramanathan M, Zhang A (2007) Semantic integration to identify overlapping functional modules in protein interaction networks. BMC Bioinformatics 8(1):1
11. Georgii E, Dietmann S, Uno T, Pagel P, Tsuda K (2009) Enumeration of condition-dependent dense modules in protein interaction networks. Bioinformatics 25(7):933–940
12. Pizzuti C, Rombo S (2012b) Experimental evaluation of topological-based fitness functions to detect complexes in PPI networks. In: Proceedings of the 14th annual conference on Genetic and evolutionary computation, pp 193–200. ACM
13. Bandyopadhyay S, Ray S, Mukhopadhyay A, Maulik U (2015) A multiobjective approach for identifying protein complexes and studying their association in multiple disorders. Algorithms Mol Biol 10(1):1
14. Ramadan E, Naef A, Ahmed M (2016) Protein complexes predictions within protein interaction networks using genetic algorithms. BMC Bioinformatics 17(7):269
15. Ding C et al (2001) A MinMaxCut spectral method for data clustering and graph partitioning. Proc IEEE Int'l Conf Data Mining 2001:107–114
16. Bara'a AA, Abdullah QZ (2018) Improving the performance of evolutionary-based complex detection models in protein–protein interaction networks. Soft Comput 1–24
17. Lancichinetti A, Fortunato S, Kertész J (2009) Detecting the overlapping and hierarchical community structure in complex networks. New J Phys 11(3):033015
18. Newman ME, Girvan M (2004) Finding and evaluating community structure in networks. Phys Rev E 69(2):026113
19. Leskovec J, Lang KJ, Mahoney M (2010, April) Empirical comparison of algorithms for network community detection. In: Proceedings of the 19th international conference on World wide web, pp 631–640. ACM
20. Park Y, Song M (1998, July) A genetic algorithm for clustering problems. In: Proceedings of the third annual conference on genetic programming, pp 568–575

21. Mewes HW et al (2002) MIPS: a database for genomes and protein sequences. Nuclc Acids Res 30(1):31–34
22. Gavin AC et al (2006) Proteome survey reveals modularity of the yeast cell machinery. Nature 440(7084):631
23. Zaki N, Berengueres J, Efimov D (2012) Detection of protein complexes using a protein ranking algorithm. Proteins Struct Funct Bioinform 80(10):2459–2468

A Comparative Analysis of Open Government Data in Several Countries: The Practices and Problems

Amirudin Syarif[1,2(✉)], Mohamad Aizi bin Salamat[2], and Rusmin Syafari[1,2]

[1] Department of Management and Information System, Universitas Bina Darma, Jl. Ahmad Yani 3, Palembang 30000, Indonesia
amirudinsyarif@binadarma.ac.id, aizi@uthm.edu.my
[2] Faculty of Computer Science and Information Technology, Universiti Tun Hussein Onn Malaysia, 86400 Parit Raja, Batu Pahat, Johor, Malaysia
Syafari.mov@gmail.com

Abstract. Open Government Data (OGD) practice for researchers are intriguing to analyze as a decade has passed and showed many differences in application from each country. This difference may occur due to diversity in culture of openness, level of openness, and the level of trust or confidence of the country in opening data. Researchers presume that there is linearity in both, in meaning if the practice of government in the real world goes well then so is in cyberspace. Researchers also assume that OGD practices in developing countries in terms of success and failure are not equal to developed countries. It is interesting for researches to study the practice of OGD member of Association of Southeast Asian Nations that is developing countries with culture of openness, level of openness, and the level of trust or confidence of the country that is similar due to its homogeneity to ASEAN nations. This paper gives out description and study of the reality of OGD practices from ASEAN countries. Qualitative Comparative Analysis (QCA) through literature review is done to see the ratio of OGD practices among developing countries especially ASEAN countries.

Keywords: Open government data · ASEAN · Openness · Qualitative comparative analysis

1 Introduction

The development of Information Technology brings great influence to many business and governmental activities. Many of said influences are in terms of open data. The development of open government data (hereinafter referred to OGD) became rapid after the inauguration speech of Barrack Obama in 2009 that mentions the importance of government to open data that is allowed to be used, and distributed freely for the stakeholder. The United States of America's open government data movement was then followed by many others countries. In the early 2010, countries in Europe became open to data after The United States. This movement was then followed by other

R. Ghazali et al. (Eds.): SCDM 2020, AISC 978, pp. 346–357, 2020.
https://doi.org/10.1007/978-3-030-36056-6_33

countries in Asia and Africa. After this movement began, OGD became a commitment to almost all countries in the world.

In 2019, the OGD movement aged a decade. The practice of OGD has recorded many successes and failures as well. Both success and failure stories of each country, if researched well enough, show many factors to its success and failure. A couple of notes from researches are as follows: [1–10] and etc.

OGD practice is intriguing to analyze for researchers as a decade has passed and many differences in application from each country is shown. [11] said these differences may occur due to diverse cultures of openness, level of openness, and the level of trust or confidence of the country in opening data and that different stakeholders also have different perspectives, as it was stated in their research. Researcher [12], who wrote on a practice of OGD in Brazil, stated that many steps need to be done in order to achieve a high level of effectiveness of OGD. Researcher [7] examined the development of components directly related to OGD in Thailand according to them, there are nine key components which include Organizations, Policies and Plans, Laws and Regulations, Innovation and participation of Citizens, Capability enhancements, Open government principles, Enterprise architecture, and Technology infrastructure.

OGD is a mirror of government practice in the real world that is brought to cyberspace. The researchers presume that there is linearity in both. Meaning, if the practice of government in the real world goes well then it will also work well in cyberspace. The governments of developed countries are those who are able to provide a high level of prosperity for its people in forms of high per capita incomes, and able to provide proper infrastructure, as well as E-government practice and OGD provisions. According to [13] there are substantial differences between developed countries and developing countries in terms of E-government development strategies that link to OGD. The researchers also assume that OGD practices in developing countries in terms of success and failure are not equal to developed countries. It is interesting for the researches to study the practice of OGD by members of the Association of Southeast Asian Nations (hereinafter referred to ASEAN) that are developing countries with the culture of openness, level of openness, and the level of trust or confidence of the country that is similar due to its homogeneity to ASEAN nations. This paper gives out description and study of the reality of OGD practices from ASEAN countries. Qualitative Comparative Analysis (QCA) through literature review is done to see the ratio of OGD practices among developing countries especially ASEAN countries. This research is done by focusing on the implementation of OGD that has been done, and then compare the success as well as failures from those countries. Literature review is done on purpose to make a systematic, factual, and accurate description through a conceptual framework. This Paper also intends to be and provide the best reference for the OGD practices in ASEAN Countries.

2 The Material and Method

2.1 The Material

OGD became a phenomenal movement relating to the openness of government data. It must be freely available for use, and free for digital distribution. The following scholars [4, 14–16] also state the same. That government data must be available and open to all stakeholders responsibly at no cost [17]. Specifically gives out the definition that open data is data that is openly available for use without restrictions and costs, and data that is published by the government is referred to as open government data.

There are various data types provided by the government. For Example: The data of citizen population, public health data, poverty data, education data, businces data, agricultural data, mining data, transportation data, and others. The implementation of OGD shows the diversity. According to [18] there is a lack of clarity in the open data context because of the diverse interests of the stakeholders. Therefore the data provided by the government must be able to adopt all the interests of the stakeholders. As mentioned by [17], implementations in various countries are diverse. According to the attention of stakeholders, the article of [19] which discusses Switzerland's OGD, says that their concern is throughout transparency, participation and collaboration.

A research written by [20] states that attitude from public servants determines the success of OGD. On the other hand [21] discusses the obstacles to implementation of OGD in China, The strengths and weaknesses of application OGD in the GCC member states: The United Arab Emirates. Bahrain, Kuwait, Saudi Arabia, Oman, and Qatar are discussed by [22].

The practice of OGD is a form of honest, responsible and dignified state administration practice. According to research [19] on OGD in Switzerland, it is stated that OGD emphasizes encouragement of transparency, collaboration and participation. Other researchers, such as [23], state that the main barrier in implementing OGD is perceptions that based on the behavior of state employees. Other significant barriers include perceived legal barriers, structuring of perceived hierarchy of authority, perceived culture of perceived bureaucratic decision making, and perceptions of organizational transparency. Researcher [24] states his opinion that countries that are not in the European group have a difference in terms of success and barrier factors. Research result of [25] states that the success factor in applying OGD is a clear responsibility and implementation of the process model, as well as the integration of the OGD platform.

2.2 The Method

Research Design

Research design is a research structure, according to [26], their research is one that seeks to advance knowledge of the impact of variables related to research design. This research is designed as Qualitative research using Qualitative Comparative Analysis (QCA), and Descriptive analytics. The aim is to make a systematic, factual, and accurate description of facts. In this study the reality of OGD will be described from several OGD practices of several ASEAN countries that have been published in indexed and reputable scientific journals. In this research, Biplot analysis is used to present the relative plot position of OGD practice from each ASEAN country towards the study variables simultaneously in one 2-dimensional graph. According to [27], Biplot is a type of analysis which uses a descriptive statistical technique that presents the plot relative position of ň objects of observation with p variables simultaneously in one 2-dimensional graph so that the characteristics and relative position of the observed variables and objects can be analyzed.

To obtain accurate data that is in accordance with the needs of this research, the steps are as follows: (1) Collecting appropriate articles which discuss the practice of OGD in ASEAN countries and are published in a reputable and indexed journal. (2) Determining the articles that are most relevant to each country. (3) Determining the variables used to make comparative analysis. The variables are from Global Open Data Index (https://index.okfn.org), (4) Concluding Biplot analysis, and relative position mapping of OGD practice from ASEAN countries. (6) Drawing conclusions from the result of research.

Object of Research
The objects of research are (1) Journal articles that discuss the practice of OGD in ASEAN member countries. It covers the country of Indonesia, Thailand, Malaysia, Philippines, Vietnam, Myanmar, Cambodia, Laos, Singapore, and Brunei. (2) Data that comes from Global Open Data Index (https://index.okfn.org).

3 Results and Discussion

The Association of Southeast Asian Nations (ASEAN) is a geo-political and economic organization of countries located in the Southeast Asian region. It was established and declared on 8 August 1967 by Indonesia, Malaysia, Philippines, Thailand and Singapore. ASEAN was founded in the spirit of building economic growth, and the cultural progress of its member countries. After standing for more than half a century, this organization has brought significant economic growth and progress for its members. In terms of OGD implementation, the ranking of ASEAN countries is still not good. Only Singapore is ranked in the world's top 20 according to the [28] Global Open Data Index (https://index.okfn.org/place/). It is interesting to see the comparison of OGD practices among ASEAN countries. Ratings issued by the Global Open Data Index are based on 15 data variables (Government Budget, National statistics, Procurement, National law, Administrative Boundaries, Draft legislation, Air quality, National maps, Weather forecasts, Company registers, Election results, Locations, Water quality, Government spending, Land ownership) weighted in percentage of data disclosure. The greater the percentage of open data, the better the country's ranking.

The discussion in this paper starts from a summary describing OGD practices from each ASEAN country that has been written by researchers in their paper. The following is the summary:

1. **Indonesian OGD practices are ranked 61st** [28]

In the papers [10, 29] it is stated that, from the initial desk research that was conducted on Indonesia open data, the Evaluation of Indonesian OGD shows that it is still in the very beginning stages of development. At the moment there is more focus on gathering awareness on the matter of open data and open government rather than integrated OGD practices. Any regulations that mention the need for specific data types, formats, licensing, or any other policy related aspects that were mentioned were not found. However, the designation of a specific taskforce to ensure the development of open government and also participating in the Open Government Partnership are signs of Indonesia's seriousness in joining the global movement.

2. **Thai OGD practices are ranked 51st** [28]

In the papers [7, 30], it is stated that Thailand government released the Digital Economic Plan of Thailand and Thailand's three years Digital Government Plan in 2016. The three years plan contains a plan for Open government data which can be used as an instrument in improving the transparency. However, there are various problems with open government data in Thailand. Open government data will have to comply with the existing degree of data privacy, and data protection. To accelerate digital government, they use four strategies which are: (1) Government integration. (2) Smart operations. (3) Citizen-centric services, and (4) Driven transformation.

3. **Malaysian OGD practices are ranked 87th** [28]

In the paper [31–33], it said that the implementation of OGD by the Malaysian government is part of an initiative to improve government performance through the use of information technology. In particular, Malaysia focuses on the public sector, with a specific focus on e-Government and open data. The "Tenth Malaysia Plan" (2011) specifically identified e-Government and open data as critical elements in the move towards more effective, transparent and accountable public service delivery. It is carried forward in the "Eleventh Malaysia Plan" (2016-2020) that clearly articulates the country's intention to use data-driven governance to improve citizen-centered service delivery, increase responsiveness, and strengthen accountability through greater transparency. The E-government initiative focuses on increasing efficiency and at the same time reducing operational costs of offered public services. Nevertheless, Malaysia's OGD ranking is still below most ASEAN countries. Only Myanmar is ranked below Malaysia.

4. **OGD Philippines practice is ranked 53rd** [28]

In the papers [29, 34], it is said that the Philippines' OGD data sets are available in user-friendly formats with a detailed description of the data set itself in the form of metadata. The portal permits data search and sharing via social media, but at the same time, the OGD portal of the Philippines has many barriers to re-use. This is a weakness of the Philippines' government's OGD practices. Another disadvantage is that the data sets are not current, and no attempts have been made at updating the records.

5. **Vietnam, Laos, and Brunei are not ranked** [28]

In the papers [35–37] it is stated that though the index currently covers 149 countries, which covers almost all countries, only three ASEAN countries namely Brunei, Laos and Vietnam could not be indexed. The reasons include the still closed and maintaining the confidentiality of data carried out by these countries. The initiatives of governments in opening their data have not been done. Access to data and information is not fully given.

6. **Cambodia's OGD practices are ranked 74th and Myanmar OGD practices are ranked 94 according to** [28]

In [36], it is stated that in 2016, the two countries have no open data portals of their own, and the websites of their national statistics offices suffer from challenges in data accessibility. It also has no national priority in Cambodia and Myanmar as well. That is why both of Open Data Index is at the level of low scores which results in low ranking.

7. **OGD Singapore practices are ranked 17th according to** [28]

Singapore is the only ASEAN country that is ranked in the top 20. This ranking is quite large, with Thailand ranking 51st in the world and rank 2 in ASEAN. This large gap according to [36] is possible because there are technological factors and the ability of Statistics to manage data. It relies on technology infrastructure, policies, and an ecosystem that enables a culture of experimentation and co-creation between citizens, industry and research institutes. The following are country data and values for each factor:

Table 1. Table Country and Factor.

Country/Factor	Indonesia	Thailand	Malaysia	Philippines	Cambodia	Myanmar	Singapore
Government budget	60	60	60	80	60	0	100
National Statistics	80	85	50	85	30	0	100
Procurement	45	70	45	65	45	0	100
National laws	45	85	0	0	30	0	60
Adminstrative boundaries	0	0	0	65	0	0	100
Draft legislation	0	45	0	45	0	0	30
Air quality	85	70	0	0	0	0	85
National maps	65	0	0	50	0	0	0
Weather forecast	0	0	0	65	45	0	85
Company register	0	100	0	0	15	15	100
Election results	0	0	0	0	0	0	0
Locations	0	0	0	0	30	0	70
Water quality	0	0	0	0	0	0	60
Government spending	0	0	0	0	0	0	0
Land ownership	0	0	0	0	0	0	15

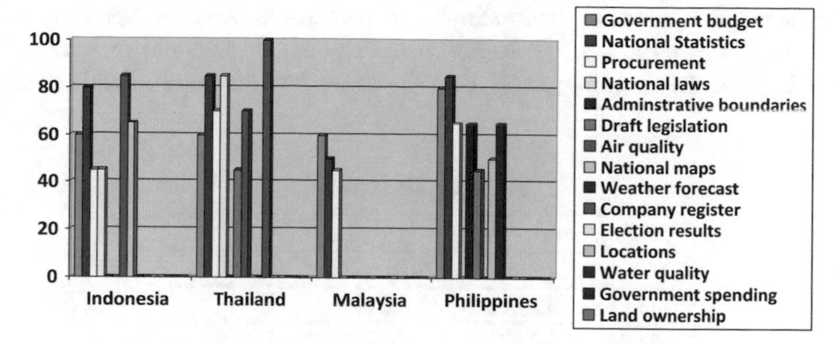

Fig. 1. Rating factors of selected countries

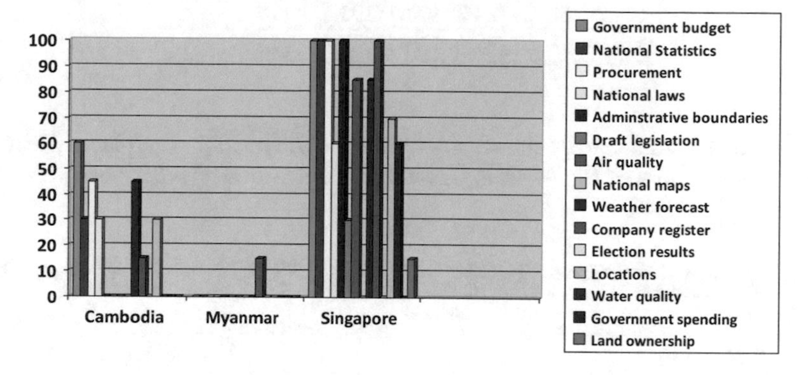

Fig. 2. Rating factors of selected countries

The following chart is the Ratings of factors of open data in selected countries in Southeast Asia.

The following are Ratings of open data factors in ASEAN member countries with coverage in the Open Data Index (Data source: Open Data Index 2018) based on [28].

The following table is the result of data processing using SPSS:

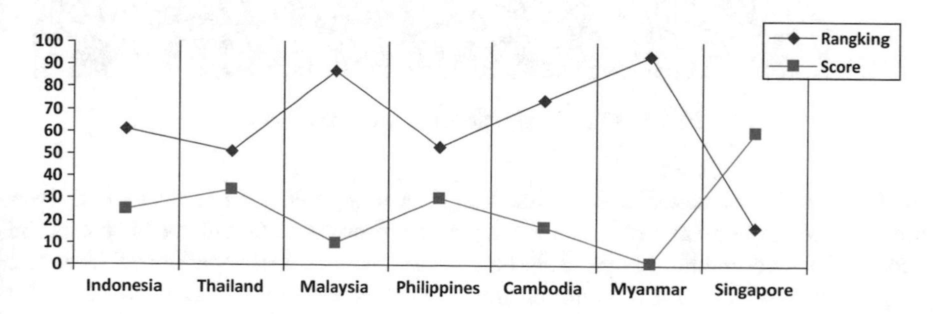

Fig. 3. Rank and scores each selected country

The descriptive statistics table shows the percentage score of openness of government data from each country based on the mean. For example in Indonesia, the mean is 25.33 if the rounding is 25 then the score for Indonesian data disclosure is 25%.

Table 2. Mean Table

Descriptive statistics			
	Mean	Std. deviation	Analysis N
Indonesia	25.3333	33.67209	15
Thailand	34.3333	39.81505	15
Malaysia	10.3333	21.58593	15
Philippines	30.3333	34.87051	15
Cambodia	17.0000	21.11195	15
Myanmar	1.0000	3.87298	15

This is in accordance with the results from https://index.okfn.org/place/. Next is the capture of the results of Fig. 4. It is likewise for the scores of other countries.

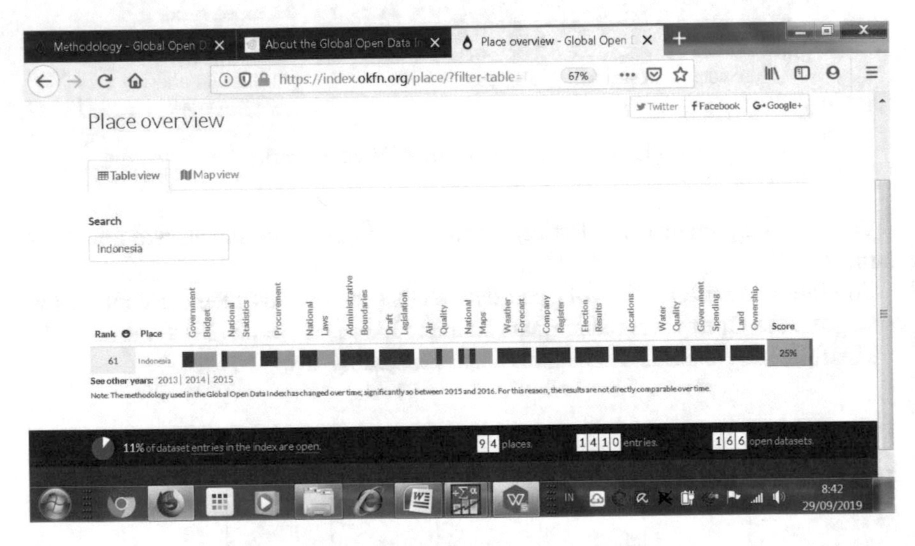

Fig. 4. Capture of Indonesian score and factors

In an effort to describe data that has been summarized in a data table, two-dimensional graphics are used. Biplot can provide information that includes objects and variables in one graphic image. This biplot analysis shows the position of closeness between factors. There are 4 groups that show the position of closeness. The first position is the National Laws and Air Quality factor group, the second position is the Procurement, National Statistics, and Government budget group, the third position is the Water quality, Land ownership, and Election results group, the fourth position is the Locations and Draft legislation. For the Company register factor, only Singapore opens the data while other countries do not.

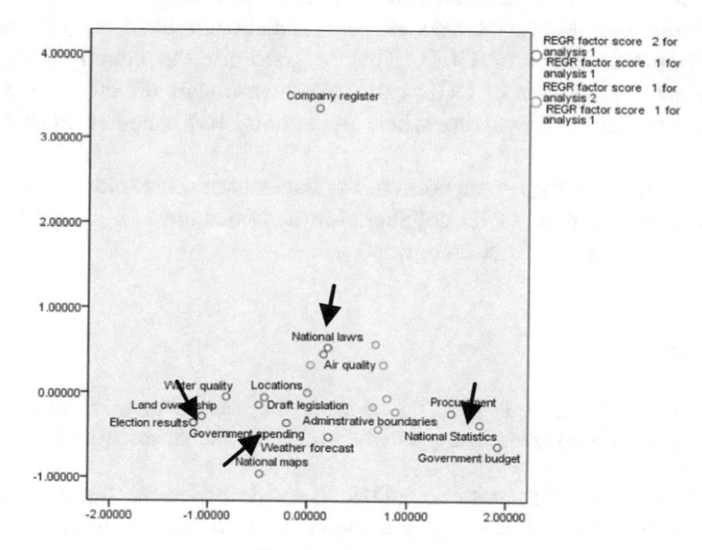

Fig. 5. Biplot graph

Of the 10 ASEAN countries, there are three countries that are not available because they cannot be assessed. The three countries are Vietnam, Laos and Brunei. In addition to the three countries there are many countries outside the ASEAN group that also cannot be given an assessment. If you look at the OGD practices discussed above, many factors are not 100% open. This shows that the issue of transparency and state security is still more dominant than the transparency factor that can benefit the country. Many countries in the practice of OGD are not prepared if this transparency is carried out one hundred percent openly. That is why data exchange and collaboration among countries cannot be applied one hundred percent honestly at this time.

The government must pay attention to many data concerning the social life of the community, such as income or salary data, paid tax data, bio data place and date of birth, mother's name, and so forth. The government must also pay attention to the interests of stakeholders. These data certainly should not be issued or provided haphazardly because it concerns the interests of the community. The security of personal data is an important thing that must be guaranteed by state law. This is a concern in designing open government data. Stakeholders must remember that the importance of data privacy is more important than data disclosure. Providing data that is suitable for the people who need data is important. Data becomes useless if it is not available and fits the needs. Data that is not suitable and incorrect in its classification will not be useful at all.

4 Conclusions

Based on the results of the analysis and discussion above, it can be concluded that the implementation of OGD in ASEAN countries is still unable to fulfill data disclosure in the information age. Many interests of each stakeholder have an influence on the successful implementation of OGD. This is good for the country's development because the implementation of OGD can make investments directly beneficial to the country. An example is Singapore, where the country has benefited from good OGD practices.

The government must pay attention to the interests of stakeholders concerned with releasing appropriate data. OGD collaboration activities are still asking for a higher level of trust from the ASEAN countries.

References

1. Cartofeanu C, Macrinici D (2017) Evaluating the success of eGovernment OpenData platform at increasing transparency in Moldova: from the perspectives of journalists and developers
2. Araújo NM, Filho FMF, Melo LA (2018) Open data from the Brazilian government : understanding the perspectives of data suppliers and developers of applications to the citizens
3. Hiramoto K (2012) e-Government and open government data in Japan, Japan
4. Geiger CP, Von Lucke J (2018) Open government and (Linked) (Open) (Government) (Data). JeDEM eJournal eDemocracy Open Gov 4(2):265–278
5. van Schalkwyk F, Willmers M, McNaughton M (2016) Viscous open data: the roles of intermediaries in an open data ecosystem. Inf Technol Dev
6. Vracic T, Varga M, Curko K (2016) Effects and evaluation of open government data initiative in Croatia. In: 2016 39th International convention on information and communication technology, electronics and microelectronics, MIPRO 2016, pp 1521–1526
7. Srimuang C, Cooharojananone N, Tanlamai U, Chandrachai A (2017) Open government data assessment model: an indicator development in Thailand
8. Roa HN, Loza-Aguirre E, Flores P (2019) A survey on the problems affecting the development of open government data initiatives. In: 2019 Sixth international conference eDemocracy eGovernment, pp 157–163
9. Aarshi S, Malik BH, Habib F (2018) Dimensions of open government data web portals : a case of Asian countries. Int J Adv Comput Sci Appl 9(6):459–469
10. Nugroho RP (2014) Comparing open data policies and their implementation in developed and developing countries. J Penelit dan Pengemb Komun dan Inform 4(3):159–172
11. Gonzalez-Zapata F, Heeks R (2015) The multiple meanings of open government data: understanding different stakeholders and their perspectives. Gov Inf Q 32(4):441–452
12. De Oliveira EF, Silveira MS (2018) Open government data in Brazil a systematic review of its uses and issues. In: Proceedings of the 19th annual international conference on digital government research, p 9

13. Chen YN, Chen HM, Huang W, Ching RKH (2006) E-government strategies in developed and developing countries: an implementation framework and case study. Glob Inf Manag 14 (1):23–46
14. Zhu X (2017) The failure of an early episode in the open government data movement : a historical case study. Gov Inf Q
15. Vetrò A et al (2016) Open data quality measurement framework: definition and application to open government data. Gov Inf Q 33(2):325–337
16. Kassen M (2017) Open data in Kazakhstan: incentives, implementation and challenges. Inf Technol People 30(2):301–323
17. Charalabidis Y, Alexopoulos C, Loukis E (2015) A taxonomy of open government data research areas and topics. J Organ Comput Electron Commer 26(1–2):41–63
18. Veljković N, Bogdanović-Dinić S, Stoimenov L (2014) Benchmarking open government: An open data perspective. Gov Inf Q 31(2):278–290
19. Neuroni AC, Riedl R, Brugger J (2013) Swiss executive authorities on open government data—policy making beyond transparency and participation. In: Proceedings of the annual Hawaii international conference on system sciences, pp 1911–1920
20. Wirtz BW, Birkmeyer S (2015) Open government: origin, development, and conceptual perspectives. Int J Public Adm 1–16(January):37–41
21. Huang R, Lai T, Zhou L (2017) Proposing a framework of barriers to opening government data in China. Libr Hi Tech 35(3):421–438
22. Saxena S (2017) Significance of open government data in the GCC countries. Digit Policy Regul Gov 19(3):251–263
23. Wirtz BW et al (2016) Resistance of public personnel to open government : a cognitive theory view of implementation barriers towards open government data. Public personnel to open, vol 9037, February 2016
24. Huber, S (2012) Indicators for the fitness of municipal open government data (ogd) to be processed into e-participation innovations with intended societal impacts. In: Electronic government and electronic participation, vol 39, pp 249–258
25. Parycek P, Hochtl J, Ginner M (2014) Open government data implementation evaluation. J Theor Appl Electron Commer Res 9(2):80–99
26. Molina-Azorin JF (2012) Mixed methods research in strategic management: impact and applications. Organ Res Methods 15(1):33–56
27. Gunarto M, Syarif MA (2014) Penggunaan Analisis Biplot pada Pemetaan Perguruan Tinggi Swasta di Kota Palembang. In: Forum Manajemen Indonesia 6, pp 1–13
28. Global Open Data Index (2019) OGD Ranking. https://index.okfn.org/place/. Accessed 25 Sept 2019
29. Pasaribu NMA (2017) A comparative study of open data policies: lessons learned from Indonesia' s and The Philippines' participation in open government partnership Asia-Europe Institute. Kuala Lumpur, Malaysia
30. Srimuang C (2018) Development of an open government data assessment model: user-centric approach to identify the weighted components in Thailand. Int J Electron Gov 10 (3):1
31. Ahmed MS, Bin Mahmuddin M, Mahat NIB (2017) The factor affecting Malaysian citizens satisfaction with open government data. J Eng Appl Sci 12(15):3843–3846
32. World Bank Group (2017) The Malaysia development experience series—open data readiness assessment (ODRA) report. Kuala Lumpur, Malaysia

33. Kaur R (2006) Malaysian e-government implementation framework, University of Malaya
34. Saxena S (2018) Drivers and barriers to re-use Open Government Data (OGD): a case study of open data initiative in Philippines. Digit Policy Regul Gov 20(4)
35. World Bank Group (2019) Digital government and open data readiness assessment, February 2019
36. Stagars M (2016) Open data in southeast Asia. Palgrave Pivot, Singapore
37. Saphangthong T, Phissamay P, Inthavong C (2017) Knowledge information and data in the Mekong Lao open data experiences, Yangon, p 25

Survey of Offline Arabic Handwriting Word Recognition

Haitham Qutaiba Ghadhban(✉), Muhaini Othman,
Noor Azah Samsudin, Mohd Norasri Bin Ismail,
and Mustafa Raad Hammoodi

Software Engineering Department, Universiti Tun Hussein Onn Malaysia,
Parit Raja, Batu Pahat 86400, Johor, Malaysia
{gi170052,gi180012}@siswa.uthm.edu.my,
{muhaini,azah,norasri}@uthm.edu.my

Abstract. The field of Arabic handwriting recognition and translation is currently experiencing rapid growth in terms of research, which is evident in the coverage of major conferences and journals that specialise in the area of handwriting recognition. Against this backdrop, a significant increase has been observed in the classification and features techniques used, as compared to some years back. Researchers have put in more efforts geared towards building a variety of databases for Arabic handwriting recognition. This article aims to provide a comprehensive survey of advances in Arabic offline handwriting recognition. We have been provided details of availability Arabic databases with limitation. Further, we focus on techniques of feature extraction and different variety of classification approaches such as ANN, HMM, SVM that used in Arabic handwriting recognition.

Keywords: Offline handwriting recognition · Arabic datasets · Feature extraction · Classifiers

1 Introduction

Handwriting recognition, which has several practical applications, is also an active area of research in Pattern Recognition. Some of these practical applications include automatic processing of bank cheques, processing of forms, postal address and zip code recognition etc. Recognition of unconstrained handwriting is an extremely challenging task due to large variations in handwriting styles. In some restricted domains, the task of handwriting recognition may become less demanding. For example, in zip code recognition, the problem may reduce to the recognition of a sequence of isolated numerical digits. However, even in this restricted domain, many challenging issues arise. Touching digits, broken characters, writing errors and noise due to scanning are some of the issues that can make the 'simpler' problem of zip code recognition much harder.

Handwriting recognition is a task that involves the transformation of a language that is represented in its spatial (off–line) and temporal (on–line) form of graphical marks into its symbolic representation [1]. Optical Character Recognition (OCR) is a

R. Ghazali et al. (Eds.): SCDM 2020, AISC 978, pp. 358–372, 2020.
https://doi.org/10.1007/978-3-030-36056-6_34

term that is related to this task; it involves translating scanned images of typewritten, printed text or handwritten text into machine-encoded text. However, the main focus of handwriting recognition is handwritten text. Figure 1 below is an example of Arabic handwriting recognition.

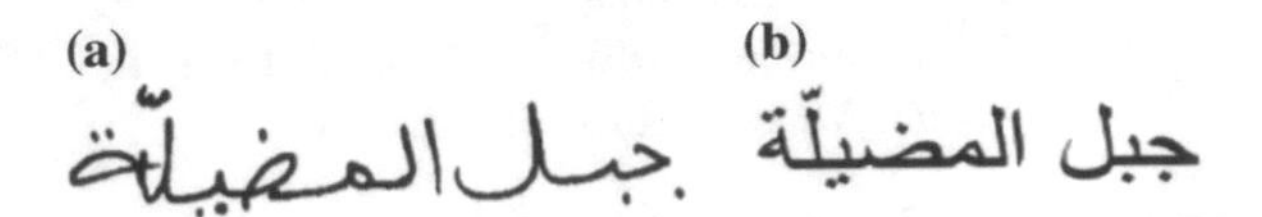

Fig. 1. Illustration of handwriting recognition: (a) scanned image of Arabic handwritten text and (b) recognized machine–readable text.

The automation of several challenging tasks have been facilitated through the advancement in handwriting recognition. There are several applications of algorithmic analysis of human handwriting, and some of them include form processing, identification and verification of writer, as well as on–line and off–line handwriting recognition etc.

For the first time in the 1950s commercial OCR systems were introduced for Latin script. The availability of commercial text recognition products has been enhanced by the significant technological advancement in document and character recognition [2]. Handwriting recognition has ever since been used in processing bank cheques, forms and postal address. However, there are new sophisticated recognition systems that process documents whose layouts are complex, multi-lingual scripts, etc. A wide range of writing modes like handwritten, hand-printed and machine-printed can be handled by such systems. Earlier works in OCR started with the recognition of numerals, but it has now been extended, thereby allowing the recognition of Chinese characters, Latin alphabets, Japanese Katakana syllabic characters, Hangul characters, Kanji (Japanese version of Chinese) characters, etc.

Al-Badr and Mahmoud (1995) noted that works in the area of Arabic OCR began in as far back as 1970s, with the first work published in 1975. It was in the 1990s the first Arabic OCR system was introduced. There are some special challenges and benefits associated with Arabic handwriting recognition [3]. Despite the fact that handwriting recognition has been in existence for over three decades, only little efforts have been made in the area of Arabic handwritten text recognition in comparison with text in other scripts. Despite the availability of few commercial Arabic OCR systems for printed text, there is no operationally accurate Arabic handwritten OCR commercial product available in the market (Cheriet 2008).

In this article, we present a comprehensive survey of research works that have been published in the different phases of Arabic handwriting recognition. The most recent survey of this topic dates back to 2013 [4, 5]. Authors tried to cover the gap back to 2006 [6]. However, the researchers in this particular field have increased significantly in recent years. This article aims to provide a comprehensive survey of Arabic datasets, features extraction and classification techniques for Arabic offline handwriting recognition to date with affirmation on date from 2013 to 2019.

2 Motivation

Approximately 280 million native speakers speak the Arabic language, while another 250 million non-native speakers also speak the language. Even though Arabic language is spoken in different forms, there is only one standard form of writing it, which is officially used for communication in Arab countries. Sometimes, this written form of Arabic is referred to as "Modern Standard Arabic" (MSA). More so, the use of Arabic language has been employed by different tribes apart from Arabs, and some of these languages are non-semantic like Kurdish, Persian and Urdu. Therefore, if the interpretation of written Arabic language can be automated, the benefits will be enormous.

In addition, with Arabic handwriting recognition, the searching and reading of Arabic manuscripts which is estimated to be over three million, can be done automatically [7]. The same techniques that have been tailored for MSA can be adopted for a large number of Arabic handwritten manuscripts because not much change has been made to written Arabic. The availability of Arabic content can be increased through the automation of Arabic manuscripts processing. Carrying out the task of recognition on manuscripts can be easier because manuscript writing is often neater than free handwriting. Nevertheless, the task of recognition may be complicated by degradation of image, previously unseen writing styles and unexpected markings [8].

3 Characteristics of Arabic Script

One plausible hypothesis states that Arabic (الْعَرَبِيَّة) alphabet was derived from the Aramaic script through Nabatean alphabets. Some Asian and African languages like Arabic and Urdu are written using the Arabic alphabet (الأبْجَدِيَّة العَرَبِيَّة). After the Latin alphabet, it is the second–most widely used alphabet around the world. One of the unique differences between Arabic script and that of other languages is that it is naturally calligraphic. More, so the direction of writing is another special feature of the Arabic text, as it is written for right to left, and has 28 basic characters. Sixteen Arabic letters have from one to three dots. Similar characters are differentiated by number and position of the dots (like خ, ح and ج). Furthermore, some characters like أ, ؤ, كئ, can have a zigzag like stroke (Hamza ء). These dots and Hamza are called secondary and they are located above the character primary part as in ALEF (أ), or below like BAA (ب), or in the middle like JEEM (ج). Arabic text appears as cursive in both hand-written and machine-printed text. In some words, there is a connection between some characters and the preceding characters, while some do not have this connection. The position of an Arabic character in a word determines the shape of that character, and often times, a character may have up to four different shapes. These shapes depend on if the character is isolated, connected from the right (ending form), connected from the left (beginning form), or connected from both sides (middle form). There may be a vertical overlapping of characters in a word. The size of Arabic characters is not fixed in terms of width and height. The size of the character varies based on its position in the word. The Arabic alphabets are listed in Tables 1 and 2 below.

Dhammah (ـُ), Shaddah (ـّ), Maddah (~), and Kasrah (ـِ). Moreover, Tanween may be formed by having double Fat–hah (ـً), Dhammah (ـٌ), or Kasrah (ـٍ). These diacritics are written as strokes and are placed either on top of or below the characters. The meaning of a word may be changed by a different diacritic on a character. Usually, when Arabic text readers read an un-discretised text, they deduce the meaning from the context.

A ligature is a character that is formed through the combination of two or more letters with horizontal or vertical overlapping between component characters. There are many standard ligatures that exist in Arabic language. For instance, „laam–alif" (لا),„laam–ha" (لحـ), „meem–haa" (محـ) etc. in addition, in Arabic cursive script there may be overlapping of two or more letter without any contact.

Table 1. Lists the alphabets of Arabic language.

Transcription	General	Contextual forms			
		Isolated	End	Middle	Beginning
ALEF	ا	ا	ـا		
BAA	ب	ب	ـب	ـبـ	بـ
TAA	ت	ت	ـت	ـتـ	تـ
THAA	ث	ث	ـث	ـثـ	ثـ
JEEM	ج	ج	ـج	ـجـ	جـ
HAA	ح	ح	ـح	ـحـ	حـ
KHAA	خ	خ	ـخ	ـخـ	خـ
DAL	د	د	ـد		
THAL	ذ	ذ	ـذ		
RAA	ر	ر	ـر		
ZAIN	ز	ز	ـز		
SEEN	س	س	ـس	ـسـ	سـ
SHEEN	ش	ش	ـش	ـشـ	شـ
SAAD	ص	ص	ـص	ـصـ	صـ
DHAD	ض	ض	ـض	ـضـ	ضـ
TTAA	ط	ط	ـط	ـطـ	طـ
TTHAA	ظ	ظ	ـظ	ـظـ	ظـ
AIN	ع	ع	ـع	ـعـ	عـ
GHAIN	غ	غ	ـغ	ـغـ	غـ
FAA	ف	ف	ـف	ـفـ	فـ
QAAF	ق	ق	ـق	ـقـ	قـ
KAAF	ك	ك	ـك	ـكـ	كـ
LAM	ل	ل	ـل	ـلـ	لـ
MEEM	م	م	ـم	ـمـ	مـ
NOON	ن	ن	ـن	ـنـ	نـ
HHAA	ه	ه	ـه	ـهـ	هـ
WAW	و	و	ـو	ـو	و
YAA	ي	ي	ـي	ـيـ	يـ
TAAM	ة	ة	ـة		

4 Databases of Arabic Text

Before the phases involved in the recognition of Arabic text are described, a discussion on the available Arabic text databases used by the researchers is provided. It is important to note that in Arabic text recognition task; there is no generally accepted database that is available to researchers. In other words, the use of different data has been employed by researchers [9], and as such, it may not be possible to carry out a comparison of the recognition rates of the various techniques. As can be seen from Table 2 summary of Arabic datasets, there is no appropriate freely accessible datasets.

Table 2. Summary of results for Arabic datasets

Database	Description	Writers	Limitation
ERIM [10]	Consists of over 750 pages of printed Arabic texts	–	No longer available
[11]	2000 samples of unconstrained Arabic handwritten characters	4 writers	– Basic character shapes without dots – Does not have handwritten character shapes
Al–ISRA [12]	Contains 37,000 Arabic words, 10,000 digits, 2,500 signatures, and 500 free–form Arabic sentences gathered	500 writers	– Until now no data has been published on the internet – Does not include handwritten text
[13]	Consists of 262,647 words with 1,126,420 characters	48 writers	Can only be used for research in isolated character recognition
AHDB [14]	10,000 words, handwriting pages	100 writers	– Limited in vocabulary – More relevant in cheque processing applications
IFN/ENIT [15]	Consists of 26,459 images of the 937 names of cities and towns in Tunisia, The database contains 115585 pieces of Arabic words (PAWs) and 212211 characters	411 writers	Limited vocabulary because it is only restricted to city names
CENPARMI [16]	3000 cheques, contains 2499 legal amounts, 2499 courtesy amounts written in Indian/Arabic digits	–	Limited vocabulary word recognition
IFHCDB [17]	Contains grayscale images of 52380 characters and 17740 numerals	–	– Lacks natural Arabic/Farsi handwritten text – Characters is not uniform

(*continued*)

Table 2. (*continued*)

Database	Description	Writers	Limitation
Applied Media Analysis [18]	5,000 images, 200 documents (forms, memos, poems, diagrams, and number lists in both English and Indic digits)	25 writers	Website to access database not available
ADBase [19]	Contains 70,000 digits, and each digit („0" to „9") was written by each writer ten times. Divided into training and testing set. There are 60,000 digits samples in the training set, while the testing set contains 10,000 digits samples	700 writers	Only database for Arabic handwritten digits
CEDAR [20]	Approximately 20,000 words, have written 10 pages of text, each includes between 150 and 200 words	10 writers	–
LMCA [21]	Contains 30000 digits, 100000 letters and 500 words	55 writers	Limited to city names and thus contains limited vocabulary
KHATT [22]	A total of 4,000 paragraphs were segmented into text lines with about 200,000 different words. In addition, 928 of the writers were right-handed and 72 were left-handed	1000 writers	– Does include writers labelling information – Low citation

5 Feature Extraction Approaches

Feature extraction is very problem dependent. Good features are those in which values are similar for objects belonging to the same category and distinct for objects in different categories. Choosing the proper type of features depends on the nature of the text, the type of the system processing which may be offline or online, and the scripts types that can be handwritten or printed.

Extracting features from images can be done in two manners. First, the most classical way is to design features especially tuned for the task and sometimes even tuned for a specific data set. These features are called handcrafted features since an algorithm was designed manually to extract them, incorporating a priori information on the specificities of the data. Some of the handcrafted features are very simple, while some more recent feature sets are very complex and generally highly-dimensional. The second category of features consists in automatically learning features from the images using machine learning. This is a solution that is used more and more since the advent of Deep Learning. These features are called learned features.

There are many techniques to learn features from a data set, either supervised learning or unsupervised learning. Historically, supervised learning was used to learn a feature extractor, either with simple dictionary learning or with an Artificial Neural

Network (ANN). There are several techniques than can be used for this task, K-means, Principal Component Analysis (PCA), auto-encoders and the family of Restricted Boltzmann Machine (RBM) models, on which this experiment focuses. Since the advent of Deep Learning, unsupervised learning has been used more and more to extract features from data, especially from images. Learned features are trying to overcome the issues of handcrafted features. Since they can be trained on any data set, they generalize very well to change. Moreover, they are generally able to cope well with unknown examples, due to the higher generalization capabilities of neural networks. Finally, they do not require expert knowledge of the data and should not require a lot of human-labor time.

Author [23] presented a new technique for dividing the image into non-uniform horizontal segments to extract the features and a new technique for solving the problems of the skewing of characters by fusing multiple HMMs. Moreover, two enhancements are introduced: the pre-processing method and feature extraction using concavity space. Author [24] proposed a novel system for the recognition of handwritten Arabic words. It is evolved based on horizontal-vertical Hidden Markov Model and Dynamic Bayesian Network Model. The strategy consists of looking for various HMM architectures and selecting those which provide the best recognition performance. They used Structural features (ascendants, descendants, loops and diacritic points) and statistical features at pixel level (pixel density distributions and local pixel configurations) are then extracted from word images.

Author [25] proposes an un-constrained recognition approach for the handwritten Arabic script. The approach starts by explicitly segment each word image into its constituent letters, then a filter-bank of Gabor wavelet transform is used to extract feature vectors corresponding to different scales and orientation in the segmented image. Author [26] focused on the improvement of the accuracy rate and the reliability of a HMM based handwriting recognition system by the use of Dempster Shafer Theory (DST). Two feature sets are proposed in this work. The first one is based on directional density features. The second one is based on foreground.

[27] explored a new model focused on integrating two classifiers; Convolutional Neural Network (CNN) and Support Vector Machine (SVM) for offline Arabic handwriting recognition (OAHR) on which the dropout technique was applied. The suggested system altered the trainable classifier of the CNN by the SVM classifier. A convolutional network is beneficial for extracting features information and SVM functions as a recognizer. It was found that this model both automatically extracts features from the raw images and performs classification. Author [28] considered two strategies CNN features HMM and Handcrafted-features-HMM. The first combination took the CNN as an automatic feature extractor and HMM as recognizer. That allows operating directly on the images and extracting relevant characteristics without much emphasis on feature extraction and pre-processing stages.

Author [29] proposed a new system based on the integration of two deep neural networks. First a Convolutional Neural Network (CNN) to automatically extract features from raw images, then the Bidirectional Long Short-Term Memory (BLSTM) followed by a Connectionist Temporal Classification layer (CTC) for sequence labelling. Author [30] presented a model CNN based HMM for Arabic handwriting word recognition. The HMM have proved a powerful to model the dynamics of handwriting.

Table 3. Summary of results for features

Authors	Features	Classifier	Data	Accuracy
[23]	Concavity features, Gradient features	Multiple HMMs	IFNIENIT database.	93.44%
[24]	Structural features, statistical features	Horizontal-vertical Hidden Markov Model, Dynamic Bayesian Network	IFNIENIT database	92.19% with HMM and 88.82% DBN
[25]	Gabor wavelet	Support Vectors Machine	IESK-arDB and IFN/ENIT databases	71% on IESK, 56% on IFN/ENIT
[26]	Densities features, Contour features	Hidden Markov Model	RIMES for Latin script and IFN/ENIT for Arabic script	87.55%
[27]	CNN	Convolutional Neural Network (CNN), Support Vector Machine (SVM)	HACDB and IFN/ENIT databases	94.17% for HACDB, 92.95% for INF/ENIT
[28]	CNN	Hidden Markov Model	IFNIENIT database	88.95%
[29]	CNN	Bidirectional Long Short-Term Memory (BLSTM), Connectionist Temporal Classification layer (CTC)	IFNIENIT database	92.21%
[30]	CNN	Hidden Markov Model	IFNIENIT database	89.23%
[31]	Structural features and CNN	Dynamic Bayesian network (DBN) and Hidden Markov model (HMM)	IFN/ENIT database	95.20%
[32]	Modified Direction Features (MDF), Global Features, Complete feature vector and Statistical Classification	Support Vector Machine (SVM)	IFN/ENIT database.	96.71%

In this model, the trainable classifier of CNN is replacing by the HMM classifier. CNN works as a generic feature extractor and HMM performs as a recognizer.

[31] described the main highlights of the dynamic Bayesian network (DBN) architecture, Features are extracted based on the word baseline which has been estimated to mainly cope with the problems of inclination and distortions. They applied deep learning architecture: a CNN that convolves learned features with input data and uses 2D convolutional layers that makes it well suited to 2D word image processing. Author [32] presented combination statistical and structural approaches for character recognition. The statistical method first recognizes the main body of a character using modified direction features and Support Vector Machines. Then in structural classification, dot descriptors are used to recognize the exact shape of an Arabic character.

6 Classification Approaches

The use of some classifiers has been employed in the recognition of Arabic handwritten numerals, words and characters. Such classifiers include Support Vector Machines (SVM), Hidden Markov Model (HMM), k-Nearest Neighbours (k-NN), Artificial Neural Networks (ANN), etc. However, there are issues when it comes to the nature of selected classifiers since it has own weakness: SVM requires huge calculation to calculate the kernels [33], Extreme Learning Machine (ELM) has non-stable performance due to the random weights between the input and hidden layer [34], Multilayer Perceptron (MLP) uses backpropagation which is slow in training and subject to local minima [35]. Obtaining strong Arabic handwriting recognition requires crucial part: which is incorporating stronger classifiers with easy calculation and stable performance.

Author [36] proposed a handwriting based biometric identification system using a large database of Arabic handwritten documents. The system first extracts, from each handwritten sample, a set of features including run lengths, edge-hinge and edge-direction features. These features are used by a Multiclass SVM (Support Vector Machine) classifier. Handcrafted feature is handled as input by the suggested method and gets going with a supervised learning algorithm. Author [37] chose the Multi-class Support Vector Machine with an RBF kernel and tested it on Handwritten Arabic Characters Database (HACDB). The suggested system performs better than the other current methods by producing better results. RBF kernel SVM is quite a promising classification method in the handwriting recognition domain.

Author [27] explored the applicability of dropout in the proposed CNN based-SVM model on Arabic handwritten recognition and demonstrated the efficiency of the system. A convolutional network is beneficial for extracting features information and SVM functions as a recognizer. They deduce that CNN based on SVM classifier offers the state-of-the-art significant results without much emphasis on feature extraction and pre-processing stages. CNN based-SVM model is indeed a full of promise classification method in the handwriting recognition domain. Author [38] presents two models for the recognition of the handwritten Arabic characters. Number one is based on deep learning approach, called Deep networks using SVM (DSVM), and dropout technique while number two insists on hand-crafted feature extraction as SIFT descriptor. The classification system is reliant on support vector machine being considered as one of the strongest classification techniques and is now largely utilized in a lot of pattern recognition applications.

Author [39] proposed system is based on a synchronous multi-stream HMM (MSHMM) which has the advantage of efficiently modelling the interaction between multiple features. These features are composed of a combination of statistical and structural ones, which are extracted over the columns and rows using a sliding window approach. In fact, two word models are implemented based on the holistic and analytical approaches without any explicit segmentation. Author [40] enhanced the HMM based reference system by using a hybrid HMM/MLP classifier. Extracted features are statistical and geometric to integrate both the peculiarities of the text and the pixel distribution characteristics in the word image. These features are modelled using hidden Markov models. The improving of HMM modelling by incorporating MLPs to estimate emission probabilities that present the major HMM problem in order to take advantage of the strength of HMM modelling and neural networks classification.

[41] proposed an efficient multiple classifier system for Arabic handwritten words recognition. First, Authors used Chebyshev moments (CM) enhanced with some Statistical and Contour-based Features (SCF) for describing word images. Then, they combined several classifiers integrated at the decision level. Authors considered the multilayer perceptron (MLP), the support vector machine (SVM) and the Extreme Learning Machine (ELM) classifiers. They proposed several combination rules between MLP, SVM and ELM classifiers trained with CM and SCF features. Author [42] presents a system for offline recognition cursive Arabic handwritten text based on Hidden Markov Models (HMMs). The system is analytical without explicit segmentation used embedded training to perform and enhance the character models. Extraction features preceded by baseline estimation are statistical and geometric to integrate both the peculiarities of the text and the pixel distribution characteristics in the word image. These features are modelled using hidden Markov models and trained by embedded training.

[30] a model CNN based HMM has been presented to solve the Arabic handwritten word recognition problem. This combination took the CNN as an automatic feature extractor and HMM as recognizer. That allows operating directly on the images and extracting relevant characteristics without much emphasis on feature extraction and pre-processing stages.

[43] proposed method for recognition handwritten Arabic word without segmentation to sub letters based on feature extraction scale invariant feature transform (SIFT) and support vector machines (SVMs). [44] proposed system for offline isolated Arabic handwriting character. Although half of the dataset used for training the Support Vector Machine (SVM) and the second half used for testing, the system achieved high performance with less training data.

The lack of database is one of the main issues associated with offline Arabic text recognition. It is important to note that there is no widely accepted database of Arabic text recognition that is freely available for researchers. Feature extraction is very problem dependent. Extracting features from images can be done in two manners. The first category called handcrafted features and the second category of features called learned features.

Table 4. Summary of results for classification

Authors	Features	Classifier	Data	Accuracy
[36]	Run Length, and Edge Direction features	Support Vector Machine (SVM)	KHATT database	92.80%
[37]	Gabor filter	Support Vector Machine with an RBF kernel	HACDB	88.77%
[27]	CNN	Convolutional Neural Network (CNN), Support Vector Machine (SVM)	HACDB and IFN/ENIT databases	94.17% for HACDB, 92.95% for INF/ENIT
[38]	SIFT descriptor	Support Vector Machine (SVM)	HACDB	91.14%
[39]	Statistical features, Topological features	MSHMM	IFN/ENIT database	91.10% on set a
[40]	Baseline estimation	HMM/MLP	IFN/ENIT database	89.03%
[41]	Chebyshev moments, Contour-based Features	Multilayer perceptron, support vector machine and the Extreme Learning Machine	IFN/ENIT database	96.82%
[42]	Baseline estimation	Hidden Markov Models (HMMs)	IFN/ENIT database	87.93%
[30]	CNN as an automatic feature extractor	CNN based HMM	IFN/ENIT database	89.23%
[43]	SIFT	Support Vector Machine (SVM)	AHDB database	99.08%
[44]	Connected components feature and Zoning features	Support Vector Machine (SVM)	560 handwriting character images	99.64%

Learned features are trying to overcome the issues of handcrafted features. Since they can be trained on any data set, they generalize very well to change. Moreover, they are generally able to cope well with unknown examples, due to the higher generalization capabilities of neural networks and they do not require expert knowledge of the data and should not require a lot of human-labor time. There is also need for the development of newer features that consider Arabic the characteristics of Arabic text. The automatic approaches have remained largely unexplored. With the growing interest in multi-classifier systems for Arabic text recognition, valuable clues that could support the improvement of recognition accuracies can be provided by automatic approaches.

Selecting the proper classifier depends on the feature distribution and task that wanted to achieve, since classifiers has own weakness such as huge calculation, not-stable performance between hidden layers and slow in training. However, obtaining strong Arabic handwriting recognition requires two crucial parts: the first one is incorporating wide range of features; the second part is incorporating stronger classifiers with easy calculation and stable performance.

7 Conclusion

The main aim of this article is to provide a comprehensive survey of research works that have been published in the different phases of Arabic handwriting recognition. Therefore, in the beginning part of this article, the characteristics of Arabic script were discussed. More so, a discussion on the available Arabic text databases which have been used by researchers is presented. It is important to note that there is no widely accepted database of Arabic text recognition that is freely available for researchers. Therefore, different data have been used by many researchers in the field of text recognition. Consequently, a comparison of the recognition rates of different techniques cannot be done. In Table 2, a summary of all the published Arabic datasets is presented.

The nature of the text determines the choice of the most appropriate kind of features. In addition to the nature if the text, other factors that influence this choice of include types of scripts (printed or machine written), as well as the type of system processing (online or offline). The use of CNN-based techniques has been employed in many researches in the area of Arabic handwritten text/word recognition. In the context of Arabic text recognition, the automatic approaches have remained largely unexplored. With the growing interest in multi-classifier systems for Arabic text recognition, valuable clues that could support the improvement of recognition accuracies can be provided by automatic approaches.

There is rarely any free database available for researchers. Even though various techniques for Arabic text recognition have been explored by researchers, the use of databases has been employed for the different techniques in Arabic for performance evaluation. These databases are mainly author generated and limited, and as such it may be impossible to make a comparison of the different results obtained by different researchers in the area of Arabic text recognition. There is a relationship between achieved recognition rates and the quality of the used database. One database can be used to achieve a technique with high recognition rates. However, it is possible for the same technique to yield low recognition rates if used with a different database. In addition, it may be impossible to directly apply techniques which recognize Arabic characters or sub-words to unconstrained Arabic handwritten text. Due to this, a comprehensive database may be required for handwritten text. One of the reasons Latin text recognition has advanced is due the availability of large databases that are freely available such as CEDAR, IRONOFF, CEDAR and IAM. Therefore, such databases must be created if Arabic handwriting recognition must advance in like manner.

It can be observed from Tables 3 and 4 that majority of the recent results for Arabic word recognition are reported for the IFN/ENIT database. Despite the fact that the

IFN/ENIT database is limited in terms of writing variations and vocabulary, the large number of reported results on the IFN/ENIT database may provide a way to compare the performance of different systems.

Acknowledgements. The author would like to acknowledge Universiti Tun Hussein Onn Malaysia (UTHM) for the support of this research under the Tier 1 Grant: vot H093.

References

1. Bahlmann C (2006) Directional features in online handwriting recognition. Pattern Recognit 39(1):115–125
2. Fujisawa H (2008) Forty years of research in character and document recognition-an industrial perspective. Pattern Recognit 41(8):2435–2446
3. Cheriet M (2008) Visual recognition of Arabic handwriting: challenges and new directions. In: Arabic and Chinese handwriting recognition. Springer, Berlin, pp 1–21
4. Parvez MT, Mahmoud SA (2013) Offline Arabic handwritten text recognition: a survey. 45 (2)
5. Tagougui N, Kherallah M, Alimi AM (2013) Online Arabic handwriting recognition: a survey. Int J Doc Anal Recognit 16(3):209–226
6. Lorigo LM, Govindaraju V (2006) Offline arabic handwriting recognition: a survey. IEEE Trans Pattern Anal Mach Intell 28(5):712–724
7. Khorsheed MSM (2000) Automatic recognition of words in arabic manuscripts, University of Cambridge, Computer Laboratory
8. Lorigo LM, Govindaraju V (2006) Offline Arabic handwriting recognition: a survey. 28 (5):712–724
9. Märgner V, El Abed H (2008) Databases and competitions: strategies to improve Arabic recognition systems. In: Arabic and Chinese handwriting recognition. Springer, pp 82–103
10. Schlosser S (1995) ERIM Arabic document database. Environmental Research Institute of Michigan
11. Abuhaiba ISI, Mahmoud SA, Green RJ (1994) Recognition of handwritten cursive Arabic characters. IEEE Trans Pattern Anal Mach Intell 16(6):664–672
12. Kharma N, Ahmed M, Ward R (1999) A new comprehensive database of handwritten Arabic words, numbers, and signatures used for OCR testing. In: 1999 IEEE Canadian conference on electrical and computer engineering, vol 2, pp 766–768
13. Khedher MZ, Abandah G (2002) Arabic character recognition using approximate stroke sequence. In: Proceedings of workshop Arabic language resources and evaluation: status and prospects and 3rd international conference on language resources and evaluation (LREC 2002)
14. Al-Ma'adeed S, Elliman D, Higgins CA (2002) A data base for Arabic handwritten text recognition research. In: 2002 Proceedings of eighth international workshop on frontiers in handwriting recognition, pp 485–489
15. Pechwitz M, Maddouri SS, Märgner V, Ellouze N, Amiri H (2002) IFN/ENIT-database of handwritten Arabic words. Proc CIFED 2:127–136
16. Al-Ohali Y, Cheriet M, Suen C (2003) Databases for recognition of handwritten Arabic cheques. Pattern Recognit 36(1):111–121
17. Mozaffari S, Faez K, Margner V, El-Abed H (2007) Strategies for large handwritten Farsi/Arabic lexicon reduction. In: 2007 Ninth international conference on document analysis and recognition, ICDAR 2007, vol 1, pp 98–102

18. A.M ANALYSIS (2007) Arabic-Handwritten-1.0 database. http://appliedmediaanalysis.com/Datasetshtm#Arabic
19. El-Sherif and EA, Abdelazeem S (2007) A two-stage system for Arabic handwritten digit recognition tested on a new large database. In: International conference on artificial intelligence pattern recognition, January 2007, pp 237–242
20. Srihari SN, Ball GR, Srinivasan H (2008) Versatile search of scanned Arabic handwriting. In: Arabic and Chinese handwriting recognition. Springer, pp 57–69
21. Kherallah M, Elbaati A, Abed HE, Alimi AM (2008) The on/off (LMCA) dual Arabic handwriting database. In: 11th International conference on frontiers in handwriting recognition (ICFHR)
22. Mahmoud SA et al (2012) KHATT: Arabic offline handwritten text database. In: International conference on frontiers in handwriting recognition, IWFHR, pp 449–454
23. Azeem SA, Ahmed H (2013) Effective technique for the recognition of offline Arabic handwritten words using hidden Markov models. Int J Doc Anal Recognit 16(4):399–412
24. Khémiri A, Kacem A, Belaïd A (2014) Towards arabic handwritten word recognition via probabilistic graphical models. In: 2014 14th International conference on frontiers in handwriting recognition, pp 678–683
25. Elzobi M, Al-Hamadi A, Al Aghbari Z, Dings L, Saeed A (2014) Gabor wavelet recognition approach for off-line handwritten arabic using explicit segmentation. In: Image processing and communications challenges, vol 5. Springer, Heidelberg, pp 245–254
26. Kessentini Y, Burger T, Paquet T (2015) A Dempster–Shafer theory based combination of handwriting recognition systems with multiple rejection strategies. Pattern Recognit 48 (2):534–544
27. Elleuch M, Maalej R, Kherallah M (2016) A new design based-SVM of the CNN classifier architecture with dropout for offline Arabic handwritten recognition. Proc Comput Sci 80:1712–1723
28. Amrouch M, Rabi M (2017) Deep neural networks features for arabic handwriting recognition. In: International conference on advanced information technology
29. Maalej R, Kherallah M (2018) Convolutional neural network and BLSTM for offline Arabic handwriting recognition. In: 2018 International Arab conference on information technology (ACIT), pp 1–6
30. Amrouch M, Rabi M, Es-Saady Y (2018) Convolutional feature learning and CNN based HMM for Arabic handwriting recognition. In: International conference on image and signal processing, pp 265–274
31. Khémiri A, Echi AK, Elloumi M (2019) Bayesian versus convolutional networks for Arabic handwriting recognition. Arab J Sci Eng 1–19
32. Siddhu MK, Parvez MT, Yaakob SN (2019) Combining statistical and structural approaches for Arabic handwriting recognition. In: 2019 International conference on computer and information sciences (ICCIS), pp 1–6
33. Karamizadeh S, Abdullah SM, Halimi M, Shayan J, Javad Rajabi M (2014) Advantage and drawback of support vector machine functionality. In: 2014 International conference on computer, communications, and control technology (I4CT), pp 63–65
34. Ding S, Zhao H, Zhang Y, Xu X, Nie R (2015) Extreme learning machine: algorithm, theory and applications. Artif Intell Rev 44(1):103–115
35. Rana A, Rawat AS, Bijalwan A, Bahuguna H (2018) Application of multi layer (Perceptron) artificial neural network in the diagnosis system: a systematic review. In: 2018 International conference on research in intelligent and computing in engineering (RICE), pp 1–6
36. Djeddi C, Meslati LS, Siddiqi I, Ennaji A, El Abed H, Gattal A (2014) Evaluation of texture features for offline arabic writer identification. In: 2014 11th IAPR international workshop on document analysis systems, pp 106–110

37. Elleuch M, Lahiani H, Kherallah M (2015) Recognizing Arabic handwritten script using support vector machine classifier. In: 2015 15th International conference on intelligent systems design and applications (ISDA), pp 551–556
38. Elleuch M, Mokni R, Kherallah M (2016) Offline Arabic Handwritten recognition system with dropout applied in deep networks based-SVMs. In: 2016 International joint conference on neural networks (IJCNN), pp 3241–3248
39. Jayech K, Mahjoub MA, Ben Amara NE (2016) Synchronous multi-stream hidden Markov model for offline Arabic handwriting recognition without explicit segmentation. Neurocomputing 214:958–971
40. Rabi M, Amrouch M, Mahani Z (2017) Contextual Arabic handwriting recognition system using embedded training based hybrid HMM/MLP models. Trans Mach Learn Artif Intell 5 (4)
41. Tamen Z, Drias H, Boughaci D (2017) An efficient multiple classifier system for Arabic handwritten words recognition. Pattern Recognit Lett 93:123–132
42. Rabi M, Amrouch M, Mahani Z (2018) Recognition of cursive Arabic handwritten text using embedded training based on hidden Markov models. Int J Pattern Recognit Artif Intell 32 (01):1860007
43. Hassan AKA, Mahdi BS, Mohammed AA (2019) Arabic handwriting word recognition based on scale invariant feature transform and support vector machine. Iraqi J Sci 60(2):381–387
44. Salam M, Hassan AA (2019) Offline isolated arabic handwriting character recognition system based on SVM. Int Arab J Inf Technol 16(3):467–472

Implementing Virtual Machine: A Performance Evaluation

Hazalila Kamaludin(✉), Muhamad Yusmaleef Jamal, Nurul Hidayah Ab Rahman, Noor Zuraidin Mohd Safar, and Suhaimi Abd Ishak

Faculty of Computer Science and Information Technology, Universiti Tun Hussein Onn Malaysia, Parit Raja Batu Pahat, Johor, Malaysia
{hazalila,hidayahar,zuraidin,suhaimiabd}@uthm.edu.my, yusmaleef@gmail.com

Abstract. A hypervisor is a hardware virtualization technique that allows multiple guest operating systems to run on a single host machine at the same time. Each Virtual Machine (VM) or known as guest operating system emulates all interfaces and resources of a real computer system. Virtualization is beneficial as one of the educational tools to facilitate students' hands-on experiences and research activities. However, the performance of VM needs to be taken into consideration. We investigate the performance of a set of VMs using Oracle VirtualBox on several host machines, each of which has its own system specifications. We observe the resource utilization of each host machine in terms of its CPU utilization, CPU speed as well as memory usage. Experimental results show that the CPU utilization averages are 51.78%, 60.7% and 62.57% for cases before memory allocation, 1/2 of memory capacity and 2/3 of memory capacity, respectively. It is indicate that the utilization of a host processor is directly proportional to the memory capacity assigned for a virtual machine.

Keywords: Virtual machine · Guest operating system · Virtualization · Performance

1 Introduction

A virtual machine (VM) uses virtualization infrastructure where it can be applied not only to subsystems (e.g. disks), but to an entire machine such as Central Processing Unit (CPU), Input/Output (I/O) and virtual resources [1]. The capability of performing tasks as a replicate machine enables VM as an educational tool to facilitate students' hands-on experiences and research activities in the virtualization of desktop, server and storage [2, 3]. It also facilitates educational activities in investigating the emerging technologies such as cloud computing and Internet of Things, as the real-world implementation requires a costly high-end infrastructure.

In the case of Operating Systems (OS) course, investigating a number of different OS types (i.e. open-source, closed-source, mobile) are useful to help undergraduate students in understanding the real-world practice [4]. One of VM practical environments are students setting up the multiple guest OS case scenario in a VM using their

R. Ghazali et al. (Eds.): SCDM 2020, AISC 978, pp. 373–381, 2020.
https://doi.org/10.1007/978-3-030-36056-6_35

own computer to ensure they can work at anywhere and anytime (not limited to lab hours).

Running multiple guest OS in a VM, essentially, requires either a high performance of computers or a good virtual resources management for standard specifications of computers. This is to avoid disruption in practical works. Inexperience undergraduate students, however, might have a poor resources management skill that can lead to computers' low performance and students are subsequently demotivated to continue the works. This study, therefore, is carried out to understand the appropriate virtual resources management for standard computer specifications case which significantly useful for undergraduate students to undertake VM practical tasks.

In this study, hands-on tasks of OS course were applied as a case study. Experiments on several host machines were conducted to study the host performance of CPU utilization, CPU speed and memory usage while using different selected guest OS. All host machines were experimented using Oracle VirtualBox on few different guest operating systems such as Ubuntu for Linux distro, Android for open source mobile operating systems and several other open source mobile operating systems (alternative to Android), for instance Phoenix OS, Lineage OS and Plasma Mobile. Oracle VirtualBox is one of the VMs existed in market that enable a host machine to run multiple operating systems in a single server. The readings of CPU utilization, CPU speed and memory usage were taken in three scenarios, namely: (1) before allocation memory size, (2) after allocates 1/2 of the host machine memory size to VM, and (3) after allocates 2/3 of the host machine memory size to VM. All data taken were summarized into table to study the differences. This paper describes a brief observation on the host machine performance before and after running the guest OS.

There were few challenges and constraints arise while conducting this virtualization study that must be aware of to ensure getting the best result such as:

1. What is the capability needed for the host machine?
2. What is the ideal host machine RAM size to run virtual machine smoothly?
3. What type of network connection should be used to allow internet connection?

Our specific contributions with reference to verify the performance of host machine are:

- We design the experiment procedure of evaluating host machine performance before and after running the guest OS.
- We evaluate the hosts performance of CPU utilization, CPU speed and memory usage by allocating 1/2 and 2/3 of the host memory size after running the guest OS.

The rest of this paper is organized as follows. Section 2 gives a survey of the related works. Section 3 defines the experiment settings. Section 4 presents the experimental results and discussion. Section 5 concludes this paper.

2 Related Works

Application in a traditional computer system runs directly on the complete physical machine. Study on virtualization however, demonstrates that virtualization techniques allow a VM to work separately from the physical host [2–5]. By using virtualization concept, applications are executed on VM and each VM typically running a single application and a different operating system [6]. The virtualization techniques also allow customized operating system and resource allocation in each of the VM.

These days, virtualization is employed extensively in data centers mostly to enable cloud computing. A lot of research have been conducted which include VM placement [10], performance [11, 12], migration and consolidation [13] as well as resource management in VM [14–18].

In learning Operating System course, the hands-on experience is crucial for helping students to experience and understand the complexity of managing a virtualized environment for example regarding the utilization of primary hardware resources [4, 5] such as CPU and memory. In order to help this, a virtual environment for instance the VM allows multiple operating systems to be tested.

The main task of VM is to virtualize the memory, I/O devices and the processor. Memory virtualization aims at mapping the physical memory of a VM to the actual machine memory. I/O device virtualization makes sure that each VM acquires a virtual device and processor virtualization pay attention of sensitive instructions [6].

A study on resource management by [5] observed that a huge element of the operational costs in a virtualized environment was related to the inherent complexity of determining good VM-to-host mappings. The difficulty is the requirement to balance the utilization of multiple resources including CPU and memory simultaneously across virtual machines and physical hosts.

According to [7], the standard personal computer processors provide inadequate support for hardware virtualization and therefore, virtualization overhead can be as high as 50% of total computing time [2].

The Xen developers, a well-known and widely used open source hypervisor evaluated their hypervisor together with VMware a commercial hypervisor that mainly used for server consolidation and User Mode Linux. The evaluation is done between the hypervisors using several large applications. Results of the experiment showed that all the three hypervisors introduced a significant slowdown with database and web applications [6]. Another experiments also measured an up to 14% runtime overhead of VMware with compute-intensive simulation applications in High Energy Physics [8].

A study in [3] examined virtual machines in different environments for education purposes. The experiment is done between Windows7 ultimate and Linux mint 15 in VirtualBox as the VM. Results of the experiment showed that there is a difference in performance between single-core and multi-core even in a single test.

In summary, the performance of applications running on VMs has been studied before. However, for implementing VMs in education, students are exposed to the case study of running multiple guest OS in VM. These experiments essentially, require either a high performance of computers or a good virtual resources management for standard specifications of computers while giving them a better experience.

3 Experiment Settings

In order to enable an Internet connection for the three guest OS in VM, a network configuration was set up. The connection used was Network Address Translation (NAT) network connection as this network connection allows guest OS to access external network. The NAT service works in a similar way to a home router, grouping the systems that use it in the network and preventing external systems from directly accessing the systems, but allowing the systems to communicate with each other and use TCP externally and UDP over IPv4 and IPv6 systems.

Following is a network configuration set up for Ubuntu (the same way is applied to every other guest OS).

1. Open the installed VM and choose the guest OS that is to be configured (e.g. Ubuntu).
2. Right click on the guest OS and choose *Settings* as in Fig. 1.

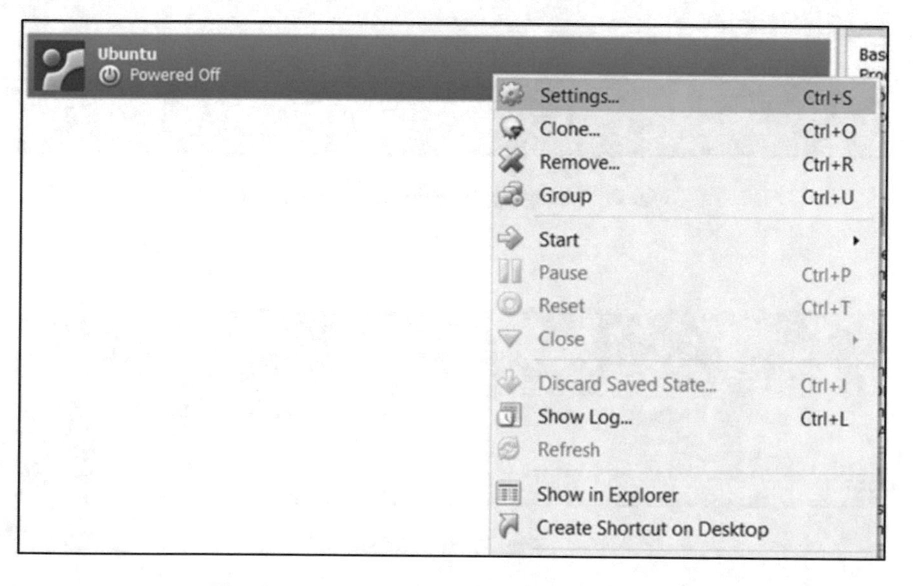

Fig. 1. Open network configuration setting panel

3. Then, at the pop-out window named Ubuntu Settings, click *Enable Network Adapter*, as shown in Fig. 2.
4. Add rule into *Port Forwarding Rule* and insert port to the added rule as in Figs. 3 and 4.
5. Click OK to save changes made to the guest OS network configuration.
6. Start the guest OS and try to access the Internet. It shows that the setting is correct as the Internet is connected as shown in Fig. 5.

Next step is to observe the host performance while running the guest OS. Each guest OS is run one after another and the CPU utilization, CPU speed and memory

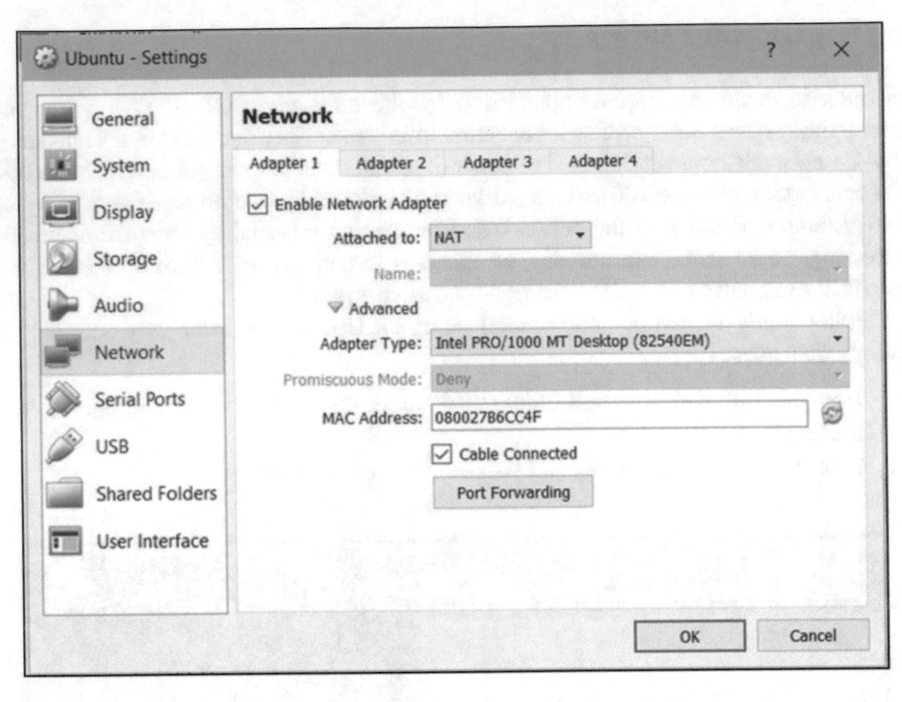

Fig. 2. Setting up the network configuration

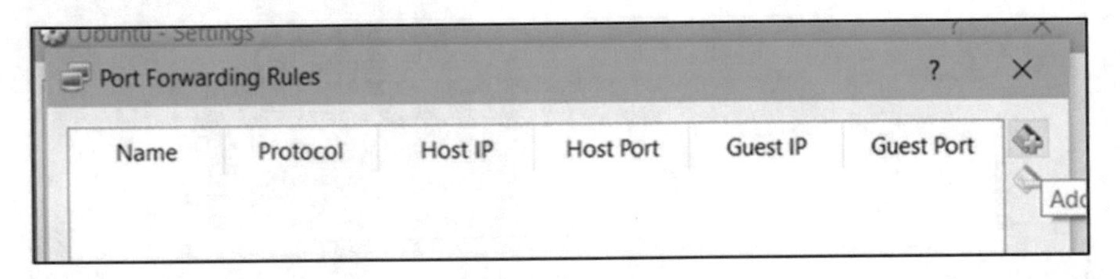

Fig. 3. Add rules

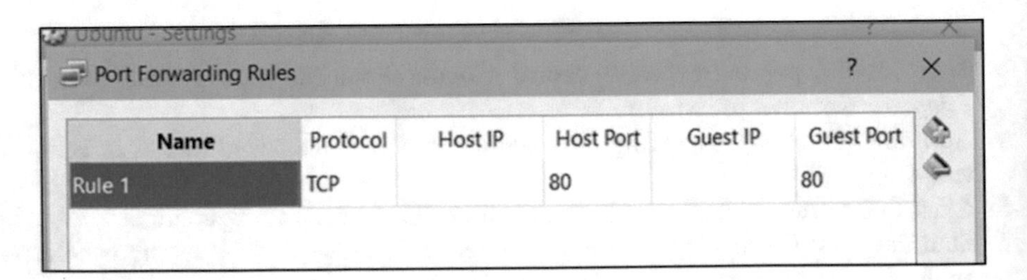

Fig. 4. Insert port

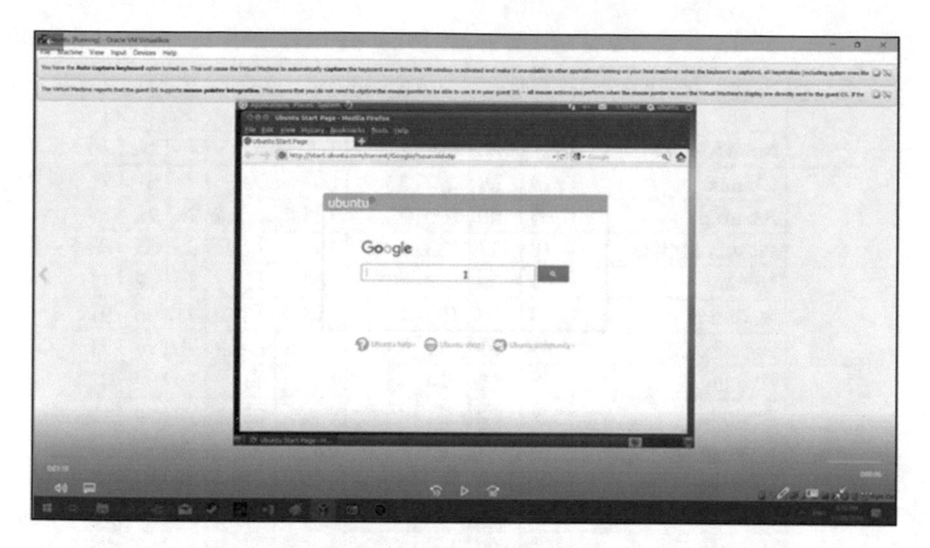

Fig. 5. Access to the internet is successful

usage were recorded. To have a more precise result, the memory size of the host was allocated to 1/2 and 2/3 and the new reading of host performance after both allocations were recorded and arranged into a table.

4 Results and Discussion

Table 1 presents the differences in performance of host machine when users undertake browsing activities in a VM using three guest OS. The performance percentage of CPU utilization, speed and memory usage were observed from 7 groups of students. The observation involves three scenarios that are: (1) before allocation memory size, (2) allocation of 1/2 memory size to VM, and (3) 2/3 allocation of memory size to VM.

Group 1 was using Ubuntu, Android and Plasma Mobile. Observation of Group 1 shows that Android has the highest workload on the resources, followed by Ubuntu and Plasma Mobile, respectively.

Group 2 was using Ubuntu, Android and Phoenix OS. Similarly, observations from Group 2 shows Android have the highest CPU utilization, speed performance and memory usage followed by Ubuntu and Phoenix OS respectively. Phoenix OS presents significant lowest memory usage in the scenario of 2/3 memory allocation compared to the other two Guest OS.

Group 3 was using Ubuntu, Android and Lineage OS. In this group, Android shows highest CPU utilization under scenario of 1/2 memory allocations and 2/3 memory allocation, while other scenario presents percentage over 50%. It was noted that Android has highest speed performance after 2/3 memory allocation.

Group 4 was using Ubuntu, Android and Phoenix OS. This group is presents full workload of CPU utilization for the three OS and for speed performance all three OS for three scenarios has exceed the host machine base speed making the percentage

Table 1. The results of experiments

Group	System performance	CPU utilization (%)			Speed (%)			Memory usage (%)		
	Memory size	1	1/2	2/3	1	1/2	2/3	1	1/2	2/3
1	Ubuntu	19	69	84	35	80	98	77	61	71
	Android	68	100	100	73.6	123.6	123.6	78	78	75
	Plasma Mobile	18	13	23	35.6	33	36.4	66	65	68
2	Ubuntu	65	50	67	123	106	116	70	72	75
	Android	89	100	100	119.2	115.2	121.6	71	90	91
	Phoenix OS	61	81	79	123.6	102.8	109.2	70	70	71
3	Ubuntu	3	20	14	45	46	49	79	92	82
	Android	3	22	21	45	63	62	79	90	92
	Lineage OS	3	15	16	45	78	49	79	92	92
4	Ubuntu	100	100	100	123.6	123.6	124.4	44	58	61
	Android	100	90	100	124	126	125.6	57	59	66
	Phoenix OS	100	100	100	123.6	125.6	123.6	67	67	75
5	Ubuntu	39	81	78	114	114	111	80	75	78
	Android	39	42	73	114	113.6	114	71	72	72
	Lineage OS	35	61	85	114	100.3	114	67	67	72
6	Ubuntu	39	34	17	110.4	112.5	86.3	97	95	92
	Android	15	33	26	82.9	97.9	112.9	92	92	92
	Phoenix OS	87	49	15	110	112.9	88.3	92	92	95
7	Ubuntu	30	55	62	110.4	112.5	86.3	79	95	92
	Android	59	59	70	94	99.6	109.9	74	85	82
	Lineage OS	49	56	65	81.2	100	102.5	79	82	85

reading become more than 100%, while Ubuntu denotes the lowest memory usage under the three scenarios.

Group 5 was using Ubuntu, Android and Lineage OS. Ubuntu and Android show same CPU utilization reading for scenario (1) and all Guest OS show similar reading for speed performance in scenario (1). Similar with group 4, workload of speed exceeding 100% was observed. Lineage OS, on the other hand, presents the lowest memory usage.

Group 6 was using Ubuntu, Android and Phoenix OS. In group 6, both Ubuntu and Phoenix OS present lowest CPU utilization and speed performance after 2/3 memory allocation, while Android presents highest speed performance after 2/3 memory allocation. All the three OS present over 90% of memory usage for the three scenarios.

Group 7 was using Ubuntu, Android and Lineage OS. Ubuntu shows the lowest CPU utilization before memory allocations, while other scenarios present more than 50% of utilization for the three OS. Ubuntu has over 90% of speed performance and memory usage in the scenario of (2) but lower than 90% in scenario (3) of speed performance and in scenario (1) of memory usage.

5 Conclusion

The percentage of CPU utilization and CPU speed will increase when opening or running an application. Task Manager in the host machine is used to observe the percentage of CPU utilization, CPU speed and memory usage. As power consumption increase when CPU utilization increase, the percentage of CPU utilization will rise temporarily during opening an application, play video or doing activities on web and the percentage of CPU usage will drop back to lower level once the in-depth process of CPU completes. The percentage of CPU utilization increases significantly when running the guest OS in the virtual machine. For example, the CPU utilization reading reach 100% when carried out intensive browsing activities in the guest OS. The computer will slow down and therefore, it has to regularly free up processing power for different processes.

Overall, the results showed that all the three VM from each group introduced a significant slowdown while opening an application, play video or web browsing. Obviously, even though virtualized machine reduces the overall number of physical systems and related recurring costs, it introduces overhead which results in a performance loss when applications run on a VM rather than directly on the physical machine. With the evolving technologies such as cloud computing, grid computing and Internet of Things, performance of virtualized machines becomes therefore a research topic.

For future work, this study plans to observe the host performance while running open source OSs targeting for IoT devices. These operating systems focus on low power IoT devices [9] (memory constraint device) such as Contiki, RIOT OS, FreeRTOS, TinyOS, OpenWSN, nuttX OS, Embedded Configurable Operating System (eCos), mbedOS and so on. Even though the device is low on its power and memory, a device should still must be able to monitor inputs, update event and as well as receiving commands from gateway/devices. Therefore an operating system for this device must have the following abilities:

- Task management and scheduling
- Interrupt servicing
- Inter process communication
- Memory management

Furthermore, the extraction of the performance levels produce by this study will benefit the development of machine learning approach to predict the performance of an open source in virtual environment.

Acknowledgments. The authors express appreciation to the Ministry of Higher Education (MOHE) and Universiti Tun Hussein Onn Malaysia (UTHM). This research is supported by the Fundamental Research Grant Scheme (FRGS) grant (Vot 1640)

References

1. Smith JE, Nair R (2007) Introduction to virtual machines. Virtual Machines. Elsevier, pp 1–26
2. Nabhen R, Maziero C (2006) Some experiences in using virtual machines for teaching computer networks. In: Education for the 21st century—impact of ICT and digital resources, pp 93–104
3. Mateljan V, Juricic V, Moguljak M (2014) Virtual machines in education. In: 37th International convention on information and communication technology, electronics and microelectronics (MIPRO), pp 603–607
4. Nieh J, Vaill C (2005) Experiences teaching operating systems using virtual platforms and linux. ACM SIGCSE Bull 37(1):520–524
5. Gulati A, Holler A, Ji M, Shanmuganathan G, Waldspurger C, Zhu X (2012) Vmware distributed resource management: Design, implementation, and lessons learned. VMware Tech J 1(1):45–64
6. Tao J, Furlinger K, Wang L, Marten H (2012) A performance study of virtual machines on multicore architectures. In: 2012 20th Euromicro international conference on parallel, distributed and network-based processing, pp 89–96
7. Chen PM, Noble BD (2001) When virtual is better than real [operating system relocation to virtual machines]. In: Proceedings eighth workshop on hot topics in operating systems, pp 133–138
8. Gilbert L et al (2005) Performance implications of virtualization and hyper-threading on high energy physics applications in a grid environment. In: 19th IEEE international parallel and distributed processing symposium, p 10
9. Hahm O, Baccelli E, Petersen H, Tsiftes N (2015) Operating systems for low-end devices in the internet of things: a survey. IEEE Internet Things J 3(5):720–734
10. Masdari M, Nabavi SS, Ahmadi V (2016) An overview of virtual machine placement schemes in cloud computing. J Netw Comput Appl 66:106–127
11. Joy AM (2015) Performance comparison between linux containers and virtual machines. In: 2015 International conference on advances in computer engineering and applications. IEEE, pp 342–346
12. Sharma P, Chaufournier L, Shenoy P, Tay YC (2016) Containers and virtual machines at scale: a comparative study. In: Proceedings of the 17th international middleware conference. ACM, p 1
13. Ahmad RW, Gani A, Hamid SHA, Shiraz M, Yousafzai A, Xia F (2015) A survey on virtual machine migration and server consolidation frameworks for cloud data centers. J Netw Comput Appl 52:11–25
14. Arianyan E, Taheri H, Sharifian S (2015) Novel energy and SLA efficient resource management heuristics for consolidation of virtual machines in cloud data centers. Comput Electric Eng 47:222–240
15. Piraghaj SF, Calheiros RN, Chan J, Dastjerdi AV, Buyya R (2015) Virtual machine customization and task mapping architecture for efficient allocation of cloud data center resources. Comput J 59(2):208–224
16. Khan MA, Paplinski A, Khan AM, Murshed M, Buyya R (2018) Dynamic virtual machine consolidation algorithms for energy-efficient cloud resource management: a review. In: Sustainable cloud and energy services. Springer, Cham, pp 135–165
17. Gupta MK, Amgoth T (2018) Resource-aware virtual machine placement algorithm for IaaS cloud. J Supercomput 74(1):122–140
18. Toosi AN, Sinnott RO, Buyya R (2018) Resource provisioning for data-intensive applications with deadline constraints on hybrid clouds using Aneka. Futur Gener Comput Syst 79:765–775

Lungs Cancer Nodules Detection from CT Scan Images with Convolutional Neural Networks

Muhammad Zubair Rehman[1], Nazri Mohd Nawi[2(✉)], Aisha Tanveer[1], Hassan Zafar[3], Hamza Munir[3], and Sher Hassan[3]

[1] Faculty of Computing, Universiti Malaysia Pahang, Pekan, Malaysia
[2] Soft Computing and Data Mining Center (SMC), Faculty of Computer Science and Information Technology, Universiti Tun Hussein Onn Malaysia, Batu Pahat, Malaysia
nazri@uthm.edu.my
[3] Machine Learning Research Group (MLRG), The University of Lahore, Islamabad, Pakistan

Abstract. Lungs cancer is a life-taking disease and is causing a problem around the world for a long time. The only plausible solution for this type of disease is the early detection of the disease because at preliminary stages it can be treated or cured. With the recent medical advancements, Computerized Tomography (CT) scan is the best technique out there to get the images of internal body organs. Sometimes, even experienced doctors are not able to identify cancer just by looking at the CT scan. During the past few years, a lot of research work is devoted to achieve the task for lung cancer detection but they failed to achieve accuracy. The main objective of this piece of this research was to find an appropriate method for classification of nodules and non-nodules. For classification, the dataset was taken from Japanese Society of Radiological Technology (JSRT) with 247 three-dimensional images. The images were preprocessed into gray-scale images. The lung cancer detection model was built using Convolutional Neural Networks (CNN). The model was able to achieve an accuracy of 88% with lowest loss rate of 0.21% and was found better than other highly complex methods for classification.

Keywords: Lungs cancer · Convolutional neural networks · Features extraction · Computed tomography · Preprocessing · Deep learning · Computer vision

1 Introduction

Lungs cancer is a major problem around the world causing millions of death every year. According to a recent survey, more than 225,000 people are diagnosed for cancer and 150,000 died because of it. If, we discuss the expense of the lung cancer treatments, not a single person will be able to pay for it as it too expensive. Almost, all the cases diagnosed for lungs cancer only a small percent survive this disease and the developed countries are facing this serious threat more than the developed ones.

R. Ghazali et al. (Eds.): SCDM 2020, AISC 978, pp. 382–391, 2020.
https://doi.org/10.1007/978-3-030-36056-6_36

The lung cancer is divided into two stages and is defined in terms of spread of the infection. In stage 1, the infection is just in a small portion of the lungs but in stage 2, cancer starts spreading from lungs to other body organs. The diagnostic methods used today are biopsies and imaging such as CT scans. The lungs cancer is the only cancer that can be treated well or can be controlled, if it is detected at an early stage. This is where the problem arrives that the detection of lungs cancer is not an easy task and even an experienced doctor may sometime fail to diagnose it properly [1–3]. That is where our Convolutional Neural Networks (CNN) model steps in to diagnose lung cancer with great proficiency. We also know that machines are more efficient and effective as compared to humans. Because humans cannot give their 100% all the time (i.e. they get tired, ill and many other problems) but machine is not effected by any problems.

The main objective of this project is to identify the lungs cancer in the CT scans of a patient. Now for the time being, we are not concerned with the stages of lungs cancer as the goal is just to identify it. The techniques like 2D CNN are used to generate an appropriate classifier model that can distinguish between cancerous nodules and non-nodules. This model can help in speeding up the learning process and greatly reduce the cost for screening test that would in return be able to detect the cancer at an early stage. Apparently, the task seems to be quite easy but indeed this is like searching for a needle in a haystack. In order to get proper results, the system would have to find whether the patient have cancer or not. The system would have to evaluate the tiny nodules that might range from 2 mm to 10 mm in diameter [3].

The main Problem with the CT scans is that there is a lot of noise in it that needs to be removed before the image can be further processed. The noise can be the tissues surrounding the lungs inside the body, rib bones, and the air filled in the lungs [4]. Thus, our prescribed pipelining uses image pre-processing and then comes the classification between nodules and non-nodules. In this paper, extensive work is done upon the dataset first to extract the images and then feeding those extracted images into CNN. The dataset contains 247 three dimension (3D) images, which after extraction were augmented using three Python libraries such as image-data-generator, Imutils and augmentation to generate enough samples for the training of the CNN [5–7].

2 Related Work

In almost 90% of the cases, in which the patients are suffering from lungs cancer, it was mainly due to smoking tobacco. In 2016, according to a study conducted in Brazil there were 28,220 cases related to lungs cancer in various patients [6]. Every year various studies are conducted globally with the aim of developing a model that can greatly increase the accuracy of the lungs cancer disease. Kumar and his fellow researchers proposed a method using Stacked Autoencoder (SAE) and deep learning techniques and were able to achieve an accuracy of 75.01% [8]. In 2014 Kuruvilla and Gunavathi suggested a method that would use texture based features using Artificial Neural Networks and achieved an accuracy of 93.30% [9]. Gupta and Tiwari developed a method using ANN that was based upon shape features and achieved an accuracy of about 90% [10]. Dandil and his mates proposed a method based on features from texture using (PCA) Principal Component Analysis and this was implemented using ANN and achieved an accuracy of 90.63% [11]. Similarly Parveen and Kavitha proposed a scheme using Support Vector Machine (SVM) and attained a Sensitivity of 91.38% and Specificity 89.56% [12].

In 2012 Nacimento and his team used SVM and Linear Discriminant Analysis (LDA) based with texture features and achieved an accuracy of 92.78% [13]. Orozco and his fellows proposed a method using K-Nearest Neighbor (KNN) and SVM with Correlation based Feature Subset Selection and gave an accuracy of 82.66% [14]. Krewer and team developed a model using texture features and shape features with Correlation-based Features Subset Selection and KNN and achieved accuracy of about 90.91% [15]. Aggarwal, Furquan and Karla developed a system that can perform classification between nodules and normal lungs their proposed model could extract geometrical, Gray level and Statistical features. The system was able to achieve 84% of accuracy, Specificity of about 53.33%. The system was able to detect nodules but this accuracy is not acceptable in classification standards of today [16]. Jin, Zhang and Jin used CNN for cancer detection in lungs scans. Their system had an accuracy of about 84.6%. This model was quite popular as it makes the use of circular filter for the Region of Interest (ROI) and the training cost was also reduced along with recognition cost [17].

In 2016, R. Golan suggested a model that can train the weights of the CNN using back-propagation and could detect lungs nodules from computed tomography (CT) scan images. The proposed system was able to achieve sensitivity of 78.9% on the nodules that were approved by at least four radiologist [18, 19]. In 2016, Gonzales and Ponomaryvo developed a system for the detection of lungs cancer that was able to classify cancer in lungs as malignant or benign. The proposed system used Housefield Unit (HU) for the calculation of region of interest (ROI). Several parameters were extracted to be used as features for detection in the Support Vector Machine (SVM) [20]. Hua et al., suggested two methodologies using deep learning algorithms, the first one was Deep Belief Network (DBN) and second was Convolutional neural network, DBN was able to achieve sensitivity rate of 73.30% and Specificity rate 82.20%, whereas CNN was able to achieve Sensitivity rate of 73.30% and Specificity rate of 78.70% [7]. The proposed model in this research will use CNN model and will try to achieve better accuracy than the previous methods used in the literature review.

The paper is organized as follows: A brief introduction to CNN algorithm is given in Sect. 3. Section 4 presents the proposed model for lung cancer detection. The experiments and implementation along with results are discussed in Sects. 5 and 6 respectively. Finally, the findings of this research are concluded in Sect. 7.

3 Architecture of the Convolutional Neural Networks

Recently, Convolutional neural networks achieved state of the art results in the field of object detection, image classification and other visual tasks. Due to the deep learning features, they hold strong grounds in the field of medical imaging, as they are able to detect various diseases in body parts using several extracted features. Their performance is far better than humans as the chances of them making mistakes can be controlled through proper network design [19]. In CNN, neurons are the basic unit that performs the calculation in the entire network. The input is fed to the neuron from various nodes or it can be from any outer source of input and after working with it the neuron learn from it to make predictions. The most significant feature of neuron is that it can learn through outputs from other neurons [21]. In Fig. 1, a simple representation of neurons activation can be seen. The numerical inputs of the from x_1 and x_2 along with their respective weights w_1 and w_2 are the baseline information source. The main function of the activation function is to bring non-linearity to the output of the respective neuron [21].

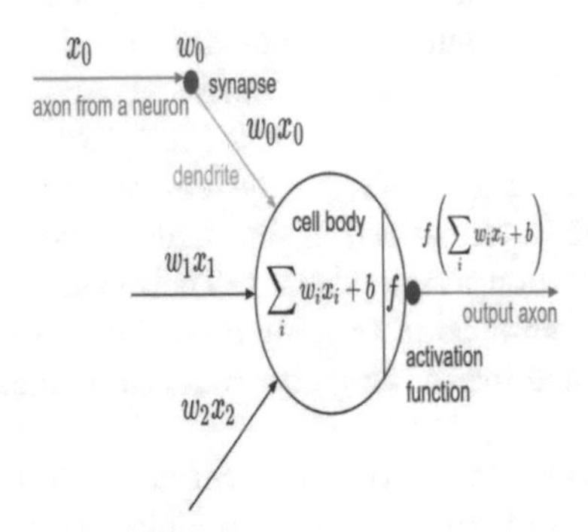

Fig. 1. A single neuron in CNN

4 The Proposed CNN Model for Lung Cancer Detection

The architecture of CNN model for lung cancer detection is simple. First CT-scan images are fed into the network as inputs and then image pre-processing is applied. In this specific case, gray scale filter is used for normalization. Then sample downsizing is performed with zero-centering. The next in the pipelining are Convolutional Layers where Relu Activation function is applied and for pooling purpose Max-Pooling is used. The same sequence is repeated in 5 hidden layers with gradual increase in its density. The filters used for convolution are 3×3 whereas the filter for the pooling layer is 2×2. The number of hidden layers can be increased as per requirement. The dense function is also used to increase the training capability of the model and it started from 32 and then it increases in a manner of 2^n fraction. The batch normalization is also performed to normalize the data stream. Finally, the dropout function is used to randomly dropout samples from the training dataset. Finally, we have our fully connected Layer in which sigmoid function is used as activation function and dropout at this level is about 50%.

4.1 Lungs Cancer Dataset

One of the major problem with computer science projects is the limited availability of resources like memory and CPU. In Deep Learning, a dataset is required to train the model. The basic goal in deep learning is the generalization issue as the model should be able to classify examples other than those from training samples. Deep Learning uses the training dataset for the tuning of parameters and perform a lot of tuning to the attained parameters. The dataset used in this practice is Japanese Society of Radiological Technology (JSRT) dataset. It was created in 1999 and since then it is regularly updated [22]. The dataset contains 154 nodules and 93 non-nodules high resolution samples of 2048 × 2048 pixels. Because of limited resources, (system and GPU memory) we just use various images library for reduction purpose and image-generator libraries given in Keras, Imutils and data-augmentation which are divided in two classes cancer nodules and non-cancer nodules. The data segmentation was 70% for training, 20% for testing, and 10% for validation. The proposed model is made using sci-kit learn as a high level library of python that can work compatibly with theano. This model contain threes major portions input layer, hidden layer and output layer. The input layer and output layer both have same parameters whereas the changes only happen in the hidden layer.

4.2 Problem Formulation

In this paper, CNN is used on CT-scan images to perform the classification of lungs cancerous nodules and non-cancerous nodules. The classification of nodules is meant to be performed upon various images from our dataset and the system should be able to classify the nodules from non-nodules correctly. For performing the classification procedure, different tasks are carried-out in the following order; scale images to same size, create labels for the input classes, save them to a file in the disk, save the designed model, perform detection of objects in the image, check class variations and perform predictions.

5 Experimentation and Implementation

The overall working of the system for performing image classification can be explained after the preprocessing of the data is discussed. For this purpose, the model is trained using an Intel Core i3 processor, 4 GB of RAM and entry level Nvidia Graphics. For preprocessing, OpenCV2 library with menpo packages is used for reducing the image size from 2048 × 2048 to 128 × 128 most significant pixels. Then, a Gray-Scale filter is applied and the resultant image of 128 × 128 is represented as an array. The training samples are labeled by the system as nodules and non-nodules in a manner [0, 1] and [1, 0] respectively. The system learns to develop model using these labels. The results from the model are compared with the test images for the calculation of accuracy. For the purpose of classification the model uses different combination of layers starting with two layer of convolution and then working our way up with five layers of convolution using various activation function to find better results. The final evaluation of the system is made by comparing the predicted results with the actual results using the images that are not used in training.

5.1 Lungs Cancer Dataset

The dataset contain 247 images of 2048 × 2048 pixels. Among them 154 are nodules and 93 are non-nodules. Each of the image from this data is labeled as file name. The test folder also contains same number of images. As the samples are not enough for the training of CNN, we have used three different libraries image data generator by Keras, Image augmentation by Keras and Imutils all these collectively fulfill the sample requirements of the system to train with. For each image in the test folder the model should predict probability of nodules and non-nodules. Output from these images will be a numpy array of size 128 × 128 for every image. Images are saved in an array of numpy as preprocessing images. The training images can be seen in the Fig. 2. The CNN model and its 5 hidden layer configurations are given in the Table 1.

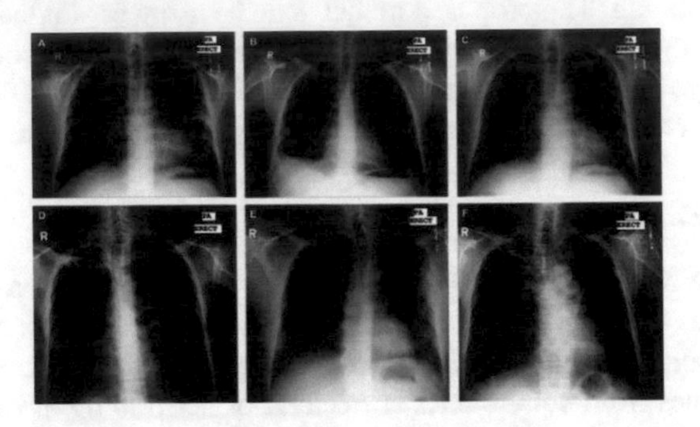

Fig. 2. Training images

On every hidden layer, there is 2D convolution and Relu is used as an activation function. In-order to prove that Relu is faster than other activation functions, we have applied same model with Tanh hidden convolution layers. After activation, Max-pooling 2D is applied and dropout is used to create randomness in the training samples. Dropping out means to remove samples of the given percentage randomly but they are not permanently removed from the network but just for the time being [23].

Table 1. CNN model with 5 hidden layer configurations

Layer	Parameter	Output
Input		128x128x1
1	Conv2D Relu MaxPool2D	3 × 3 × 32 2 × 2 × 32
2	Conv2D Relu MaxPool2D	3 × 3 × 64 2 × 2 × 64
3	Conv2D Relu MaxPool2D	3 × 3 × 128 2 × 2 × 128
4	Conv2D Relu MaxPool2D	3 × 3 × 256 2 × 2 × 256
5	Conv2D Relu MaxPool2D	3 × 3 × 512 2 × 2 × 512
Output	Fully connected Dropout Fully connected	1 × 1 × 1024 0.30 Sigmoid

6 Results and Discussions

In this research, Adam optimizer is with a very small learning rate of 0.001 with sklearn library. The network was trained with different number of epochs to find better results. The network was trained on 120 epochs to 1000 epochs. As the number of hidden layers are increased, the accuracy also increased and the same case was observed with an increase in the number of epochs.

Several tests were performed upon the model to get the best out of it. To prove the model's working, the model was tested using unseen images and check for the prediction whether it is correct or not. All the epochs have been ran on same and different version of same model. And as per results the model with 5 convolutional Relu layer gave the best results and will be used for further study. Different Accuracies with different activation function and epochs are shown in Table 2.

Table 2. Epochs comparison for different layers of CNN

Epoch	(2CNN) loss/accuracy Relu	(5CNN) loss/accuracy Relu	(6CNN) loss/accuracy tanh
120	0.89/0.60	0.512/0.71	0.62/0.75
200	0.80/0.61	0.55/0.72	0.65/0.77
300	0.60/0.75	0.71/0.75	0.70/0.79
400	0.60/0.92	0.20/0.88	0.72/0.80
1000	0.60/0.95	0.27/0.91	0.60/0.82

The confusion matrix given in the Fig. 3 represent the earliest results in this study for 120 and 200 epochs. All the given below results are not quite good in terms of accuracy and the rate of losses is quite high.

After 200 epochs, a positive trend of CNN converging to global minima with a slightly better accuracy was observed. Therefore, maximum number of epochs were increased to 400 and 1000 respectively. From the Fig. 4(a), it can be easily observed that false negatives were rapidly declining and CNN is showing better convergence with an increase in the number of epochs. Even after training the model with 400 epochs, it did not give satisfactory results. True Negatives have reduced to zero, so these results are unacceptable and there is a radical need of improvement in the model. Therefore, the epochs were finally set to 1000 and were tested on the model. The system was able to achieve state-of-the-art results on five convolutional layers, a filter size of 3×3, and pooling layer of 2×2. The confusion matrix can be seen in the Fig. 4(b). While training our CNN model up to 1000 epochs, we are able to successfully train our model with 80 plus percent of validation accuracy. The model actually gave a low rate validation loss after completing its testing phase. The lowest loss rate was about 0.27, which was the most acceptable under the given circumstances. There were several limitation we faced, such as insufficient dataset and less powerful GPU during simulations. Despite all limitations, this model performed well.

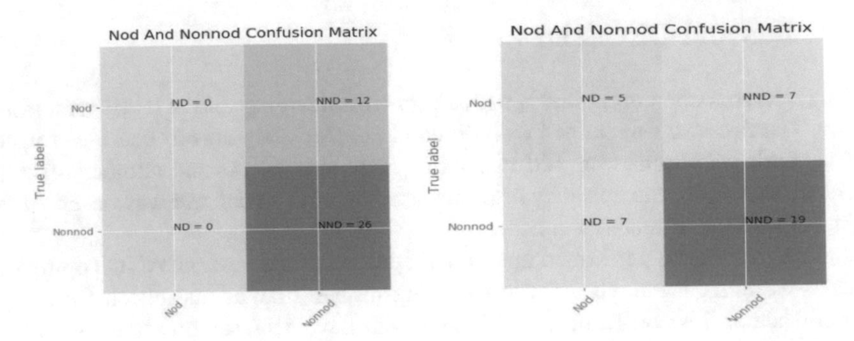

Fig. 3. Confusion matrix for (a) 120 and (b) 200 epochs

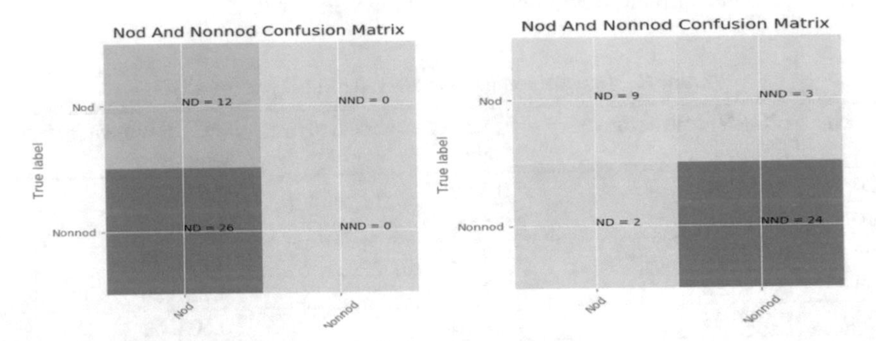

Fig. 4. Confusion matrix for (a) 400 and (b) 1000 epochs

7 Conclusions

CNN can be a helpful tool in lung cancer classification. Using CNN, can significantly improve the accuracy of the model. In addition, increasing the size of the dataset and number of epochs can greatly improve a system's accuracy. In this paper, the proposed CNN model deployed with 5 convolution layers, Relu hidden activations, Max-pooling, and Sigmoid function in the final layer was able to achieve high accuracy in detecting lung cancer. However, this model can be made more reliable with the provision of large number of CT scan images and powerful machine for diagnosing lung cancer.

Acknowledgments. The authors would like to thank Universiti Tun Hussein Onn Malaysia (UTHM) and Ministry of Higher Education (MOHE) Malaysia for financially supporting this Research under IGSP grants note U420 and under Trans-displinary Research Grant Scheme (TRGS) vote no. T003.

References

1. Gindi A, Attiatalla TA, Sami MM (2014) A comparative study for comparing two feature extraction methods and two classifiers in classification of earlystage lung cancer diagnosis of chest x-ray images. J Am Sci 10(6):13–22
2. Choi Wook-Jin, Choi Tae-Sun (2013) Automated pulmonary nodule detection system in computed tomography images: a hierarchical block classification approach. Entropy 15 (2):507–523
3. Chon A, Balachandar N, Lu P (2017) Deep convolutional neural networks for lung cancer detection, Standford University
4. Suzuki K et al (2006) Radiologic classification of small adenocarcinoma of the lung: radiologic-pathologic correlation and its prognostic impact. Ann Thorac Surg 81(2):413–419
5. Kanarek NF et al (2014) Survival after community diagnosis of early-stage non-small cell lung cancer. Am J Med 127(5):443–449
6. Dieguez C (2016) Passive smoking: the importance of 100% smoke-free environments. https://www2.inca.gov.br/wps/wcm/connect/tiposedecancer/site/home/pulmao. Accessed 09 Feb 2016
7. Hua KL et al (2015) Computer-aided classification of lung nodules on computed tomography images via deep learning technique. OncoTargets Ther 8
8. Kumar D, Wong A, Clausi DA (2015) Lung nodule classification using deep features in CT images. In: 2015 12th conference on computer and robot vision. IEEE
9. Kuruvilla J, Gunavathi K ((2014)) Lung cancer classification using neural networks for CT images. Comput Methods Program Biomed 113(1): 202–209
10. Gupta B, Tiwari S (2014) Lung cancer detection using curvelet transform and neural network. Int J Comput Appl 86(1)
11. Dandıl E et al (2014) Artificial neural network-based classification system for lung nodules on computed tomography scans. In: 2014 6th International conference of soft computing and pattern recognition (SoCPaR). IEEE
12. Parveen SS, Kavitha C (2014) Classification of lung cancer nodules using SVM Kernels. Int J Comput Appl 95(25)
13. Nascimento LB, de Paiva AC, Silva AC (2012) Lung nodules classification in CT images using Shannon and Simpson diversity indices and SVM. In: International workshop on machine learning and data mining in pattern recognition. Springer, Berlin
14. Orozco HM et al (2015) Automated system for lung nodules classification based on wavelet feature descriptor and support vector machine. Biomed Eng Online 14(1): 9
15. Krewer H et al (2013) Effect of texture features in computer aided diagnosis of pulmonary nodules in low-dose computed tomography. In: 2013 IEEE International conference on systems, man, and cybernetics. IEEE
16. Aggarwal T, Furqan A, Kalra K (2015) Feature extraction and LDA based classification of lung nodules in chest CT scan images. In: 2015 International conference on advances in computing, communications and informatics (ICACCI). IEEE
17. Jin XY, Zhang YC, Jin QL (2016) Pulmonary nodule detection based on CT images using convolution neural network. In: 2016 9th International symposium on computational intelligence and design (ISCID), vol 1. IEEE
18. Golan R, Jacob C, Denzinger J (2016) Lung nodule detection in CT images using deep convolutional neural networks. In: 2016 International joint conference on neural networks (IJCNN). IEEE

19. Milletari F, Navab N, Ahmadi SA (2016) V-net: Fully convolutional neural networks for volumetric medical image segmentation. In: 2016 Fourth international conference on 3D vision (3DV). IEEE
20. Rendon-Gonzalez, E, Ponomaryov V (2016) Automatic lung nodule segmentation and classification in CT images based on SVM. In: 2016 9th International Kharkiv symposium on physics and engineering of microwaves, millimeter and submillimeter waves (MSMW). IEEE
21. Andrej C (2018) Convolutional neural networks for visual recognition. https://cs231n.github.io/convolutional-networks/Accessed. Accessed 09 July 2018
22. Shiraishi J et al (2000) Development of a digital image database for chest radiographs with and without a lung nodule: receiver operating characteristic analysis of radiologists' detection of pulmonary nodules. Am J Roentgenol 174(1):71–74
23. Srivastava N et al (2014) Dropout: a simple way to prevent neural networks from overfitting. J Mach Learn Res 15(1):1929–1958

Conditional Generative Adversarial Networks for Data Augmentation in Breast Cancer Classification

Weng San Wong(✉), Mohammed Amer, Tomas Maul, Iman Yi Liao, and Amr Ahmed

School of Computer Science, University of Nottingham Malaysia, Semenyih, Malaysia
khcy5wws@exmail.nottingham.edu.my

Abstract. Automatic breast cancer classification benefits pathologists in obtaining fast and precise diagnoses and improving early detection. However, the performance of deep learning models depends greatly on the quality and quantity of the datasets used. Due to the complexity and high costs of patient data collection, many medical datasets, particularly for pathological conditions, suffer from small sample sizes. Hence, developing a deep learning solution for breast cancer classification is still challenging. Data augmentation is one of the popular approaches to bridge this gap. In this work, we propose to use Conditional Generative Adversarial Networks (CGANs) for data augmentation. The aim of training CGANs is to generate a new set of realistic synthetic images and combine these together with real images to form a new augmented training set. The experiments show that most of the images produced by CGAN are reliable and classification performance with CGAN-based data augmentation can achieve good results. This method, unlike traditional data augmentation, can produce histopathological images that are completely different from the existing data. Therefore, this technique has the potential to address data scarcity and to directly benefit the training of deep learning models.

Keywords: Breast cancer classification · Deep learning · Histopathological images · Data augmentation · CGANs

1 Introduction

Breast cancer (BC) is a massive public health problem in the world today. It is the most prevalent form of cancer among women, impacting 2.1 million women every year [1]. Recent breast cancer statistics published by the World Health Organization (WHO) reported that, in 2018 alone, 627,000 women died due to breast cancer, which represents around 15% of all cancer related deaths in women [1].

R. Ghazali et al. (Eds.): SCDM 2020, AISC 978, pp. 392–402, 2020.
https://doi.org/10.1007/978-3-030-36056-6_37

Early detection is critical for treatment. If an abnormal growth of cancer-causing breast cells is detected at an earlier, more curable stage, there is a high probability of preventing the growth becoming life-threatening [2]. However, the identification of morphological features and immunohistochemical markers can be complicated and thus time consuming. Manual interpretation of histopathological images is difficult and laborious [14], which consequently creates the urgent need of Computer-aided diagnosis (CADx) systems to assist pathologists in being more accurate and efficient. CADx systems promise to relieve the heavy workload of pathologists by providing useful information as a second opinion in the diagnostic process.

Machine learning approaches to automatic classification of BC have been frequently researched, and initial results have been promising. However, collecting image datasets in the medical domain is usually difficult and sometimes impossible due to privacy concerns [15]. Only a limited number of datasets are publicly available, and the sizes of these datasets are mostly insufficiently large to train more powerful deep learning models, consequently often tend to over-fit. A common approach is to increase the size of datasets through data augmentation by applying various simple image transformations (e.g. affine transformations and noise) to the original images. Data augmentation is well-known in deep learning as it is a simple technique that can nevertheless effectively reduce over-fitting.

The success of data augmentation techniques motivates an interest in researching more advanced methods to further generate new images for training. Besides the traditional data augmentation methods, potential avenues could involve the deep generative models that have made recent breakthroughs in the field of Computer Vision. Generative Adversarial Networks (GANs) constitute a powerful method for training generative models with neural networks and have proven effective for image generation tasks. This method has become popular for augmenting image data in recent works due to its capability in generating new and realistic images that are similar to those in the training set.

In this paper, we propose to augment the dataset using a conditional GAN (CGAN) model, using the class label as the conditioning information to generate new training samples according to class. We then present the classification performances of different data augmentation methods on the BreaKHis dataset [24] and compare the results. CGAN for data augmentation confers clear benefits to the chosen classification model as it could overcome the small size of the dataset, balance the proportion of benign and malignant class, and most importantly, help avoid misdiagnosis in order to improve the quality of human health.

2 Related Works

In the literature, research related to automatic image processing for cancer diagnosis has been explored for more than four decades. Despite the long interest and large number of approaches directed towards it, cancer detection is still challenging due to the inherent complexity of the underlying images used. In recent years, most of the approaches pertaining to BC classification focus on histopathology images. In [24], the authors introduced a database called BreaKHis and provided a baseline pattern recognition system for BC classification. Approaches used consisted of traditional feature descriptors, achieving a classification accuracy ranging from 80% to 85%.

In BC classification, the current state-of-the-art for distinguishing benign and malignant breast tumors can be divided into three categories: feature extraction [5,24], deep learning and the combination of both. In [6], the authors proposed a Vectors of Locally Aggregated Descriptors (VLAD) encoding for each image in Euclidean space and this method resulted in high detection rates. In contrast, deep learning has been widely applied to develop solutions for BC classification where a CNN is trained to classify BC images. Works such as [3,23] reported results where CNNs can clearly achieve better classification performance as compared to traditional techniques.

Recent studies have also introduced transfer learning in the context of different models such as Inception [18], ResNet [12], VGG [18] and DenseNet [20]. These works showed that the classification performance can be further improved even though such techniques, when applied from scratch, require longer time to train. In [12] the authors utilized a pretrained ResNet to achieve an accuracy of 98.30% for binary BC classification and outperformed the state-of-the-art. In recent years, works such as [11,19,21] have utilized both global and local image information to train models. The method proposed in [19] utilizes CT transform, LBP and Histogram information to extract local features and the experiments obtained a best accuracy of 97.19% and an average accuracy of 95.88%.

The availability of large training sets is an important requirement for achieving good performance in deep learning. As mentioned earlier, due to the high cost and complexity of patient data collection, medical image datasets tend to be small, especially for pathological conditions. There are also concerns about privacy, making it almost impossible to obtain a sufficiently large dataset for training. Small datasets tend to cause over-fitting whereby the model remembers all of training samples but performs poorly on test samples. In other words, the trained model does not generalize well on the data from test set.

Data augmentation has been widely used to cope with data limitations over the years, especially in the medical domain. It works well in machine learning because it acts as a regularizer, preventing the model from over-fitting [22]. Data augmentation can be performed via online or offline processes. In offline data augmentation, transformations are applied before training and increase the size of datasets by a factor which is usually equal to the number of transformations performed. Conversely, in online data augmentation, transformations are applied to mini-batches in real time as they are fed into training.

In recent developments, Generative models based on deep CNNs have become a hot research topic and have achieved several interesting breakthroughs for data augmentation in recent years [4,7]. GANs are a popular framework that typically utilize deep neural networks as image generators. GANs are extremely powerful in generating new images through the transformation of latent variables into data instances. The basic approach and the theoretical foundation of GANs were introduced in [10]. GANs consist of two models, a generator and a discriminator. The training of these models is called adversarial because both are competing against each other in terms of the training objective. Both models are typically implemented using neural networks, and training involves a min-max strategy. The generator is trained to generate better counterfeit images that can maximally fool the discriminator. The role of the discriminator is to learn to distinguish counterfeit samples from real data. GANs have been widely applied in many data generation tasks. Some popular applications of GANs include anime character generation [13], human pose estimation [17], 3D object generation [8,25], image inpainting (hole-filling) [16] and music generation [26].

Table 1. Structure of BreaKHis dataset

Magnification	Benign	Malignant	Total
40×	652	1370	1995
100×	644	1437	2081
200×	623	1390	2013
400×	588	1232	1820
Total of images	2480	5429	7909

3 Methods

3.1 Dataset

The data used in this paper was obtained from the BreaKHis database [24], where the biopsy images were collected and labelled by pathologists of the Pathological Anatomy and Cytopathology (P&D) Lab. Table 1 summarizes the structure of this dataset at different magnification levels.

3.2 Standard Data Augmentation

By investigating previous work in BC classification, it was found that the data augmentation techniques used were mostly based on rotations and flipping. It was therefore decided that the standard geometric transformations to perform in this work would include horizontal and vertical flips, and rotation at 90°, with a 50% probability of being applied to any training image. These transformations were applied online at every batch generation, during training, and were implemented using standard functions in the *torchvision* library in PyTorch.

3.3 Conditional Generative Adversarial Network (CGAN)

GAN is a framework that consists of a pair of neural networks for training generative models. [10] Unlike most generative models that are trained to maximize log likelihood, the learning of GANs is based on game theory. The two models in a GAN are designed to solve a zero-sum game [9], where the goal of the generator is to sample new data that looks like it has been sampled from *the target distribution, and the goal of the discriminator is to distinguish fake data from real data. Both networks are tasked to optimize an opposing loss function, competing and pushing their losses against each other. Since GAN is a game-theoretic method, the objective function is expressed as a mini-max function:

$$\min_G \max_D V(D,G) = E_{x\sim p_{data}(x)}[logD(x)] + E_{z\sim p_z(z)}[log(1-D(G(z)))] \quad (1)$$

where $x \in X$ is data variable, $z \in Z$ is latent variable, p_{data} is data distribution.

This function is used to train both models jointly with opposite goals. During training, discriminator D tries to maximize the probability of the real data and minimize the probability of the generated data. Meanwhile, generator G is trained to minimize the objective function, thus encouraging the underlying network to generate instances that fool the discriminator.

CGAN is a natural extensions of GAN whereby both generator G and discriminator D are conditioned on additional information y. The conditioning information y is fed into both the generator and the discriminator as an additional input, allowing the generator to generate samples with certain attributes. In our case, since we are interested in generating images for specific classes (for supervised learning), we use class labels as the conditioning information. Therefore, the optimization objective function becomes:

$$\min_G \max_D V(D,G) = E_{x\sim p_{data}(x)}[logD(x \mid y)] + E_{z\sim p_z(z)}[log(1-D(G(z \mid y)))] \quad (2)$$

In this paper, we use the CGAN model from *TorchFusion* for the image generation. It consists of two CNNs: one for generator and another for discriminator. The input and output space for the model is defined as below:

- Z is a latent space used to seed the generator network. Some noise z is sampled from a noise distribution p_z, where p_z is a normal distribution with 0 mean and a standard deviation of 1.
- Y is an embedding space for the prior class distribution that is drawn from the training data. The class information is combined with the input noise in the hidden representation, which determines how the model generates images for a specific class.
- X corresponds to the image/data space, which encompasses the output produced by the generator or the input of the discriminator.
- The generator is defined as a function $G : (Z, Y) \to X$. The generative network transforms the data $z \in Z$ from the latent space along with the class attribute $y \in Y$ into a sample in data space X.

- Discriminator $D : (X, Y) \to [0, 1]$ estimates the probability that the input data $x \in X$ under condition Y comes from the data distribution instead of the generator.
- Training details. The training of the generator and the discriminator is done separately using the Adam optimizer with $\beta_1 = 0.5$, $\beta_2 = 0.999$, latent space of size 128, a learning rate of 0.0002 and batch size of 4.

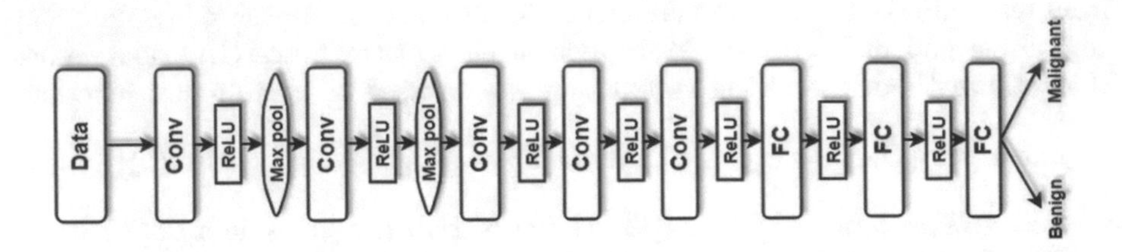

Fig. 1. Schematic presentation of CNN network architecture for BC classification

3.4 Convolutional Neural Network (CNN) Classifier

The architecture of CNN used for classifying BC images in this paper is presented in Fig. 1. The parameters are as follow:

- Input layer. It loads the images with dimensions 64×64 pixels and 3 channels.
- Convolutional layers and pooling layers. There are four convolutional layers in our model. The kernel is of size 5×5 and the stride is set to 1. The first two convolutional layers learn 16 filters and 32 filters respectively, each layer followed by a rectified linear operator (ReLU) and a max pooling layer that is set to use a 2×2 receptive field. Each feature map from a convolutional layer is sub-sampled with max pooling. In the third and last convolutional layers, a set of 64 and 100 filters of size 5×5 are used respectively, each layer being followed by a ReLU non-linearity.
- Fully connected layers. This network contains three fully-connected layers where each layer adopts a ReLU function at the output. The final layer is the output layer, whose size depends on the number of classes in the classification problem. In our case, the final layer has two output nodes corresponding to our intended BC binary classification.
- Training details. The training of CNN classifier is done using Stochastic Gradient Descent (SGD) optimizer at learning rate of 0.0001, batch size of 4 and momentum of 0.9.

4 Experiment and Results

4.1 Experiment

In this work, experiments were conducted on two datasets with different numbers of instances. For the larger dataset, experiments were carried out on the full set of image data, which comprised all BreaKHis magnification levels (i.e. 40×, 100×, 200×, and 400×). For the smaller dataset, experiments were carried out on the subset of images corresponding only to the 400× magnification level. Images from BreaKHis were grouped into benign and malignant classes. The data was split into 70% for training and 30% for testing. To increase efficiency and feasibility, each image was re-sized to 64 × 64 and each channel was normalized using the data mean and standard deviation.

The experimental design consisted of two main parts. In the first part, the CGAN model was trained, and in the second the trained CGAN was used to augment the training set in order to improve classification performance on the target problem. During CGAN training, samples were generated every 500 iterations. At the beginning of training, generated images were indistinguishable from noise. Subsequently, as training progressed, generated images gradually started to show certain similarities with the training images. Figure 2 clearly shows that after 275,500 iterations the CGAN generator is capable of outputting good quality synthetic images. After completion of CGAN training, the generator is ready for generating a set of images according to the class condition. The existing training data was doubled using the following process:

- For the malignant class, we generated malignant images totaling the number of real training images from the benign class.
- For the benign class, we generated benign images totaling the number of real training images from the malignant class.

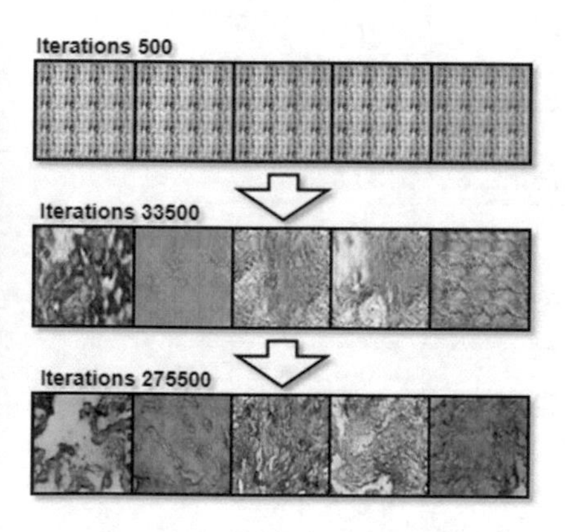

Fig. 2. Generated samples by CGAN in the learning process

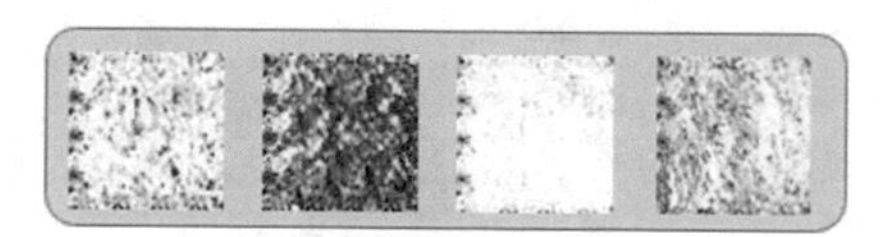

Fig. 3. Some generated samples by CGAN that are noisy

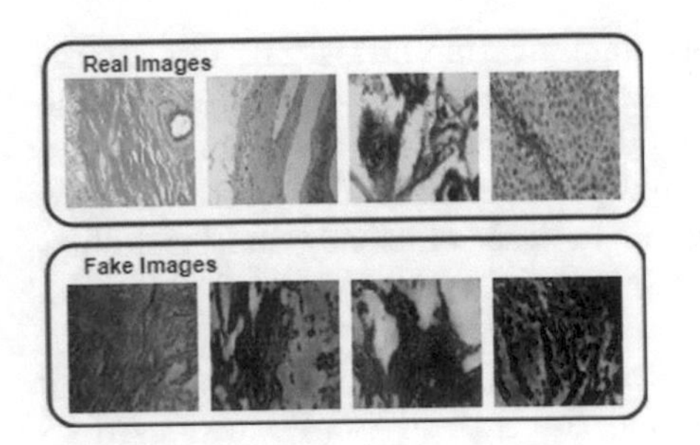

Fig. 4. Generated benign images

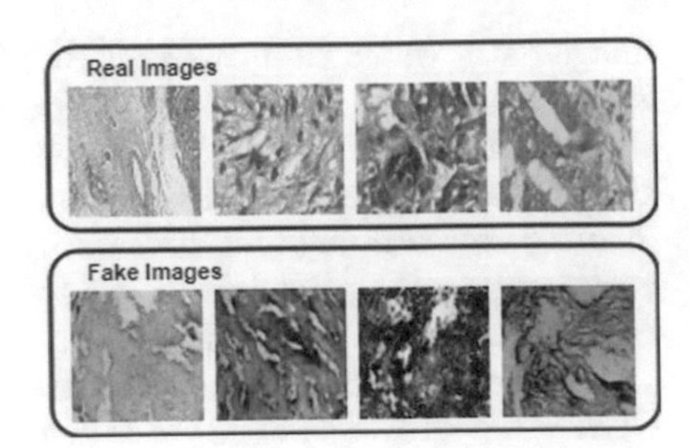

Fig. 5. Generated malignant images

Figures 4 and 5 depicts a comparison between real images from BreaKHis and the synthetic images produced by the CGAN generator. The types of features and patterns found in the generated images are virtually indistinguishable from those found in real images. Even though CGAN is able to produce images with highly similar content, judging by the image color it is evident that the brightness of the generated images is lower, while the contrast is higher. Moreover, some CGAN generated images can exhibit some subtle but noisy artifacts, as depicted in Fig. 3.

Table 2. Classification accuracy for full BreaKHis dataset

Methods	Average accuracy
CNN (No augmentation)	77.167
Data augmentation	76.667
CGAN-based data augmentation	77.345
Data augmentation & CGAN augmentation	75.225

Table 3. Classification accuracy for 400× dataset

Methods	Average accuracy
CNN (No Augmentation)	75.408
Data augmentation	75.932
CGAN-based data augmentation	78.520
Data augmentation & CGAN Augmentation	78.700

4.2 Results

Table 2 contains the classification results for four different data augmentation techniques. CGAN-based data augmentation, achieved an average accuracy of 77.345%, which is better than classification with standard augmentation (i.e. rotations and flips) by 0.678%. Table 3 depicts the results of the same set of experiments applied to the smaller dataset containing only images at the 400× magnification level. CGAN, combined with standard augmentation, achieved an average accuracy of 78.7%, which outperforms experiments with standard augmentation alone by 2.768%, and outperforms experiments with no data augmentation by 3.292%.

5 Conclusion

In this work, we proposed the utilization of CGANs for data augmentation in the context of BC histopathology image classification. We hypothesized that the images automatically generated by a properly trained CGAN, would improve the performance of another model trained on our target dataset, compared to no augmentation and standard augmentation conditions. Our experiments confirmed our hypothesis, especially in the case of smaller datasets. Although some generated images might still contain subtle noisy artefacts, the experiments show that CGANs, even when designed and trained without major modifications, can be used for extending training sets and improving deep learning solutions for classification problems with limited and imbalanced data. For the limited data case, as in our small dataset condition containing only 400× magnification images, CGANs can extend the size of the dataset with new instances, whereas in the imbalanced case, CGANs can generate samples to compensate for the under-represented classes. Given these positive results, we expect future extensions of the CGAN architecture and training protocols, to contribute even more effectively to data augmentation, and therefore classification performance, in data scarce and imbalanced domains.

References

1. Cancer: Breast cancer. https://www.who.int/cancer/prevention/diagnosis-screening/breast-cancer/en/
2. Why is early diagnosis important? (2018). https://www.cancerresearchuk.org/about-cancer/cancer-symptoms/why-is-early-diagnosis-important
3. Bayramoglu N, Kannala J, Heikkil J (Dec 2016) Deep learning for magnification independent breast cancer histopathology image classification. In: 2016 23rd international conference on pattern recognition (ICPR), pp 2440–2445. https://doi.org/10.1109/ICPR.2016.7900002
4. Bowles C, Chen L, Guerrero R, Bentley P, Gunn R, Hammers A, Dickie DA, Valdés Hernández M, Wardlaw J, Rueckert D (2018) GAN augmentation: augmenting training data using generative adversarial networks. CoRR abs/1810.10863. http://arxiv.org/abs/1810.10863
5. Chattoraj, S., Vishwakarma, K.: Classification of histopathological breast cancer images using iterative VMD aided zernike moments & textural signatures. CoRR abs/1801.04880. http://arxiv.org/abs/1801.04880

6. Dimitropoulos K, Barmpoutis P, Zioga C, Kamas A, Patsiaoura K, Grammalidis N (2017) Grading of invasive breast carcinoma through grassmannian vlad encoding. PLOS One 12(9):1–18. https://doi.org/10.1371/journal.pone.0185110
7. Frid-Adar M, Klang E, Amitai M, Goldberger J, Greenspan H (2018) Synthetic data augmentation using GAN for improved liver lesion classification. CoRR abs/1801.02385. http://arxiv.org/abs/1801.02385
8. Gadelha M, Maji S, Wang R (2016) 3d shape induction from 2d views of multiple objects. CoRR abs/1612.05872. http://arxiv.org/abs/1612.05872
9. Gauthier J (2015) Conditional generative adversarial nets for convolutional face generation
10. Goodfellow IJ, Pouget-Abadie J, Mirza M, Xu B, Warde-Farley D, Ozair S, Courville A, Bengio Y (2014) Generative adversarial nets. In: Proceedings of the 27th international conference on neural information processing systems, NIPS 2014, vol 2. MIT Press, Cambridge, MA, USA, pp 2672–2680. http://dl.acm.org/citation.cfm?id=2969033.2969125
11. Gupta V, Bhavsar A (June 2018) Sequential modeling of deep features for breast cancer histopathological image classification. In: The IEEE conference on computer vision and pattern recognition (CVPR) workshops
12. Habibzadeh Motlagh N, Jannesary M, Aboulkheyr H, Khosravi P, Elemento O, Totonchi M, Hajirasouliha I (2018) Breast cancer histopathological image classification: a deep learning approach. https://www.biorxiv.org/content/early/2018/01/04/242818
13. Jin Y, Zhang J, Li M, Tian Y, Zhu H, Fang Z (2017) Towards the automatic anime characters creation with generative adversarial networks. CoRR abs/1708.05509. http://arxiv.org/abs/1708.05509
14. Kårsnäs A (2014) Image analysis methods and tools for digital histopathology applications relevant to breast cancer diagnosis. PhD thesis, Uppsala University, Division of visual information and interaction, computerized image analysis and human-computer interaction
15. Kohli MD, Summers RM, Geis JR (2017) Medical image data and datasets in the era of machine learning whitepaper from the 2016 C-MIMI meeting dataset session. J Digit Imaging
16. Li Y, Liu S, Yang J, Yang M (2017) Generative face completion. CoRR abs/1704.05838. http://arxiv.org/abs/1704.05838
17. Ma L, Jia X, Sun Q, Schiele B, Tuytelaars T, Gool LV (2017) Pose guided person image generation. CoRR abs/1705.09368. http://arxiv.org/abs/1705.09368
18. Myung Jae L, Da Eun K, Dong Kun C, Hong L, Young Man K (2018) Deep convolution neural networks for medical image analysis. Int J Eng Technol 7(3.33). https://doi.org/10.14419/ijet.v7i3.33.18588
19. Nahid A, Kong Y (2018) Histopathological breast-image classification using local and frequency domains by convolutional neural network. Information 9:19. https://doi.org/10.3390/info9010019
20. Nawaz M, Sewissy AA, Soliman THA (2018) Multi-class breast cancer classification using deep learning convolutional neural network. Int J Adv Comput Sci Appl 9(6):316–332. https://doi.org/10.14569/IJACSA.2018.090645
21. Nazeri K, Aminpour A, Ebrahimi M (2018) Two-stage convolutional neural network for breast cancer histology image classification. CoRR abs/1803.04054. http://arxiv.org/abs/1803.04054
22. Perez L, Wang J (2017) The effectiveness of data augmentation in image classification using deep learning. CoRR abs/1712.04621. http://arxiv.org/abs/1712.04621

23. Spanhol FA, Oliveira LS, Petitjean C, Heutte L (July 2016) Breast cancer histopathological image classification using convolutional neural networks. In: 2016 international joint conference on neural networks (IJCNN), pp 2560–2567. https://doi.org/10.1109/IJCNN.2016.7727519
24. Spanhol FA, de Oliveira LES, Petitjean C, Heutte L (2016) A dataset for breast cancer histopathological image classification. IEEE Trans Biomed Eng 63:1455–1462
25. Wu J, Zhang C, Xue T, Freeman WT, Tenenbaum JB (2016) Learning a probabilistic latent space of object shapes via 3d generative-adversarial modeling. CoRR abs/1610.07584. http://arxiv.org/abs/1610.07584
26. Yang L, Chou S, Yang Y (2017) Midinet: a convolutional generative adversarial network for symbolic-domain music generation using 1d and 2d conditions. CoRR abs/1703.10847. http://arxiv.org/abs/1703.10847

Reducing the Effects of Time Cheating on the Performance of Divisible Load Scheduling Using Analytical Hierarchy Process

Shamsollah Ghanbari[1,3(✉)] and Mohamed Othman[2,3(✉)]

[1] Islamic Azad University, Ashtian Branch, Iran
myrshg@gmail.com

[2] Department of Communication Technology and Network, Faculty of Computer Science and Information Technology, Universiti Putra Malaysia, UPM Serdang, 43400 Seri Kembangan, Selangor D.E., Malaysia
mothman@upm.edu.my

[3] Institute for Mathematical Research, Universiti Putra Malaysia, UPM Serdang, 43400 Seri Kembangan, Selangor D.E., Malaysia

Abstract. The divisible load theory (DLT) is a paradigm in the area of distributed and parallel computing. As a matter of fact, the computations and communications can be divided into some independent part in which each part can be executed separately by a processor. The problem that the processors may cheat the algorithm has examined the divisible load theory. However, the computation rate cheating issue may appear if the processors accomplish their fraction of loads with various rates. According to the literature, if the processors do not report their true computation rates, they can not obtain optimal performance. This paper focuses on this problem. This paper proposes an AHP-based divisible load scheduling method aiming to decrease the impacts of cheating on the efficiency of divisible load scheduling. The experimental results indicate the proposed method considerably reduce the impacts of cheating on the startup time, speedup, and makespan specially when a huge number of processors cheat the algorithm.

Keywords: Analytical hierarchy process (ahp) · Scheduling · Divisible load theory · Multi-level tree network

1 Introduction

Divisible Load Scheduling (DLS) is a class of scheduling techniques that is suitable for using in the field of parallel and distributed computing with big data. The first articles regarding to the divisible load scheduling theory (DLT) were appeared in 1988, [1,2]. The DLT has found an expansive range of applications over the past three decades e.g., image and vision processing [3], data grid applications [4], wireless sensor networks [5], cloud computing [6], and so on.

R. Ghazali et al. (Eds.): SCDM 2020, AISC 978, pp. 403–416, 2020.
https://doi.org/10.1007/978-3-030-36056-6_38

An extensive survey on the divisible load scheduling, containing applications, approaches, procedures and challenges can be found in [8,9]. Basically, the DLT is due to the fact that the communication and computation can be divided into some arbitrary sizes fractions, and each fraction can be processed separately in parallel. Assume p_0, p_1, p_2, ..., p_m are $m+1$ independent processors. The p_0 is called originator and others are called worker processor. It is assumed that, preliminary amount $V=1$ of load is delivered to the originator. The originator holds α_0 fraction of load for itself and distributes the $1-\alpha_0$ fraction of load to the worker processors. Consider p_i $(0 \leq i \leq m)$ receives α_i fraction of load from the originator. The condition for obtaining optimal performance is that all processors stop proceeding simultaneously [10]. The goal is to calculate optimized value of $\alpha_0, \alpha_1, \ldots, \alpha_m$ in the DLS system. Equation (1.1), indicates a close form for a general DLS model [10].

$$\alpha_j = \left(\frac{z_{j-1}T_{cm} + w_{j-1}T_{cp}}{z_jT_{cm} + w_jT_{cp}}\right)\alpha_{j-1} \tag{1.1}$$

and

$$\alpha_0 = \left(\frac{z_1T_{cm} + w_1T_{cp}}{w_1T_{cp}}\right)\alpha_1 \tag{1.2}$$

where $\alpha_0 + \alpha_1 + \cdots + \alpha_m = v$.

In Eq. (1.3), We have:

$$T = \alpha_0 w_0 T_{cp} \tag{1.3}$$

In a multi level tree form of divisible load scheduling the load is spread from a high-level to the lower one, by catching through each level in a multi-level tree. An optimal makespan can be achieved by traversing the tree from the higher level to the lower subtrees with single equivalent processors (denoted by w_j^{eq}) until T is eliminated to one processor. The details about multi level tree divisible load scheduling is available at [7]. The existing DLS models assume that the processors announce their real communication and computation rates, it means, the processors do not cheat the algorithm. In the real applications of the DLS, the algorithm may be cheated by the processors. It means that the processors may not announce the real computation rates to the originator. This problem was considered in [11] first time. The results of that research demonstrate that the cheating decreases the makespan in the divisible load scheduling model. Few years later, the cheating problems of the divisible load scheduling has been investigated by the other researchers [7].

The method proposed in [7] was able to considerably reduce the effects of computation rate-cheating on the performance of DLS. However, that method has some limitations for instance it increases the start-up time when we have a huge number of processors that cheat the algorithm.

In this paper, we propose a solution for limitations of [7]. The proposed method is called "AHP-Based" method. The results indicate that the proposed method considerably reduces the startup time and makespan particularly when we have a number of processors that cheat the algorithm.

We apply the AHP-based method on a binary tree network (multi-level tree) in two different scenarios. In the first scenario we consider a binary tree networks comprising fifteen processors labeled p_0, p_1, ..., p_{14}. In the second scenario we consider the same binary tree networks comprising 63 processors labeled $p_0, p_1, \ldots, p_{62}$. The results indicated that the proposed method has considerable performance in the both scenarios.

We use the Analytical Hierarchy Process (AHP)for this purpose. Generally, the AHP contains three major levels, including objective, attributes and alternatives levels. Each level uses the comparison matrices to compare the priorities [13,14]. Assume that $A = [a_{rt}]$ is a comparison matrix. Every entry in matrix A is positive. In this occasion, A is a square matrix ($A_{n \times n}$). There is only one vector of weights such as u=(u_1, u_2, ..., u_n) in associate with any arbitrary comparison matrix including A. This vector also is named *priority vector*. The relationship between the elements of comparison matrix (A) and its vector of weights (u) is demonstrated by Eq. (1.4).

$$a_{rt} = \begin{cases} \frac{u_r}{u_t} & r \neq t \\ 1 & r = t \end{cases} \tag{1.4}$$

According to [13,14] the priority vector of matrix A can be computed with the following equation:

$$Av = \lambda_{max}.v \tag{1.5}$$

where λ_{max} and v are the principal eigenvalue and the equivalent priority vector of A respectively. If A is wholly consistent then $\lambda_{max} = n$. There are diverse methods for clarify the consistency of a comparison matrix [12–14]. A simple method for computing vector of weights(v) is indicated by Eq. (1.6).

$$u = \lim_{\mu \to \infty} \frac{1}{\mu} \sum_{t=1}^{\mu} \frac{A^t e}{e^T A^t e} \tag{1.6}$$

where $e = (1, 1, \ldots, 1)^T$ [13].

2 Definitions and Notations

The basic notions used in this paper is indicated in Table 1. Furthermore, the following definition has been used in the proposed method.

3 Proposed Method

We propose an analytical hierarchy process method (AHP-based) for the divisible load scheduling. As Fig. 1 shows, the proposed method consists of the following four phases:

Table 1. Definitions and notations.

Notation	Description	Notation	Description
$\hat{w}_j$	One-line computation rate	w_j	True computation rate
w_{ij}^{rep}	Reported rate for p_j in i^{th} probing	w_{ij}^{ch}	Online rate for p_j in i^{th} probing
m	Number of processors	k	Number of probing
α_j	The initial fraction of load for p_j	$\tilde{\alpha}_j$	The estimated fraction load for p_j
T_i^e	Expected finish time in i^{th} probing	T_i^o	Observed makespan of ith probing

3.1 Computation-Based Probing (Phase 1)

The originator gather information regarding to the performance and behavior of the worker processors, containing the on-line computation rates and diversity of makespan. In fact, the originator understands the on-line computation rates of the worker processors after carrying out the task of allocated fractions of load. The collected information helps the originator to estimate the best approximated fraction of load depend on the on-line computation rate of the worker processors. This phase includes the coming next six steps:

- **Step 1:** In the first level the originator proposes a computation rate (w_{ij}^{rep}) for the worker processor one by one, i.e., stated computation rate in the ith probing for the jth processor.
- **Step 2:** The originator later allocates the v ($v \ll V$) quantity of load on the worker processors. At that point, any processor obtains the fraction of load based on its stated computation rate.
- **Step 3:** Subsequently, the worker processors execute their fraction of load depend on their on-line computation rates.
- **Step 4:** The originator understands the real computation rate of every processor that is denoted by w_{ij}^{ch}.
- **Step 5:** In this step, the originator calculates the anticipated makespan using [7], along with the noticed makespan, which are designated by T_i^e and T_i^o respectively.
- **Step 6:** The computed information, containing w_{ij}^{ch}, w_{ij}^{rep}, T_{ij}^o, T_{ij}^e, T_i^e, and T_i^o is documented. In fact, T_{ij}^o and T_{ij}^e are the noticed implementation time and anticipated implementation time of the jth processor in the ith probing.

Table 2 demonstrate the population which are collected by the probing phase.

3.2 AHP-based Approximation for the Level of Cheating (Phase 2)

In this phase we use the analytical hierarchy process in order to approximate the level of cheating. This phase consists of the following five steps:

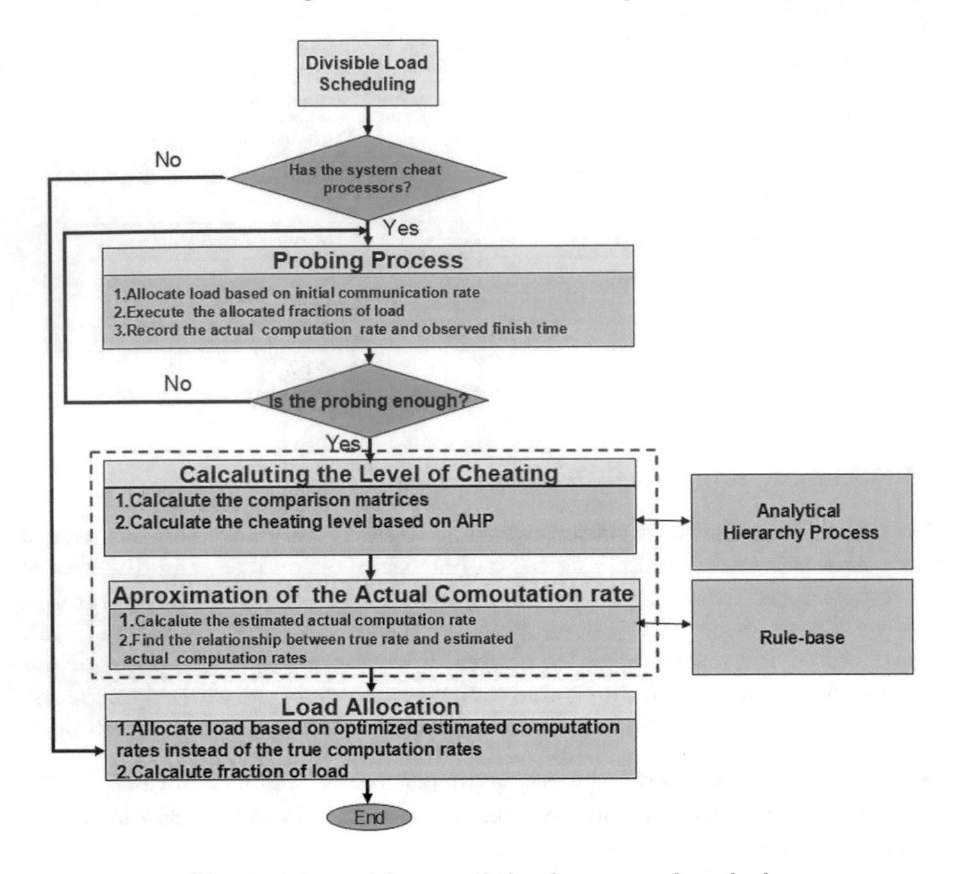

Fig. 1. A general framework for the proposed method.

- Step 1: (Calculation of the number of cheat-processor). The goal of this step is to determine the number of processors that cheat the algorithm. If the number of cheat-processor is reasonable (it has been defined in [7]) then we use the multi-objective method [7], otherwise we start from the next step. The number of reasonable cheat-processors (ℓ) can be computed using Eq. (3.1).

$$d^{\ell} \times R \leq T^{ch} \tag{3.1}$$

where T^{ch} is the cheated time and can be calculated by comparing the multi-objective method [7] and Carroll's method [11]. Furthermore, d and R were defined in [7]. The details of this step is sketched in Algorithm 1.
- Step 2: (Computing the Comparison Matrix of Criteria). This step, finds the processor that has the highest effect on the cheating. Every probing process can be considered as a criterion in this step. The difference of observed finish time (T_i^o) and expected finish time (T_i^e) can be computed for comparing the criteria. The details of this step is sketched in Algorithm 2. Furthermore, Ψ is a function that changes a real number to a ranking number. A method For calculation of Ψ can be found in [12].

Table 2. A pattern of data gathered by the computation-based probing

Probing	Cheated time	The diversities of computation rates			
		p_1	p_2	...	p_m
1	$T_1^o - T_1^e$	$w_{11}^{ch} - w_{11}^{rep}$	$w_{12}^{ch} - w_{12}^{rep}$	...	$w_{1m}^{ch} - w_{1m}^{rep}$
		$T_{11}^o - T_{11}^e$	$T_{12}^o - T_{12}^e$	...	$T_{1m}^o - T_{1m}^e$
2	$T_2^o - T_2^e$	$w_{21}^{ch} - w_{21}^{rep}$	$w_{22}^{ch} - w_{22}^{rep}$	...	$w_{2m}^{ch} - w_{2m}^{rep}$
		$T_{21}^o - T_{21}^e$	$T_{22}^o - T_{22}^e$	...	$T_{2m}^o - T_{2m}^e$
⋮	⋮	⋮	⋮	...	⋮
$i-1$	$T_{i-1}^o - T_{i-1}^e$	$w_{i-1,1}^{ch} - w_{i-1,1}^{rep}$	$w_{i-1,2}^{ch} - w_{i-1,2}^{rep}$	...	$w_{i-1,m}^{ch} - w_{i-1,m}^{rep}$
		$T_{i-1,1}^o - T_{i-1,1}^e$	$T_{i-1,2}^o - T_{i-1,2}^e$	...	$T_{i-1,m}^o - T_{i-1,m}^e$
i	$T_i^o - T_i^e$	$w_{i1}^{ch} - w_{i1}^{rep}$	$w_{i2}^{ch} - w_{i2}^{rep}$	...	$w_{im}^{ch} - w_{im}^{rep}$
		$T_{i1}^o - T_{i1}^e$	$T_{i2}^o - w_{i2}^e$	...	$T_{im}^o - T_{im}^e$
$i+1$	$T_{i+1}^o - T_{i+1}^e$	$w_{i+1,1}^{ch} - w_{i+1,1}^{rep}$	$w_{i+1,2}^{ch} - w_{i+1,2}^{rep}$	...	$w_{i+1,m}^{ch} - w_{i+1,m}^{rep}$
		$T_{i+1,1}^o - T_{i+1,1}^e$	$T_{i+1,2}^o - T_{i+1,2}^e$	...	$T_{i+1,m}^o - w_{i+1,m}^e$
⋮	⋮	⋮	⋮	...	⋮
k	$T_k^o - T_k^e$	$w_{k1}^{ch} - w_{k1}^{rep}$	$w_{k2}^{ch} - w_{k2}^{rep}$	...	$w_{km}^{ch} - w_{km}^{rep}$
		$T_{k1}^o - T_{k1}^e$	$T_{k2}^o - T_{k2}^e$	...	$T_{km}^o - T_{km}^e$

Algorithm 1. Calculation of Number of Cheat-processor()

Input: A population of recorded data i.e., Table 2.
Output: Number of cheat-processor(ℓ)

```
1: ℓ=0
2: for r=1 to m do
3:     for s=1 to k do
4:         if w_sr^ch − w_sr^rep ≠ 0
5:         then {
           ℓ++;
           Exit;
           }
6:     end for
7: end for
8: if ℓ satisfies Eq. (3.1) then
9:     the system has reasonable cheat processors
10: else
11:     the system has not reasonable cheat processors
12: end if
```

- Step 3: (Making Comparison Matrices for the Attributes). In this step, comparison matrices for the processors based on the criteria must be computed. The comparison matrices indicate the effects of each processor on the other processors. In this case the comparison matrix based on the ith criterion can be defined by the following equation:

Algorithm 2. Making Comparison Matrix of Criteria()

1: **for** x=1 to k **do**
2:     **for** y=x+1 to k **do**
3:         $C[x,y] = \frac{\Psi(T_y^0, T_y^e)}{\Psi(T_x^0, T_x^e)}$
4:         $C[y,x] = \frac{\Psi(T_x^0, T_x^e)}{\Psi(T_y^0, T_y^e)}$
5:     **end for**
6: **end for**
7: **for** r=1 to k **do**
8:     C[x, x]=1
9: **end for**

$$Q_{rs}^i = \begin{cases} \frac{\Psi(T_{ir}^o, T_{ir}^e)}{\Psi(T_{is}^o, T_{is}^e)} & r \neq s \\ 1 & r = s \end{cases} \tag{3.2}$$

where Q^i is the comparison matrix of attributes in the ith *(i=1, 2, ..., k)* probing process.

- *Step 4: (Calculation of the Principal Eigenvector).* For each comparison matrix principal eigenvector must be calculated. The principal eigenvector of comparison matrices can be calculated using Eq. (1.6).
Assume that u^1, u^2, ..., u^k are the principal eigenvectors of Q^1, Q^2, ..., Q^k. Therefore, we have k principal eigenvector associated with the k comparison matrices.
- *Step 5: (Calculation of the Level of Cheating).* The level of cheating denoted by δ can be computed by Eq. (3.3).

$$\delta = \Delta \times \Gamma \tag{3.3}$$

where Δ can be calculated using Eq. (3.4).

$$\Delta = [u^1 u^2 \dots u^k] = \begin{pmatrix} u_1^1 & u_1^2 & \dots & u_1^k \\ u_2^1 & u_2^2 & \dots & u_2^k \\ \vdots & \vdots & \vdots & \vdots \\ u_m^1 & v_m^2 & \dots & u_m^k \end{pmatrix} \tag{3.4}$$

In addition, the Γ in Eq. (3.3) is the principal eigenvector of C which was calculated by Algorithm 2.

3.3 Estimation of the Actual Computation Rate (Phase 3)

This phase estimates the on line computation rate of the processors. For this purpose we use the following steps:

- Step 1: Find the level of cheating (δ) from the previous phase. Clearly δ is an array with m elements. We have also $0 \leq \delta_j \leq 1$ for all $1 \leq j \leq m$.

- Step 2: Determine the type of cheat-processor. The type of cheat-processor denoted by s_j, indicates the behaviour of the jth processor to report its computation rate. The type of cheat-processor can be defined as the following equation.

$$s_j = \begin{cases} 1 & r_j > 0 \\ 0 & r_j = 0 \\ -1 & r_j < 0 \end{cases} \tag{3.5}$$

In Eq. (3.5) the parameter r_j can be defined as the following equation:

$$r_j = \sum_{i=1}^{k} (w_{ij}^{rep} - w_{ij}^{ch}) \tag{3.6}$$

where k is the number of probing processes.

- Step 3: Calculate critical rate (w_j^{cr}) using Algorithm 3. In this algorithm the *"Transfer"* is a function that maps real number into [−1,1].

Algorithm 3. Critical Rate (processor p_j)

Input: The recorded population

1: **For** j^{th} processor **Find** the dth probing with the lowest $|T_d^e - T_d^o|$

2: w_j^{cr} = Transfer($w_{dj}^{ch} - w_{dj}^{rep}$,[0,1])

3: **Return** (w_j^{cr})

- Step 4: Estimate the actual computation rate using Algorithm 4.

In Algorithm 4 the *classify* is a function that produces an output of positive integer value for an input of a real number between −1 and 1. A typical output for Algorithm 4 can be formed as a rule-based which is shown in Table 3. In the table we classified the level of cheating (δ_j) into five classes, including N(no cheat), L(low cheat), LL(very low cheat), H(high cheat) and HH (very high cheat). Clearly, increasing the number of classes increases the accuracy of the solution.

3.4 Load Allocation (Phase 4)

Now, by replacing $\tilde{w}_j$ with w_j in Eqs. (1.1) and (1.2) the estimated fraction of load can be calculated.

Algorithm 4. Estimation of Actual Computation Rate ()

Input: δ_j, s_j, w_j^{cr}.

1: $j \leftarrow 1$
2: **while** $(j \leq m)$ **do**
3: $n_1 = classify(\delta_j \times s_j)$
4: $n_2 = classify(w_j^{cr})$
5: $\tilde{w}_j = w_j + (n_1 + n_2) \times \tau$
6: $j \leftarrow j + 1$
7: **end while**
8: **Return** $\tilde{w}_j = w_j$

Table 3. Estimated Actual Computation Rate($\tilde{w}_j$)

Class of cheating($\tilde{w}_j$)	The range of computation rate-cheating				
	$w_j \ll w_j^{cr}$	$w_j < w_j^{cr}$	$w_j = w_j^{cr}$	$w_j > w_j^{cr}$	$w_j \gg w_j^{cr}$
LL	$w_j + 4\tau$	$w_j + 3\tau$	$w_j + 2\tau$	$w_j + \tau$	w_j
L	$w_j + 3\tau$	$w_j + 2\tau$	$w_j + \tau$	w_j	$w_j - \tau$
N	$w_j + 2\tau$	$w_j + \tau$	w_j	$w_j - \tau$	$w_j - 2\tau$
H	$w_{+}\tau$	w_j	$w_j - \tau$	$w_j - 2\tau$	$w_j - 3\tau$
HH	w_j	$w_j - \tau$	$w_j - 2\tau$	$w_j - 3\tau$	$w_j - 4\tau$

4 Experimental Results and Discussion

4.1 Experimental Setup

For comparing the proposed method with the existing works [7,11], we consider the same assumptions including the topology of network (binary tree), the cases of examining (8 cases in [7,11] and data. We also consider two different scenarios. In the first scenario which is indicated in Fig. 2, we consider a binary tree network comprising fifteen processors labeled $p_0, p_1, \ldots, p_{14}$. In the second scenario which is indicated in Fig. 3, we consider a binary tree networks comprising 63 processors labeled $p_0, p_1, \ldots, p_{62}$. We generate a number of random population in the eight mentioned cases (Table 4.). We explore the effects of computation rate-cheating on makespan, speedup, and startup time in the eight different cases.

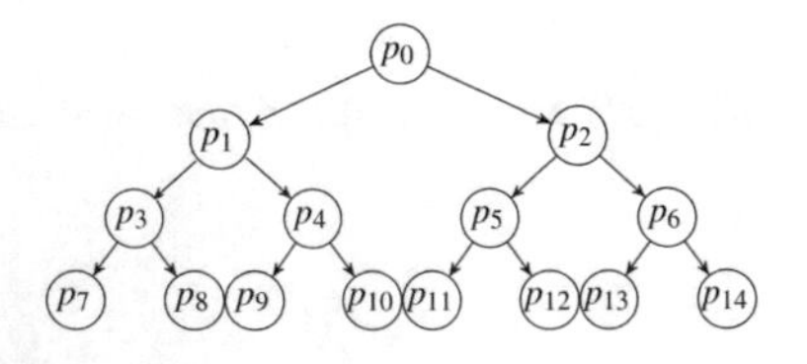

Fig. 2. Binary tree network contains 15 processors

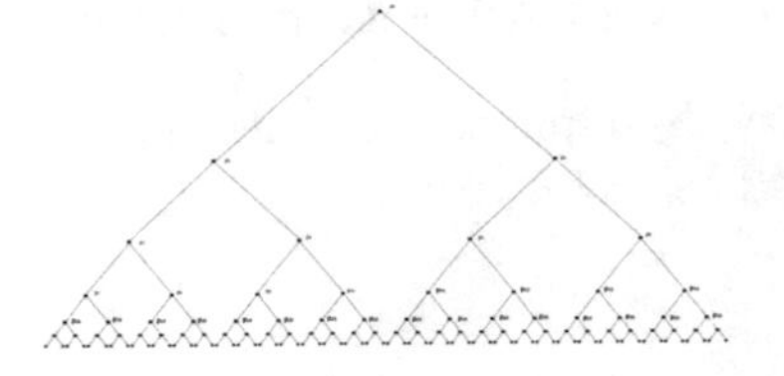

Fig. 3. Binary tree network with 63 processors

Table 4. The reported, on-line and estimated computation rates

Rates	Cases							
	1	2	3	4	5	6	7	8
$w_1\prime$	250	250	750	750	750	500	100	100
$\hat{w}_1$	250	500	750	250	500	750	250	500
$\tilde{w}_1$	250	360	750	501	621	626	172	309

4.2 First Scenario: Impacts of the Proposed Method on Makespan

By changing the anticipated computation rates for the eight above mentioned cases contrary to the real computation rates, the total finish time in both first and second scenarios can be calculated. The results are shown in Fig. 4.

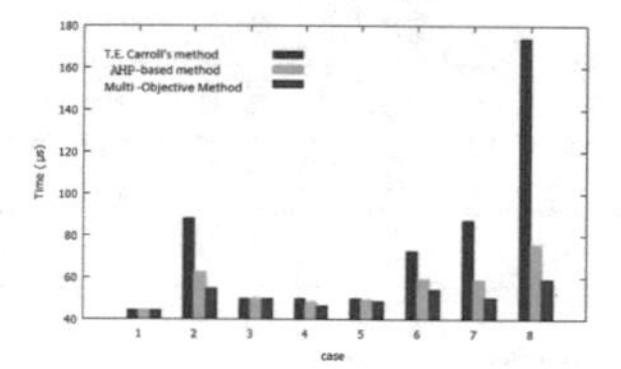

Fig. 4. Effects on Makespan (Scenario 1)

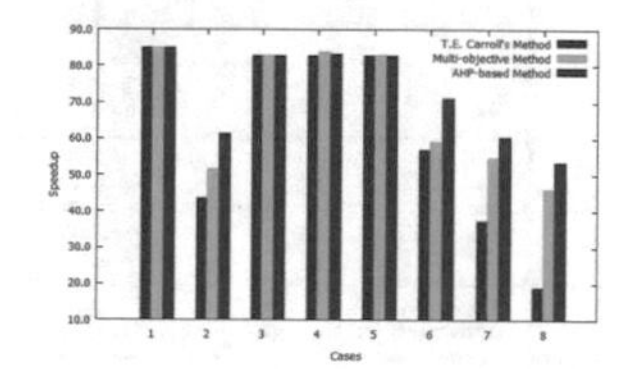

Fig. 5. Effects on speed-up (Scenario 2)

As Fig. 4 shows the AHP-based method can reduce the makespan almost as well as multi-objective method [7]. Comparing [11], the AHP-based method reduces the makespan having reduction of 28.98%, 4.07%, 2.18%, 18.36%, 32.63% and 56.47% in cases 2, 4, 5, 6, 7 and 8 respectively. The results show that the multi-objective [7] method, in this scenario is slightly more effective than the AHP-based method.

4.3 Second Scenario: Effects of Proposed Method on Speedup & Startup Time

An important problem is that in the real application we may have a huge numbers of processors that cheat the algorithm. The second scenario investigates the problem with 63 processors. Now, we consider the impacts of computation rate-cheating on speed-up in a binary tree divisible load scheduling. The speed-up is the ratio of computation time of one processor to computation time of the entire tree with m children. It is a measure of parallel processing advantage. Based on [10] the speedup in a multi-level tree network can be calculated by the following equation.

$$Speedup = \frac{T_{f,0}}{T_{f,m}} = \frac{w_0}{w^{eq}} \tag{4.1}$$

Now we calculate the speed-up for a binary tree networks. We consider a binary tree with 63 processors. Comparing with [11], the proposed AHP-based method increases the speed-up 24.67% of case 2, 15.04% of case 6, 48.00% of case 7 and 94.86% of case 8. The results show that the proposed method does not increase the speed-up in cases 1 and 3. The main reason is that, the rate-cheating problem has not been occurred in these two cases. It can also be seen that in cases 4 and 5 the proposed method increases the speedup 2.08% and 3.09% respectively. The detail of improvement of the proposed method has been indicated in Fig. 5 and Table 5.

Table 5. The speed-up of proposed method and existing works

Cases	Results of [12]	Results of [7]	Proposed results	Comparing with (%)	
				[11]	[7]
1	83.26	83.26	83.26	0.00	0.00
2	57.04	59.13	71.12	24.67	20.27
3	82.71	82.71	82.71	0.00	0.00
4	82.71	83.05	80.90	2.18	1.86
5	64.43	68.96	74.12	3.09	2.55
6	64.43	66.62	74.12	15.04	11.26
7	43.59	54.31	64.51	48.00	18.77
8	29.27	47.48	57.04	94.86	20.13

In the *second* part of this section, we investigate the start-up time in the AHP-based method and compare it with the existing methods. The start-up time in the multi-objective method was computed in [7]. Now we compute the start-up time equation in the AHP-based method. It is shown in Eq. (4.2).

$$T^{start} = t_{prob} + t_{ahp} \tag{4.2}$$

The t_{prob} has also been explained in [7]. The other parameter is calculated using Eq. (4.3).

$$t_{ahp} = [max(m,k)]^{2.81} \times \mu \times R \tag{4.3}$$

where k, R, m and μ are the number of probing process, speed of originator, the number of processors and number of steps to obtain a consistent comparison matrix respectively. Comparing the t_{ahp} and t_{obj} which calculated in [7] indicates that the AHP-based method has considerably less start-up time than the multi-objective method. We investigate the problem in the worst state. In this situation, we consider the eight mentioned cases. Therefore, the start-up time have been shown in the Figs. 6, 7, 8, 9, 10 and 11. However, Fig. 6 shows the problem in case 1 which is an optimal case. In this case the processors report their true computation rates. As the figure show the three methods have the

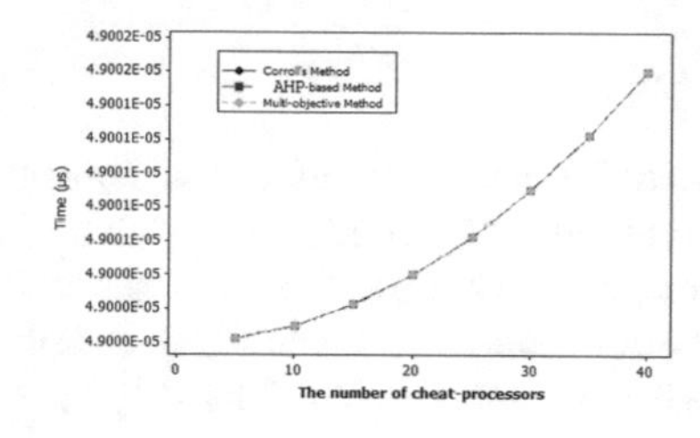

Fig. 6. Case 1

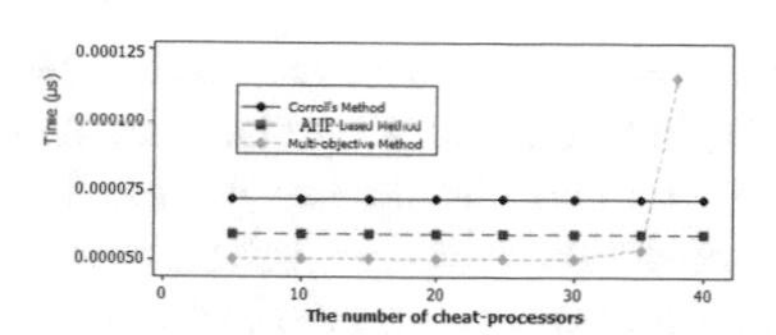

Fig. 7. Case 2

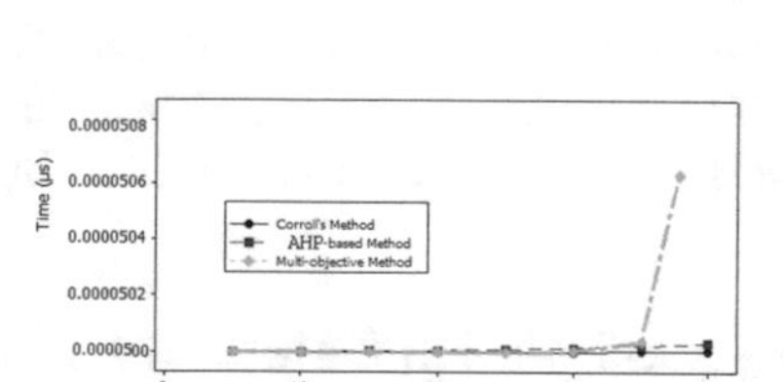

Fig. 8. Case 3

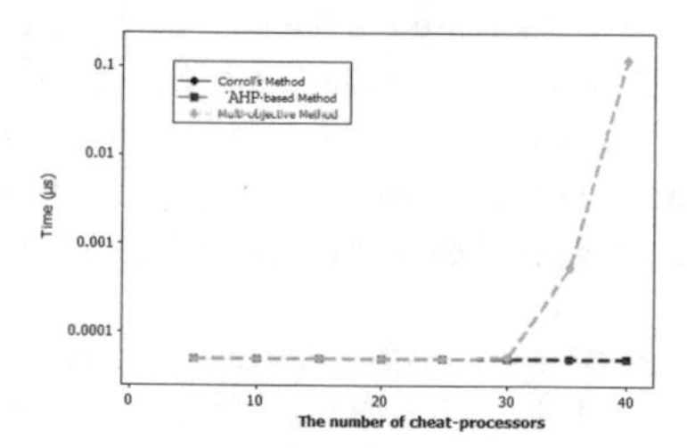

Fig. 9. Cases 4&5

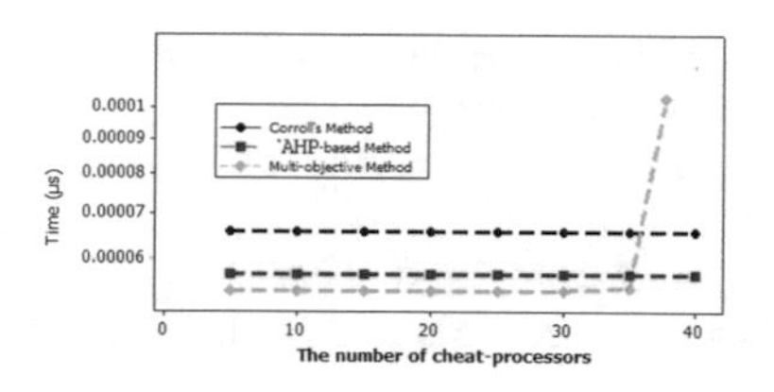

Fig. 10. Case 6

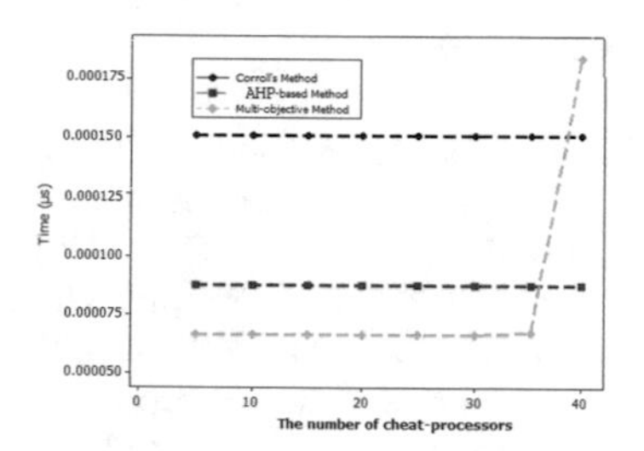

Fig. 11. Cases 7&8

same behaviour. We also investigate the problem in case 2. As Fig. 7 shows, the proposed multi-objective method has the lowest total time when the system has less than 35 cheat-processors. The figure also shows that the AHP-based method has lower total time comparing the other methods when the system has more than 35 cheat-processors. The same behavior can also be seen in cases 6, 7 and 8. Furthermore, Fig. 8 shows that not only the multi-objective method but also the AHP-based method have slightly more time than [11] in case 3 when the system has more than 35 cheat-processors. It is because in this case the start-up time of the AHP-based method is more than the method proposed in [11]. In fact, in this case the time for both the muti-objective and AHP-based methods can be improved in the future works. Lastly, we discuss the problem in cases 4 and 5. As Fig. 9 show, in cases 4 and 5, the proposed AHP-based method and the method proposed in [11] have the same total time. It means these two methods have the same behaviour when the system have considerable numbers of cheat-processors.

5 Conclusion

This paper presented an AHP-based method to decrease the effects of computation rate-cheating on the execution of the divisible load scheduling model. We discussed the simulation results in two different scenarios. In the first scenario we considered that a limited range of processors cheat the system. The results indicated that demonstrated that the AHP-based method has considerably lower start-up time comparing the other methods. In the second scenario we considered that a huge number of the processors cheat the system (63 processors). The results indicated that in this scenario the AHP-based method has considerably lower startup time and make span than the other methods. In fact, in the real applications the systems may have several processors that cheat the algorithms (second scenario). It means the AHP-based method is considerably more applicable than the other methods.

Acknowledgments. This research work is partially supported by the Malaysian Ministry of Education under Research Management Center, Universiti Putra Malaysia, Putra Grant with High Impact Factor. Grant no. UPM/700-2/1/GPB/2017/9557900.

References

1. Yuan-Chieh C, Robertazzi TG (1988) Distributed computation with communication delay. IEEE Trans Aerosp Electron Syst 24(6):700–712
2. Agrawal R, Jagadish HV (1988) Partitioning techniques for large-grained parallelism. IEEE Trans Comput 37(12):1627–1634
3. Bharadwaj V, Li X, Ko CC (2000) Efficient partitioning and scheduling of computer vision and image processing data on bus networks using divisible load analysis. Image Vis Comput 18(11):919–938
4. Korkhov VV, Moscicki JT, Krzhizhanovskaya V (2009) Dynamic workload balancing of parallel applications with user-level scheduling on the Grid. Futur Gener Comput Syst 25(1):28–34
5. Shi H, Kwok N, Wang W, Chen S (2014) Divisible load theory based active-sleep workload assignment schemes for wireless sensor networks. Int J Distrib Sens Netw, Hindawi Publishing Corporation. https://doi.org/10.1155/2014/929416
6. Suresh S, Huang H, Kim HJ (2015) Scheduling in compute cloud with multiple data banks using divisible load paradigm. IEEE Trans Aerosp Electron Syst 51(2):1288–1297
7. Ghanbari S, Othman M, Leong WJ, Bakar MRA (2016) Multi-objective method for divisible load scheduling in multi-level tree network. Futur Gener Comput Syst 54:132–143
8. Ghanbari S, Othman M (2018) Time cheating in divisible load scheduling: sensitivity analysis, results and open problems. Procedia Comput Sci 125:935–943
9. Ghanbari S, Othman M (2014) Comprehensive review on divisible load theory: concepts, strategies, and approaches. Hindawi Publishing Corporation, Math Probl Eng. https://doi.org/10.1155/2014/460354
10. Robertazzi TG (2003) Ten reasons to use divisible load theory. Computer 36(5):63–68

11. Carroll TE, Grosu D (2012) An incentive-based distributed mechanism for scheduling divisible loads in tree-networks. J Parallel Distrib Comput 72:389–401
12. Ghanbari S, Othman M, Bakar MRA, Leong WJ (2015) Priority-based divisible load scheduling using analytical hierarchy process. Appl Math Inf Sci 9(5):2541
13. Saaty TL (1988) What is the analytic hierarchy process. J Math Models Decis Support 48:109–121
14. Saaty TL (2013) The modern science of multicriteria decision making and its practical applications: the AHP/ANP approach. Oper Res 6(15):1101–1118

Emerging Trends in Information Technology and Data Science

Enhanced Bat Algorithm for Solving Non-Convex Economic Dispatch Problem

Kashif Hussain[1(✉)], William Zhu[1], Mohd Najib Mohd Salleh[2], Haseeb Ali[2], Noreen Talpur[3], Rashid Naseem[4], Arshad Ahmad[5], and Ayaz Ullah[5]

[1] Institute of Fundamental and Frontier Sciences, University of Electronic Science and Technology of China, Chengdu, Sichuan, China
k.hussain@uestc.edu.cn

[2] Faculty of Computer Science and Information Technology, Universiti Tun Hussein Onn Malaysia, Batu Pahat, Malaysia

[3] Faculty of Science and Information Technology, Universiti Teknologi PETRONAS, Perak, Malaysia

[4] Department of Computer Science, City University of Science and Information Technology, Peshawar, Pakistan

[5] Department of Computer Science, University of Swabi, Swabi, Pakistan

Abstract. Bat algorithm lags behind other modern metaheuristic algorithms in terms of search efficiency, due to premature convergence. Once trapped in any sub-optimal region, the algorithm is unable to escape because of deficiency in population diversity. To address this, an enhanced Bat Algorithm (EBA) is introduced in this paper. The EBA algorithm comes with adaptive exploration and exploitation capability, as well as, additional population diversity. This enables EBA improve its convergence ability to find even better solutions towards the end of search process, where standard BA is often trapped. To illustrate effectiveness of the proposed method, EBA is applied on non-linear, non-convex economic dispatch problem with a power generation system comprising of twenty thermal units. The experimental results suggest that EBA not only saved power generation cost but also reduced transmission losses, more efficiently as compared to original BA and other methods reported in literature. The EBA algorithm also showed enhanced convergence ability than BA towards the end of iterations.

Keywords: Bat algorithm · Economic dispatch · Power generation dispatch · Optimization · Non-convex

1 Introduction

Economic dispatch (ED) problem comes in the category of engineering optimization problems that are solved subject to certain practical constraints. The objective is to minimize the cost of power generation by economically (or optimally) allocating power generation values to different power producing units, while satisfying specific operational constrains at the same time. Because power generation is performed using costly non-renewable energy resources like petroleum or

R. Ghazali et al. (Eds.): SCDM 2020, AISC 978, pp. 419–428, 2020.
https://doi.org/10.1007/978-3-030-36056-6_39

coal, it is important to achieve optimum output from power plants. The problem is mathematically formulated in the form of an optimization problem which is solved by mathematical programming models; such as, linear and non-linear programming algorithms. Optimization methods, as an efficient approach to solving optimization problems, have been employed to solve ED problem in literature [1–4].

Bat algorithm (BA) is one of the popular swarm-based metaheuristic algorithms [5], which has been employed to solve numerous challenging optimization problems, as it can solve both continuous and discrete optimization problems [6]. Applications of BA in engineering sciences include Biomedical Engineering, Computer Science Engineering, Electrical and Electronics Engineering, Industrial and Production Engineering, Mechanical Engineering, etc. [7]. Since its introduction by Yang in 2010 [8], it has attracted researchers from different fields of sciences, who not only used it but also developed various variants for improved performance [7]. Because of efficiency and simplicity of the algorithm, along with other metaheuristic algorithms; such as Particle Swarm Optimization (PSO), Firefly Algorithm (FA), Cuckoo Search (CS), etc., researchers have attempted BA on ED [9]. Despite of wider acceptance in research community, BA still has to overcome its infancy stage by improving performance on hard optimization problems [7]. It is, therefore, this research is aimed at modifying BA for enhanced performance, while solving non-convex constrained ED problem.

Rest of the paper is organized as follows. The subsequent section briefly describes ED problem, as well as, recalls literature on BA implementations on solving this problem. Section 3 introduces enhancement proposed in BA, hence called EBA in this paper. This section also explains how EBA is employed on ED problem. Results of the implementation are discussed in Sects. 4, and 5 concludes this study.

2 Literature Review

2.1 Economic Dispatch Problem

In a power plant, a number of power generation units are installed. Here, the power generation is always subject to several different operational constraints which are also to be satisfied. For example, the thermal units should be run in a way that maximum demand is met with lowest possible total cost. This is mainly concerned with determining how much power to be dispatched by each unit, while minimizing operational costs expressed in nonlinear form. Hence, this becomes an optimization problem, termed as economic dispatch (ED) problem [9]. Its objective function f can be defined as (1):

$$\min_{P \in R^N} = \sum_{i=1}^{N} F_i(P_i) = \sum_{i=1}^{N} (a_i + b_i P_i^2 + c_i P_i^2) \tag{1}$$

where N is the number of thermal units, $F_i(P_i)$ is the ith thermal unit's fuel cost. a_i, b_i, c_i are cost coefficients of the ith thermal unit. The subjected constraints attached to objective function are expressed via (2):

$$P_D = \sum_{i=1}^{N}(P_i) - P_L,$$
$$P_L = \sum_{i=1}^{N}\sum_{j=1}^{N}(P_i B_{ij} P_j) + \sum_{j=1}^{N}(B_0 P_i + B_{00}) \tag{2}$$
$$\text{where } P_L^{min} \leq P_i \leq P_L^{max}$$

where P_D and P_L are the demand load and transmission loss in megawatt. B_{ij}, B_{0i} and B_{00} are loss coefficients. Whereas, P_i^{min} and P_i^{max} are minimum and maximum power generation limits of the ith thermal unit.

2.2 BA Applications on ED

In literature, BA has been applied on ED problem to lower the per hour cost and transmission losses. It can be said that BA is a promising technique in this area. In this section, we review some of the recent literature in this area. Biswal *et al.*, [10] implemented BA to solve ED problem with valve-point loading effect and few more constraints. It was found that BA was useful and easy to implement technique than PSO and Intelligent Water Drop (IWD) algorithm. To achieve optimum total generation cost of a power plant, Wulandhari *et al.* [11] used BA. The research achieved 1.23% cost reduction per hour as compared to 0.12% by FA. Dinh *et al.* [12] modified BA and applied on ED problem with combined heat and power generation using three different systems (7, 24, and 48 thermal unit systems). According to the research, it is considered as hard optimization problem as it takes care of both heat and power generation from co-generation system. The modified BA proved to be efficient solution to the problem, as compared to various other counterparts. Adarsh *et al.* [9] employed chaotic BA on five different ED problems comprising of 6, 13, 20, 40, and 160 thermal unit systems. The research used sinusoidal chaotic map for updating the loudness parameter of BA and achieved considerably improved performance.

Based on a brief literature review, it can be inferred that BA is a useful approach to solving ED problem. Motivated from existing applications, this research also employed BA on ED problem after modification for improved results. The proposed medication is discussed in the following section.

3 Methodology

3.1 Basic Bat Algorithm (BA)

The BA is a swarm-based metaheuristic algorithm inspired from the echolocation phenomenon of bats. These blind-by-nature animals use frequency and pulse rate to locate the prey and update their positions accordingly. A bat flies randomly with a certain velocity at a position. It searches for prey and communicates with other bats by varying its frequency, loudness, and pulse rate. Based on this description, BA is structured by Yang [8], where each virtual bat i (swarm agent) maintains four properties: velocity v_i, position x_i, loudness A_i, and pulse rate r_i. The position of ith bat is updated using (3):

$$\begin{aligned} f_i &= f_{min} + (f_{max} - f_{min})\beta, \\ v_i^{t+1} &= v_i^t + (x_i^t - x_{best}) f_i, \\ x_i^{t+1} &= x_i^t + v_i^{t+1} \end{aligned} \tag{3}$$

where $\beta \in [0, 1]$ is a randomly generated value from uniform distribution. As the iterations t proceed, the ith bat varies its loudness and pulse emission rate using (4):

$$\begin{aligned} A_i^{t+1} &= \alpha A_i^t, \\ r_i^{t+1} &= r_i^0[1 - exp(-\gamma t)] \end{aligned} \tag{4}$$

where $0 < \alpha \leq 1$ and $\gamma > 0$ are constants. The Algorithm 1 presents step by step schema of BA algorithm.

Algorithm 1. Pseudo-code of basic BAT algorithm

```
procedure BA(r, A)
    Initialize bat swarm with position x_i, velocity v_i, pulse emission rate r_i, and
loudness A_i
    Define frequency limit [f_min − f_max]
    while termination criteria not met do
        Evaluate each position and assign objective function value f_i
        Find best position x_best
        Use equation (3) to update velocity and position of each bat
        if random < r_i then    ▷ random=randomly generated value between [0,1]
            Generate new position x_new around x_best
            Evaluate x_new and assign f_new
        end if
        if f_new < f_i and rand < A_i then    ▷ Conditionally update best position
            x_i = x_new, f_i = f_new
        end if
    end while
    return x_best as best solution found
end procedure
```

It is repeatedly reported in literature that BA suffers from lack of exploration hence prematurely converges early in search process [13–15]. Moreover, BA produces efficient solutions on problems with small dimension-size. When dealing with high dimensional problems, its performance deters significantly. Similar to other metaheuristic algorithms like PSO and GA, BA performance is also sensitive to its parameters (A and γ). Proper assignment and control of these parameters during iterations (increasing or decreasing) helps enhance its search efficiency [6].

Several researchers have proposed modification and hybrids of BA. These modifications can be categorized into four strategies: replacing loudness property of BA with chaotic maps [16,17], applying different selection schemes [15], Lévy flights [18], inertia weight and other modification [19,20]. Other than these modifications, BA has also been hybridized with other heuristic and metaheuristic techniques. Few of the latest hybrids of BA with PSO, ABC, CS, FA can be found in [21–24]. All these modifications and hybrids of BA have shown their efficiencies in specific scenarios, however there still exists no universally accepted version of BA. It motivates this research to perform enhancement in the standard BA; especially while applying on ED problem.

3.2 Enhanced Bat Algorithm (EBA)

Generally speaking, the standard BA is simplest implementation of swarm behavior in bats. It lacks self-adaptive capability as in other latest swarm-based metaheuristic algorithms. Provided this capability in BA can make it able to mold search behavior according to search status and solution space. Moreover, BA lacks population diversity due to insufficient randomization or information from outside the local environment. To address the issue, this paper proposes two simple yet effective modifications in the standard BA: (a) use of inertia weight in velocity update equation to address self-adaptability problem, and (b) injecting additional but controlled randomization in position update step (line 7 and 9) in Algorithm 1.

The first modification is introducing inertia weight to velocity equation, as in PSO [25]. This controls diversity from being initially high in the beginning of iterations to lower towards the end of iterations, when the algorithm is about to converge onto a global optimum. As expressed in the Eq. (5), the inertia weight $(0 < w_t < 1)$ reduces as iteration counter t approaches maximum number of iterations t_{max}. This parameter is put in the boundary of w_{min} and w_{min} values.

$$w_i = (w_{max} - w_{min}) \times \frac{t_{max} - t}{t_{max}} + w_{min}, \quad (5)$$

where w_t with value near to 1 enforces exploration to new regions at the beginning of iterations, while its value approaching to 0 boosts exploitation. It is worth noticing that the use of inertia weight factor in the velocity update equation brings balance between exploration and exploitation in BA. The inertia weight factor is incorporated in the velocity update equation as in (6):

$$v_i^{t+1} = w_t v_i^t (w_i^t - x_{best}) f_i, \quad (6)$$

This modification is not enough to significantly enhance BA performance. It is because, when the search status is towards the end of iterations, the population will mainly revolve around the best solutions found so far, due to small scale of diversity at this stage. This best solution found so far might be the undesirable local optimum, hence it is necessary to inject ample diversity at this stage also. That said, we modify position update equations in the steps 7 and

9 in Algorithm 1. The modified position update equation in line 7 is expressed in (7):

$$x_i^{t+1} = x_i + v_i^t \times \cos(w_t \times \frac{\pi}{x_i}), \tag{7}$$

Furthermore, the enhanced diversity factor added to line 9 in Algorithm 1 is expressed via (8):

$$x_i^{t+1} = x_{best} + 0.01 \times rand \cos(0.5 \times \frac{\pi}{x_i}), \tag{8}$$

where x_best is the best solution found so-far.

3.3 EBA Implementation on ED

Algorithm 2 outlines general steps to implement any metaheuristic algorithm on ED problem. EBA is also implemented in similar way. Here, a population individual represents a thermal unit.

Algorithm 2. Schema of BA implementation on ED problem

1: Initialize lower and upper bounds of thermal units, maximum number of iterations, and population-size. Also, initialize population with random positions.
2: Evaluate population fitness by objective function defined in (1) and constraints given in the system.
3: Calculate total generation cost.
4: Update positions of candidate solutions using the relevant update equation.
5: Memorize the best solution found so-far.
6: If stopping criteria is met, output the best solution; otherwise, return to Step 2.

Following are the parameter settings adopted in the standard BA and EBA for solving ED problem in this research. Population-size = 30, Maximum Iterations = 1000, Minimum Inertia Weight = 0.1, Maximum Inertia Weight = 0.9.

4 Results and Discussion

To evaluate the proposed EBA, it is applied on ED problem with 20-unit power generation system having loss coefficients. The power generating demand of the system is 2500 MW. The system information is taken from [26] and presented in Table 1. The code is implemented in MATLAB®on a computer system with Windows 10 operating system, 8 GB of RAM, and 3.80 GHz processor. Since, metaheuristic algorithm, by nature are based on randomization technique, so it is important to execute them on problem more than once. Hence, the EBA was run 10 times and the average results are given in this section. The experimental presented are listed in Table 2, whereas convergence graphs of BA and

Table 1. 20-Unit Power Generation System with Loss Information

Unit	a ($/MW)	b ($/MW)	c ($/MW)	P_i^{min}	P_i^{max}
1	1000	18.19	0.00068	150	600
2	970	19.26	0.00071	50	200
3	600	19.80	0.00650	50	200
4	700	19.10	0.00500	50	200
5	420	18.10	0.00738	50	160
6	360	19.26	0.00612	20	100
7	490	17.14	0.00790	25	125
8	660	18.92	0.00813	50	150
9	765	18.27	0.00522	50	200
10	770	18.92	0.00573	30	150
11	800	16.69	0.00480	100	300
12	970	16.76	0.00310	150	500
13	900	17.36	0.00850	40	160
14	700	18.70	0.00511	20	130
15	450	18.70	0.00398	25	185
16	370	14.26	0.07120	20	80
17	480	19.14	0.00890	30	85
18	680	18.92	0.00713	30	120
19	700	18.47	0.00622	40	120
20	850	19.79	0.00773	30	100

EBA are given in Fig. 1. According to the results presented in this section, EBA achieved better power generation cost in dollars per hour than standard BA. EBA achieved 62456.6182 $/hour against 62456.8042 $/hour achieved by BA. Comparison with the results reported in literature, EBA also managed to outperform Backtracking Search Algorithm (BSA) [27], Gravitational Search Algorithm (GSA) [28], and Biogeography-Based Optimization (BBO) [29]. Apart from generation cost, as shown in Table 2, EBA also achieved the lowest power loss against the counterparts.

When analyzing convergence graphs given in Fig. 1, it is obvious that EBA converged relatively faster than BA and achieved better cost function value. Moreover, from focused later part of iterations in Fig. 1, it can also be seen that EBA was able to escape suboptimal region in the later part of iterations and found even better solution towards the end of search process. It is because EBA was specifically modified to keep on searching for better regions even in the last stage of optimization.

Table 2. Power Generation Cost ($/hr) and Power Loss (MW) Comparison

Method	Generation cost($/hr)	Power loss(MW)
BA	62456.8042	92.544
EBA	62456.6182	91.860
BSA [27]	62456.6920	91.893
GSA [28]	62456.6330	91.965
BBO [29]	62456.7790	92.101

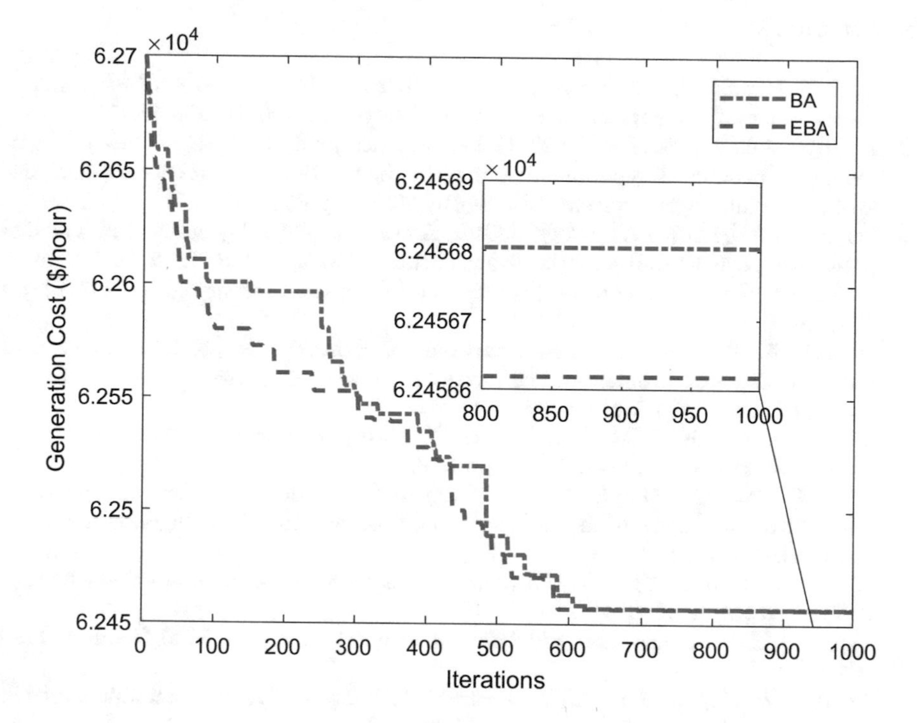

Fig. 1. BA and EBA convergence on ED

5 Conclusion

In this paper, Bat Algorithm (BA) was modified to improve its performance, hence called enhanced Bat Algorithm (EBA). The proposed modification incorporated inertia weight factor for adaptive search strategy. Also, additional but controlled randomization was introduced the algorithm, which enabled it to escape suboptimal region even at the later stage of search process – when the standard BA was unable to improve search results. The EBA algorithm was successfully applied on electronic dispatch problem with 20-unit test system for finding optimum power generation cost, in terms of fuel cost in dollars per hour. The methods were also evaluated on reducing power loss of the system. From comparative analysis, it can be inferred that the proposed EBA not only outperformed standard BA, but also other counterparts reported in literature. Apart from improvements, however, EBA introduced additional control parameters like minimum and maximum values for inertia weight factor.

Acknowledgments. The authors would like to thank University of Electronic Science and Technology of China (UESTC) and National Natural Science Foundation of China (NSFC) for supporting this research under Grant No. 61772120.

References

1. Abdi H, Fattahi H, Lumbreras S (2018) What metaheuristic solves the economic dispatch faster? a comparative case study. Electr Eng 100(4):2825–2837
2. Pal HK, Jain K, Pandit M (2011) Performance analysis of metaheuristic techniques for nonconvex economic dispatch. In: International conference on susttainable energy intelligent systems (SEISCON 2011), pp 396–402
3. Fergougui AE, Ladjici AA, Benseddik A, Amrane Y (2018) Dynamic economic dispatch using genetic and particle swarm optimization algorithm. In: 2018 5th International conference on control, decision and information technologies (CoDIT), pp 1001–1005
4. Habachi R, Touil A, Boulal A, Charkaoui A, Echchatbi A (2019) Resolution of economic dispatch problem of the morocco network using crow search algorithm. Indones J Elect Eng Comput Sci 13(1):347–353
5. Hussain K, Salleh MNM, Cheng S, Shi Y (2018) Metaheuristic research: a comprehensive survey. Artif Intell Rev, pp 1–43
6. Fister I, Yang XS, Fong S, Zhuang Y (2014) Bat algorithm: recent advances. In: 2014 IEEE 15th International symposium on computational intelligence and informatics (CINTI), pp 163–167
7. Chawla M, Duhan M (2015) Bat algorithm: a survey of the state-of-the-art. Appl Artif Intell 29(6):617–634
8. Yang XS (2010) A new metaheuristic bat-inspired algorithm. Stud Comput Intell 284:65–74
9. Adarsh BR, Raghunathan T, Jayabarathi T, Yang XS (2016) Economic dispatch using chaotic bat algorithm. Energy 96:666–675
10. Biswal S, Barisal AK, Behera A, Prakash T (2013) Optimal power dispatch using bat algorithm. In: 2013 International conference on energy efficient technologies for sustainability, pp 1018–1023
11. Wulandhari LA, Komsiyah S, Wicaksono W (2018) Bat algorithm implementation on economic dispatch optimization problem. Procedia Comput Sci 135:275–282
12. Dinh B, Nguyen T, Quynh N, Dai L (2018) A novel method for economic dispatch of combined heat and power generation. Energies 11(11):3113
13. Meng X, Gao XZ, Liu Y, Zhang H (2015) A novel bat algorithm with habitat selection and doppler effect in echoes for optimization. Expert Syst Appl 42(17):6350–6364
14. Liang H, Liu Y, Shen Y, Li F, Man Y (2018) A hybrid bat algorithm for economic dispatch with random wind power. IEEE Trans Power Syst 33(5):5052–5061
15. Al-Betar MA, Awadallah MA, Faris H, Yang XS, Khader AT, Alomari OA (2018) Bat-inspired algorithms with natural selection mechanisms for global optimization. Neurocomputing 273:448–465

16. Gandomi AH, Yang XS (2014) Chaotic bat algorithm. J Comput Sci 5(2):224–232
17. Mitić M, Vuković N, Petrović M, Miljković Z (2018) Chaotic metaheuristic algorithms for learning and reproduction of robot motion trajectories. Neural Comput Appl 30(4):1065–1083
18. Tu D, Wang E, Zhang F (2019) An intelligent wireless sensor positioning strategy based on improved bat algorithm. In: 2019 International conference on intelligent transportation, big data and smart city (ICITBS) (2019)
19. Reddy MP, Ganguli R (2018) Enhancement structures for the bat algorithm. In: 2018 IEEE symposium series on computational intelligence (SSCI), pp 601–608
20. Cui Z, Li F, Zhang W (2019) Bat algorithm with principal component analysis. Int J Mach Learn Cybern 10(3):603–622
21. Ghosh S, Kaur M, Bhullar S, Karar V (2019) Hybrid abc-bat for solving short-term hydrothermal scheduling problems. Energies 12(3):551
22. Ferdowsi A, Farzin S, Mousavi SF, Karami H (2019) Hybrid bat and particle swarm algorithm for optimization of labyrinth spillway based on half and quarter round crest shapes. Flow Meas Instrum 66:209–217
23. Gunji B, Deepak BBVL, Saraswathi MBL, Mogili UR (2019) Optimal path planning of mobile robot using the hybrid cuckoo-bat algorithm in assorted environment. Int J Intell Unmanned Syst 7(1):35–52
24. Ponmalar PS, Kumar JS, Harikrishnan R (2017) Bat-firefly localization algorithm for wireless sensor networks. In: 2017 IEEE international conference on computational intelligence and computing research (ICCIC) (2017)
25. Kennedy J, Eberhart R (1995) Particle swarm optimization (pso). In: Proc. IEEE international conference on neural networks, Perth, Australia, pp 1942–1948
26. Su CT, Lin CT (2000) New approach with a hopfield modeling framework to economic dispatch. IEEE Trans Power Syst 15(2):541–545
27. Modiri-Delshad M, Rahim NA (2014) Solving non-convex economic dispatch problem via backtracking search algorithm. Energy 77:372–381
28. Udgir M, Dubey HM, Pandit M (2013) Gravitational search algorithm: a novel optimization approach for economic load dispatch. In: 2013 Annual international conference on emerging research areas and 2013 international conference on microelectronics, communications and renewable energy, pp 1–6
29. Bhattacharya A, Chattopadhyay PK (2010) Solving complex economic load dispatch problems using biogeography-based optimization. Expert Syst Appl 37(5):3605–3615

A Hybrid Social Mining Approach for Companies Current Reputation Analysis

Falwah AlHamed(✉) and Aljohara AlGwaiz

King Saud University, Riyadh, Saudi Arabia
falhamed@kacst.edu.sa, aalgwaiz@ksu.edu.sa

Abstract. This paper presents an approach for company's reputation analysis using data mining techniques. It obtains knowledge from huge data written about these companies and available publicly on the internet. It is done by extracting data from social media, such as twitter, containing relevant company's mentions. Then, data is then injected into the first layer where it is classified into a positive and negative classes using machine learning, specifically artificial neural network. It takes tweets after preprocessing as an input, then outputs the sentiment of each tweet. The result will be the general reputation and perception of the company for a given timeframe. To further understand these results, the analyzed data is then transferred to a second layer of analysis where consumers to identify products, services, and announcements that lead to the positive of negative perception. Using Term Frequency (TF), this will result to ranked list of most mentioned words in each of the negative and positive classes. This will be valuable for companies to identify points of weakness and strength, advertisement impressions, and strategic decisions impact.

Keywords: Sentiment analysis · Social mining · Twitter analysis · Data mining · Reputation analysis

1 Introduction

Nowadays, social media is widely used by different people. Enormous number of users are posting news, opinions, diaries, feelings and lifestyle to various social media platforms such as Twitter, Instagram, Snapchat and others. The volume of data generated on these platforms is very large. For instance, Twitter statistics showed that there are more than one million tweets published per day [1]. This huge growth of content arises a demand to utilize information in order to help companies in decision- making and understanding market trends depending on consumers' opinions. Therefore, data mining approaches must be used to be capable of transferring vast amount of data into meaningful information. This work aims to develop a sentiment analysis system that targets retrieving companies' reputation from Twitter. Specifically, it addresses tweets containing predetermined large companies containing different departments or producing multiple well-known products. Hybrid sentiment analysis methodology is used, this includes using lexicon and machine learning tweets classification. First, the data is collected and injected into the first layer where it is classified to positive and negative

R. Ghazali et al. (Eds.): SCDM 2020, AISC 978, pp. 429–438, 2020.
https://doi.org/10.1007/978-3-030-36056-6_40

data. Each data class is then transferred into the second layer where the most frequent words are extracted. This information is very valuable to companies for decision making and evaluation. For instance, a word such as "delivery" at the top of the positive class indicates a that the delivery department performance is good. It may also be a product such as "croissant" which can be a new successful product or the best seller in the store. Also, news relevant to the company and executive decision impression on consumers can be better understood. For instance, if the company made a major rebranding to the company, words such as "logo" or "colors" might pop up in the positive or the negative list based how the changed was perceived by consumers. This provides valuable information in marketing, decision making, and reputation management.

The rest of this paper is organized as follows: Sect. 2 overviews some work related to reputation analysis in social media. Section 3 describes the methodology of our system. Section 4 illustrates the implementation. Section 5 emphasizes results. Section 6 concludes this paper with an overview of contributions and future work.

2 Related Work

There have been numerous studies to investigate user's sentiment by analyzing public content. Some researchers studied people's opinions by utilizing data from online review websites to classify positive or negative feedback about a product. Others used data on social media to gather people's thoughts about particular subject. However, we can categorize previous studies by the methodology used into three categories, Lexicon-based sentiment analysis, machine-learning based sentiment analysis and hybrid approach of both.

2.1 Lexicon-Based Sentiment Analysis

In this section, the paper will go through some studies that used lexicons only to obtain sentiments. Anwar Hridoy et al. [2] developed a location based Twitter sentiment analysis system. The aim of this work is to determine the popularity of a given product in several locations. Their methodology utilizes the lexicon source of word from predefined list or dictionary. After pulling all tweets containing words in the list, they filtered out obtained tweets based on the Geographical Co-ordinates tagged within the tweet. They only include tweets with specific locations. Then, another filter is applied to get extract tweets with predefined grammatical dependencies that are known to be useful in information analysts, whereas the rest are excluded. Next, a positive or negative score is assigned to each word. The model was tested on tweets related to "iPhone 6". Model Accuracy was not illustrated clearly in this work. However, authors state that quality of filtered tweets was low even after applying filters. There could be room for enhancement in this direction.

Some prior researches suggest that degree of polarity is more important than polarity classification alone. For example, recent research conducted by Jurek et al. in [3] designed a lexicon-based sentiment analysis algorithm that consists of two main components: sentiment normalization and evidence-based combination function.

This algorithm is targeting Twitter content analysis. Sentiment normalization aims at gathering the degree of polarity rather than Positive/negative decision, while evidence-based function improves decision accuracy when a tweet with mixed sentiments occurs. To evaluate the algorithm, authors conducted an experiment on Stanford Twitter test set with 177 negative, 182 positive and 139 neutral manually labelled tweets. Results showed that the two new functions increase the performance of the standard lexicon-based sentiment analysis by about 8% (from 69% classification accuracy using standard lexicon to 77% using authors' proposed algorithm). Mostafa [4] analyzed customers' sentiments towards major worldwide brands such as IBM, Nokia, and DHL. He collected and analyzed 3516 tweets for 16 different brands. A predefined opinion lexicon with positive and negative words has been used. However, this study did not cover the reason which lead to the positive or negative feedback.

2.2 Machine Learning Based Sentiment Analysis

There exists a considerable body of literature on using machine learning in sentiment analysis for different blogging and social media platforms. The greater part of these studies used supervised learning techniques. For example, Narayanan et al. in [5] used Naïve Bayes classifier on Internet Movies Data-Base (IMDB) to gather people's opinions on movies. Their approach classifies opinions to positive, negative and neutral. They focused on improving accuracy of Naïve Bayes for sentiment analysis. This led to combining multiple methodologies including: negation handling, improved feature selection and words N-grams. This resulted in significant improvement in the accuracy and speed of sentiment classifier. Their methodology could be generalized to many text categorization problems.

Other study conducted by Shukla and Misra [6] used K-means besides Naïve Bayes to obtain customers' opinion about mobile products. They created dataset containing 2000 mobile reviews retrieved from Amazon, Flipkart and Review Centre, with 1000 positive and 1000 negative labelled reviews. Their contribution was utilizing K- means in feature selection to remove noise and reduce complexity for the classifier. Then, they applied Naïve Bayes to classify mobile reviews. To evaluate their proposed algorithm, a comparison was conducted between Naïve Bayes alone, Support Vector Machine (SVM) and proposed algorithm on the same dataset, results showed that their proposed algorithms outperforms other mentioned classifiers.

When it comes to mining social media in order to get sentiments from users, Ghiassi et al. in [7] addressed the challenges associated to the nature of Twitter language. For this reason, their study focused on feature engineering to deal with these challenges, including but not limited to, frequency analysis, affinity analysis and negation analysis. Their proposed methodology used Dynamic Architecture for Artificial Neural Network (DAN2) as a machine learning model to create 3-class and 5-class sentiment analysis classifiers. DAN2 is known to be sensitive to mild expressions. For their evaluation, they targeted Starbucks brand sentiments in Twitter, to obtain meaningful results, they compared their methodology using SVM model, and Sentiment140 [8] and Repustate [9] Twitter sentiment analysis systems, results showed that proposed methodology achieved noticed higher accuracy than other classifiers.

2.3 Hybrid Approach Based Sentiment Analysis

A number of authors have recognized that combining Lexicon and machine learning may produce better approaches. Mudinas et al. [10] created pSenti, a sentiment analysis system developed by combining lexicon-based and machine learning-based approaches. Their approach utilized Lexicon for initial sentiment detection. The output of this stage is used to train Space Vector Machine classifier. They used two datasets to evaluate their approach: software reviews collected from CNET's software download website, and movies reviews collected from Internet Movies Database IMDB. Results showed that hybrid approach gives much better accuracy than using lexicon or machine learning separately.

Another study conducted by [11] proposed a hybrid approach that combined a lexicon-based approach and machine learning. First, they applied a lexicon classifier to classify tweets' sentiments. Then, Support Vector Machine was trained to identify sentiments of new tweets. Data set was retrieved from Twitter with 5 key-words, for each key-word, 500 tweets used for test set and the rest for training. Authors compared their approach with four different approaches including lexicon based, maximum entropy and two other approaches, results clearly showed that their approach achieved higher accuracy than others.

It can be inferred from that hybrid approach allows automatic labeling for tweets, by this way, time for labelling training data is reduced and adding more training data is achievable, therefore, this work hybrid approach is used, that is using lexicon classification and Neural network as a machine learning model, to our knowledge, this is the first work that uses neural network in hybrid company reputation analysis system based on Twitter.

3 Methodology

This work utilizes hybrid approach since it proved to enhance accuracy in sentiment analysis as described in Chap. 2, hybrid approach combines lexicon and machine learning in classifying tweets. One of its benefits is eliminating manual labeling of data for training since the collected data will pass through a lexicon-based classifier for labelling. The labelling data are used as training data for a machine learning classifier to build the classification model. This model is used to predict the polarity of the new data set (testing data). For the machine learning classifier, Artificial Neural Networks will be used since it outperforms other classifiers in Twitter sentiment analysis based on [7].

Proposed system consists of two phases: building Artificial Neural Network model for tweets sentiment analysis and using this validated model in company reputation along with viewing most and least liked products/departments. Figures 1 and 2 illustrates further system methodology.

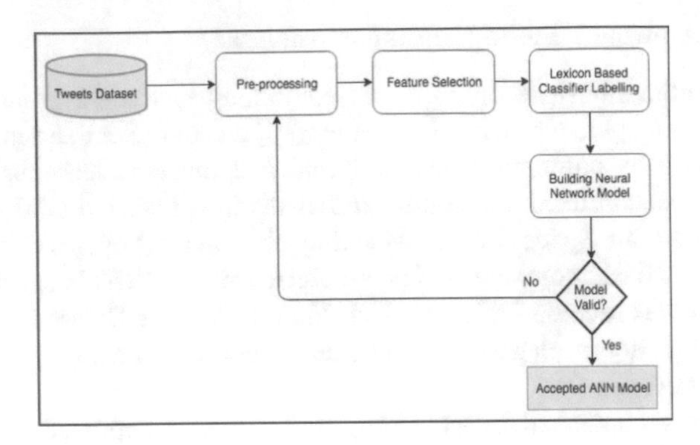

Fig. 1. ANN sentiment classification model

3.1 Data Collection

Official Twitter API is used to search and retrieve tweets. A company's name is entered as search keywords for tweets retrieval. The API returns then relevant tweets based on the given word.

3.2 Data Preprocessing

Different preprocessing techniques is applied on the tweets in sequential manner:

- **Tweet cleaning**
  In this process, tweets are cleaned from irrelevant information and Twitter special characters and words such as: "RT", "Fav", "Mentions".
- **Word boundaries**
  Some characters were defined to be word boundaries characters which include comma, period, colon, question mark and semicolon. All these characters are replaced with a white space
- **Stemming**
  Light stemmer is used to proposed system in order to map an input word to the shortest possible form while maintaining its meaning. Light stemmers have proved effective in a number of information retrieval systems and are considered simple to implement. Stemming is done using Scikit learn light stemmer [12] to remove redundant letters in a word.
- **Tokenization**
  The tokenization process breaks each tweet into a bag of words. These words are named tokens.
- **Stop word removal**
  Stop words are very common words. They are usually words which help build ideas but do not carry any significance themselves. Stop words include words such as conjunctions, articles, prepositions and so on [13]. Stop word filtering means the

elimination of these words, such as (in, of, on, the…etc.). Official Natural Language Toolkit stop words list [14] was used in this work. Customization was made on the list by removing some sentiments from the list such as "not, below, above' since it affects the tweet sentiment.

- **Negation Handling**
  Since negation words such as "not" may change the sentiment to the opposite, this work handles these situations by detecting negation words and reversing the sentiment of the word next.

3.3 Sentiment Lexicon Building

- **Words Enrichment**
  Predefined lexicon containing list of positive and negative words taken from work by Liu et al. in [15] was used in this work. However, after inspecting some tweets words and lexicon classification, it was found that the lexicon needs enriching with some words. Therefore, some frequently used words in social media, such as "lovin", were added to the positive words lexicon. The negative list was also modified. The word "ignoring" was added to the negative lexicon and the word "cold" was removed.
- **Emoji Consideration**
  It was noted that many people are posting tweets with only Emoji representing their sentiments towards the topic or product. So, this work includes all Emoji characters that represent opinion or sentiment in the proposed lexicon.

3.4 Lexicon Classifier

After cleaning tweets, it is passed to lexicon classifier. The main goal of this process is labelling tweets into positive, negative, and neutral before passing them to NN based on tweet content. Methodology of labelling is parsing through each word in the tweet and look for it in both positive and negative lookup tables from work in [15]. If the word is found in positive lookup table, it is given +1 score, if found in the negative list it is given -1 score, if the word is not found the tables, then it is given score of 0. After scoring all words in tweet, the words are summed up to give the sentiment of the whole tweet. This methodology is previously done by the work in [16]. By the end of this step, a data frame of all tweets with corresponding sentiment labels is ready to be passed to ANN.

3.5 Neural Network Classifier

Neural network is powerful machine learning classifier that is composed of an input layer, hidden layers, and an output layer [17]. In this work, multi-layer perception (MLP) neural networks from Scikit learn framework is used [18]. Since neural networks only works on numerical values, all text was transformed into [0, 1] vectors, that passed to neural networks as an input layer (features), the output layer is the sentiment [1, 0, −1] while hidden layers are conducted and assigned internally by the model.

Neural networks algorithms have lot of parameters to set, such as number of hidden layers, number of neurons in each layer, maximum number of iterations, weight optimization function …etc. in order to find the best parameters set to achieve maximum accuracy, a grid search with range of possible parameters was conducted, the grid search is done with the following set of parameters ranges: [number of hidden layers: (1–5), number of neurons: (32–1000), Alpha range: (0.5–1.5), solvers: (adam, sgd), Learning rate: (constant, adaptive, invscaling), number of epochs: (0–3000)], and best parameters found by grid search are: Number of hidden layers: 3.

Number of neurons in each layer: 64, 128, 256 respectively, Learning rate: Adaptive, Number of epochs: 128, Alpha: 1.005, Solver: Adam.

3.6 Term Frequency

In this stage, the goal is to understand the reason behind the negative or positive perception of the company. After classifying the tweets to negative and positive using ANN, words that indicate the product or department is identified. It is determined using term frequency after tweets classification. Once sentiment is detected for company, term frequency is counted on positive tweets to get most recurring word in the positive class. Words are ranked based on their frequency. This is done by finding the most recurring words in the whole document of positive tweets of the company, while excluding stop words, special characters and some frequently used words and verbs that does not represent a department, product and service, as an example (eat, drink, love, thanks, hello ..etc.). The resulting words can be words such as “delivery” which indicates a that the delivery department performance is good. It may also be a product such as “croissant” which can be a new successful product or the best seller in the store. Some words can be related to recent news.

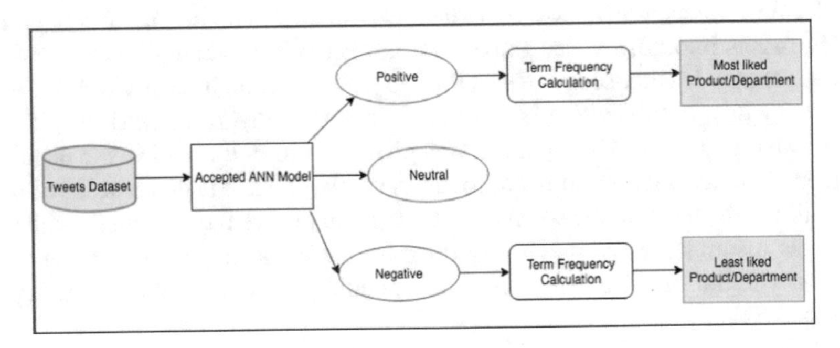

Fig. 2. Using validated model in companies reputation analysis

4 Implementation

To test our approach, four well-known companies were chosen from different sectors; “Starbucks”, “Aramex”, “Domino’s Pizza” and “Uber”. Data was retrieved for the period Jan 2019 – April 2019. More than 20,000 tweets were retrieved, 14,200 tweets only were selected, and the remaining were eliminated by data cleaning process. Companies distribution of selected tweets are: 4600 tweets for Starbucks, 2100 for Aramex, 4400 for Uber and 3100 for Domino's Pizza. Overall data contains around 4590 positive tweets, 4070 negative tweets and the remaining are neutral. companies' tweets reputation results are illustrated in Fig. 3, when it comes to ranked services, top ranked services for Aramex are customs clearing, agents and delivery, while delay and customer services were least ranked services. For Starbucks, people are positively mentioning espresso, mocha and Wi-Fi, while tea, music selection and Machiatto are least ranked products. For Domino's Pizza top ranked products are crunchy crust, mobile application and "points for pies" program while least ranked products are: delivery, driver and smell. Tweets of Uber showed that people like their mobile application, free-rides and "Uber-cool" service the most while they dislike "annoying" drivers, "asking" drivers and dirty cars.

5 Results

In order to evaluate proposed classifier, hold-out split approach was used, this approach depends on splitting the data into two datasets, training and testing, in our case, 75% of data were used for training (10,650 rows) while 25% for testing (3550 rows), testing dataset was labelled manually in order to get accurate results. Accuracy of classifier is calculated using accuracy function as follows:

NT = number of testing examples

NC = number of correctly classified testing examples (which includes True Positive and True Negatives).

$$\text{Classification accuracy} = \frac{NC}{NT} \Rightarrow \frac{3164}{3550} = 89\%$$

This work is compared with Zhang et al., work in [11] which used Hybrid approach with SVM as a machine learning classifier, the same approach was applied to our dataset using same settings and data, it gave the accuracy of 87%, this illustrates that proposed classifier outperforms other hybrid SVM approaches in classifying companies tweets datasets into natural, positive, and negatives sets.

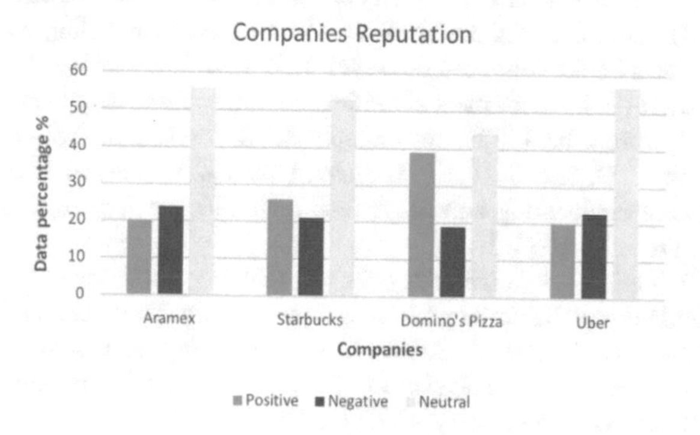

Fig. 3. Companies reputation based on retrieved tweets

6 Conclusion

In conclusion, this paper describes Twitter sentiment analysis system for perceiving company's reputation along with service/product reputation. It is intended to assist companies to identify customers' perceived value and study the company's points of weakness and strength by utilizing unbiased available data on social networks. The methodology is using hybrid approach including lexicon classification to label data and neural networks approach to classify reputation from pulled tweets. Since proposed system depends on text analysis and mining, sometimes it is hard to extract the desired meaning of the word. Specifically, if a company name has another meaning, it is difficult to pull tweets related to this specific company and ignore other tweets with the same word. For example, Amazon is well-known company, however it is also name of a river, so when retrieving tweets about Amazon (the company) it happened to get tweets mixed of Amazon company and Amazon river, accordingly, proposed system will work properly with unique company names. Importantly, proposed system was able to successfully classify unique companies tweets by 89% accuracy.

Looking forward, investigations are necessary to extract the context of the word in order to be able to analyze companies' reputation that have common names. Historical data could be considered to find the progression of company over years. Moreover, this approach can be implemented on other customers feedback platforms, such as customer reviews on food delivery websites/applications/blogs, customers reviews on shopping sites (e.g. ebay, amazon, ..etc.), and also on other social media platforms such as Facebook and Instagram.

References

1. Van Gerven M, Bohte S (2018) Twitter power: tweets as electronic word of mouth
2. Anwar Hridoy SA, Ekram MT, Islam MS, Ahmed F, Rahman RM (2015) Localized twitter opinion mining using sentiment analysis. Decis Anal 2(1):8
3. Jurek A, Mulvenna MD, Bi Y (2015) Improved lexicon-based sentiment analysis for social media analytics. Secur Inform 4(1):9
4. Mostafa MM (2013) Expert systems with applications more than words : social networks' text mining for consumer brand sentiments. Expert Syst Appl 40(10):4241–4251
5. Narayanan V, Arora I, Bhatia A (2013) Fast and accurate sentiment classification using an enhanced Naive Bayes model. Lect. Notes Comput. Sci. (including Subser. Lect. Notes Artif. Intell. Lect. Notes Bioinformatics), vol 8206 LNCS, pp 194–201
6. Shukla A, Misra R (2015) Sentiment classification and analysis using modified K-means and naïve Bayes algorithm. Int J Adv Res Comput Sci Softw Eng 5(8):80–85
7. Ghiassi M, Zimbra D, Lee S (2016) Targeted twitter sentiment analysis for brands using supervised feature engineering and the dynamic architecture for artificial neural networks. J Manag Inf Syst 33(4):1034–1058
8. Sentiment 140. [Online]. Available: http://www.sentiment140.com/
9. Repustate. [Online]. Available: https://www.repustate.com/sentiment-analysis/
10. Mudinas A, Zhang D, Levene M (2012) Combining lexicon and learning based approaches for concept-level sentiment analysis. In: Proceedings first international working issues sentiment discovery opinion mining - WISDOM '12, pp 1–8

11. Zhang L, Ghosh R, Dekhil M, Hsu M, Lui B (2011) Combining lexicon-based and learning-based methods for twitter sentiment analysis. In: Proceedings First international working issues sentiment discovery opinion mining - WISDOM '12, pp 1–8
12. Scikit Learn Lemmatizer
13. Srivastava J, Desikan P, Kumar V (2002) Web mining: accomplishments & future directions. In: National science foundation workshop on next generation data mining (NGDM'02), pp 51–69
14. Natural Language ToolKit
15. Liu B, Street SM, Street SM, Street SM, Opinion observer analyzing and comparing opinions .pdf
16. Han H, Zhang Y, Zhang J, Yang J, Zou X (2018) Improving the performance of lexicon-based review sentiment analysis method by reducing additional introduced sentiment bias, pp 1–12
17. Kubat M, Introduction to machine learning
18. Scikit-learn developers (BSD License). Scikit Learn MLP. [Online]. Available: https://scikit-learn.org/stable/modules/generated/sklearn.neural_network.MLPClassifier.html

Mathematical Modeling of Multimodal Transportation Risks

Vitalii Nitsenko[1(✉)], Sergiy Kotenko[2], Iryna Hanzhurenko[2], Abbas Mardani[3], Ihor Stashkevych[4], and Maksym Karakai[4]

[1] Private Joint-Stock Company "Higher Education Institution "Interregional Academy of Personnel Management", Frometivska Str., 2, Kiev 03039, Ukraine
vitaliinitsenko@gmail.com
[2] Institute of Market Problems and Economic-Ecological Research, National Academy of Sciences of Ukraine, French Boulevard 29, Odessa 65044, Ukraine
[3] Azman Hashim International Business School, Universiti Teknologi Malaysia, 81310 Johor Bahru, Johor, Malaysia
[4] Donbas State Engineering Academy, Mashinobudivnykiv blvd., 39, Kramatorsk 84313, Ukraine

Abstract. Research has shown that the risks of multimodal transportation depend as both on stochastic and fuzzy parameters.

Mathematical vehicles for the stochastic and fuzzy quantities are different. Therefore, a mathematical model is suggested to evaluate for the integral risk of cargo transportation. This makes it possible to use this model in support systems while making decisions on logistics of multimodal transportation. The use of a mathematical model requires careful analysis of all risks attributed to the multimodal transportation chain, possible overload options, and taking into account the entire spectrum of control activities.

After determining the most appropriate, from the point of view of risk minimization, the mode of transportation and its first links, the next stage of dynamic risk management is recursive review of the status vector of the chosen variant of the specified transportation route. For this information system it is necessary to process large data sets, while the suggested model economically uses computer resources and reduces the calculation time. The given mathematical model allows real-time changes in the transportation risk at specific stage to offer options for reducing integral risk, leverage it, in particular, choosing other routes and types of transport.

Keywords: Mathematical model · Multimodal transportation · Risks · Big data · Fuzzy variables · Stochastic parameters · Time-discretization · Dynamic system

1 Introduction

Multimodal transport can be considered as a system of separate subsystems. In this case, the subsystems are: the types of transport for the carriage of goods, points of overload, temporary warehousing, information support of transportation. From the point of view of multimodal transportation, as dynamic system for each level of cargo

R. Ghazali et al. (Eds.): SCDM 2020, AISC 978, pp. 439–447, 2020.
https://doi.org/10.1007/978-3-030-36056-6_41

transportation, it is possible to consider different variants of use of separate subsystems, in particular: vehicles, routes of transportation, etc., changing them in real time.

The risks of transportation at each transportation stage for each of the subsystems can lead to an increase in the total cost of transportation, in an increase in the time of transportation, in the damage or loss of cargo. Therefore, it is advisable to choose variants of transport ways, in general, and variants of their stages with the lowest risks.

The mathematical formalization of multimodal traffic begins with a description of risks within a complex multilevel hierarchical system.

Determining the risks of multimodal traffic across Ukraine is complicated by the fact that these risks depend not only on deterministic values, but also on non-deterministic, stochastic ones. However, the task become even more complicated due to the fact that the distribution of risk probability could not be evaluated or risk could not be identified. Therefore, in the risk analysis, it is necessary to operate fuzzy values as well. The use of the apparatus of mathematical statistics and probability theory, suitable for the processing of stochastic data, for fuzzy values is not entirely correct. There is a logical question on which mathematical means to apply for the definition and prediction of integral risks of multimodal transportation of cargoes.

There are studies devoted to the analysis of risks during transportation, in particular, during multi-modal transportation, both in the presence of purely vague variables of these risks [1–4], and in the presence of strictly viable parameters of transportation risks [5–8]. In addition, there are well-known studies on finding and forecasting risks for specified routes or certain types of transport [9–14]. While, for example, logistic companies, which are planning multimodal transportation through Ukraine, in view of the presence of systemic risks in their practical work, must also take into account the backup route options at certain stages of transportation, so that they can be changed, if necessary. Therefore, mathematical models for computerized decision support systems (DSS) for the transportation of goods, their transshipment, the selection of the most efficient routes should be based upon dynamic mathematical models. Such models should include not only possible changes in the modes of transportation or in the routes of goods delivery at certain their parts, but also, it is desirable to point out to the main controlling factors to reduce the possibility or consequences of certain issues arising from the transportation of multimodal cargo. Therefore, the purpose of this work was to construct a mathematical model for a dynamic system of support and decision making of multimodal transportation taking into account both the risks depending on stochastic parameters and the risks determined by fuzzy variables.

Therefore, the purpose of this work was to build a mathematical model for a dynamic system of support of and decision making in multimodal transportations with the use of an effective method of finding integral risk in the conditions of stochastic and fuzzy local risk parameters.

2 The Mathematical Modeling of Multimodal Transport Risks

Since the task of multimodal transportation is hierarchical, at the first stage it is possible, using the principle of decomposition, to identify those units where the risk parameters are fuzzy values, which, do not require information from other parts of the transport chain to evaluate the risks of the traffic. Then the risks of transportation of cargoes at such links can be determined only with the use of methods of fuzzy logic. The principle of decomposition consists in the formal replacement of the task of finding the transportation risk of the selected option of the transport chain by the equivalent set of tasks of transportation by individual links of the specified chain. The execution of decomposition is carried out according to a certain algorithm. The first of the decomposition stages is the transformation of the selected route of cargo transportation into a formalized system suitable for the decomposition, the separation of individual subsystems according to selected features. The criterion for the relevance of the selection of individual subsystems will be such risk values, which will be in line with the risks of systems that include the whole set of subsystems. This can be determined without calculating the specific values of the risks of both systems and their respective subsystems. To do this we use the following method. When the risk-determining parameters are represented by a universal set X, then the fuzzy set A will be a set of sets $(x|\mu_{\mathrm{A}}(x))$ for $x \in \mathrm{X}$. Value of the function of belonging μ_{A} comply with $\mu_{\mathrm{A}} : \mathrm{X} \to [0, 1]$, and for a separate set $(x|\mu_{\mathrm{A}}(x))_i$ it shows the degree of affiliation of a particular value $\mu_{\mathrm{A}i}$ to the fuzzy set A. Then the statement of the problem of finding the minimum risk value, which is determined by the fuzzy set of parameters A, will take the form:

$$\min_{x} U(x) \;\; \text{provided}\, x \in X \tag{1}$$

An implicit set can be used to solve a problem:

$$\{x^* : U(x^*) \geq U(x)\} \text{ provided } x \in \mathrm{X} \text{ and } x^* \in \mathrm{X} \tag{2}$$

This expression indicates that there may not be a single solution to the problem, but a set of such solutions, which corresponds to the purpose of the study. Since this expression formalizes the division of the system into subsystems, but does not determine the composition of sub-elements of subsystems, this expression can be considered in terms of the formation of compressed sets. The compression of a set X can be made in a tangent manner, so that the global minimum of the target function corresponds not only to the set X, but also to the compressed set for which the following is true:

$$V(X) \subset X \tag{3}$$

where $V(X)$- compressed set.

Moreover,

$$V(X) = \{x^* : U(x^*) \geq U(x)\} \text{ provided } x \in \mathrm{X} \tag{4}$$

As shown by the analysis of the multimodal transportation risks, which are determined by fuzzy variables, these variables can be described as fuzzy numbers of (L-R)-type. The membership function μ_A of such numbers is given using functions $L(x)$ and $R(x)$. These are functions of real variables which do not increase within the set of nonnegative numbers and have the following properties:

$$1.\ L(-x) = L(x); R(-x) = R(x); \tag{5}$$

$$2.\ L(0) = R(0)$$

where $x \in [c, d]$. Dots c - left limit and d - right limit of interval of fuzziness.

Accordingly, $[a, b] \in [c, d]$, where a, b the left limit and the right limit are the tolerance interval, respectively. Then the membership function can be reduced to an algebraic expression as follows:

$$\mu_{\mathrm{A}}(x) = \begin{cases} L[^{(a-x)}/_c & \text{for the case } x \in [(a-c), a]] \\ R[^{(x-b)}/_d & \text{for the case } x \in [b, (b+d)] \\ 1, & \text{for the case } x \in [a, b] \\ 0, & \text{for the case } x \notin [a, b] \end{cases} \tag{6}$$

Using the theory of graphs, multimodal transportation can be described as a coherent indicative graph. Each of its vertices can be divided into sets of vertices in such a way that the generated by the specified procedure subgraphs are also connected. Thus, it is possible to form the mathematical basis for the decomposition of multimodal transportation. An approximate graph can be represented in analytical form by the tensor equation of the following form:

$$WQ = V \tag{7}$$

where the tensor $\boldsymbol{W}$ - the probability of a certain traffic flow at certain stages of the entire route of carriage in a given time interval, the tensor $\boldsymbol{Q}$ - bandwidth of each individual section of the chain of transportation, tensor $\boldsymbol{V}$ - stochastic characteristic of the possibility of passing goods by each individual section of the chain of transportation.

The possibility of passing the cargo by a separate section of the chain of transportation is characterized by a matrix of vectors

$$\vec{\vartheta} := \varphi(\mathrm{P}, \vec{\tau}) \tag{8}$$

where $\vec{\vartheta}$ - the vector for estimating each of the set of event risk, can be represented by a linear matrix; P - the significance of the probability of the relevant risk factor; $\vec{\tau}$ - vector of consequences of specified risk.

It should be taken into account, that the consequences of this risk may vary, depending on its specific value. The risks of multimodal transportation are a dynamically system dispersed in time, which is characterized by the fact that the dimension of the vector of traffic graph is bigger than the vector of input parameters, on which the risk of transportation of cargo depends during the whole chain of transportation. After determining the most appropriate, from the point of view of risks minimization, the mode of transportation and its first links, the stage of dynamic risk management comes through. It includes the recursive reviewing of the status vector of the chosen variant of the specified transportation route. In our opinion, the most expedient mathematical tool for this, is the Kalman filter, because it is, unlike other filters, suitable not only and not so much for finding the vector of the state of the transport graph, but for controlling the uncertainty of the vector. Kalman's filter uses the Bayes theorem, which organically describes the dependencies between risks of each of the stages of multimodal transportation. That is, the probability of an event on the previous section of the chain of the carriage rout, denote it as ***A***, and the probability of the risk of transportation on it, respectively, denote ***P(A)***, affects the probability value on the next section of the chain, denote the section as ***B***, and the probability of transportation risk on it, respectively, ***P(B)***. Then, as it is known, according to Bayes' theorem, when the probability of transportation risk is $\boldsymbol{P(A) \neq 0}$, $P(B|A) = P(A|B)P(B)/P(A)$. Then, according to the Bayes theorem, in the case when the probability of the transportation risk $\boldsymbol{P(A) \neq 0}$. It is advisable to use the Kalman filter, which is widely used in economics for DSGE-simulation [9]. In this case, the mathematical model of multimodal transportation should take into account the fact inherent to the multimodal transportation of goods for which the state of carriage process in stage ***B*** depends on the state of the carriage process in the preceding stage ***A***. Then for the state vector of the transportation stage $\vec{Y}_A$, for which a certain static set is characteristic risks, can be written in form:

$$\overrightarrow{Y_A} = \overrightarrow{U_A}\,\overrightarrow{X}_B + \overrightarrow{W_A}\,\overrightarrow{s_A} + \overrightarrow{z_A} \tag{9}$$

where $\overrightarrow{U_A}$ - vector of change of the transportation process, inherent to the process of transportation at the stage B, $\overrightarrow{W_A}$ - control vector, which is accompanied by a matrix (set) of control actions $\overrightarrow{s_A}$, $\boldsymbol{w_k}$ - the Gaussian set of transportation risks, which is characterized by the risk matrix of the entire transport chain $\boldsymbol{\delta}$, the diagonal of which is the dispersion of the components of the specified risk vector, and outside of the diagonal there are the covariances of the risk components. For the risk matrix of the entire transport chain, $\mathbf{cov(X)} \geq 0$. The initial state and the vectors of individual risk components of the entire transportation chain are independent values. The use of this method requires careful analysis of all risks on the multimodal transportation chain, possible overloads, and taking into account the entire spectrum of control activities. Without such an analysis, the use of the Kalman filter is not appropriate. This method allows to construct a system of control of the process of multimodal transportation with the dynamic change of individual parts of the chain of cargo transportation, depending on the risks increase at certain stages or changes in transportation conditions. Thus, one can in real time, taking control over the effects on the transportation process $\overrightarrow{s_A}$, reduce the risks, and achieve the minimization of the target function of cargo transportation - the cost of transportation, time or damage to the cargo. In the case, when functions $L(x)$

and $R(x)$ are linear functions, the fuzzy number is described as a trapezoidal, in particular, a triangular number.

The weight of each risk or the risk of each of the stages of transportation is the proportion of its impact on thc value of the integral risk, that is, in an analytical form:

$$f_{sum} = \sum a_i \times f_i \quad (10)$$

where f_i - the value of i-th risk, or the risk of the i-th stage of transportation, a_i - weight of i-th risk.

Graphic interpretation of the search for integral risk consisting of two components is shown in Fig. 1.

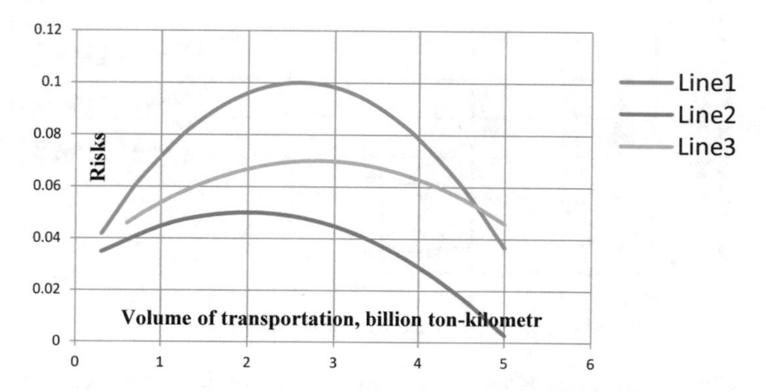

Fig. 1. Graphical interpretation of the search for integral risk consisting of two components. Row 1 - Transshipment risk, Row 2 - Tariff Risk, Row 3 - Integral Risk

The analysis of finding the integral risk of transportation including two risks, which have the biggest weight, related to tariff and transshipment of multimodal cargoes (Fig. 1), is characterized by the fact that the first one is stochastic and the second one is fuzzy. That is, this analysis clearly demonstrates the workability of the mathematical model, since the mathematical apparatus for calculating these risks are different, however, according to the suggested approach, they allow finding the integral risk of multimodal transportation. Handling risks were calculated for the ports of Yuzhkiy, Odessa, Chernomorsk, Kherson and Mariupol. The factors, that affect the risk of transshipment through each of these ports, are different and the risk magnitude for the ports varies significantly. After the risk assessment of routes passing through the named ports this allowed choosing the most efficient route in real time. For demonstration purposes (Fig. 1), the total risk of freight re-roll over the country's Black Sea ports is calculated as a sub-integral risk that combines local stochastic risks of transshipment through each of the named ports. Practical implementation of the model foresees the separate assessment of the risks of routes passing through so-called ports. A characteristic feature of local risks (Fig. 1) is their maximization for certain values of the multimodal turnover of the country. Reducing the value of integral risk by reducing its component - the risk of transshipment through the ports with an unacceptable risk level and the subsequent reorientation of cargo to other ports- made it possible to find levels of risk acceptable for maintaining the volume of cargo flows.

As the test of the DSS based on the suggested mathematical apparatus showed, increase in integrated risk of multimodal transportation (see Fig. 2) leads to an increase in the cost of transportation (line 1) at the interval of its real change, which, in turn, due to the competition between the Black Sea ports of other countries, gives the foreseen result - a decrease in the volume of bulk cargo (line 3) and container transportation (line 2). Thus, the integral risk of transportation makes it possible to predict not only the cost indicators, but also the volumes of total freight turnover and freight turnover by individual types of cargo.

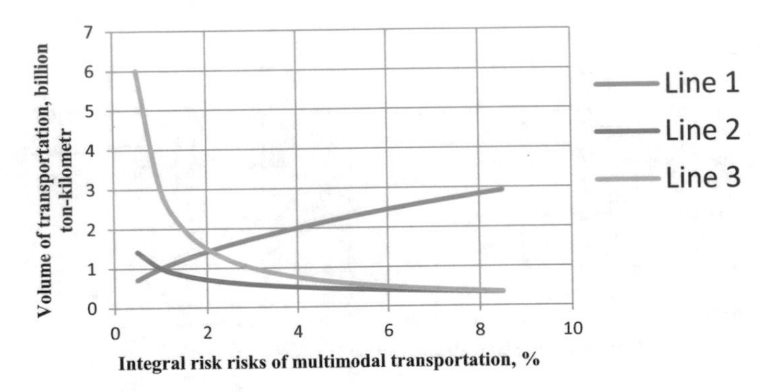

Fig. 2. Dependence of transportation costs and volume of bulk and container cargo on integrated risk

For this information system it is necessary to process large data arrays and the suggested model economically uses computer resources, reduces the calculation time. The given mathematical model allows real-time changes in the risk of transportation at some stage to offer options for reducing integral risk, leverage for this, in particular, to indicate other routes and modes of transportation.

The mathematical model was tested for the analysis of the transport of grain from the elevators of Kremenchuk to the ports of Odesa and the city of Chornomorsk. This made it possible to identify the risks at each stage of grain handling. It is revealed that at the stage of using the services of the railway company "Ukrzaliznytsya", risks associated with transportation, tariff policy of the company-monopolist, regulatory risks of transportation can take undesirable values. The alternative option is suggested - use river transport at separate stages, this will increase the cost of transportation of grain by around 5%, but will reduce the consequences of the risk fluctuations for the integral risk value in their range of changes.

An alternative option is recommended - use river transport at separate stages, this will increase the costs by 5%, but will reduce the impact of tariff risk on the value of integrated risk. The practical results obtained confirmed the adequacy of the suggested mathematical model.

3 Conclusion and Future Work

The study showed that the risks of multimodal transport depend both on stochastic quantities and fuzzy parameters. Since it is not possible to use a single mathematical device to handle such risks, a mathematical model for calculating the integral risk of multimodal transportation of goods is suggested, which allows using different mathematical approaches for calculating different types of local risks at individual stages of transportation. The application of this mathematical model implies the need for a thorough risk analysis at each stage of the entire multimodal transportation chain. In addition, we must take into account all possible overload options, develop a list of all spectrum of control actions. This allows, first, to establish the most expedient way of transportation, based on minimal integral risk and the accompanying economic consequences from the point of view of minimizing risks. It is also possible at each link of the entire route of transportation to manage risks dynamically, by recursively reviewing the status vector of the selected version of the specified transportation path.

Since the task involves the processing of large data sets, it is extremely useful, when applying the developed mathematical model, to have the possibility of economic use of computer resources, to reduce the calculation time. Pilot application of the given mathematical model, as the core of the system of support and decision making for finding the best solutions when choosing a variant of the way of multimodal transportation, is promising. For the practical work of the logistics center, there is an opportunity with the means of suggested model to choose the variant of the transportation route with the lowest integral risk, as well as to manage its levers, and to choose other routes and modes of transportation. The test trials of the core of the system of support and decision-making proved the correlation between the increase in the integral risk of multimodal transportation and the cost of transportation, reduced transportation of bulk cargo and container traffic for the seaports of Odesa and Chornomorsk.

References

1. John A, Paraskevadakis D, Bury A (2014) An integrated fuzzy risk assessment for seaport operation. Safety Sci 68:180–194
2. Liu Y, Fan ZP, Yuan Y (2014) A FTA-based method for risk decision making in emergency response. Comput Oper Res 42:49–57
3. Ferdous R, Khan F, Veitch B (2009) Methodology for computer aided fuzzy fault tree analysis. Process Saf 87:217–226
4. Bansal A (2011) Trapezoidal Fuzzy Numbers (a, b, c, d): arithmetic behavior. Int J Phys Math Sci 2(1):39–44
5. Vilko JPP, Hallikas JM (2012) Risk assessment in multimodal supply chains. Int J Product Econ 140(2):586–595, https://doi.org/10.1016/j.ijpe.2011.09.010
6. Frazila RB, Zukhruf F (2017) A stochastic discrete optimization model for multimodal freight transportation. network design. Int J Oper Res 14(3):107–120
7. Steadie Seifi M, Dellaert NP, Nuijten W, Van Woensel T, Raoufi R (2014) Multimodal freight transportation planning. Eur J Oper Res 233:1–15

8. Yamada T, Febri Z (2015) Freight transport network design using particle swarm optimization in supply chain–transport super network equilibrium. Transp Res Part E 75:164–187
9. Andrease MM (2008) Non-linear DSGE Models, The Central Di⁄erence Kalman Filter, and The Mean Shifted Particle Filter 46, (ftp://ftp.econ.au.dk/creates/rp/08/rp08_33.pdf)
10. Wang Y, Yeo G-T, A study on international multimodal transport networks from Korea to Central Asia
11. Litman T (2017) Introduction to multi-modal transportation planning principles and practices victoria transport policy Institute 19
12. Sossoe K (2018) Modeling of multimodal transportation systems of large networks. Automatic Control Engineering. University Paris-Est, 187
13. Jian Z (2017) Multimodal freight transportation problem: model, algorithm and environmental impacts. A dissertation submitted to the graduate school Newark Rutgers, The State University of New Jersey. 117
14. Liu Y, Chen J, Wu W, Ye J (2019) Typical combined travel mode choice utility model in multimodal transportation network, https://www.mdpi.com/2071-1050/11/2/549

Economic Load Dispatch Problem via Simulated Annealing Method

Jamaluddin Mir[1], Maria Imdad[1](✉), Junaid A. Khan[2], Nurul Aswa Omar[1], Shahreen Kasim[1], and Tauseef Sajid[3]

[1] Faculty of Computer Science & Information Technology, Universiti Tun Hussein Onn Malaysia, Parit Raja, Johor, Malaysia
mirjamal70@gmail.com

[2] Department of Computer Science & Engineering, HITEC University, Taxila, Pakistan

[3] Department of Computer Science, University of Lahore, Lahore, Islamabad, Pakistan

Abstract. In the presented research work an economic load dispatch problem based on various thermal units is processed by a derivative free method based on an analogy of metals annealing. The method is heuristic in nature and has minimum probability to get stuck in the local minima with better accuracy than that of classical schemes used solve economic dispatch problem. The test data has been incorporated from IEEE bus system of thermal generators and being observed the minimum cost ($) for power generation by maximizing the power utilization and minimizing the power losses. The simulations of the various scenarios are performed for different number of thermal units and constraints applicable in economic load dispatch. The author will evaluate the effects of the applied scheme in term of applicability, accuracy, cost and the computational complexity.

Keywords: Economic load dispatch problem ELD · Power generation · Simulated annealing SA

1 Introduction

The ever increasing differences of supply and demand in electric power generation systems require adequate policies to be adopted that ensure minimum cost of operation. One of the primary objectives of the power generation systems is to overcome the unremitting variations in power demands. Therefore, minimizing the operational cost of power generation becomes very important factor. Economic Load Dispatch (ELD) problem is a scheduling technique of power generation in regard to the load demand. Here the goal is to operate the power generating units at most economical cost while meeting all the constraints of the system [1]. In power generation, our main aim is to generate the required amount of power with minimum possible cost. The term ELD means that in order to achieve a specified power output, the generator's real and reactive powers are allowed to vary between certain limits while at the same time, incurring minimal fuel cost. Simulated Annealing algorithm has been able to finding a

R. Ghazali et al. (Eds.): SCDM 2020, AISC 978, pp. 448–459, 2020.
https://doi.org/10.1007/978-3-030-36056-6_42

solution of all such optimization problems as it is not susceptible of getting stuck in local optima or slow convergence towards a solution. Literature review is presented in Sect. 2, Sect. 3 discusses the used algorithm, methodology is discussed in Sect. 4, Sect. 5 discusses the results of simulations and Sect. 6 discusses the conclusion and future works.

2 Related Work

Researchers have applied several different mathematical programming as well as optimization methods for solving the ELD Problem. Early solutions used mathematical methods like Gradient Method, Linear programming [2], Newton-Raphson method, Base point method, Lambda iteration method, Participation Factor method, etc. [3]. It was observed that these traditional methods needed the incremental cost curves to be piece-wise linear or monotonically increasing [4]. However, it has been noticed that the modern power generating units are fundamentally highly nonlinear in nature and resultantly they have numerous local minima in the objective function. Therefore, the traditional techniques resulted in 'sub-optimal' solutions to the problem i.e. either increase in cost or reduced production. As a result, the nonlinear nature of the problem demanded such robust algorithms which will not face the problem of getting stuck in local minima or maxima [5]. To overcome this problem, certain algorithms based on the stochastic search prove to be more efficient to overcome problem under discussion. A few examples of such algorithms are particle swarm optimization (PSO) algorithm [6–8], evolutionary programming (EP) algorithm [9, 10], Hop field neural network [11], genetic algorithm (GA) [12], Chaotic Bat algorithm [13], Backtracking search algorithm [14], Grey wolf Optimizer [15], Ant lion optimizer [16], Group search optimizer [17], Artificial Bee Colony algorithm [18] and simulated annealing (SA) algorithm [19]. However, the load assignment must be such that the power is generated at lowest possible cost. It is evident from literature that the stochastic optimization techniques fail to achieve a balance between exploration and the exploitation of results which causes them to converge towards a local minima [18]. Traditionally, ABC algorithm is good when it comes to exploration and less efficient in exploitation. Sinha, et al. [20] have examined evolutionary programs for optimizing economic load dispatch problem. Fang and Hua [21] have discussed an improved particle swarm optimization algorithm (PSO) which incorporates the penalty functions as well.

The Simulated Annealing approach in this paper aims to find such a configuration of the power generating nodes which incur the least possible cost. Total power generation is kept as a hard constraint i.e. no compromise is made on the output of the system. However, there is no uniform distribution of load among all nodes of the system. The power generated by each node varies in each of the discussed system.

3 Simulated Annealing

The SA was introduced in the early 1980s by Kirkpatrick et al. and Cerny [22–24]. SA is a general purpose robust algorithm which is based on probabilistic methods. This algorithm has been vastly applied to several fields of sciences and is not limited to any particular problem. The application areas include network design and layout, circuit design, neural networks, component placement in the electric circuitry, image processing and many more. This algorithm is single objective optimization techniques which is motivated by the process of annealing of metals. The algorithm starts with extremely large search space i.e. ultra high initial temperature and subsequently moves towards those areas in search space that seem to have good solutions. The effective exploration and exploitation techniques of the method enable it to overcome the problem of getting stuck in local minima.

3.1 Analogy of Method

The SA algorithm is analogous to the physical process of annealing of metals. In the physical process of annealing, a solid is heated to extremely high temperature and is allowed to cool down very slowly. In the physical process of annealing, when the metals are heated to extremely high temperatures, the metals change their solid state and move freely in their newly adopted molten state. However, with the gradual cooling of the metals, the movements of the atoms of the material get restricted and they arrange themselves in crystalline form with minimum possible energy, solely depending upon the rate of cooling. It must be noted that the cooling process is very slow and gradual. If the metal is allowed to cool down quickly, the metal may not attain the perfect crystalline state and it is quite possible that the metal would end up in poly-crystalline form which has higher energy as compared to a perfect crystalline form [25]. In relation to optimization problem, this physical process of annealing is analogous to finding out a global optima. Energy of the atoms can be considered as the objective function whereas the final state can be considered as the global minima for that particular objective function. The analogy that is drawn between the optimization problem and the actual physical process of annealing is shown in Table 1.

Table 1. Simulated annealing and physical process of annealing

Physical annealing process	Optimization problem
State	Feasible solution
Energy	Cost function
Ground state	Optimal solution

The objective is to attain such a state which has minimum internal energy. Slow cooling of the solids allows the matter to achieve the state of thermal equilibrium at every temperature. This state is characterized by the Boltzmann distribution

$$PTX = x = e - Ex/kbT\Sigma e - Ei/kbt \tag{1}$$

- Where X indicates the current state,
- Ex is energy of the current state,
- Kb is Boltzmann's constant and
- T is the temperature.

4 Proposed Methodology

On the basis of the problem formulation presented in the previous section, a Matlab code of SA algorithm was developed. Here we will discuss the practical implementation of SA i.e. the components necessary for application of simulated annealing to overcome particular optimization problem. These essential components are: Search space, Objective/cost function, Perturbation technique and Cooling schedule. Here search space and objective are specific to problem formulation while perturbation technique and cooling schedule are specific to algorithm used (Fig. 1).

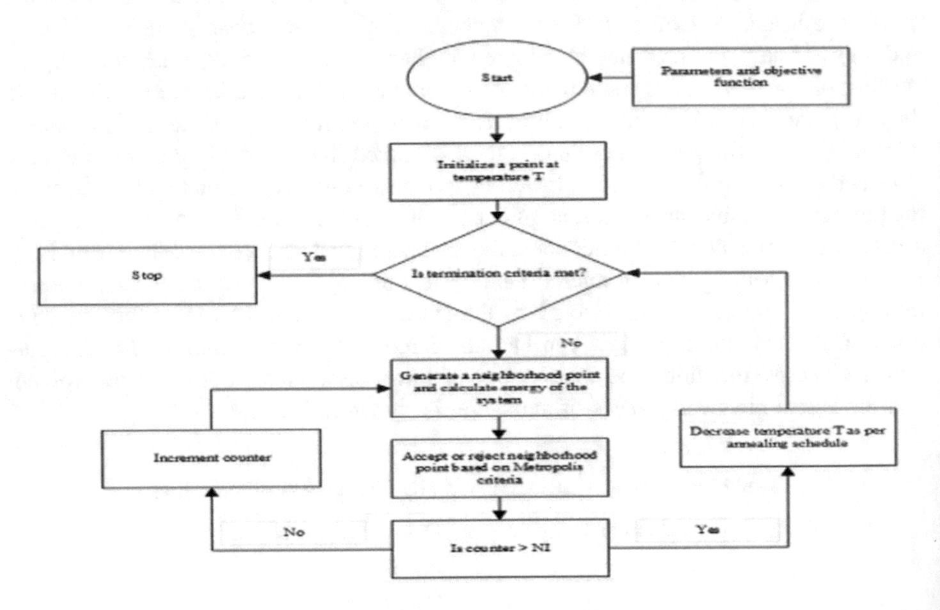

Fig. 1. Algorithm methodology

Search Space is the collection of all possible solutions to the problem. ***Initial Population*** is generated randomly in accordance to the search space. ***Initial Temperature*** parameter is extremely important for the algorithm to return global optimum values as it is responsible for the acceptance criteria defined by the Botlzmann's distribution. ***Objective*** is to produce certain amount of power while minimizing the overall cost. The ***Perturbation technique*** generates new solutions, ensuring that the optimal configuration is possible. The algorithm produces certain solutions that are not feasible at system configuration level such solutions are discarded and will not be discussed here. Several variants of SA have different ***Cooling schedules*** and selecting an efficient one is very important for the algorithm to run efficiently. The algorithm must be stopped upon finding the optimal solution and upon reaching the temperature which is low enough. The algorithm considers such a state to be as frozen; which remains unchanged for at least 4 iterations leading to search termination.

5 Simulation and Results

5.1 Case I: Convex System

The power generation values for three machine convex system, six machine convex system and eleven machine convex system are shown in the Table 2 given below. It is evident that the optimal solution here is to divide the load among all nodes but this load division is necessarily not equal while keeping the net power generated fix.

Table 2. Power generation values of convex system

SysTyp	P1	P2	P3	P4	P5	P6	P7	P8	P9	P10	P11
3	335.45	323.38	191.17								
6	372.30	306.97	192.59	275.69	410.16	242.30					
11	180.72	209.82	216.84	298.19	182.90	130.04	164.41	259.45	286.80	251.11	319.72

Table 3 shows the cost in dollars of the systems under discussion. As can be seen from the table, the cost of power generation varies from system to system. As we introduce more power generating nodes into the system, it will not only produce more power but the overall cost increases as well. However, the standard deviation of all three convex systems remains fairly uniform across all three models.

Table 3. Cost in dollars of convex systems

Cost in dollars				
System type	Minimum	Maximum	Mean	Standard deviation
3	1.6232E+06	8.8788E+06	3.1938E+06	1.3503E+06
6	6.0228E+06	2.3224E+07	1.1124E+07	3.7589E+06
11	1.3955E+07	3.3107E+07	2.3331E+07	5.3525E+06

Type 1: Three Machine System.
In Fig. 2, shows the graph of error and fitness values for three machine convex system. Due to smaller size of the system and comparatively smaller size of dataset, the values shown in the graph are quite closer to each other. The x-axis has the number of

independent runs or iterations. Error and fitness values for the system are shown at y-axis.

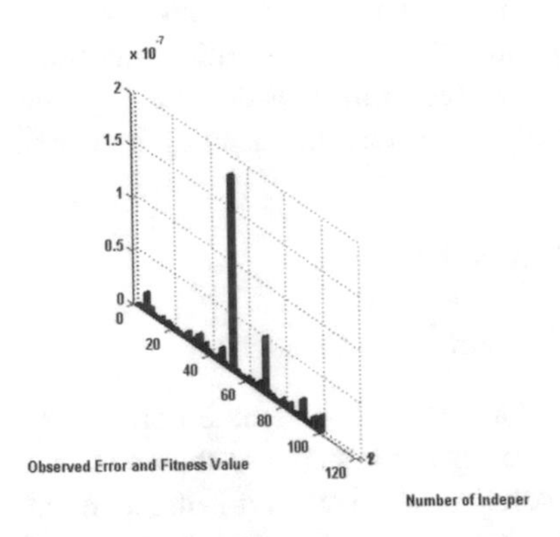

Fig. 2. Error and fitness value of 100 independent runs; 3 machine convex system

Type II: Six Machine Convex System.

In Fig. 3, error and fitness values of six machine convex system are shown during the course of hundred independent iterations. The graph for six machine convex system shows distinct error and fitness values. The behavior of the system and simulations is quite predictable as shown in the graph.

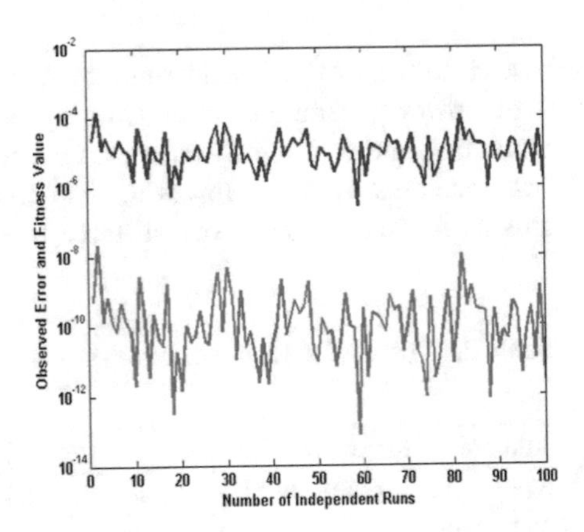

Fig. 3. Error and fitness value of 100 independent runs; 6 machine convex system

Type III: Eleven Machine Convex System.

Figure 4. shows the graph of fitness value and error for eleven machine convex system. The graph shows that the 20th and 39th iteration have produced the minimum values for fitness and error. This shows that during these iterations, the results of the simulation were closest to desired values. As a result, the simulation produced best possible fitness values during these iterations. It can be seen that as the data set increases, the simulation produces more clear and distinct results as well as becomes able to produce results with better fitness values

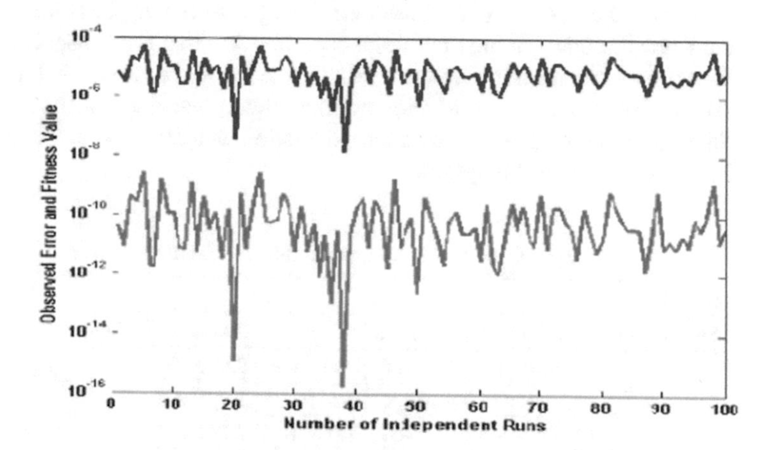

Fig. 4. Eleven machine convex system

Case II: Non convex System.
Table 4 shows the power generation values of each of the non-convex system. It must be noted that the total power generation is considered to be a hard constraint i.e. no compromise on net power. The individual machines in each of the discussed system are allowed to vary their amount of power generation, thereby running at a certain configuration which is most economical in terms of cost and producing the adequate amount of power which adds up to the desired amount when whole of the system is considered. In case of the 40-machine Korean model, it can be seen that the machines

Table 4. Power generation values of non-convex systems

System type	P1	P2	P3	P4	P5	P6	P7	P8	P9	P10
3	271.12	387.59	191.29							
6	67.99	76.24	43.63	32.64	23.78	39.11				
13	157.66	149.27	179.08	178.18	103.54	176.65	144.21	146.21	139.83	117.87
40	112.47	101.21	101.42	180.59	77.01	117.24	293.89	299.08	288.5	286.01
	P11	P12	P13	P14	P15	P16	P17	P18	P19	P20
	105.45	91.65	110.39							
	229.55	304.79	279.85	418.47	428.45	348.6	485.59	451.84	242.14	489.56
	P21	P22	P23	P24	P25	P26	P27	P28	P29	P30
	350	542.56	434.99	402.53	466.31	500.69	134.94	149.16	135.24	91.08
	P31	P32	P33	P34	P35	P36	P37	P38	P39	P40
	183.72	180.5	183.92	168.27	187.43	173.64	49.85	91.92	78.25	458.74

in the overall system produce varied amount of power which is as low as 77 MW and on the other hand as high as 540 MW.

The comparison of cost for non-convex systems is given in Table 5. The table shows that cost of power generation increases as we introduce more and more machines/nodes into the system, with three machine system having least cost and the forty machine Korean model having maximum cost. It seems that forty machine system is costlier than the three machine system, but it must also be noted that the forty machine system also generates more power than three machine system. Another important thing to note is that the standard deviation actually decreases as more machines are introduced into the system.

Table 5. Cost in dollars of non-convex systems

Cost in dollars				
System type	Minimum	Maximum	Mean	Standard deviation
3	1.46E+06	5.68E+06	2.89E+06	9.69E+05
6	1.54E+05	9.82E+05	4.14E+05	1.79E+05
13	6.59E+06	1.55E+07	1.09E+07	2.18E+06
40	1.99E+08	2.45E+08	2.21E+08	1.08E+07

Type I: Three Machine Non-convex System.

Figure 5 shows the graph of Error and fitness values for three machine convex system. As the dataset in this simulation was fairly smaller i.e. there were only three machines in the system, the values for fitness and error are fairly close to each other. The x-axis of the graph shows the number of iteration whereas y-axis shows the values of error and fitness during their respective iterations.

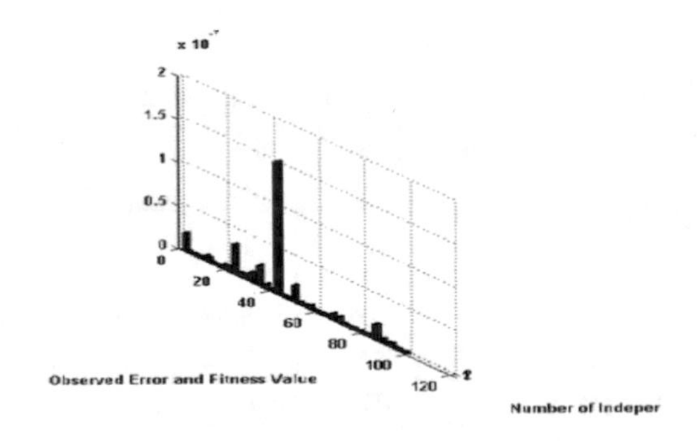

Fig. 5. Error and fitness values for three machine non-convex system

Type II: Six Machine Non-convex System.
In Fig. 6. error and fitness values for six machine non-convex system are shown for the hundred independent runs. The minimum error achieved for this system is as low as 10-14. As error is the value that shows how close the algorithm is to thc result, the smaller exponent means that the algorithm is able to achieve good results.

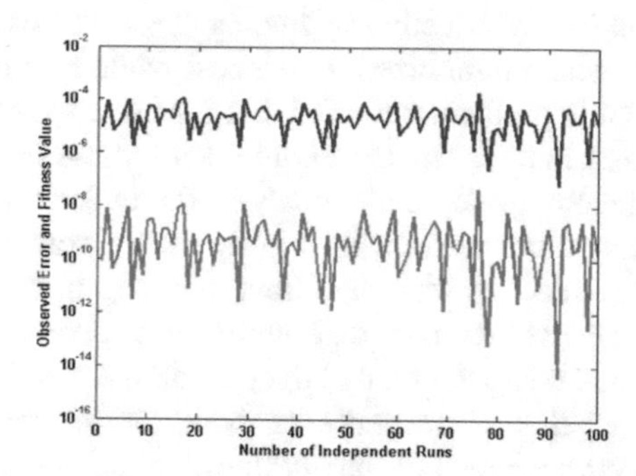

Fig. 6. Six machine non-convex system

Type IV. 40 Machine Non-convex Korean Model.
In Fig. 7 error and fitness values for forty machine non-convex system are shown. The graph shows that the error of the simulation to be lower than 10−14. This observation can be interpreted as that the algorithm is able to produce results that are extremely close to the desired output. It has been able to minimize the difference between the actual power produced and amount of power that was set to be the goal of the model. The lower values for error shows the closeness of attained results as compared to the goals that were set. In the same way, the fitness value shows the overall performance of the optimizer i.e. simulated annealing algorithm.

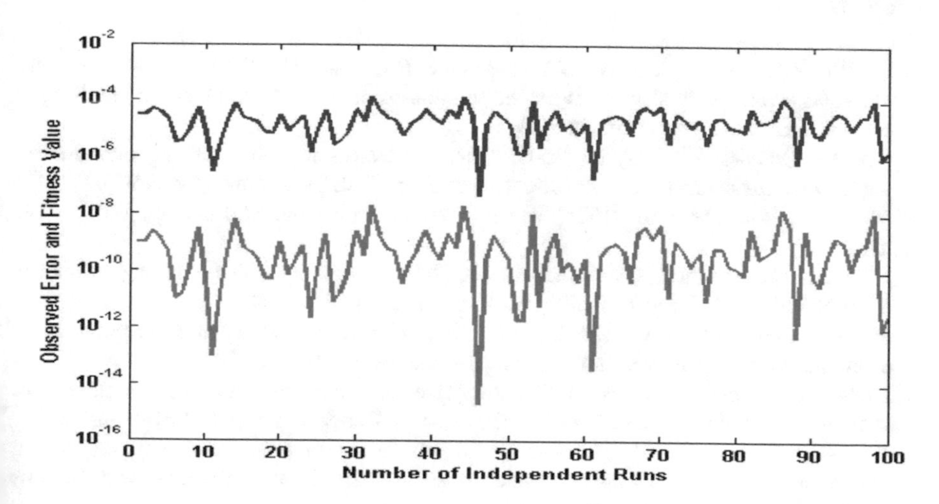

Fig. 7. Forty machine non-convex system

6 Conclusions and Future Works

ELD problem is an important issue with respect to scheduling and generating units to facilitate the ever-increasing power demand. The conventional and soft computing methods are explored from literature keeping emphasis on minimal cost of power generation using the ELD optimal solutions. The objective of this research work is to demonstrate that SA is highly effective in finding the best possible configuration of power generating units which minimize the fuel cost while keeping all constraints. SA was used to solve the ELD problem with 3, 6, 11, machine system and on 40-machine Korean power generation model. The algorithm explored the efficiency of convex and non-convex models, thereby giving authenticity to the undertaken study. SA depicted good results of superior quality, showing highly stable convergence characteristics moreover solutions provided by the algorithm are superior to several traditional/Newtonian methods and evolutionary algorithms. We have simulated the result of convergence behavior and computational budget for above mentioned convex and non-convex systems showing that convergence of the algorithm tends to improve as the complexity of the system increases by adding more nodes into the system. The reliability of the algorithm over several iterations is fairly good, indicating algorithm produces good results every time the simulation is run.

As a future direction ELD problem can be solved by using Neural Networks techniques like ANFIS, Deep learning and several other new methods. Comparative studies focusing on the convergence behavior and computational budget of the above mentioned can be a way forward.

Acknowledgments. We would like to say thank you to Universiti Tun Hussein Onn Malaysia (UTHM) and Research Mangement Centre (RMC) for kindly providing us with the internal funding Tier 1 (Grant Vot: H107).

References

1. Qu BY, Zhu YS, Jiao YC, Wu MY, Suganthan PN, Liang JJ (2018) A survey on multi-objective evolutionary algorithms for the solution of the environmental/economic dispatch problems. Swarm Evolut Comput 38:1–11
2. Jabr RA, Coonick AH, Cory BJ (2000) A homogeneous linear programming algorithm for the security constrained economic dispatch problem. IEEE Trans Power Syst 15(3):930–936
3. Wood AJ, Wollenberg BF (1984) Power generation, operation and control. Wiley, New York
4. Nidul S, Chakrabarthi R, Chattopadhyay PK (2003) Evolutionary programming techniques for economic load dispatch. IEEE Trans Evolut Comput 7:83– 94
5. Sinha N, Chakrabarti R, Chatopadhyay PK (2004) Improved fast evolutionary program for economic load dispatch with non-smooth cost curves. IE (I) J. EL 85
6. Panigrahi BK, Pandi VR, Das S (2008) Adaptive particle swarm optimization approach for static and dynamic economic load dispatch. Energy Convers Manag 49(6):1407–1415
7. Jiang S, Ji Z, Shen Y (2014) A novel hybrid particle swarm optimization and gravitational search algorithm for solving economic emission load dispatch problems with various practical constraints. Int J Electr Power Energy Syst

8. Jiang S, Ji Z, Shen Y (2014) A novel hybrid particle swarm optimization and gravitational search algorithm for solving economic emission load dispatch problems with various practical constraints. Int J Electr Power Energy Syst
9. Nidul S, Chakrabarthi R, Chattopadhyay PK (2003) Evolutionary programming techniques for economic load dispatch. IEEE Trans Evolut Comput 7:83–94
10. Sinha N, Chakrabarti R, Chatopadhyay PK (2004) Improved fast evolutionary program for economic load dispatch with non-smooth cost curves. IE (I) J. EL 85
11. Vo DN, Ongsakul W (2012) Economic dispatch with multiple fuel types by enhanced augmented Lagrange Hopfield network. Appl Energy 91(1):281–289
12. Chiang CL (2007) Genetic-based algorithm for power economic load dispatch. IET Gener Transm Distrib 1(2):261–269
13. Adarsh BR, Raghunathan T, Jayabarathi T, Yang XS (2016) Economic dispatch using chaotic bat algorithm. Energy 96:666–675
14. Modiri-Delshad M, Kaboli SHA, Taslimi-Renani E, Rahim NA (2016) Backtracking search algorithm for solving economic dispatch problems with valve-point effects and multiple fuel options. Energy 116:637–649
15. Kamboj VK, Bath SK, Dhillon JS (2016) Solution of non-convex economic load dispatch problem using Grey Wolf Optimizer. Neural Comput Appl 27(5)
16. Kamboj VK, Bhadoria A, Bath SK (2017) Solution of non-convex economic load dispatch problem for small-scale power systems using ant lion optimizer. Neural Comput Appl 28 (8):2181–2192
17. Daryani N, Zare K (2018) Multiobjective power and emission dispatch using modified group search optimization method. Ain Shams Eng J 9(3):319–328
18. Dixit GP, Dubey HM, Pandit M, Panigrahi BK (2011) Economic load dispatch using artificial bee colony optimization. Int J Adv Electr Eng 1(1):119–124
19. Basu M (2005) A simulated annealing-based goal-attainment method for economic emission load dispatch of fixed head hydrothermal power systems. Int J Electr Power Energy Syst 27 (2):147–153
20. Nidul S, Chakrabarthi R, Chattopadhyay PK (2003) Evolutionary programming techniques for economic load dispatch. IEEE Trans Evolut Comput 7:83–94
21. Ting-Fang YU, Chun-Hua Peng (2010) Application of an improved particle swarm optimization to economic load dispatch in power plant. In: 3rd international conference on advanced computer theory and engineering (ICACTE), Vol 2, pp 619–624
22. Kirkpatrick S, Gelatt Jr CD, Vecchi MP (1982) Optimization by simulated annealing. IBM Res Report RC 9355
23. Kirkpatrick S, Gelatt CD Jr, Vecchi MP (1983) Optimization by simulated annealing. Science 220:671–680
24. Cerny V (1985) Thermodynamic approach to the travelling salesman problem: an efficient simulation algorithm. J Optim Theory Appl 45:41–51
25. Wang L, Li N, Zhang XN, Wei T, Chen YF, Zha JF (2018) Full parameters inversion model for mining subsidence prediction using simulated annealing based on single line of sight D-InSAR. Environ Earth Sci 77(5):161
26. Alcantar V, Ledesma S, Aceves SM, Ledesma E, Saldana A (2017) Optimization of type III pressure vessels using genetic algorithm and simulated annealing. Int J Hydrogen Energy 42 (31):20125–20132
27. Matai R (2015) Solving multi objective facility layout problem by modified simulated annealing. Appl Math Comput 261:302–311

28. Zaretalab A, Hajipour V, Sharifi M, Shahriari MR (2015) A knowledge-based archive multi-objective simulated annealing algorithm to optimize series–parallel system with choice of redundancy strategies. Comput Ind Eng 80:33–44
29. Cakir B, Altiparmak F, Dengiz B (2011) Multi-objective optimization of a stochastic assembly line balancing: a hybrid simulated annealing algorithm. Comput Ind Eng 60 (3):376–384

A Comparison of Weighted Support Vector Machine (WSVM), One-Step WSVM (OWSVM) and Iteratively WSVM (IWSVM) for Mislabeled Data

Syarizul Amri Mohd Dzulkifli(✉), Mohd. Najib Mohd. Salleh, and Ida Aryanie Bahrudin

Universiti Tun Hussein Onn Malaysia, 86400 Batu Pahat, Parit Raja, Johor, Malaysia
syzm81@gmail.com, {najib, aryanie}@uthm.edu.my

Abstract. Labeling error can occur for various reasons such as the subjective nature of the labeling task, the lack of information to determine the true label of a given example and data entry error. Labeling errors were categorized as mislabeled, unlabeled, partially labeled, incompletely labeled and illegible label. In this study, the focus will be on mislabeled data. The problem of dealing with mislabeled data and in particular of constructing a classifier from such data has been approached from a number of different directions. Therefore, developing learning algorithms that effectively and efficiently deal with mislabeled data is a great practical importance and key aspect in machine learning. Support Vector Machine (SVM) has been widely accepted to be one of the most effective techniques in machine learning algorithms. One of the main drawbacks of SVM is it depends on only a small part of the data points (support vectors) and it treats all training data of a given class equally. To address this problem, one of the solution is the Weighted Support Vector Machines (WSVM). Wu & Liu proposed two different WSVM namely one-step WSVM (OWSVM) and iteratively WSVM (IWSVM). In this paper, a comparison of Weighted Support Vector Machine (WSVM), One-step WSVM (OWSVM) and Iteratively WSVM (IWSVM) for mislabeled data has been done to see the classification accuracy of each of the method. The three methods were compared based on correctly labeled, mislabeled data, data within margin, mislabeled data within margin and classification accuracy for eight KEEL repository datasets using 20% noise in training data. Based on the experimental results, the performance of OWSVM is better than both WSVM and IWSVM based on the correctly labeled, mislabeled data, data within margin, mislabeled data within margin and classification accuracy.

Keywords: Weighted support vector machine (WSVM) · One-step WSVM (OWSVM) · Iteratively WSVM (IWSVM) · Mislabeled data

R. Ghazali et al. (Eds.): SCDM 2020, AISC 978, pp. 460–469, 2020.
https://doi.org/10.1007/978-3-030-36056-6_43

1 Introduction

Supervised learning is the data mining task where the input data or the training data come with a label. The goal of this learning is to be able to predict the label for new, unforeseen examples. Labeling the data is expensive and error prone [1]. Labeling error can occur for various reasons such as the subjective nature of the labeling task, the lack of information to determine the true label of a given example and data entry error [2]. A subjective labeling error may arise if some experts disagree with the general consensus. Labeling errors arises when there is a shortage of information to determine the label. For example, a physician might fail to make a 100% accurate diagnosis if some expensive medical procedures are not available. In addition, labeling errors may come from communication or encoding problems and databases of real-word are deem to include around five percent of encoding errors [3]. The most frequent type of error is mistakes made during data-entry [2]. Labeling errors were categorized as mislabeled, unlabeled, partially labeled, incompletely labeled and illegible label [4]. This study will focus on mislabeled data. Mislabeled data is ever-present in real-word datasets and has different effects [3]. Firstly, the prediction performances of mislabeled data will reduce which has been proved for simple models such as linear classifiers or quadratic classifiers. Secondly, the number of necessary training data may raise up as well as the complexity of inferred models such as the number of support vectors (SV) in Support Vector Machine (SVM). Thirdly, the observed frequencies of the possible classes may be changed which is of particular importance in medical contexts. Thus, there exist three kinds of approaches to deal with mislabeled data namely label noise robust models, data cleansing methods and label noise tolerant learning algorithms. Label noise robust models are inherently robust to mislabeled data. The main purpose of this approach is to mitigate the effect of the mislabeled data. For example, a label noise robust extension of the widely used Bayesian logistic regression classifier was proposed by [5] and the same approach was used for the classification of mislabeled microarrays in [6]. A generalized label noise model which can resist the negative effects of labeling errors for classification in presence of labeling errors was proposed by [7]. Data cleansing methods are to remove data that appear to be mislabeled. This method is similar to outlier detection and anomaly detection. Model predictions may also be used to filter instances although this may remove too many instances. Many label noise tolerant learning algorithms are probabilistic methods such as Bayesian and frequentist methods. Clustering or belief functions also are in the same category as probabilistic methods. An important issue that is highlighted by these methods is the identifiability of mislabeled data [8]. Apart from probabilistic methods, specific strategies have been developed to obtain label noise-tolerant variants of popular learning algorithms such as SVM, neural networks and decision trees. The problem of dealing with mislabeled data and in particular of constructing a classifier from such data has been approached from a number of different directions. Therefore, developing learning algorithms that effectively and efficiently deal with mislabeled data is a great practical importance and key aspect in machine learning [9]. SVM has been widely accepted to be one of the most effective techniques in machine learning algorithms designed for classification problem [10] was introduced by Boser et al. [11] and then initiated by Vapnik [12]. SVM is also one of the best known margin based learner models [13]. One of the main drawbacks of SVM is it depends on only a small part of the data points (support vectors) and it treats all training data of a given class equally. To address this problem, one of the solution is the Weighted Support Vector Machines (WSVM). WSVM which uses a function of weights treats each data point differently and correctly classify more important data whether or not they are mislabeled. Yang et al. [14] proposed WSVM that could effectively mitigate the outlier sensitivity problem (i.e., WSVM-outlier). In this research, kernel-based possibilistic c-means (KPCM) was used to generate weights but the weights that generated directly by KPCM were not quite suitable to the algorithm. Fan & Ramamohanarao [15] proposed the same algorithm as [14] but they employ Emerging Patterns (EP) to allocate a weight to each data point. They stated that the important issue in training WSVM is how to initiate a dependable weighting strategy to contemplate the true noise distribution in the training data. Another interesting research proposed by [16] that combines PCA and WSVM. In this research, the purpose of PCA is to extract features from the dataset for KPCM to produce suitable weights. Wu & Liu [17] proposed two different WSVM namely one-step WSVM (OWSVM) and iteratively WSVM (IWSVM). Both methods produce more accurate classifiers than the standard SVM when there are noise in the training data. In this paper, OWSVM and IWSVM will be used as a comparison for the WSVM. The paper is organized in the following manner: Sect. 2 describes the theoretical background of the WSVM, OWSVM and IWSVM. Section 3 describes the experimental results and discussion. Finally, Sect. 4 gives the conclusion.

2 Theoretical Background

In this section, a description of the WSVM, OWSVM and IWSVM are briefly presented.

2.1 Weighted Support Vector Machine (WSVM)

WSVM is the SVM adaptation that solves the following optimization problem:

$$\min_{\mathrm{w,b},\xi} \frac{1}{2}\|w\|^2 + \mathrm{C}\sum_{i=1}^{n} W_i H_1(y_i f(x_i)) \tag{1}$$

$$\text{Subject to } : y_i(\langle w, x_i \rangle + b) \geq 1 - \xi_i, \xi_i \geq 0 \; for \; i = 1, \ldots n \tag{2}$$

where W_i is a weight of the i-th data point. When $W_i = 1$, it became the standard SVM. The hinge loss $H_1(u) = (1 - u)_+$, where $(u)_+ = u$ if $u \geq 0$ and 0 otherwise. The dual problem of WSVM is:

$$\max_{a_i} \sum_{i=1}^{n} \alpha_i - \frac{1}{2}\sum_{i=1}^{n}\sum_{j=1}^{n} \alpha_i \alpha_j y_i y_j (\langle x_i, x_j \rangle) \tag{3}$$

$$\sum_{i=1}^{n} \alpha_i y_i = 0$$

$$\text{Subject to } : 0 \leq \alpha_i \leq CW_i, \, for\, i = 1, \ldots, n \tag{4}$$

where α_i is the Lagrange multiplier and the data points with $\alpha_i > 0$ is called the SV. SV are classified into two groups such as margin SV and error SV. The margin SV are firm on the margin $(y_i f(x_i) - 1 = 0)$ while the error SV surpass the margin $(y_i f(x_i) - 1 < 0)$. Some error SV may be mislabeled $(y_i f(x_i) < 0)$. Most of the noises would become the error SV. Therefore, noise data should be assigned lower weights. From calculated α_i, the number of SV is M, then the decision function is obtained by

$$f(x) = sign\left(\sum_{i=1}^{M} \alpha_i y_i \langle x_i, x \rangle + b\right) \tag{5}$$

Figure 1 shows the flowchart process of WSVM.

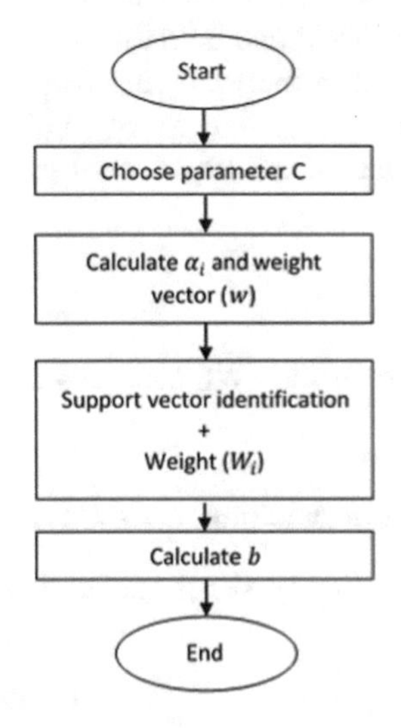

Fig. 1. The flowchart process of WSVM

2.2 One-Step WSVM (OWSVM)

OWSVM only uses one step weighting. The inspiration of the weight function draw closer from the form of hinge loss function. When $u \leq 0$, the hinge loss $H_1(u)$ is $(1 + |u|)$. To append with the $0 - 1$ loss $I(u \leq 0)$, the weight of OWSVM have been assigned in the form of $W_i = 1/(1 + |f(x_i)|)$ for the i-th data point. The dual problem of OWSVM is similar with Eqs. (3) and (4). OWSVM method imply two steps. The first step is to train the standard SVM with weight ($W_i = 1$) for all training data points. The second step is to obtain the weighted learning using the weight ($W_i = 1/(1 + |f(x_i)|)$) for all the training data points. Figure 2 shows the flowchart process of OWSVM.

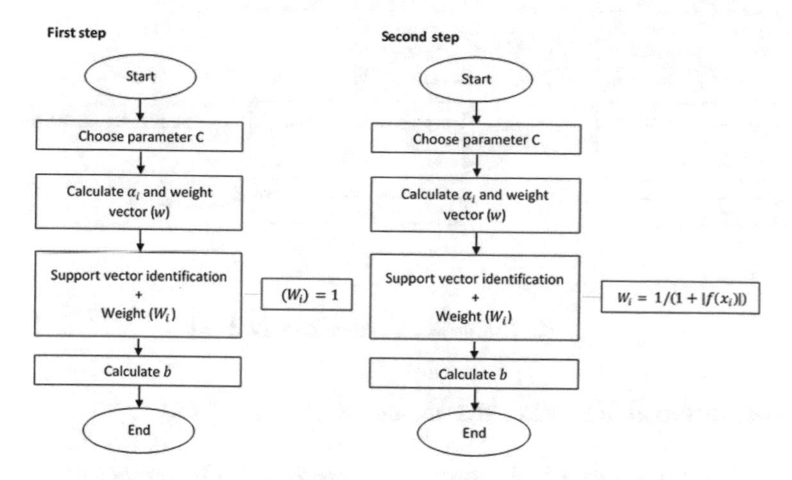

Fig. 2. The flowchart process of OWSVM

2.3 Iteratively WSVM (IWSVM)

IWSVM solves the corresponding WSVM by iteratively updating the weights. IWSVM truncate the hinge loss function and outlined the truncated hinge loss $T_s(u) = H_1(u) - H_s(u)$, where $H_s(u) = [s - u]_+$ for $s \leq 0$. The use of s, $s = -1/(k-1)$ where k denotes the number of classes. IWSVM solves the following optimization problem:

$$\min_{w,b,\xi} \frac{1}{2}||w||^2 + C \sum_{i=1}^{n} W_i T_s(y_i f(x_i)) \tag{6}$$

$$\text{Subject to} : y_i(\langle w, x_i \rangle + b) \geq 1 - \xi_i, \xi_i \geq 0 \; for \; i = 1, \ldots n \tag{7}$$

The dual problem of IWSVM is similar with Eqs. (3) and (4). IWSVM method imply three steps. The first step is to train the standard SVM with weight ($W_i = 1$) for all training data points. The second step is the iteration step. At the i-th data point iteration with $t \geq 1$, the weight will be outlined as $W_i^t = 1 \; if \; s \leq y_i f^{(t-1)}(x_i) \leq 1$ and 0 otherwise. The WSVM will be conclude with the upgrade weights W_i^t. The third step is to repeat the second step until convergence. The convergence will be claim when $W_i^{(t)} = W_i^{(t-1)}$. Figure 3 shows the flowchart process of IWSVM.

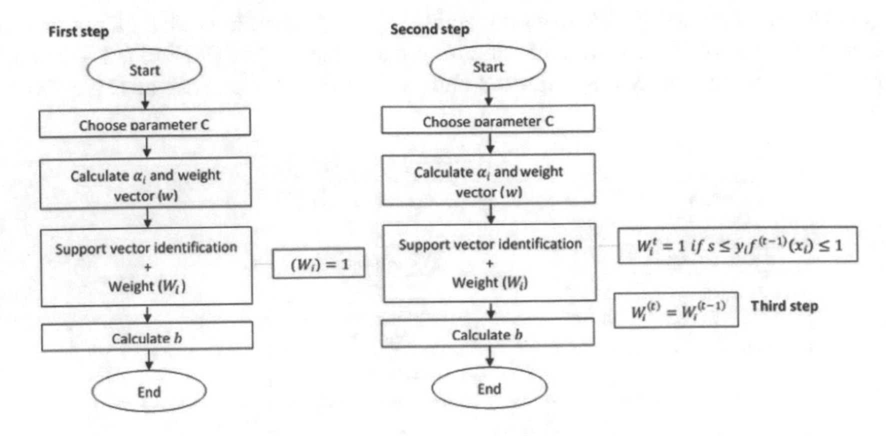

Fig. 3. The flowchart process of IWSVM

3 Experimental Results and Discussion

In this section, the experimental results and discussion will be presented.

3.1 Experimental Setup

The experimental setup is based on the following phases. First, eight datasets drawn from the Knowledge Extraction based on Evolutionary Learning (KEEL) repository datasets as shown in Table 1. The preprocessing data for the eight datasets is important to make sure that all the datasets will produces clean and well-curated data that leads to more practical and accurate model outcomes. Secondly, the datasets will be divided into two sets namely training data and test data. For each datasets, the training data will be partitioned by means of a 10-folds cross validation procedure as the final result. The programming algorithm was developed based on MATLAB software that build up around vectors and matrices. Thirdly, 20% of noise will be included in each of the dataset for the training data. Finally, the performance of the WSVM, OWSVM and IWSVM will be evaluated based on correctly labeled, mislabeled data, data within margin, mislabeled data within margin and classification accuracy.

Table 1. Characteristics of datasets used in the experiment

Datasets	Description	Training data	Attributes
Statlog (Heart)	Classification of heart disease as absence or presence	243	13
Ionosphere	Classification of radar returns from the ionosphere	351	33
Spambase	Classifying email as spam or non-spam	4601	57

(*continued*)

Table 1. *(continued)*

Datasets	Description	Training data	Attributes
Pima indians diabetes	Classifying of diabetes as negative or positive	691	8
Ringnom	Classifying 20 dimensional of 2 class classification problem. Class 1 has mean zero and covariance 4 times the identity	6660	20
Twonorm	Classifying 20 dimensional of 2 class classification problem. Class 1 has mean (a,a,..a)	6660	20
Breast cancer wisconsin (Diagnostic)	Classifying of tumor as benign or malignant	515	30
Connectionist bench (Sonar, Mines vs. Rocks)	Classifying of object as rock or mine	187	60

3.2 Evaluation Measurements

The classification accuracy (CA) was calculated to evaluate the performance of the WSVM, OWSVM and IWSVM. As shown in Eq. (9), the classification accuracy which is used to evaluate the overall performance of the model as follows

$$\text{CA} = (\text{Number of correctly labeled / Overall training data}) * 100\,\% \tag{8}$$

3.3 Experimental Results

This section shows the results for the eight KEEL datasets that include 20% of noise in training data using WSVM, OWSVM and IWSVM. The parameter C is set to 1. Linear kernel is the only kernel used in this study. All the tables below show the result for the 20% of noise in training data using WSVM, OWSVM and IWSVM for eight KEEL datasets based on correctly labeled, mislabeled data, data within margin, mislabeled data within margin and classification accuracy.

Tables 2 to 4 presents the correctly labeled, mislabeled data, data within margin, mislabeled data within margin and classification accuracy results using WSVM for eight KEEL datasets. From Table 2, the highest classification accuracy using WSVM is 97.90% while the lowest classification accuracy using WSVM is 61.08%. From Table 3, the highest classification accuracy using OWSVM is 98.00% while the lowest classification accuracy using OWSVM is 61.08%. From Table 4, the highest classification accuracy using IWSVM is 97.97% while the lowest classification accuracy using IWSVM is 55.51%.

Table 2. Results for 20% of noise in training data using WSVM for eight KEEL datasets

Datasets	Training data (20% noise)				
	Correctly labeled	Mislabeled data	Data within margin	Mislabeled data within margin	Classification accuracy (%)
Statlog (Heart)	216	27	58	14	88.89
Ionosphere	296	20	72	15	93.67
Spambase	3874	263	1249	190	93.64
Pima indians diabetes	537	154	361	122	77.71
Ringnom	4068	2592	6349	2394	61.08
Twonorm	6520	140	323	92	97.90
Breast cancer wisconsin (Diagnostic)	491	21	54	19	95.90
Connectionist bench (Sonar, Mines vs. Rocks)	160	28	103	23	85.11

Table 3. Results for 20% of noise in training data using OWSVM for eight KEEL datasets

Datasets	Training data (20% noise)				
	Correctly labeled	Mislabeled data	Data within margin	Mislabeled data within margin	Classification accuracy (%)
Statlog (Heart)	220	23	50	8	90.54
Ionosphere	297	19	62	10	93.99
Spambase	3043	1094	282	168	73.56
Pima indians diabetes	543	148	299	105	78.58
Ringnom	4068	2592	6349	2394	61.08
Twonorm	6527	133	264	73	98.00
Breast cancer wisconsin (Diagnostic)	491	21	60	17	95.90
Connectionist bench (Sonar, Mines vs. Rocks)	166	22	99	14	88.30

3.4 Discussion

In this section, the results from Tables 2 to 4 will be discussed. From all the tables above, it can be seen that WSVM and OWSVM did not perform well on Ringnom dataset because the numbers of data within margin are higher than the numbers of

Table 4. Results for 20% of noise in training data using IWSVM for eight KEEL datasets

Datasets	Training data (20% noise)				
	Correctly labeled	Mislabeled data	Data within margin	Mislabeled data within margin	Classification accuracy (%)
Statlog (Heart)	213	30	19	8	87.65
Ionosphere	295	21	35	4	93.35
Spambase	3873	264	267	71	93.62
Pima indians diabetes	536	155	15	4	77.57
Ringnom	3697	2963	23	6	55.51
Twonorm	6525	135	100	38	97.97
Breast cancer wisconsin (Diagnostic)	493	19	9	1	96.29
Connectionist bench (Sonar, Mines vs. Rocks)	156	32	74	20	82.98

correctly labeled. IWSVM did not perform well on most of the dataset except Spambase and Breast Cancer Wisconsin (Diagnostic) dataset because IWSVM rely heavily on noise data that located outside the margin. The dataset that produce the highest classification accuracy for WSVM, OWSVM and IWSVM are Twonorm dataset. The highest classification accuracy between WSVM, OWSVM and IWSVM show that OWSVM have the highest compared to WSVM and IWSVM. This is because OWSVM only uses one-step weighting procedure and more computationally efficient than WSVM and IWSVM. OWSVM also have the lowest classification accuracy compared to WSVM and IWSVM simply because the numbers of mislabeled data are higher than the numbers of data within margin. However, OWSVM has some drawbacks such as is it a one-step weighting procedure and did not have iteration process to select the best hyperplane that have the optimal margin.

4 Conclusion

Tables 2 to 4 reports the 20% noise in training data for eight datasets using WSVM, OWSVM and IWSVM based on correctly labeled, mislabeled data, data within margin, mislabeled data within margin and classification accuracy. The presence of mislabeled data in training dataset can severely affect the performance of machine learning classifiers. Based on the experimental results, the performance of OWSVM is better than both WSVM and IWSVM based on the correctly labeled, mislabeled data, data within margin, mislabeled data within margin and classification accuracy. Future work might focus on modifying WSVM, OWSVM or IWSVM to solve the presence of mislabeled data especially mislabeled data within margin.

References

1. Reddy M (2018) Ground Truth Gold—Intelligent data labeling and annotation. The Hive
2. Brodley CE, Friedl MA (1999) Identifying mislabeled training data. J Artif Intell Res 11:131–167
3. Frénay B, Kabán A (2014) A comprehensive introduction to label noise. In: European Symposium on Artificial Neural Networks. Comput Intell Mach Learn 23–25
4. Wagar EA, Stankovic AK, Raab S, Nakhleh RE, Walsh MK (2008) Specimen labeling errors: a Q-probes analysis of 147 clinical laboratories. Arch Pathol Lab Med
5. Bootkrajang J, Kabán A (2012) Label-noise robust logistic regression and its applications. In: Lecture Notes in Computer Science (including subseries Lecture Notes in Artificial Intelligence and Lecture Notes in Bioinformatics)
6. Bootkrajang J, Kabán A (2013) Classification of mislabelled microarrays using robust sparse logistic regression. Bioinformatics
7. Bootkrajang J (2016) A generalised label noise model for classification in the presence of annotation errors. Neurocomputing
8. Frénay B, Verleysen M (2014) Classification in the presence of label noise: a survey. IEEE Trans Neural Networks Learn Syst 25(5):845–869
9. Liu T, Tao D (2015) Classification with noisy labels by importance reweighting. IEEE Trans Pattern Anal Mach Intell 38(3):447–461
10. Almasi ON, Rouhani M (2016) Fast and de-noise support vector machine training method based on fuzzy clustering method for large real world datasets. Turkish J Electr Eng Comput Sci 24(1):219–233
11. Boser BE, Guyon IM, Vapnik VN (1992) A training algorithm for optimal margin classifiers. In: Proceedings of the Fifth Annual Workshop On Computational Learning Theory-COLT '92, pp 144–152
12. Vapnik VN (1995) The nature of statistical learning theory, vol 8
13. Sabzevari M (2015) Ensemble learning in the presence of noise
14. Yang X, Song Q, Wang Y (2007) A weighted support vector machine for data classification. Int J Pattern Recognit Artif Intell 21(5):961–976
15. Fan H, Ramamohanarao K (2005) A weighting scheme based on emerging patterns for weighted support vector machines. In: 2005 IEEE International Conference on Granular Computing, pp 435–440
16. Tian J, Gu H, Liu W, Gao C (2011) Robust prediction of protein subcellular localization combining PCA and WSVMs. Comput Biol Med 41(8):648–652
17. Wu Y, Liu Y (2013) Adaptively weighted large margin classifiers. J Comput Graph Stat 22 (2):37–41

eSNN for Spatio-Temporal fMRI Brain Pattern Recognition with a Graphical Object Recognition Case Study

Norhanifah Murli[1(✉)], Nikola Kasabov[2], and Nurul Amirah Paham[1]

[1] Universiti Tun Hussein Onn, Parit Raja, Malaysia
hanifah@uthm.edu.my, amerpaham@gmail.com
[2] Auckland University of Technology, Auckland, New Zealand
nkasabov@aut.ac.nz

Abstract. This paper describes an experiment involving visual object fMRI brain data and the NeuCube [1] architecture. fMRI spatio- and spectro- temporal data (SSTD), apart from EEG, audio and video data, comprises both space and time information, that requires a specific and specialized architecture to process, interpret and visualize the data for better understanding and interpretation of the information it may carries. At the same time, any patterns can be better recognized and thus new knowledge that may be embedded within the pattern can be extracted. From the experiment with the case study of Haxby fMRI data, NeuCube^{B} has accomplished better accuracy in recognizing the brain patterns compared with the standard machine learning techniques (i.e. SVM and MLP). In addition, the NeuCube method assists deep learning of the SSTD and deeper analysis of the spatio-temporal characteristics and patterns in the fMRI SSTD.

Keywords: NeuCube · Spatio- spectro- temporal pattern · Functional magnetic resonance imaging (fMRI) · Evolving spiking neural network · Deep learning

1 Introduction

Evolving spiking neural networks (eSNN) was originally developed for visual pattern recognition system [1]. The networks progress their functionality and structure increasingly in an on-line manner in such a way that for every incoming fresh input pattern, a new spiking neuron is allocated and connected dynamically to the input neurons [1–4]. In certain application, if the spiking neurons acquired the same patterns (i.e. class) and acquired the same connection weights, these neurons are consolidated [3, 4]. Consequently, this enables for the networks to achieve a very fast learning both in supervised or unsupervised mode. With these features, NeuCube was proposed as a unified computational framework for learning and understanding complex data for instance EEG, fMRI, genetic, audio and video data.

A NeuCube^{B} model is introduced that specifically deals with fMRI of brain data. The model is experimented with StarPlus dataset [5]. This paper presents NeuCube^{B} model that map, learn, classify and interpret another fMRI dataset known as the Haxby dataset. Haxby fMRI dataset is very different from the StarPlus – it is in NII format that

R. Ghazali et al. (Eds.): SCDM 2020, AISC 978, pp. 470–478, 2020.
https://doi.org/10.1007/978-3-030-36056-6_44

is in principal, is associated with NIfTI-1 format. The format is developed by Neuroimaging Informatics Technology Initiative, specifically to handle fMRI data in volume. The data consists of a header (.hdr), the actual data (.img) and other additional information.

The Haxby's data is about visual object recognition and was downloaded from OpenfMRI.[1] The same NeuCube^B framework [6] was used. The brain data is mapped differently from the previous experiment because of the variance in the voxels number, as well as in the size and dimension of the brain.

2 NeuCube^B Methodology for Modelling the Spatio-Temporal fMRI Data

The basis of this research is NeuCube [5] spiking neuron networks. Earlier experiments on NeuCube and fMRI data had produced successful results and it is very desirable to conduct more research especially to study its capability (i.e. to learn and classify) other brain data. We used the same NeuCube unsupervised and supervised learning mode with major modification on data mapping, data visualization and data interpretation. This implementation is to verify that the model is able to recognize, learn and classify the fMRI brain data that has time and space components embedded in it. The NeuCube model that specifically developed and experimented to learn spatio- temporal brain data is known as NeuCube^B (Fig. 1).

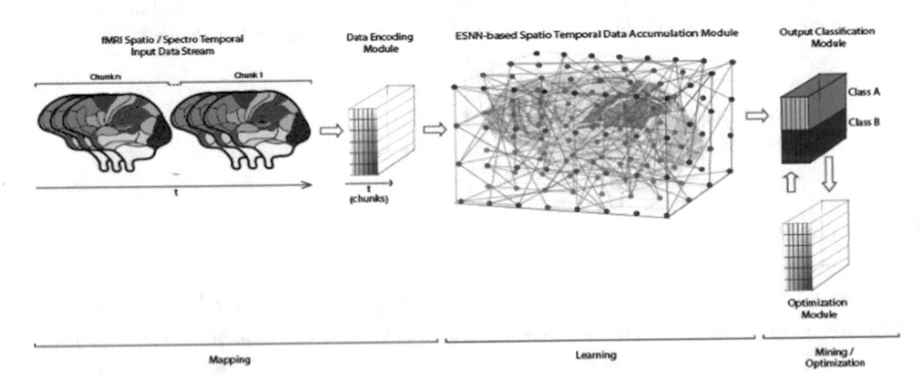

Fig. 1. NeuCube^B framework with phases of mapping, learning and mining [6].

2.1 Visualization

NeuCube^B and fMRI input data neurons are mapped within the same environment. For instance, as depicted in Fig. 2, mapping of neurons are displayed in different z-slice view (right figure) and in three-dimensional view (left figure).

[1] www.openfmri.org/dataset/ds000105.

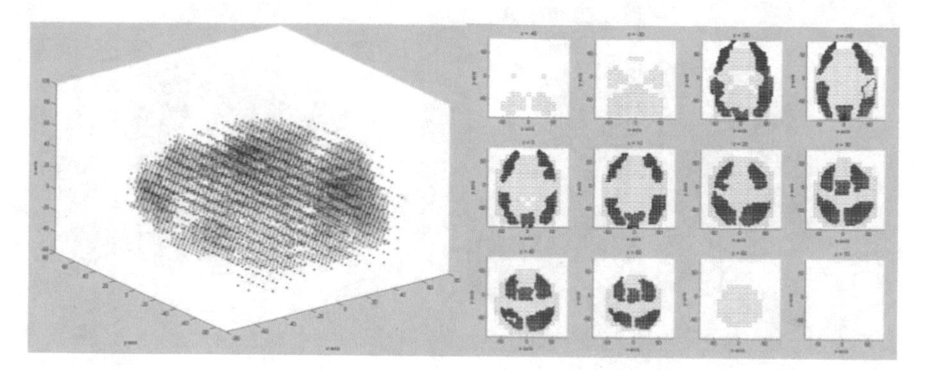

Fig. 2. 3D mapping of NeuCubeB neurons (dots) and fMRI input neurons (crosses) in the initialization phase (left) and its 2D mapping of z-slice fMRI input neurons (right).

NeuCube^{B} visualizes its neurons' state either in its initial, after unsupervised or after supervised learning state. The neurons' state are visualized as a massive inter-connected neurons. These connected neurons are created in every training cycle based on small world recurrent connections. In eSNN, connections are evolved i.e. it learns the spatio- temporal brain data incrementally, propagated in a single pass, in which the spiking neurons are created and merged.

In STDP learning paradigm, weight of synapses' is increasing or decreasing based on the spike time of the post-synaptic neurons in relation to the spike time of pre-synaptic neurons. The weight of connection between two neurons will be increased if a pre-synaptic neuron spikes first (the spike time difference is negative) or else connection weight will be decreased (the spike time difference is positive). Neurons that are connected learn the consecutive temporal association from the brain data and these connections evolved over time.

We also proposed that these spatio-temporal brain data are input into deSNN [9], in which weights will be calculated and adjusted again, by using rank order (RO) and STDP. RO learning initializes the weights of connection and STDP adjusts the weights to the connection based on the following spikes occurrence.

2.2 Model Interpretation

In this study, a new method to interpret connections of neuron created in the eSNN model in association with the activity of fMRI spiking during the learning, is proposed. Referring to Fig. 3, the activity of spiking neurons (before and after) in unsupervised or supervised learning are taken and visualized using blue and red lines that actually represented the connections between fMRI neurons. Positive connections are in blue lines (positive spikes with connection weight > 0) and negative connections are in red lines (negative spikes with connection weight < 0). During each training, neurons' spatio-temporal connections will be initialized, created and calculated. For visualization purposes, active neurons and inactive neurons are displayed with blue and red dots respectively.

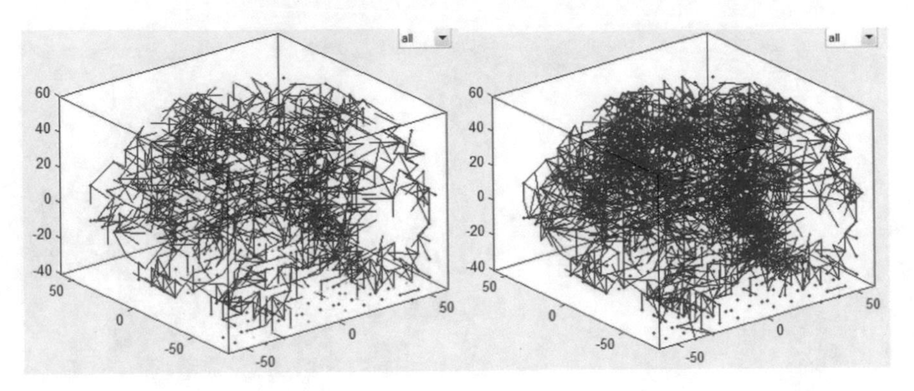

Fig. 3. 3D visualization of neurons connectivity of before (left) and after (right) STDP learning

In initialization phase, connections and their connection weights in the NeuCube^B will be randomly initialized. More interconnected neurons are generated in different areas of the NeuCube^B resulting from the increase in positive spikes activity during the training phase. Connections with stronger spikes can also be displayed in the model. Spots with higher number of connections visualizes the spots in the brain which are more stimulated when presented with some kind of stimulus. Spikes that are stronger (indicated by connections with maximum weight) can be displayed and the corresponding brain areas that are affected can be labelled and recognized. The extraction and identification of brain areas that are activated is certainly a new knowledge impossible to be extracted from the traditional machine learning methods.

3 Experiment Setting

There are 5 different subjects downloaded from OpenfMRI (labelled as SUB001, SUB002, SUB003, SUB004 and SUB006) involved in the experiment. The subjects were presented with grey-scaled images of House (COND001), Scrambled pictures (COND002), Cat (COND003), Shoe (COND004), Bottle (COND005), Scissors (COND006), Chair (COND007) and Face (COND008). 12 time-series of images were acquired for each subject, during which begins and ends with a rest period of 12 s. It consisted of 8 stimulus blocks of 24 s interval, i.e. a stimulus for each category. A stimulus is presented for 0.5 s, followed by 1.5 s interval before the next stimulus is presented. The same image is captures in 4 different directions and each category consisted of 12 different patterns.

The image data is extracted from the.img structure specified by $40 \times 64 \times 64$ 121 from the body.nii file. The image structure from the 12 runs for the 5 subjects were then extracted and tabulated.

For each run, a subject is displayed with a stimulus of different patterns for 0.5 s and 1.5 s for an inter-stimulus break, therefore generating 24 s of a specific stimulus block. In between stimulus, there are 12 s rest breaks. For example, the stimulus sequence for subject 2 is as the following: Chair (COND007) → Scissors (COND006) → Cat (COND003) → Face (COND008) → Bottle (COND005) → Scrambled pictures

(COND002) → House (COND001) → Shoe (COND004). A unique ID that represents a different time- series is used to identify each stimulus block, thus producing 108 time-series (9 IDs x 12 experiment runs) for every stimulus for every subject.

The study is carried out in two approaches: (1) Face against Scrambled Pictures classification and (2) Face against Not Face and Scrambled Pictures against Not Scrambled Pictures) classification. The first approach was a classification within the same subject i.e. Face (Class 1) against Scrambled Pictures (Class 2), that had 12 samples for each class. Another approach was a classification of 12 samples of Face (Class 1) against 84 samples of Not Face (Class 2).

3.1 fMRI Neuron Mapping

Figure 4 shows the 3D mapping of Haxby's SUB001 fMRI neuron coordinates into NeuCube^{B} neurons. The threshold value is fixed to 1000, thus producing more than 19,000 s of significant voxels. Reducing the threshold value will generate more voxels to be determined. These 19,000 s coordinate locations can be identified from the cube. To reduce the number of neurons mapping, each coordinate that nearby (radius is specified to 7 mm) the coordinates of NeuCubeB were identified. This will reduce the number of neurons' coordinates to only 1471 that actually follows the number of locations specified in NeuCube^{B}. In this experiment, with radius set to 7 mm, only 854 locations were identified, i.e. some of Haxby's fMRI coordinates were not mapped into the cube. Voxels in these locations were the actual input to the NeuCube^{B} model, and these inputs were transformed into sequence of spikes.

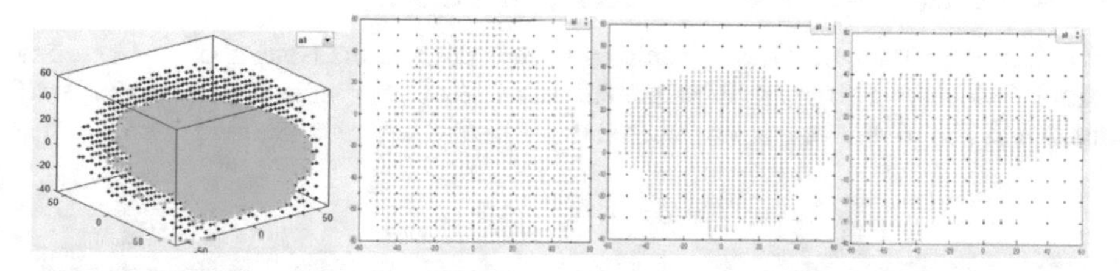

Fig. 4. 3D visualization mapping of Haxby coordinates into NeuCube^{B} model, coordinates of Haxby's fMRI data is in cyan dots while NeuCube^{B} coordinates are in blue dots.

3.2 Data Learning

In NeuCube^{B}, during the unsupervised learning phase, STDP is used to initialize, adjust and preserve the connection weights (i.e. memory). Classification is between Face against the other seven stimuli brain patterns. This experiment approach was based on the experiment of distributed and overlapped activation in the human brain ventral temporal and occipital cortex [4, 5].

Face (Class 1) versus Not Face (Class 2) and Scrambled Picture (Class 1) versus Not Scrambled Picture (Class 2)

Classification of fMRI patterns was based on 1 stimulus (i.e. Face) against the other 7 stimuli (all other stimuli but Face). There were 10 different experiments produced. For illustration, parameters shown in Table 1 are used in the experiment in determining the classification accuracies for each class for both experiments.

Table 1. Parameter's setting for the experiment

Parameter	Face stimulus against the other 7 stimuli
Size of the NeuCube^B	$10 \times 10 \times 10$
AER threshold	4.25
SWC	0.15
LIFM neurons threshold	0.8
LIFM neurons leak parameter	0.002
STDP learning rate	0.01
Number of training	2
Mod parameter & drift	0.4 and 0.25

4 Results and Discussion

Face (Class 1) versus Not Face (Class 2) and Scrambled Picture (Class 1) versus Not Scrambled Picture (Class 2) Classification Result

To preserve the temporal details, samples were input into the NeuCube^B one at a time. Figure 5 displays a software implementation of NeuCube^B after a successful classification and Fig. 6 shows neurons connectivity before and after training.

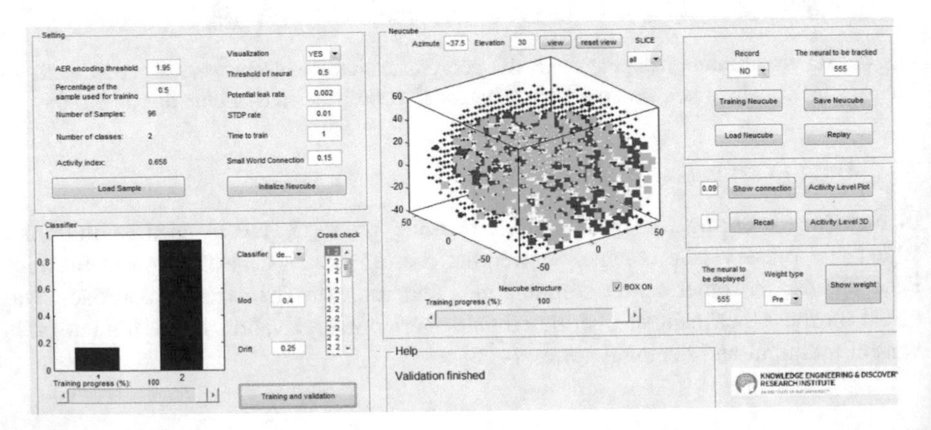

Fig. 5. A sample of parameter setting in the NeuCube^B with deSNN as the classifier. Accuracy percentage acquired is 17% (Class 1) and 97% (Class 2).

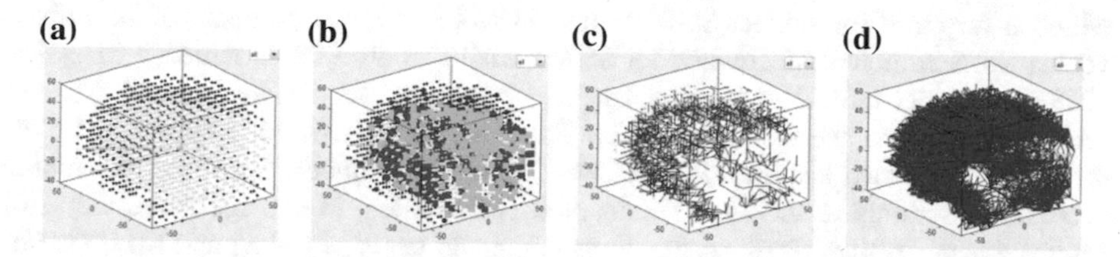

Fig. 6. 3D visualization of fMRI brain data and its connectivity of eSNN for subject SUB001 for Face (Class 1) against Not Face (Class 2) experiment: (a) no spike, blue inactive neurons are in blue, yellow fMRI input neurons; (b) spiking neurons activity: red active neurons, blue inactive neurons, magenta positive input neurons, cyan negative input neurons and yellow zero input neuron; (c) neurons connection prior to training, blue positive connections and red negative connection; (d) neurons connection once training is completed

The results as shown in Table 2 were acquired using 50% of training data and 50% of testing data, for the modeled and trained NeuCube^B versus Support Vector Machine (SVM) and Multi-layer Perceptron (MLP) standard classifiers for SUB001, SUB002, SUB003, SUB004 and SUB006. The average percentage achieved for Class 1 and Class 2 from NeuCube^B all the 5 experiment runs are in brackets. From these experiments, it can be concluded that has better accuracies as compared to SVM and MLP. The average NeuCube^B accuracy is 85% while SVM 66% and MLP 76%. MLP's performance is better than SVM. Face against Not Face experiment for SUB001 achieved the 90% (highest percentage) in which 23% identify correctly patterns of brain acquired from Class 1 and 99% identify correctly patterns of brain acquired from Class 2.

Table 2. Classification results for SUB001, SUB002, SUB003, SUB004, and SUB006 for Face (Class 1) against Not Face (Class 2) and Scrambled Pictures (Class 1) against Not Scrambled Pictures (Class 2).

Subject/classifier	MLP	SVM	NeuCube^B
SUB001 - Scrambled against Not Scrambled	87 (0, 87)	76 (12, 87)	87 (13, 97)
SUB001 - Face against Not Face	86 (10, 88)	74 (12, 87)	90 (23, 99)
SUB002 - Scrambled against Not Scrambled	71 (0, 73)	55 (17, 73)	83 (33, 91)
SUB002 - Face against Not Face	71 (6, 75)	62 (23, 75)	86 (17, 96)
SUB003 - Scrambled against Not Scrambled	77 (49, 80)	70 (31, 75)	87 (50, 92)
SUB003 - Face against Not Face	74 (27, 76)	64 (33, 79)	76 (17, 85)
SUB004 - Scrambled against Not Scrambled	74 (8, 75)	69 (30, 77)	81 (20, 90)
SUB004 - Face against Not Face	72 (3, 74)	68 (31, 78)	83 (17, 92)
SUB006 - Scrambled against Not Scrambled	73 (3, 75)	68 (27, 75)	85 (33, 93)
SUB006 - Face against Not Face	71(21, 76)	58 (20, 73)	90 (17, 100)

From all experiments, smaller accuracy percentage is achieved for Class 1 as compared to Class 2. For example, the least result for Class 1 of MLP classifier is 0%

which is acquired for subject SUB001 and SUB002 in the experiment of Scrambled versus Not Scrambled. Meanwhile for SVM classifier, only 12% is achieved Class 1 is 12% for subject SUB001 in the experiment of Scrambled versus Not Scrambled and in the experiment of Face versus Not Face. Small accuracy (13% for SUB001 and 17% for SUB002) is also acquired in NeuCube^B in the experiments involving Scrambled versus Not Scrambled and Face versus Not. Nonetheless, in average NeuCube^B still achieved better results (30% accuracy) as compared to SVM (24%) and MLP (13%).

It can be concluded that the low percentage results for Class 1 are acquired from the small sample number (6 samples each for training and testing). The spiking neurons networks has insufficient samples (and time) to learn the fMRI brain pattern. The network learns better (more brain patterns were identified) when more samples were extracted and used throughout the unsupervised learning. This will result in a better accuracy percentage – more brain patterns were classified correctly.

Higher percentage results achieved for Class 2 that had 96 samples (48 samples each for training and testing). For instance, for subject SUB006 the achieved accuracy result was as high as 100%. It can be concluded that the network of spiking neurons could recognized perfectly between the Face and Not Face samples and generally produced better accuracies as opposed to the SVM and MLP classifiers.

This study also has the ability to visualize neurons connections when it is initialized, after it completed an unsupervised training as well as when it has been through a supervised training. NeuCube^B is capable in visualizing brain areas with the most neurons activity.

5 Concluding Remarks and Future Work

The experiment conducted in this study involved the use of different set of optical fMRI Haxby's dataset incorporated into the NeuCube^B model. fMRI brain patterns could be classified better using the NeuCube^B model as opposed to standard classifier methods. Further experiments could be tested to the dataset especially to further analyze the neurons connections before/after the training. Different spots of the brain areas are stimulated when the subjects were presented with different stimulus and this can be visualized directly through the model. The most activated areas are visualized using highly interconnected neurons. This visualization is impossible to be done in the standard black-box machine learning methods. In general, we can also conclude that NeuCube^B offers not only better accuracy performance but also a better analysis and understanding of the fMRI Haxby's brain dataset as compared to the earlier experiments involving the similar set of data.

Acknowledgements. This experiment is part of a project supported by the EU FP7 Marie Curie project EvoSpike PIIF-GA-2010-272006, supported by the Institute for Neuroinformatics at ETH/UZH Zurich (http://ncs.ethz.ch/projects/evospike), as well as by the Knowledge Engineering and Discovery Research Institute (KEDRI, http://www.kedri.info) of the Auckland University of Technology and the New Zealand Ministry of Science and Innovation. This work also is supported by TIER 1 No. H100 as Graduate Research Assistant (GRA) of University Tun Hussein Onn Malaysia (UTHM).

References

1. Gholami Doborjeh M, Gholami Doborjeh Z, Gollahalli A, Kumarasinghe K, Breen V, Sengupta N et al (2018) From von neumann architecture and Atanasoff's ABC to neuromorphic computation and Kasabov's NeuCube. Part II: applications. Stud Syst Decis Control 17–36. https://doi.org/10.1007/978-3-319-78437-3_2
2. Muscinelli S, Gerstner W, Schwalger T (2019) How single neuron properties shape chaotic dynamics and signal transmission in random neural networks. PLoS Comput Biol 15(6): e1007122. https://doi.org/10.1371/journal.pcbi.1007122
3. Kasabov N (2018) Evolving spiking neural networks. Springer series on bio- and neurosystems, pp 169–199. https://doi.org/10.1007/978-3-662-57715-8_5
4. Kasabov N (2019) Evolving and spiking connectionist systems for brain-inspired artificial intelligence. Artif Intell Age Neural Netw Brain Comput 111–138:2019. https://doi.org/10.1016/b978-0-12-815480-9.00006-2
5. Kasabov N (2014) neucube: a spiking neural network architecture for mapping, learning and under-standing of spatio-temporal brain data. Neural Netw 52:62–76
6. Murli N, Kasabov N, Handaga B (2014) Classification of fMRI data in the neucube evolving spiking neural network architecture. In: International conference on neural information processing. Springer, Cham, pp 421–428
7. Kasabov NK (2019) Artificial neural networks. evolving connectionist systems. In: Time-Space, spiking neural networks and brain-inspired artificial intelligence. Springer series on bio- and neurosystems, vol 7. Springer, Berlin, Heidelberg (2019)
8. Oosterhof N, Connolly A, Haxby J (2016) CoSMoMVPA: Multi-modal multivariate pattern analysis of neuroimaging data in matlab/GNU Octave. Front Neuroinformatics 10. https://doi.org/10.3389/fninf.2016.00027
9. Kasabov N (2018) Deep learning of multisensory streaming data for predictive modelling with applications in finance, ecology, transport and environment. Springer series on bio- and neurosystems, pp 619–658. https://doi.org/10.1007/978-3-662-57715-8_19

Author Index

R. Ghazali et al. (Eds.): SCDM 2020, AISC 978, pp. 479–481, 2020.
https://doi.org/10.1007/978-3-030-36056-6

Printed in the United States
By Bookmasters